INTERSTELLAR DUST AND RELATED TOPICS

INTERNATIONAL ASTRONOMICAL UNION

UNION ASTRONOMIQUE INTERNATIONALE

SYMPOSIUM No. 52

HELD AT THE STATE UNIVERSITY OF NEW YORK AT ALBANY,
ALBANY, N.Y., U.S.A.

INTERSTELLAR DUST AND RELATED TOPICS

EDITED BY

J. MAYO GREENBERG

State University of New York at Albany and Dudley Observatory, U.S.A.

AND

H. C. VAN DE HULST

Sterrewacht, Leiden, The Netherlands

D. REIDEL PUBLISHING COMPANY

DORDRECHT-HOLLAND / BOSTON-U.S.A.

1973

First printing: December 1973

Published on behalf of
the International Astronomical Union
by
D. Reidel Publishing Company, P.O. Box 17, Dordrecht, Holland

Sold and distributed in the U.S.A., Canada, and Mexico
by D. Reidel Publishing Company, Inc.
306 Dartmouth Street, Boston,
Mass.02116, U.S.A.

Library of Congress Catalog Card Number 72–88590

ISBN-13: 978-94-010-2663-5   e-ISBN-13: 978-94-010-2661-1
DOI: 10.1007/978-94-010-2661-1

TABLE OF CONTENTS

PART III / REFLECTION NEBULAE AND DIFFUSE GALACTIC LIGHT

PART IV / INTERSTELLAR POLARIZATION

PART V / DISTRIBUTION OF DUST AND GAS

PART VI / PHYSICAL PROCESSES, THEORY AND EXPERIMENT

PART VII / MOLECULES, THEORY AND OBSERVATIONS

CONCLUDING REMARKS

PREFACE

IAU Symposium Number 52 on Interstellar Dust and Related Topics was held at Albany, N.Y., on the campus of the State University of New York at Albany from May 29 to June 2, 1972. The members of the Organizing Committee were:

Dr A. D. Code, University of Wisconsin, Madison, Wis., U.S.A.

Dr B. D. Donn, Goddard Space Flight Center, Greenbelt, Md., U.S.A.

Dr A. Elvius, Stockholm Observatory, Saltsjöbaden, Sweden.

Dr T. Gehrels, Lunar and Planetary Laboratory, University of Arizona, Tucson, Ariz., U.S.A.

Dr J. M. Greenberg (Chairman), State University of New York at Albany, Albany, N.Y., U.S.A.

Dr H. C. van de Hulst, Sterrewacht, Leiden, Holland.

Dr S. B. Pikel'ner, Sternberg Astronomical Institute, Moscow, U.S.S.R.

Dr E. E. Salpeter, Cornell University, Ithaca, N.Y., U.S.A.

Dr B. E. Turner, National Radio Astronomy Observatory, Charlottesville, Va., U.S.A.

The suggestion was first made in 1971 that a symposium on interstellar grains would be timely. The response to the first preliminary announcement, which was sent out on November 29, 1971, was well beyond our expectations. The meeting was locally sponsored by the State University and by Dudley Observatory. The National Aeronautics and Space Administration and the National Science Foundation contributed along with the IAU. There were 158 participants of whom 49 were from 15 countries outside the United States. A total of 92 papers were presented.

An important feature of the meeting was the attendance by a sizable number of physicists, chemists, and geologists. The principal topics discussed were: (1) extinction and polarization, (2) diffuse interstellar features, (3) dust around and in close association with stars, (4) reflection nebulae and other aspects of dust scattering properties, (5) alignment mechanisms, (6) dust and gas, (7) distribution of molecules and processes of molecule formation, (8) radiation effects on dust, (9) physical and chemical interactions of dust with the ambient medium – dust formation and destruction, (10) gas and dust in H II regions. There was a clearly indicated need for additional data from laboratory and theoretical studies of the chemical, physical, and optical properties of potentially important dust materials. Identification of both silicate and ice in grains was inferred from infrared absorption data. In general the value of the infrared as well as the ultraviolet observations was amply demonstrated. New techniques in polarimetry appear to be providing important grain optics criteria. The physical and chemical activities of the dust grains in the interstellar medium and their interactions in association with stars dominated in interest over the more classic passive properties of dust.

One of the principal purposes for holding a symposium is to provide the opportunity for active workers in a field to confront one another with their ideas and most current results. This usually plays an important role in channeling the research efforts into the most challenging and fruitful direction. During the course of the symposium there were numerous active discussions. Unfortunately, there were no recording capabilities in the lecture hall and consequently our permanent records were somewhat limited. We are grateful to those conscientious participants who provided us with written statements of their questions and comments. These have been placed at the end of the volume and after some editing, for the purposes of continuity, have been generally ordered according to the Section (Part) in which they originated. The papers themselves have been somewhat reordered from their original sequence of presentation so that the material can be more naturally divided into chapters.

We were very grateful for the invaluable assistance of Mrs Pat Rudd in organizing and handling the local arrangements. Ample assistance was provided by the graduate students and Mrs Betty Sterrett, Secretary of the Department of Astronomy and Space Science, during the course of the meeting. Early on we had been helped out by Mrs Chris Bain at Dudley Observatory. Finally, and this is not at all to be considered pro forma, recognition is due Mrs Naomi Greenberg who played such a creative role in providing for the personal comfort of the participants.

J. MAYO GREENBERG

LIST OF PARTICIPANTS

BELGIUM

P. Cugnon – Observatoire Royal de Belgique, (Brussels)

BRAZIL

J. A. M. Machado – Instituto Astronomico e Geofisico (Sao Paulo)
G. R. Quast – Instituto Astronomico e Geofisico (Sao Paulo)

CANADA

W. W. Duley – York University (Ontario)
G. Fahlman – University of British Columbia (Vancouver)
G. L. Floyd – York University (Ontario)
J. D. McCullough – York University (Ontario)
R. H. Prince – York University (Ontario)
S. van den Bergh – David Dunlop Observatory (Ontario)

CZECHOSLOVAKIA

V. Vanýsek – Charles University (Prague)

FRANCE

E. Bussoletti – Observatoire de Paris
G. Courtes – Observatoire de Marseille
P. Lena – Observatoire de Paris
R. Louise – Observatoire de Lyon
G. Stazinska – Observatoire de Meudon

GERMANY

H. D. Breuer – Universität des Saarlandes (Saarbrücken)
F. F. Gardner – Max Planck Institut für Radioastronomie (Bonn)
P. G. Mezger – Max Planck Institut für Radioastronomie (Bonn)
Y. K. Minn – Max Planck Institut für Radioastronomie (Bonn)
Th. Schmidt – Max Planck Institut für Astronomie (Heidelberg)
K. Voelcker – Max Planck Institut für Astronomie (Heidelberg)

GREAT BRITAIN

R. D. Davies – Nuffield Radio Astronomy Laboratory (Cheshire)
F. A. Goldsworthy – University of Leeds

Great Britain (continued)
A. Martin – University of Cambridge
P. Martin – Institute of Theoretical Astronomy (Cambridge)
D. McNally – University of London Observatory
N. C. Wickramasinghe – Institute of Theoretical Astronomy (Cambridge)

HOLLAND

J. Borgman – Kapteyn Sterrewacht (Roden)
M. de Vries – Kapteyn Sterrewacht (Roden)
B. Fitton – European Space Research Organization (Noordwijk)
H. Habing – Sterrewacht te Leiden
H. C. van de Hulst – Sterrewacht te Leiden
P. R. Wesselius – Sterrenkundig Laboratorium 'Kapteyn' (Groningen)
R. Willis – European Space Research Organization (Noordwijk)

INDIA

K. S. Krishna-Swamy – TATA Institute for Fundamental Research (Bombay)
T. K. Menon – TATA Institute for Fundamental Research (Bombay)

ITALY

G. V. Coyne, S.J. – Vatican Observatory
P. J. Treanor, S.J. – Vatican Observatory

JAPAN

S. Hayakawa – Nagoya University
S. Isobe – Tokyo Astronomical Observatory
F. Kamijo – University of Tokyo
H. Okuda – Kyoto University
M. Shimizu – University of Tokyo

MEXICO

I. M. Peimbert – Insti. de Astron. UNAM
S. T. Peimbert – Insti. de Astron. UNAM

SCOTLAND

K. Nandy – Royal Observatory

SWEDEN

A. Elvius – Stockholm Observatorium
L. O. Lodén – Astronomiska Observatory

SWITZERLAND

A. Behr – European Southern Observatory (Geneva)

UNITED STATES OF AMERICA

P. Aannestad – NASA Goddard Research Institute (New York)
M. A'Hearn – University of Maryland
L. Aller – University of California at Los Angeles
B. J. Bok – Steward Observatory (University of Arizona)
J. H. Cahn – University of Illinois
A. G. W. Cameron – Yeshiva University (New York)
L. Caroff – NASA Ames Research Center (California)
B. Carroll – Rutgers University (New Jersey)
E. Chappelle – Goddard Space Flight Center (Greenbelt, Md.)
A. D. Code – University of Wisconsin
A. J. Cohen – University of Pittsburgh (Pennsylvania)
S. Cohen – Massachusetts Institute of Technology
C. C. Dahn – U. S. Naval Observatory (Washington, D.C.)
K. L. Day – Ohio State University
L. Denoyer – Cornell University (New York)
H. R. Dickel – University of Illinois Observatory
B. D. Donn – Goddard Space Flight Center (Greenbelt, Md.)
M. Dubin – NASA (Washington, D.C.)
J. W. Erkes – State University of New York at Albany
M. Feldman – University of California at Berkeley
R. L. Ferch – Cornell University (New York)
T. Gehrels – University of Arizona
D. P. Gilra – University of Wisconsin
P. Goldsmith – University of California at Berkeley
W. R. M. Graham – University of Florida
J. M. Greenberg – State University of New York at Albany
L. Greenberg – University of California at Berkeley
J. Hackwell – University of Wyoming
M. S. Hanner – Dudley Observatory (New York)
D. Harris – University of Arizona
M. Harwit – Cornell University (New York)
D. S. Hayes – Rensselaer Polytechnic Institute (New York)
C. E. Heiles – University of California at Berkeley
C. L. Hemenway – Dudley Observatory and State University of New York at Albany
R. C. Henry – Johns Hopkins University (Maryland)
R. K. Honeycutt – Indiana University
S. S. Hong – State University of New York at Albany
D. R. Huffman – University of Arizona
J. H. Hunter – University of South Florida
Y. Itikawa – Argonne National Laboratory (Illinois)
E. B. Jenkins – Princeton University (New Jersey)

U.S.A. (continued)

F. M. Johnson – California State College at Fullerton
H. M. Johnson – Lockheed (California)
M. Kaftin-Kassim – State University of New York at Albany
D. W. Kammer – Albion College (Michigan)
F. J. Kerr – University of Maryland
B. N. Khare – Cornell University (New York)
L. W. King – Cornell University (New York)
S. Kleinmann – Massachusetts Institute of Technology
W. Klemperer – Harvard University (Massachusetts)
C. K. Kumar – Carnegie Institute (Washington, D.C.)
C. F. Lillie – University of Colorado
C. C. Lin – Massachusetts Institute of Technology
S. Lloyd – Cornell University (New York)
J. H. Lutz – Washington State University
A. S. Milman – University of Maryland
N. Y. Misconi – State University of New York at Albany
M. R. Morris – University of Chicago (Illinois)
P. E. Palmer – University of Chicago (Illinois)
P. Parrish – University of California at Berkeley
S. E. Persson – Harvard College Observatory (Massachusetts)
P. Pesch – Warner and Swasey Observatory (Ohio)
V. Petrosian – Stanford University (California)
A. G. D. Philip – State University of New York at Albany
J. L. Pipher – University of Rochester (New York)
M. Price – Massachusetts Institute of Technology
E. M. Purcell – Harvard University (Massachusetts)
R. Reeves – Rensselaer Polytechnic Institute (New York)
L. J. Rickard – University of Chicago (Illinois)
F. E. Roach – Honolulu, Hawaii
J. Roach – Boulder, Colorado
T. P. Roark – Ohio State University
K. Rohlfs – University of Maryland
E. E. Salpeter – Cornell University (New York)
B. D. Savage – University of Wisconsin
M. P. Savedoff – University of Rochester (New York)
D. C. Schmalberger – State University of New York at Albany
K. Serkowski – University of Arizona
S. J. Shawl – University of Kansas
F. H. Shu – State University of New York at Stony Brook
S. C. Simonson – University of Maryland
G. Sistla – State University of New York at Albany
T. Snow – University of Washington

U.S.A. (continued)

L. E. Snyder – University of Virginia

L. Spitzer, Jr. – Princeton University (New Jersey)

T. P. Stecher – Joint Institute for Laboratory Astronomy (Colorado)

C. Sturch – University of Rochester (New York)

L. G. Taff – University of Rochester (New York)

S. Tapia – University of Arizona

S. C. Temesvary – State University of New York at Albany

L. E. Tifft – State University of New York at Albany

M. Tomasko – University of Arizona

B. E. Turner – National Radio Astronomy Observatory (Virginia)

G. L. Verschuur – National Radio Astronomy Observatory (Virginia)

D. Wallace – Cornell University (New York)

W. D. Watson – Cornell University (New York)

J. L. Weinberg – Dudley Observatory (New York)

M. W. Werner – University of California at Berkeley

J. A. Williams – Albion College (Michigan)

A. N. Witt – University of Toledo (Ohio)

F. C. Witteborn – NASA Ames Research Center (California)

R. D. Wolstencroft – University of Hawaii at Honolulu

N. J. Woolf – University of Minnesota

J. P. Wright – NSF (Washington, D.C.)

C. C. Wu – University of Wisconsin Experimenter's Office (Greenbelt, Md.)

C. G. Wynn-Williams – California Institute of Technology

A. J. Yencha – State University of New York at Albany

T. T. Zajdel, Jr. – University of Pittsburgh (Pennsylvania)

B. Zellner – University of Arizona

PART I

EXTINCTION

SOME SCATTERING PROBLEMS OF INTERSTELLAR GRAINS*

J. MAYO GREENBERG

State University of New York at Albany

and

Dudley Observatory

Abstract. Some general effects of size, shape, and material on the scattering properties of small particles are considered with a view to determining their relative importance to interstellar dust interpreatations.

1. Introduction

Although this was originally intended to be a broad introduction to the symposium it seemed more useful, after seeing the excellent review papers as well as the contributed papers, to limit my discussion to a few basic scattering problems. A substantial portion of this paper on the nonspherical particle scattering consists of previously unpublished results which have not yet been fully exploited toward understanding the interstellar dust problem but which are clearly relevant and ultimately must be applied.

The most general way of defining the scattering problem is in terms of size, shape, and index of refraction. The index of refraction may be simple as in the case of an isotropic homogeneous particle or it may be complicated as in the case of such inhomogeneities as occur when the particle accretes a mantle of one material on a core of another. A somewhat more subtle physical phenomenon is involved if the particles are sufficiently small that bulk optical properties are no longer applicable. Very little has been done on this latter problem as regards conducting particles. In any case it is a problem which, as we shall see in the discussion of the far ultraviolet extinction, may be of considerable importance and deserve a great deal of attention.

Disregarding size effects, the character of the index of refraction is determined primarily by the chemical constituents of the grains. However, even here there may be significant modifications produced by the physical state of the particle; e.g., temperature, radiation damage, frozen free radicals.

The interstellar particle shapes are obviously mostly nonspherical as evidenced by the amount of interstellar polarization relative to the interstellar extinction.

A full definition of the interstellar grain optics involves therefore, a knowledge not only of the composition, size and shape of the particles but also their physical state.

2. Indices of Refraction

Current evidence leads us to consider silicates, ices, and carbon or combinations of these as the principal grain ingredients.

* Work supported in part by NASA grant #NGR-33-011-043

Greenberg and Van de Hulst (eds.), Interstellar Dust and Related Topics, 3–9.
All Rights Reserved. Copyright © 1973 by the IAU.

The most direct evidence for the silicates and to a certain extent the ices comes from the infrared spectral characteristics of the extinction curve. The limited direct comparison which can be made or the relative amount of silicates and ices is based on the observations of Gillett and Forrest (1973). From their curve the absorption at 3.1 μ is very close to that at 10 μ. It is normal to attribute the 3.1 μ absorption to H_2O ice and the 10 μ absorption to a silicate. The mass of absorbing material may then be estimated for spherical particles assumed to be small enough relative to the wavelength so that the Rayleigh approximation, $x = 2\pi/\lambda \ll 1$ is valid but not so small that the bulk refractive index does not apply. The expression for the optical depths is

$$\tau = \frac{18\pi}{\lambda} V \frac{y}{(\varepsilon+2)^2 + y^2},$$ (1)

where V is the total volume of the material $\varepsilon = m'^2 - m''^2$, $y = 2 m'm''$, m' = real part of refractive index, m'' = imaginary part.

For $\tau_{ice} = \tau_{sil}$ we get

$$\frac{V_{ice}}{V_{sil}} = \frac{3.1}{10} \frac{y_{sil}/[(\varepsilon_{sil}+2)^2 + y_{sil}^2]}{y_{ice}/[(\varepsilon_{ice}+2)^2 + y_{ice}^2]} = 0.21,$$ (2)

where we have used the values $m_{ice} = 1.375-0.815$ (Bertie et al., 1969) and $m_{sil} = 1.7-0.71$ i (Greenberg, 1972, as deduced from Launer, 1952) as the respective complex indices for ice and silicate at their absorption maxima. The ratio of volumes given in Equation (2) leads to a *mass* ratio of silicate to H_2O ice of 10:1 which is somewhat larger than the mass ratio of 4:1 derived by Gillett and Forrest, this difference being probably due to the fact that we have used an updated value of m for ice which gives greater absorptivity. We note that in the dirty ice model made up of water, methane, and ammonia, the above results really imply a somewhat larger relative volume of dirty ice. If we roughly assume that O, C, and N ices contribute relative volumes to the total according to their cosmic abundance the value 0.21 should be raised by a factor of

$$\frac{[O]+[C]+[N]}{[O]} = 1.5.$$

If further we follow the argument that photolysis by ultraviolet and other radiation in space may be shifting the 3.07 band (Greenberg, 1972) it would appear that there is a substantial amount of ices in interstellar space. A resonable assumption is that the ice forms by accretion from the interstellar gas as a mantle on a silicate core, thus leading to inhomogeneous particles.

On the basis of the infrared observation, we are thus led to consider such dielectric particles as silicates, ices, and inhomogeneous combinations of these.

The evidence for other materials is somewhat less direct. In particular, in Section 4, we shall indicate the most convincing basis for consideration of graphite as an interstellar grain constituent.

3. Shape Effects on Mass Estimates

The estimate of ice and silicate masses is subject to errors, not only in the assumption of bulk optical properties but also in the shape assumed. For the former there is no definite answer but for the latter it may readily be shown that spheroidal particles may be 10–75% more efficient *absorbers* in Rayleigh approximation thus implying that mass estimates based on spherical shape are too large. It should also be kept in mind that estimates of visual extinction based on material mass (and hence volume) should properly include consideration of shape effects.

Let us compare prolate and oblate spheroids to spheres. Other elongated and flat particles produce the same effects. Assuming equal volumes of spheres and nonspheres producing the given IR absorption we have

prolate
or $$n_1 \tfrac{4}{3}\pi a^3 = n_2 \tfrac{4}{3}\pi B^2 A, \tag{3}$$
oblate

where n_1 and n_2 are the spatial number densities of spheres and spheroids respectively and where A is taken as the rotation semi-axis of the spheroids.

For needles $A/B \to \infty$ and for plates $A/B \to 0$. In these limits the ratios of sphere to spheroid extinction are

$$\text{prolate:} \qquad \frac{4}{\pi}\,\frac{B}{a} \tag{4}$$

$$\text{oblate:} \qquad 2\,A/a.$$

These ratios clearly depend on the relative 'size' chosen for the spherical and nonspherical particles. Equal volumes (of individual particles) and equal thcikness produce quite different effects. There is no obvious a priori way of making a choice other than by matching the observed extinction curve with models of each particle type and thus defining the appropriate relative sizes based on shape. The two-fold dimension dominates and it turns out that less than 20% difference is formed between the radii of very long cylinders and of spheres used to produce similar extinction curves (Greenberg, 1968).

4. Particle Sizes

The general shapes of the curves for the wavelength dependence of interstellar extinction and polarization and the choice of index of refraction are the means of defining typical particle sizes. The shape of the extinction curve in a particular wavelength region gives information primarily about the particles whose size is in the same region. This is due to the universal character of the dependence of extinction by individual particles on the dimensionless parameters $x = 2\pi a/\lambda$ and/or $\rho = 2x(m'-1)$. This has been demonstrated numerous times in the literature.

We should examine the representative extinction curve shown in Figure 1 as it has

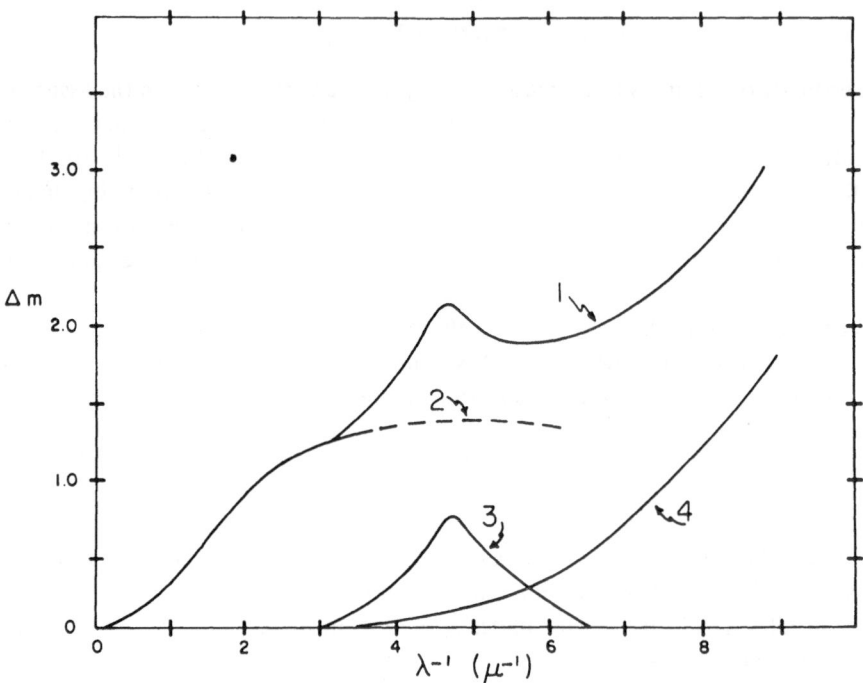

Fig. 1. Representative extinction curve and schematic separation into several contributions. Curves are labeled as follows: (1) Typical OAO extinction curve, (2) Dashed extension showing contribution by classical sized particles (as discussed in text), (3) Contribution of absorption in the 0.22μ band, (4) Contribution by very small particles.

been observed by Bless and Savage (1972). The 'visual' portion ($1 \leq \lambda^{-1} \leq 3\mu^{-1}$) is characterized by the classical particles whose size (radius) is $\bar{a}_{ice} = 0.165\ \mu$ radius for dirty ice and $\bar{a}_{sii} = 0.082$ for silicates. Core mantle particles with silicate core of about $0.05\ \mu$ radius and ice mantles of about $0.10\ \mu$ would probably also be satisfactory. In these dielectric particles it is the parameter ρ which seems to apply. Much work has been done on the portion labeled '2'. The part labeled '3' is often attributed to graphite although other suggestions involving silicates (Huffman and Stapp, 1971) have been made. The particle sizes associated with either of these models are much smaller than the classical sizes. If the contribution to the extinction labeled '4' is due to solid particles, their sizes would be less than 100 Å (Greenberg, 1973). An anomaly with respect to the far ultraviolet albedo of these grains has not yet been resolved (Witt and Lillie, 1971).

In order to account for the observed extinction in terms of solid particles we must therefore include particles from smaller than $0.01\ \mu$ to as large or larger than $0.15\ \mu$.

Since the mass contribution to the interstellar medium is proportional to the size (all other factors being equal) we see that even for equal contributions to the total extinction in the far ultraviolet the small particles add only the order of 10% to the total interstellar dust material.

We have implicitly in all the above, assumed that the small particles react optically in the same way as their classical counterparts. There are a number of reasons why this may not be the case. In the next section we consider one of these.

5. Small Particle Temperatures

It has been noted (Greenberg, 1968) that the definition of the temperature of very small particles is subject to fluctuations which may become sufficiently large that the 'average' temperature loses significance.

We first calculate the grain temperature as if it were properly defineable and then compare the likely fluctuations in the particle's heat content with the average value which is given by the temperature.

Radiation equilibrium may be described by the equation

$$\bar{\varepsilon}_{ab} \int_0^\infty R(\lambda) \, d\lambda = \bar{\varepsilon}_{em} \int_0^\infty B(\lambda, T_g) \, d\lambda, \tag{5}$$

where $\bar{\varepsilon}_{ab}$ and $\bar{\varepsilon}_{em}$ are the average efficiencies for grain absorption from the interstellar radiation field $R(\lambda)$ and grain emission at temperature T_g respectively. For $\bar{\varepsilon}_{em} = \varepsilon_{eff}$ $= 33.3 \, aT_g$ (the maximum microwave emissivity as defined in Greenberg (1971), we find a lower bound on the small particle temperature to be given by

$$\bar{\varepsilon}_{ab} W T_R^4 = \varepsilon_{eff} T_g^4 = 33.3 a T_g^5, \tag{6}$$

where we have let the interstellar field be at a temperature T_R diluted by the factor W.

The values of $\bar{\varepsilon}_{ab}$ and $\bar{\varepsilon}_{em}$ are strongly size dependent. For example, we may note that $\bar{\varepsilon}_{ab} = 0.1$ for a 0.1 μ particle and is $\bar{\varepsilon}_{ab} = 1.0$ for a 1.0 μ particle (Greenberg, 1968). For $W = 10^{-14}, T_R = 10000°, a = 10^{-5}$ cm, $\bar{\varepsilon}_{ab} = 0.1$ we get $T_g = 8°$. Noting that $\bar{\varepsilon}_{ab}$ for a 10^{-6} cm particle is certainly less than that for a 10^{-5} cm particle we may immediately note that the temperature of a 10^{-6} cm particle would be very close to that of a 10^{-5} cm particle. This could have been expected from extrapolation of the curve showing the variation of grain temperature with size (Greenberg, 1968). However, the temperature fluctuations will be substantial.

Following standard procedures (Kittel, 1956) we find the heat content of a low temperature particle to be

$$U = (3\pi^4/5) N(kT_g)(T_g/\Theta)^3,$$

where Θ is the Debye temperature and N is the total number of molecules in the grain. Using a representative value of $\Theta = 300$ and obtaining N by dividing the volume of the grain by the volume per molecule assumed to be $\sim 8 \times 10^{-24}$ cm^3 we get $U = 1.55 \times 10^{-12}$ ergs or about 1 eV. Thus a 5eV photon ($\lambda = 2480$ Å) would give rise to a grain internal energy of about five times the average and the instantaneous grain 'temperature' would be 15°. If the accretion of condensible atoms is a sensitive function of grain temperature, frequent temperature spikes would inhibit grain mantle growth. One possible consequence could be that if we were to inject a size spectrum of silicate particles into interstellar space, there could be a size below which dirty ice mantles

could not accrete so that one would find a mixture or very small 'pure' silicate particle
with a distribution of silicate core-ice mantle particles. This would be a possible simultaneous way of accounting for the classical portion '2' of the extinction curve by
core-mantle particles and the ultraviolet portion, '4', by small silicate particles. In dark
regions of interstellar space we might than expect that all particles would be at a
sufficiently low temperature and/or subject to so few high energy photons that they
could accrete the condensible gas material and that in such regions the ultraviolet rise
in the extinction should disappear. It is rather suggestive that this combination of
core-mantle and small bare particles is the most consistent with silicate absorption
observations and cosmic abundance arguments.

Moesta (private communication) has suggested another mechanism which would
lead to lack of accretion on very small particles.

6. Nonspherical Particle Scattering

This section is inteded to show that we are now capable of performing essentially
precise theoretical calculations on the scattering by smooth nonspherical particles up
to and beyond the resonance region. The details are much too cumbersome to report
here but the results as obtained by Reilly (1969) are convincing evidence on the reliability of the method. It should be mentioned that although we limit ourselves to presenting only extinction efficiencies, the method produces *complete* scattering information
analogous to Mie theory.

Extinction efficiencies for prolate spheroids of 2:1 ratio and refractive index $m =$
$1.33 - 0.05 i$ are shown in Figure 2 for three orthogonal orientations. Experimental
results from the microwave scattering laboratory are seen to be quite reliable in the size
range. For comparison are shown the E and H extinction efficiencies of perfectly

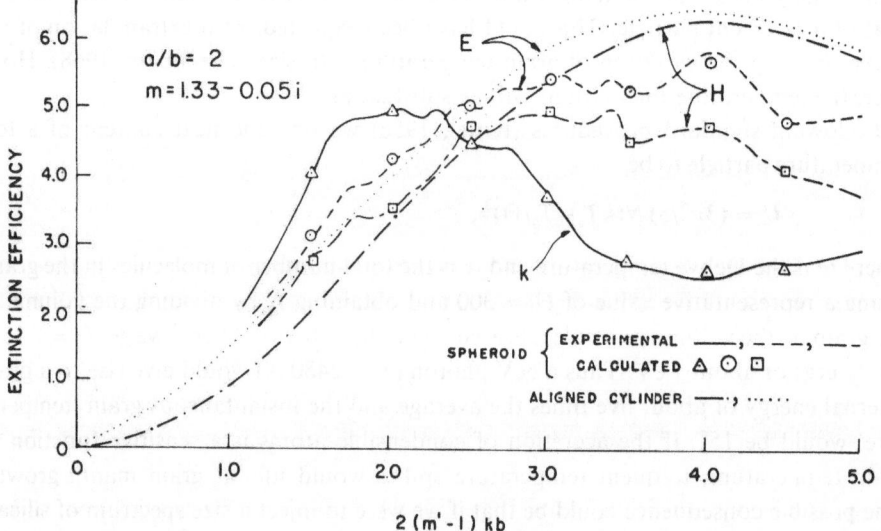

Fig. 2.   Theoretical and experimental extinction curves for prolate spheroids and cylinders.

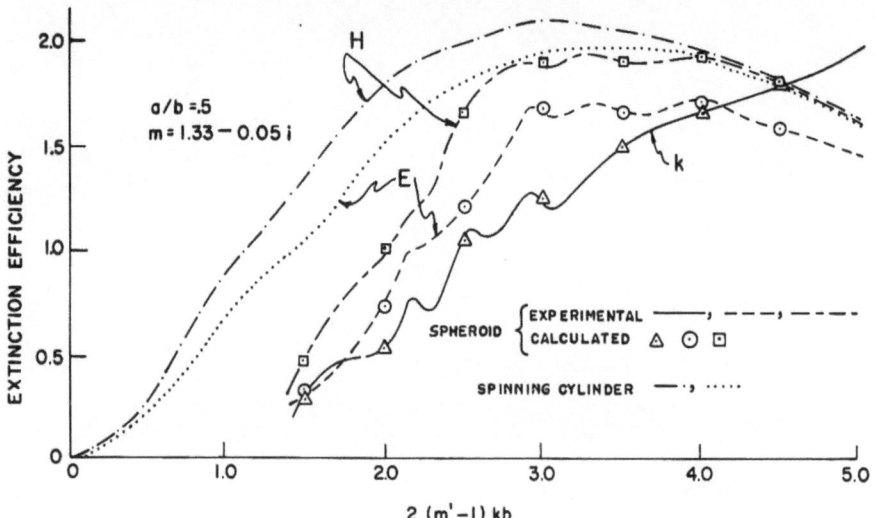

Fig. 3. Theoretical and experimental extinction curves for oblate spheroids and spinning cylinders.

aligned infinite cylinders normalized to correspond to the same asymptotic $(x \to \infty)$ value of 4 for the E and H prolate spheroids. Although the results differ in detail, it appears that semi-quantitative inferences based on cylinders are realistic. In particular, the degree of polarization is not very different between the finite spheroid and the infinite cylinder.

In Figure 3 are shown some new theoretical results for $1:2$ oblate spheroids. Again comparison is made with experimental results. We have calculated the average extinction from a set of cylinders spinning about an axis as a reasonable basis for comparison with the oblate spheroid. For small sizes the results are quite disparate but in the resonance region there is some similarity. Of particular note is that the degree of polarization by the spinning cylinders is both comparable with the oblate spheroid as well as the perfectly aligned cylinders.

References

Bertie, J. E., Labbé, A. J., and Whalley, E.: 1969, *Chem. Phys.* **50**, 4501.
Bless, R. C. and Savage, R. D.: 1972, *Astrophys. J.* **171**, 293.
Gillett, F. C. and Forrest, W. J.: 1973, *Astrophys. J.* **179**, 483.
Greenberg, J. M.: 1968, in B. M. Middlehurst and L. H. Alter (eds.), *Nebulae and Interstellar Matter*, Univ. of Chicago Press, Chicago, p. 221.
Greenberg, J. M.: 1971, *Astron Astrophys.* **12**, 240.
Greenberg, J. M.: 1972, *J. Colloid Interface Sci.* **39**, 513.
Greenberg, J. M.: 1973, in M. A. Gordon and L. E. Snyder (eds.), *Molecules in the Galactic Environment*, J. Wiley & Sons, p. 94.
Huffman, D. R. and Stapp, J. L.: 1971, *Nature Phys. Sci.* **229**, 45.
Launer, P. J.: 1952, *Am. Minerologist* **37**, 764.
Reilly, E. D., Jr.: 1969, Thesis, R.P.I.
Witt, A. N. and Lillie, C. F.: 1971, *OAO Symposium*, Amherst, Mass.

VARIATIONS OF INTERSTELLAR ULTRA-VIOLET
ABSORPTION*

SYUZO ISOBE

Tokyo Astronomical Observatory, University of Tokyo, Tokyo, Japan

Abstract. It is shown that mixtures of graphite core-ice mantle grains with large and small mean size and graphite grains provide good fits to the observed interstellar extinction curve.

Since the existence of interstellar absorption was found in 1930 by Trumpler (1930), many types of dust grains have been proposed to explain the absorption. However, as the observed wavelengths were very limited before 1965, almost any shape of reddening curve could be fitted to the observed ones by using any type of grains, if the size distribution of grains was adjusted in proper ways. Since 1965, the observed wavelengths have been extended to longer and shorter wavelengths. And in these infra-red and ultra-violet wavelengths, various types of interstellar reddening curves and circumstellar infra-red emission curves have been observed.

Stecher (1965) observed several stars at wavelengths extended to $1/\lambda \simeq 8.3\ \mu^{-1}$ and pointed out the existence of interstellar absorption bands at $1/\lambda \simeq 4.4\ \mu^{-1}$. It was already pointed out by Greenberg at this time and subsequently (Greenberg, 1966, 1967, 1968, 1969) that no model of interstellar grains which included dielectric materials with 'realistic' indices of refraction in the far ultraviolet could be used to represent the continued rise of the interstellar extinction in the far ultraviolet. All the models which presume to match the far ultraviolet extinction suffer from this defect. Carruthers (1970) and Stecher (1970) confirmed the results. Recently, Bless and Savage (1972) obtained the ultra-violet interstellar reddening curve for several stars (ζ Oph O9.5V, β^1 Sco B0.5V, $\theta^{1,2}$ Ori O8.5, σ Sco B1III, and etc.) at wavelengths extended to $1/\lambda \simeq 9.1\ \mu^{-1}$. It is found by these observations that the positions of the wavelengths of maximum interstellar absorption around $1/\lambda \simeq 4.6\ \mu^{-1}$ and the features of the reddening curves at the ultra-violet wavelengths are quite different from star to star.

Wickramasinghe and Nandy (1970, 1971a, 1971b) found that graphite-iron-silicate grain mixtures provide good fits to the observed interstellar extinction curve including the characteristic graphite feature at $1/\lambda - 4.4\ \mu^{-1}$ if one ignores the variation of the absorptivity of the silicate in the ultraviolet. Mixtures of enstatite silicate grains $((MgFe)\ SiO_4)$ with three types of size distribution (Huffman and Stapp, 1971) and mixtures of graphite-silicate-silicon carbide grains (Gilra, 1971) are also shown to provide good fits to the observed curves. The Gilra paper assumes constant indices of refraction for all but the graphite and therefore its predictions in the far ultraviolet are misleading. With this *basic* reservation it is explained by these grain mixtures that

---

* Detailed discussions of these problems will be given in *Pub. Astron. Soc. Japan* **25**, No. 2, 1973.

Greenberg and Van de Hulst (eds.), Interstellar Dust and Related Topics, 11–19.

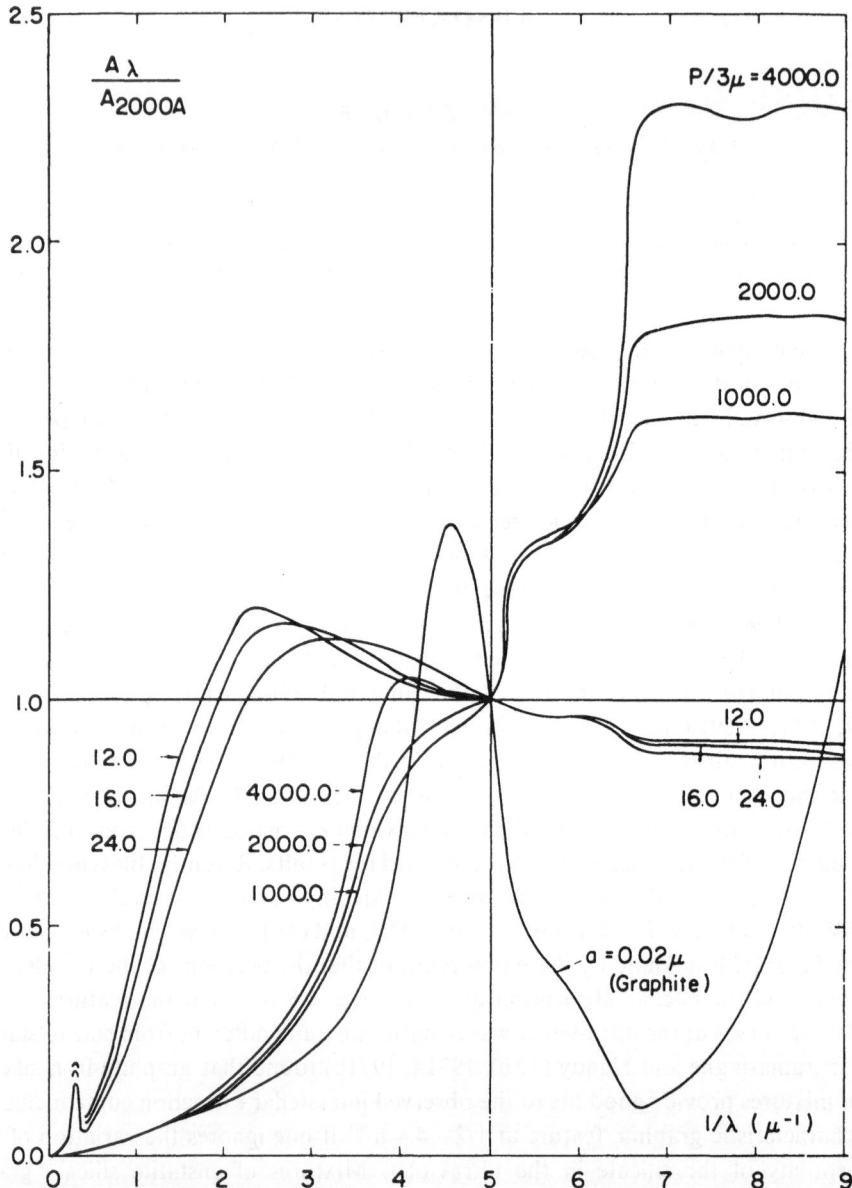

Fig. 1.   The normalized extinction curves of graphite grains and graphite core-ice mantle grains. They are normalized so that the extinction at the wavelength, 2000 Å, is 1.0 mag. The size distribution of the graphite core ice mantle grains is given by $n(a) = n(0) \exp(-P/3\mu \, a^3)$.

the varieties of the reddening curves at the ultra-violet wavelengths may be fitted by adjusting the abundance ratio, and or, the size distribution of each grain. However, every type of grain shown in the previous papers is stable in the interstellar space and therefore the varieties of the abundance ratio and the size distribution must be made before the grains are ejected from stars.

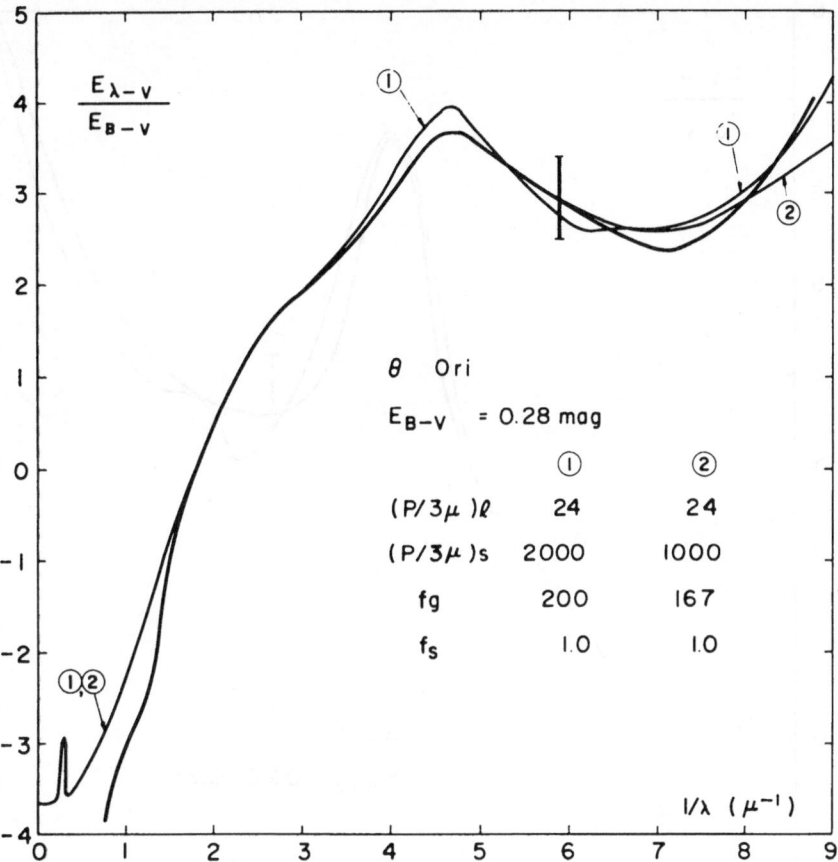

Fig. 2a. The observed and calculated extinction curves for the star $\theta^{1,2}$ Ori. The normalization is made by $A_v = 0.0$ magnitude and $E_{B-V} = 1.0$ mag. Thick line shows the observed one and thin lines show the calculated ones. The parameters of grains are shown. The vertical bar is the magnitude of the observational error.

In order to provide good fits to the observed interstellar extinction curve, we need grain mixtures with three or more types of grains and or size distributions, each one of which contributes to the extinction at the optical wavelengths, at the ultra-violet wavelength, and around $1/\lambda - 4.6\ \mu^{-1}$. Figure 1 shows the extinction curve of graphite grains with radius, 0.02 μ, and the graphite core-ice mantle grains, which are normalized so that the extinction at the wavelength, 2000 Å, is 1.0 mag. The size distribution of grains is given by $n(a) = n(0) \exp(-P/3\mu\, a^3)$, where P and μ are the destruction probability and the growth rate of ice mantle, respectively (Greenberg, 1966).

The graphite core-ice mantle grains with relatively large mean grain radius ($P/3\mu - 12-24$) mainly contribute to the extinction at optical and infra-red wavelengths and those with relatively small mean radius ($P/3\mu - 1000-4000$) mainly contribute to that at the ultraviolet wavelengths including the ultraviolet absorption by ice. Figures 2a and 2b show the normalized extinction curves observed by Bless and Savage (1972)

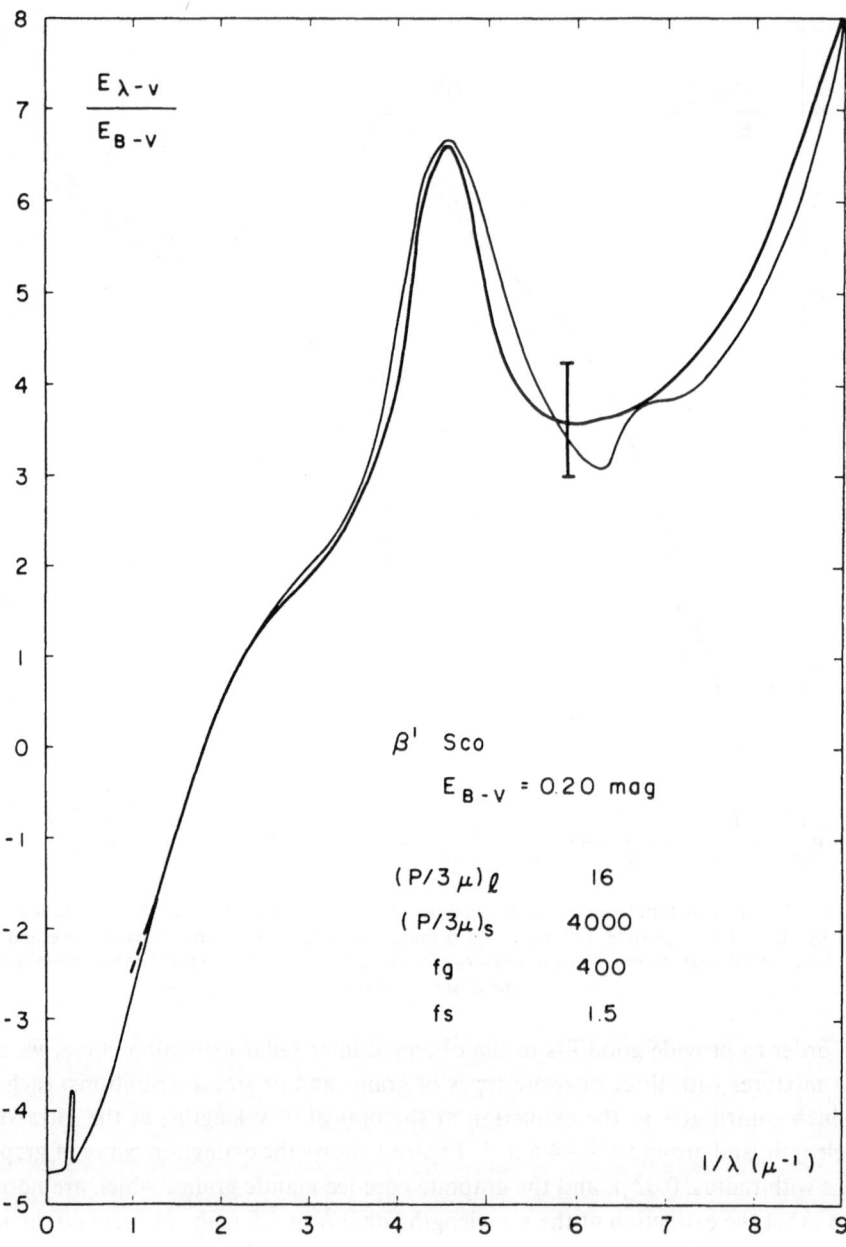

Fig. 2b.   Same figure as Figure 2a, but for the star β^1 Sco.

and calculated by the author for $\theta^{1,2}$ Ori and β^2 Sco respectively. And Figures 3a and 3b show the contribution of each type of grains to the interstellar absorption. The same figures for the other stars as shown in Figure 2 are shown by Isobe (1973) and the parameters of the size distributions and the abundances of each grain are shown

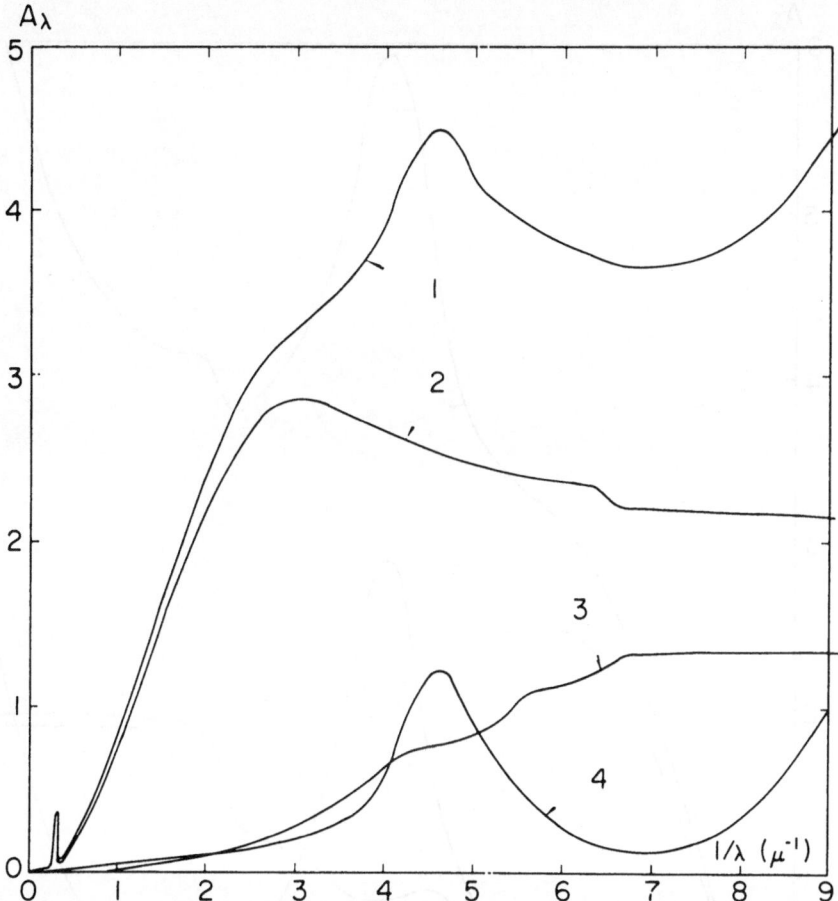

Fig. 3a. The contribution of each type of grains to the total extinction on the line of sight to $\theta^{1,2}$ Ori. The curves 2 and 3 show the extinction curves of the graphite core-ice mantle grains with large and small mean radius, respectively, and the curve 4 shows that of the graphite grains with radius of 0.02 μ. All curves are normalized so that A_v for the curve 2 is 2.0 mag. The curve 1 shows the total extinction curve.

TABLE I

Parameter of grains

	ζ Oph	σ Sco	θ Ori	β^1 Sco	Average
$(P/3\mu)_l$	24	24	24	16	20
$(P/3\mu)_s$	2000	1000	1000	4000	1000
f_g	300	300	167	400	333
f_s	3.0	1.0	1.0	1.5	2.0
$(n_d)_l$	2.8×10^{-13}	2.8×10^{-13}	2.8×10^{-13}	1.9×10^{-13}	2.3×10^{-13}
$(n_d)_s$	8.0×10^{-12}	9.9×10^{-13}	9.9×10^{-13}	1.1×10^{-11}	2.0×10^{-12}
$(n_d)_g$	2.8×10^{-11}	2.8×10^{-11}	1.8×10^{-11}	2.6×10^{-11}	2.7×10^{-11}
(A_v)	$0.^m32$	$0.^m10$	$0.^m10$	$0.^m18$	$0.^m20$

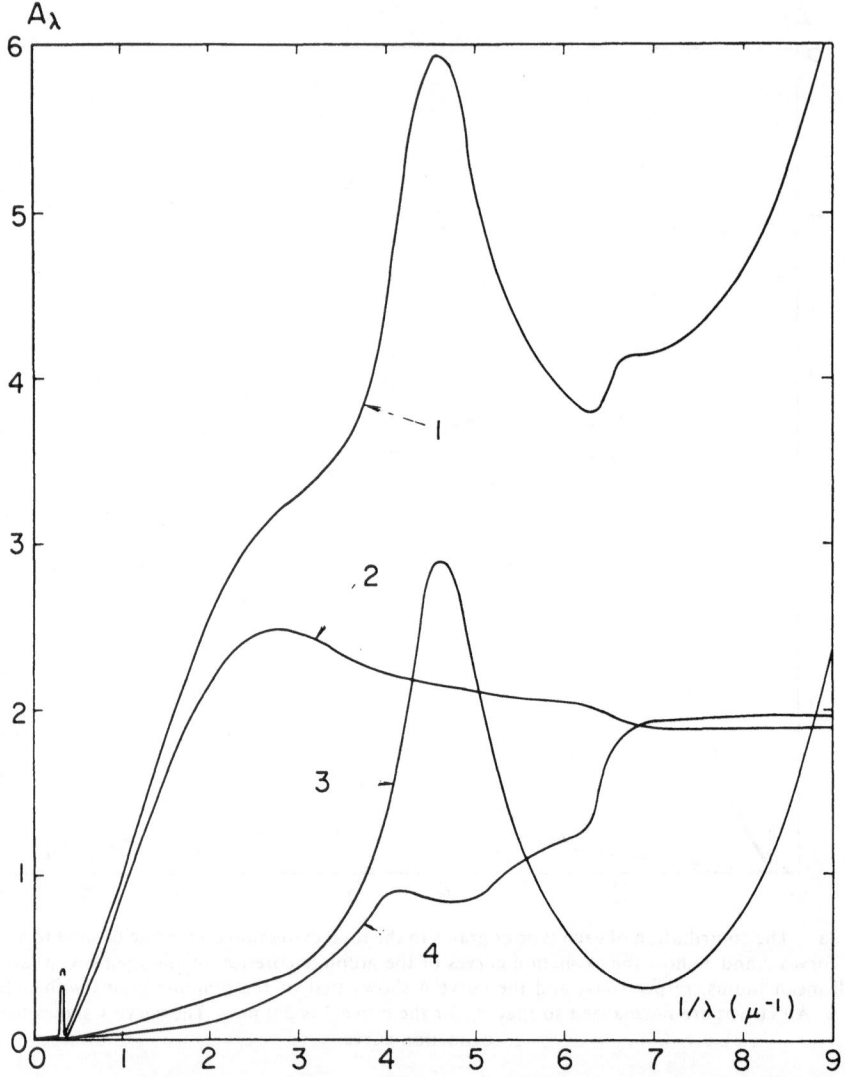

Fig. 3b. Same figure as Figure 3a, but for the case of β^1 Sco.

TABLE II
Maximum radius of grain

n_g/n_H	a_{max}
7.7×10^{-10}	$0.02\ \mu$
7.7×10^{-11}	$0.04\ \mu$
7.7×10^{-12}	$0.09\ \mu$
7.7×10^{-13}	$0.20\ \mu$
7.7×10^{-14}	$0.44\ \mu$

TABLE III

The mean number density of interstellar grains

$P/3\mu$	n_a[a]	n_b[b]
10	8.45×10^{-14} cm^{-3}	1.19×10^{-13} cm^{-3}
20	1.61×10^{-13} cm^{-3}	2.34×10^{-13} cm^{-3}
40	2.47×10^{-13} cm^{-3}	3.47×10^{-13} cm^{-3}
60	3.75×10^{-13} cm^{-3}	6.57×10^{-13} cm^{-3}
80	7.55×10^{-13} cm^{-3}	8.91×10^{-13} cm^{-3}
100	9.91×10^{-13} cm^{-3}	1.10×10^{-12} cm^{-3}
200	2.39×10^{-12} cm^{-3}	2.21×10^{-12} cm^{-3}
400	5.95×10^{-12} cm^{-3}	4.03×10^{-12} cm^{-3}
600	1.02×10^{-11} cm^{-3}	5.84×10^{-12} cm^{-3}
800	1.50×10^{-11} cm^{-3}	7.55×10^{-12} cm^{-3}
1000	2.02×10^{-11} cm^{-3}	9.20×10^{-12} cm^{-3}
2000	5.01×10^{-11} cm^{-3}	1.68×10^{-11} cm^{-3}
4000	1.20×10^{-10} cm^{-3}	3.00×10^{-11} cm^{-3}

[a] The mean number density, when the interstellar absorption is 2.0 mag. kpc^{-1}.
[b] The mean number density, when the mean density of hydrogen atoms is 0.7 cm^{-3} in interstellar space.

in Table I. As shown in Figure 2, the observed normalized extinction curves fit well to the calculated normalized extinction curve.

It is shown by Isobe (1971) that the growth rate of ice grains is proportional to the number density of oxygen atoms in the interstellar space. However, the maximum size of ice grains in the interstellar space depends on the ratio of the number density of grains, n_g, to that of oxygen atoms, n_O, by the exhaustion of the adsorbed oxygen atoms. Table II shows the relations between the ratio, n_g/n_H, and the maximum grain radius when the ratio, n_O/n_H, is 6.76×10^{-4} (Allen, 1963, p. 30). Giving the distribution function for grain radius, $n(a) = n(0) \exp(-P/3\mu a^3)$ and considering the facts that the interstellar absorption is 2.0 mag. kpc^{-1} (Allen, 1963, p. 251), we can obtain the mean density of grains, n_a, in the interstellar space, which is shown in Table III. The mean number density, n_b, is also shown in Table III for the case that the mean number density of hydrogen atoms, n_H, is 0.7 cm^{-3} and the ratio of number density of oxygen atoms, n_O, to that of hydrogen atoms is 6.76×10^{-4} (Allen, 1963, p. 30). When $P/3\mu$ is 12–24, the number density of grain is 1.42–2.79×10^{-13}. The number density of graphite grains, which produce the observed 2200 Å absorption band, is f_g times of number density of the graphite core-ice mantle grains and is shown in Table I. As the gas densities in the interstellar clouds are ten times of the mean interstellar gas densities, the ratio of number density of grains to that of hydrogen is about 10^{-11}. Therefore, the ice mantle can grow up to the size of 0.09 μ, which is smaller than the mean grain of 0.2 μ in the interstellar clouds. Table IV shows the relations between the values of $P/3\mu$ and the mean grain size. However, de Jong and Kamijo (1972) show that the destruction rate of interstellar ice grains by sub-cosmic rays is independent of the grain size and is 10^{-14} cm y^{-1}, which is the same order of destruction rate as the mutual

TABLE IV

Mean radius of grain

$P/3\mu$	$\bar{a}$	$R(=A_V/E_{B-V})$
14.6	0.22 μ	6.4
23.2	0.19 μ	4.1
40.0	0.16 μ	2.9
78.2	0.13 μ	2.2
185	0.10 μ	1.9
625	0.07 μ	1.6

impacts of grains with radius of 0.1 μ at the cloud-cloud encounters. Therefore, the adsorbed oxygen atoms are supplied to interstellar space and the ice mantle can grow up to 0.2 μ or more. From these results it follows that the interstellar reddening curves at the optical wavelengths are nearly the same in all regions of interstellar space.

If the number density of gas atoms is 0.2 times the mean interstellar gas density in the inter-cloud regions, the values of $P/3\mu$ are about 1500, which is consistent with the values of $P/3\mu$ for the grains with relatively small mean radius shown in Table I. In Table I are also shown the visual interstellar absorption by the small graphite core-ice mantle grains if the absorption by the large grains is 2.0 mag. The small grains hardly contribute to the visual absorption. In the inter-cloud regions, only 5% of interstellar atoms condense to the grains, which means that most of the atoms remain in the gas phase.

If we can somehow rationalize the assumptions regarding the ultraviolet absorptivity of ice (Greenberg, 1968) it is possible to arrive at the following conclusions: (1) that there are relatively large grains in the interstellar clouds, (2) that there are relatively small grains in the inter-cloud regions, and (3) that there are graphite grains with a size of about 0.02 μ as the kernels of the graphite core-ice mantle grains in all regions of interstellar space. Therefore, the variety of the interstellar ultra-violet absorption depends on the effective path-length ratio of the cloud regions to the inter-cloud regions along the line of sight.

References

Allen, C. W.: 1963, *Astrophysical Quantities*, 2nd ed., Athlone Press, London.

Bless, R. C. and Savage, B. D.: 1972, *Astrophys. J.* **171**, 293.

Carruthers, G. R.: 1970, in L. Houziaux and H. E. Butler (eds.), 'Ultraviolet Stellar Spectra and Related Ground-Based Observations', *IAU Symp.* **36**, 100.

De Jong, T. and Kamijo, F.: 1973, *Astron. Astrophys.* **25**, 363.

Gilra, D. P.: 1971, *Nature* **229**, 237.

Greenberg, J. M.: 1966, in K. Lodén, L. O. Lodén and U. Simerstad (eds.), 'Spectral Classification and Multicolour Photometry', *IAU Symp.* **24**, 291.

Greenberg, J. M.: 1967, in J. M. Greenberg and T. P. Roark (eds.), 'Interstellar Grains', NASA SP-140, p. 224.

Greenberg, J. M.: 1968, *Stars and Stellar Systems* **7**, 221.

Greenberg, J. M. and Shah, G. A.: 1969, *Physica* **41**, 92.

Huffman, D. R. and Stapp, J. L.: 1971, *Nature Phys. Sci.* **229**, 45.

Isobe, S.: 1971, *Publ. Astron. Soc. Japan* **24**, 27.
Isobe, S.: 1973, *Publ. Astron. Soc. Japan* **25**, 253..
Stecher, T. P.: 1965, *Astrophys. J.* **142**, 1683.
Stecher, T. P.: 1970, in L. Houziaux and H. E. Butler (eds.), 'Ultraviolet Stellar Spectra and Related Ground-Based Observations', *IAU Symp.* **36**, 24.
Trumpler, R. J.: 1930, *Lick Obs. Bull.* **14**, 154.
Wickramasinghe, N. C. and Nandy, K.: 1970, *Nature Phys. Sci.* **227**, 51.
Wickramasinghe, N. C. and Nandy, K.: 1971a, *Nature Phys. Sci.* **230**, 16.
Wickramasinghe, N. C. and Nandy, K.: 1971b, *Monthly Notices Roy. Astron. Soc.* **153**, 205.

OAO-2 OBSERVATIONS OF THE
NEAR ULTRAVIOLET EXTINCTION FEATURE

BLAIR D. SAVAGE

Space Astronomy Laboratory, Washburn Observatory, University of Wisconsin,
Madison, Wis. U.S.A.

Abstract. OAO-2 near UV spectral scans of 75 early-type stars have been examined for evidence of the UV extinction bump. The bump has been detected in 40 of these stars. The position of the maximum of the bump is within ± 40 Å of 2175 Å in all cases. A color excess between the wavelengths 2175 Å and 3500 Å was determined for all 40 stars. This color excess, which is a measure of the bump strength, correlates very well with $E(B-V)$ except for stars associated with nebulosity.

In a recent paper Bless and Savage (1972, hereafter referred to as Paper I) presented ultraviolet extinction curves for 17 stars. These curves were derived from Orbiting Astronomical Observatory 2 (OAO-2) data and covered the spectral region λ 1100–3600. The most significant feature of these curves was the well defined bump in extinction which peaked at 2175 Å. The purpose of the present paper is to briefly discuss new observations of this very important feature.

The solid curve in Figure 1 shows the 'average' ultraviolet extinction curve obtained in Paper I. We have plotted extinction in magnitudes vs $1/\lambda$ in μ^{-1}. The extinction is normalized to zero at the V bandpass and to $E(B-V) = 1.0$. The curves for ζ Oph and θ^{1+2} Ori illustrate the range of variability in extinction found so far. These curves were obtained from an analysis of low resolution ($\Delta\lambda \sim$ 10–20 Å) spectral scans of reddened early-type stars. OAO-2 contains two objective grating spectral scanners; one covers the near ultraviolet and the other the far ultraviolet. The relationship between this spectral coverage and the ultraviolet extinction is illustrated in Figure 1. As can be seen, Spectrometer 1 covers the region of the ultraviolet extinction bump while Spectrometer 2 covers the minimum and subsequent rise in extinction toward short wavelengths. Unfortunately Spectrometer 1 operated successfully for only $1\frac{1}{2}$ yr. In comparison, at the time of writing ($3\frac{1}{2}$ yr after launch) Spectrometer 2 continues to operate perfectly. This paper mainly concerns an analysis of all the good Spectrometer 1 scans for normal stars of spectral type B3 or earlier.

Since large variations are seen in the normalized ultraviolet extinction curves it is of interest to investigate the systematics of these variations in a large sample of stars. With this project in mind, we have defined the color excess, $E(2175–3500)$, which is a measure of the extinction between 2175 Å and 3500 Å. 2175 Å is the average position of the maximum in the ultraviolet extinction bump. This color excess is illustrated in Figure 1. One can see that this quantity gives a good measure of the extinction in the near ultraviolet bump. Fortunately, this quantity can be derived from individual near ultraviolet spectrometer scans of reddened stars whereas other measures of the bump, for example the height above the dashed line in Figure 1, require both near and far ultraviolet data. Both sets of data are available for only a limited number of reddened stars.

Greenberg and Van de Hulst (eds.), Interstellar Dust and Related Topics, 21–27..

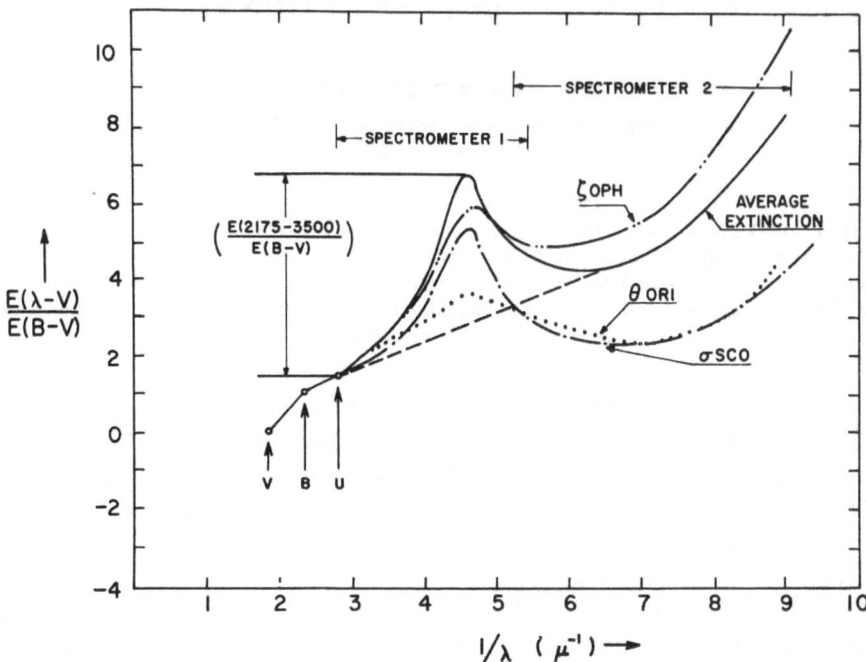

Fig. 1.   Ultraviolet extinction curves from Bless and Savage (1972). The curves for ζ Oph and θ^{1+2} Ori illustrate the range of variability in ultraviolet extinction found so far. The wavelength coverage of each OAO-2 spectrometer is indicated above the extinction curves. $E(2175-3500)$ refers to the color excess which gives a measure of the amount of additional extinction in the near ultraviolet bump.

In order to derive an extinction curve or a color excess one must of course know the intrinsic behavior of stars of different spectral type and luminosity class. A detailed discussion of this subject was given in Paper I. The basic result of that analysis was that the intrinsic ultraviolet energy distributions of normal early-type stars correlates well with $(B-V)_0$, the intrinsic $(B-V)$ color. Furthermore this result does not depend sensitively on luminosity class. Emission line stars often deviate from normal stars due to the presence of continuous Balmer emission in the near ultraviolet. Except where noted stars with Be characteristics have been omitted from this program. In this paper we have continued the investigation of intrinsic properties of stars. Figure 2 shows a color-color plot for normal early-type stars with $E(B-V) \leq 0.10$. The ultraviolet color spans the wavelengths 2175 and 3500 Å and was obtained from near ultraviolet scanner data. The colors plotted have been corrected to intrinsic colors by application of the 'average' extinction curve shown in Figure 1. One can see that the near ultraviolet color correlates well with $(B-V)_0$ and there are no *pronounced* luminosity class effects. 95% of the stars are contained in a band $\pm\ 0\overset{m}{.}15$ from the dashed line (a visual estimate of the mean relation). We can conclude that a star's $(B-V)_0$ is a good measure of its intrinsic near ultraviolet energy distribution. The amount of scatter in Figure 2 ($\pm\ 0\overset{m}{.}15$) can be considered to be a conservative estimate of the uncertainty in measures of $E(2175-3500)$ to be discussed shortly. As will be seen this

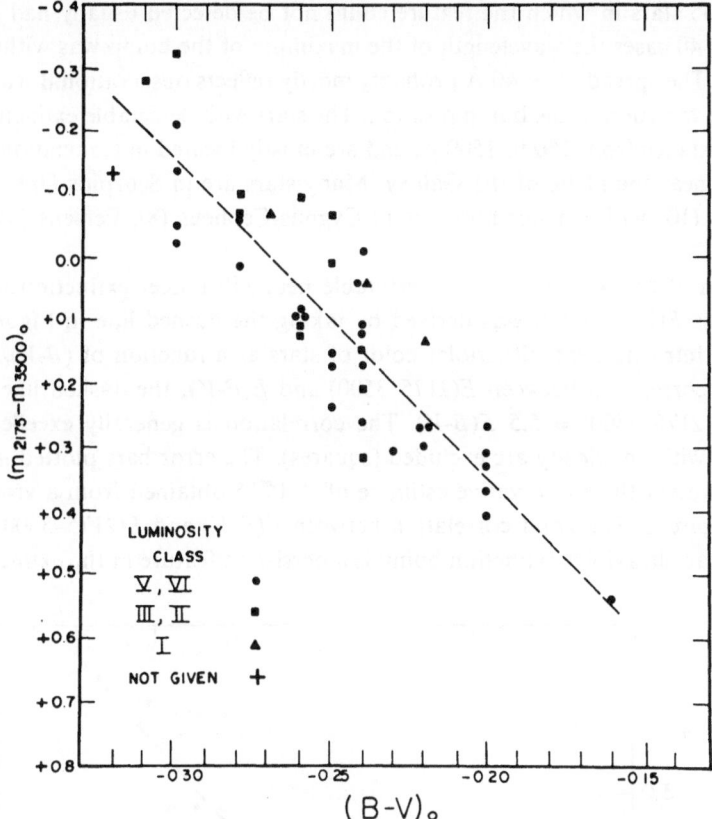

Fig. 2. A color-color plot obtained from OAO-2 Spectrometer 1 scans of slightly reddened stars $(E(B-V) < 0.10)$. $m_{2175}-m_{3500}$ refers to a color that spans the wavelengths 2175 and 3500 Å. The colors plotted have been corrected to zero reddening by using the average extinction curve seen in Figure 1. The symbols indicate stars of various luminosity classes. The dashed line represents the mean relation between the near ultraviolet color and $(B-V)_0$.

uncertainty is much smaller than the variations of normalized near ultraviolet extinction observed from object to object. Underhill (1972) has suggested that differences in line blanketing between stars with the same visual spectra could introduce large errors into the derived extinction curves, particularly in the vicinity of 2200 Å. On the basis of the good correlation in the observed sample of stars seen in Figure 2 and also in Figure 4 of Paper I we must conclude that Underhill's suggestion is not supported by the observations.

During its lifetime, the near ultraviolet scanner on OAO-2 obtained spectral scans of 75 normal stars of spectral type B3 or earlier. We have searched all these scans for evidence of the near ultraviolet extinction feature. The method was simply to intercompare all the spectrometer scans with those of a few very slightly reddened stars $(E(B-V) < 0.02)$. By this process the extinction bump was detected in 40 of 75 stars. Those stars with a detectable near ultraviolet feature had $E(B-V)$ in the range 0.04 to

0.61. The 35 stars in which the feature could not be detected usually had $E(B\text{-}V) <$ 0.05. In all 40 cases the wavelength of the maximum of the bump was within $\pm$ 40 Å of 2175 Å. The spread of $\pm$ 40 Å probably mostly reflects observational scatter rather than a true variation of the bump position. The stars with detectable extinction bumps range in distance from 150 to 1500 pc and are mostly located in associations confined to regions near the plane of the Galaxy. Many stars are in Scorpius-Ophiuchus (11) and Orion (10) with smaller numbers in Cygnus-Cepheus (8), Perseus (3) and elsewhere (8).

For each of the 40 stars with a detectable near ultraviolet extinction feature the color excess $E(2175\text{-}3500)$ was derived by taking the dashed line in Figure 2 to represent the intrinsic near ultraviolet color of stars as a function of $(B\text{-}V)_0$. Figure 3 shows the correlation between $E(2175\text{-}3500)$ and $E(B\text{-}V)$; the dashed line is for the relation $E(2175\text{-}3500) = 5.5\ E(B\text{-}V)$. The correlation is generally excellent if stars associated with nebulosity are excluded (squares). The error bars plotted in Figure 3 simply represent the conservative estimate of $\pm\ 0.^m15$ obtained from a visual inspection of Figure 2. The good correlation between $E(B\text{-}V)$ and $E(2175\text{-}3500)$ indicates that the near ultraviolet extinction bump is a persistent feature in the extinction curve

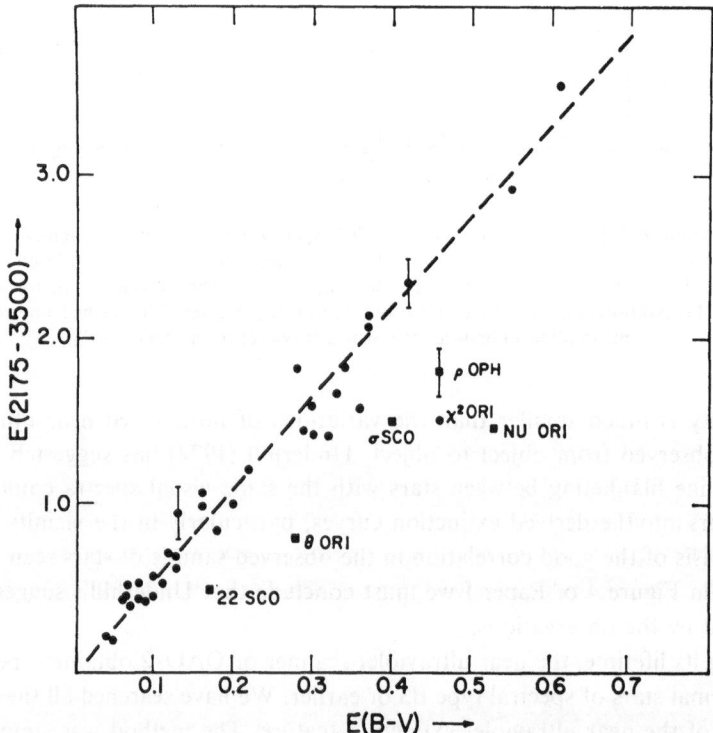

Fig. 3.   A correlation between the near ultraviolet color excess and $E(B-V)$ for 40 stars. $E(2175-3500)$ is basically a measure of the excess extinction in the near ultraviolet bump. Stars associated with nebulosity are marked with squares (■) and other stars are marked with circles (●). The dashed line represents the relation $E(2175-3500) = 5.5\ E(B-V)$.

TABLE I

Stars with abnormal near ultraviolet extinction

Object	S.T.	$E(B-V)$	Comments
θ^{1+2}	06_p	0.36	Associated with emission nebula.
			Visual and infrared extinction abnormal (Lee, 1968).
	$09.5V_p$	0.21	θ^2 Ori exhibits large P_V/P_B (Serkowski, 1968) and variable polarization (Coyne and Gehrels, 1967). The variable polarization indicates circumstellar shell activity and the possibility of contamination from continuous Balmer emission in the near ultraviolet. The extinction measurement for this star should be treated with caution.
NU Ori	B1V	0.53	Associated with nebulosity.
			Visual and infrared extinction abnormal (Lee, 1968).
			Large P_V/P_B (Serkowski, 1968).
χ^2 Ori	B2Ia	0.46	No associated nebulosity.
			Variable polarization (Vitrichenko and Efimov, 1965) indicates shell activity and possibility of contamination by continuous Balmer emission. The extinction measurement should be treated with caution.
σ Sco	B1III	0.40	Associated with reflection nebula (Hubble, 1922).
ϱ Oph	B2IV	0.46	Associated with reflection nebula (Hubble, 1922).
	B2V		Large P_V/P_B (Serkowski, 1968).
22 Sco	B2V	0.17	Associated with reflection nebula (Hubble, 1922).

of stars. Furthermore Figure 3 gives additional support to our contention that Underhill's suggestion about *large* errors being introduced into extinction measurements by variable ultraviolet line blanketing is wrong.

Six stars deviate significantly from the generally good correlation seen in Figure 3. Table I lists these stars and gives relevant information about their characteristics. A significant fact is that five of these six stars are associated with nebulosity. In addition a few of these stars are known to have peculiar visual and infrared extinction and abnormally large ratios of visual to blue polarization (P_V/P_B). As pointed out in Paper I it's possible that the acceptance of both stellar and nebular radiation by the OAO scanners produces the abnormal ultraviolet extinction measurements. In the near ultraviolet the important possibilities for nebular contamination are continuous Balmer and two photon emission from atomic hydrogen, and scattering by local dust into the line of sight. Contamination by two photon and Balmer emission should be negligible for all these objects except θ^{1+2} Ori because the other nebulae shine mostly by reflected light. Code (1973) has discussed the effect local scattering by dust might have on extinction measurements. For particle albedos similar to those presented by Witt and Lillie (1972) one would expect local scattering if important to make the bump appear stronger than normal. The observations reported here, however, indicate the opposite effect.

For several objects with abnormal near ultraviolet extinction additional data are available which can set some limits on the possibility of nebular contaminations. With a spatial resolution of 2' on direct far ultraviolet images of θ^{1+2} Ori Henry and

Carruthers (1970) found no evidence for nebular radiation and an analysis of the stellar images produced a far ultraviolet extinction determination (Weber *et al.*, 1971) in reasonable agreement with the far ultraviolet portion of the θ^{1+2} Ori curve presented in Paper I (also shown in Figure 1). From a far ultraviolet spectrum of σ Sco with an objective spectrograph Jenkins (private communicction) was able to conclude that the observed far ultraviolet radiation comes from a region having an angular diameter of less than 1'. Furthermore direct photographs of σ Sco in the visual were made with the University of Wisconsin 16" *f*/19 telescope. Using Kodak Tri-X film (the only film readily available) the star was first detected with an exposure of $^1/_{50}$ s. An increase of exposure to 60 s (a factor of 3000 increase neglecting reciprocity failure) failed to reveal nebulosity. In addition photometry observations of σ Sco plus nebulosity in a 40 Å band centered at $\sim$4500 Å with diaphragms of 2' and 1' were essentially identical to an accuracy of about 1%. From these data one can conclude that nebular contamination for these objects is probably unimportant unless it arises from a region very close to the stars.

The abnormal extinction for stars associated with nebulosity is most likely due to the local modification of the grains by the intense stellar radiation fields of the illuminating stars. Preferential destruction or ejection of small particles could explain the low near ultraviolet extinction evident in Figure 3. Paper I presented complete extinction curves for four of the six objects having peculiar near ultraviolet extinction. Figure 1 shows two of those curves which deserve additional comment. The near ultraviolet bump is much less pronounced in θ^{1+2} Ori compared to σ Sco. It appears in the case of σ Sco that the component mainly responsible for the far ultraviolet extinction rise has been destroyed whereas for θ^{1+2} Ori both the far ultraviolet component and the component responsible for the near ultraviolet bump have been destroyed. The absence of a strong extinction bump in θ^{1+2} Ori implies that one should be cautious in interpreting determinations of particle albedos from measurements of reflection nebulae. The results found here and in Paper I strongly suggest that particles which can exist in the general interstellar medium may be quite different from particles which can exist in the vicinity of hot stars.

The star χ^2 Ori is the only object listed in Table I that is not associated with nebulosity. This star does, however, exhibit variable polarization (Vitrichenko and Efimov, 1965) which indicates circumstellar shell activity and the possibility that the near ultraviolet data are contaminated by continuous Balmer emission from this shell. The extinction measurement for this star should therefore be treated with caution. A similar criticism applies to θ^2 Ori for which Coyne and Gehrels (1967) also detected variable polarization.

As is evident in Table I there appears to be a close relation between stars with peculiar ultraviolet extinction and peculiar polarization in the visual. This characteristic also holds for HD 37903 which has a very large P_V/P_B (Serkowski, 1968) and a peculiar far ultraviolet extinction (Weber *et al.*, 1971). Furthermore HD 37903 is associated with nebulosity.

A number of other correlations can be explored with the new near ultraviolet color excess. For example, one very interesting problem is that of finding the carriers of the diffuse interstellar features. Wu (1972) has investigated the strength of the diffuse features at 5780 Å and 4430 Å in stars with measured near ultraviolet excesses and has found that the correlation of the diffuse features with $E(2175-3500)$ shows as much scatter as the correlation of these features with $E(B-V)$. Therefore there is no convincing evidence to indicate that the diffuse features are carried by the material producing the near ultraviolet extinction bump. A possible difficulty with this conclusion is that $E(2175-3500)$ is significantly influenced by the behavior of the extinction curve in the far ultraviolet. This effect is evident in Figure 1 by comparing the extinction curve of ζ Oph with that of σ Sco. Although ζ Oph has a larger $E(2175-3500)/E(B-V)$ than σ Sco, the bump is more pronounced in the extinction curve of σ Sco. For the purpose of investigating the relationship between diffuse interstellar features and the near ultraviolet bump alternate measures of the bump strength are desirable.

References

Bless, R. C. and Savage, B. D.: 1972, *Astrophys. J.* **171**, 293.

Code, A. D,: 1973, this volume, p. 505.

Coyne, G. V. and Gehrels, T.: 1967, *Astron. J.* **72**, 887.

Henry, R. C. and Carruthers, G. R.: 1970, *Science* **170**, 527.

Hubble, E.: 1922, *Astrophys. J.* **56**, 416.

Lee, T. A.: 1968, *Astrophys. J.* **152**, 913.

Serkowski, K.: 1968, *Astrophys. J.* **154**, 115.

Underhill, A.: 1972, in A. D. Code (ed.), *Proceedings of OAO Symposium*, NASA Publication SP-310.

Vitrichenko, E. A. and Efimov, Y. S.: 1965, *Izv. Krymsk. Astrofiz. Obs.* **34**, 114.

Weber, S. V., Henry, R. C., and Carruthers, G. R.: 1970, *Astrophys. J.* **166**, 543.

Witt, A. and Lillie, C.: 1972, in A. D. Code (ed.), *Proceedings of OAO Symposium*, NASA Publication SP-310.

Wu, C. C.: 1972, *Astrophys. J.* **178**, 681.

THE RATIO OF TOTAL TO SELECTIVE EXTINCTION IN
THE DIRECTIONS OF SELECTED PLANETARY NEBULAE

JULIE H. LUTZ

Washington State University, Pullman, Washington, D.C., U.S.A.

Abstract. The ratio of total to selective extinction has been determined for the lines-of-sight to six planetary nebulae.

Values of the ratio of total to selective extinction (R) were derived for the lines-of-sight to six planetary nebulae by using statistical distances to the nebulae (Cahn and Kaler, 1971) in conjunction with diagrams of color excess vs distance (hereafter called ce − d diagrams). The ce − d diagrams were constructed using early-type stars within 1°.5 of each planetary (see Lutz and Lutz (1972) and Lutz (1973) for details of the observing program).

A value of R was assumed for the line-of-sight to a given planetary. The color excess of the planetary was calculated from

$$E(B - V) = 2.03 \frac{c}{R},$$

where c is the extinction constant of the planetary and the constant 2.03 results from adopting the reddening law given by Whitford (1958). The extinction constant was derived by comparing the observed ratio of the radio and Hβ flux with a theoretically computed ratio. This requires an assumption of electron temperature. For this study the electron temperature was assumed to be 7000 K. Raising the electron temperature from 7000 K to 10000 K would lower the value of R ultimately derived by 0.2. The radio fluxes were evaluated at 3 GHz, except in the cases of NGC 7026 (8 GHz) and NGC 6741 (10 GHz). The distance to each early-type program star was calculated using the assumed R, and a ce − d diagram was constructed. The distance to the planetary was found from the ce − d diagram. If the distance did not agree with the Cahn and Kaler distance, then another value of R was tried. This method introduced an uncertainty of ± 0.1 to ± 0.2 into the value of R derived.

TABLE I

Ratio of total to selective extinction

Nebula	l	R
NGC 6741	34°	>2.0
NGC 6894	70°	3.2
NGC 7026	89°	3.0
NGC 7354	108°	4.0
IC 1747	130°	>3.1
IC 289	139°	4.6

Greenberg and Van de Hulst (eds.), Interstellar Dust and Related Topics, 29–30.

The value of R for the line-of-sight to each planetary is shown in Table I. NGC 6741 and IC 1747 are optically thick, so the values of R for these nebulae are lower limits to the true values of R. A 50% change in any of the Cahn and Kaler distances would produce a change of about 1.0 in the value of R derived.

References

Cahn, J. H. and Kaler, J. B.: 1971, *Astrophys. J. Suppl.* **22**, 319.
Lutz, J. H.: 1973, *Astrophys. J.* **181**, 135.
Lutz, J. H. and Lutz, T. E.: 1972, *Astron. J.* **77**, 376.
Whitford, A. E.: 1958, *Astron. J.* **63**, 201.

THE RATIO OF TOTAL TO SELECTIVE EXTINCTION
FROM THE CLUSTER DIAMETER METHOD

DANIEL H. HARRIS

Steward Observatory and Institute of Atmospheric Physics, University of Arizona, Ariz., U.S.A.

Abstract. One hundred and fifty six open clusters with modern photometry, and apparent diameter measures by Trumpler, have been used to measure the mean ratio of total to selective extinction, $\bar{R}$. It is found that cluster diameter is related to both concentration class and richness class. Diameter also seems to depend on the background stellar density and to decrease with reddening. With such effects measured and removed $\bar{R} = 3.15 \pm 0.20$. This result is not significantly dependent on the cluster's position in space.

Of the methods available for measuring the interstellar extinction, only the cluster diameter method can measure the grey extinction of large particles. This method is therefore potentially very useful. To fully realize this potential we must collect accurate diameter measures and photometry for a large homogeneous sample of open clusters.

Trumpler (1930) and Wallenquist (1959) have published the only large and uniform sets of open cluster diameter measures. I have made an effort to assemble the necessary photometry of these clusters. Photometric data tabulated in Becker (1963), Johnson *et al.* (1961), and Hagen (1970) have been examined and inter-compared. Where possible the cluster color-magnitude diagrams were studied. Only 156 of the clusters had well determined distance modulii and color excesses. These were transformed to the UBV system. The photometric data were then used to compute the cluster distances for various assumed values of the ratio of total to selective absorption, **R**. For the purposes of this paper we consider clusters with $E_{B-V} < 0.15$ to be unreddened. Such small color excesses with normal values of **R** change the distance modulus by less than one-half magnitude.

Figures 1, 2, 3, and 4 present the log distance-log apparent angular diameter data for Trumpler's diameters. It is evident that the reddened and unreddened clusters follow nearly the same distance-apparent diameter relation. This is confirmed by the near agreement of the last squares fits also shown. Note that with the large scatter in apparent diameter the general distribution of points doesn't change rapidly with **R**. The best agreement of the reddened and unreddened clusters seems yo be near **R** = 3.2.

The distance-apparent diameter plots for Wallenquist's diameters look nearly the same as the Trumpler plots except that there are fewer points. Figure 5 presents the distance-apparent diameter data for Wallenquist's diameters and **R** = 3.2.

There are two noteworthy peculiarities in the distance-apparent diameter relations. First, the steep slope and noticable curvature of the least squares fits suggest that effects are present other than simply the reduction of apparent size with distance.

Greenberg and Van de Hulst (eds.), Interstellar Dust and Related Topics, 31–41.
All Rights Reserved. Copyright © 1973 by the IAU.

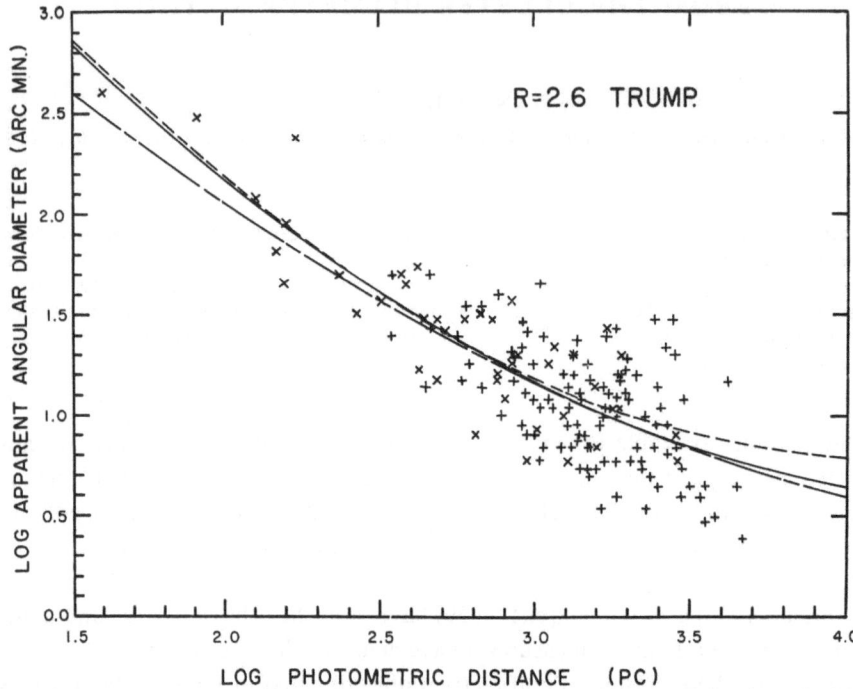

Fig. 1. The apparent angular diameter-photometric distance relations for open clusters with Trumpler diameters and $R = 2.6$. The ($\times$) points are clusters with $E_{B-V} < 0.15$ and the ($+$) points are clusters with $E_{B-V} \geq 0.15$. The curves are least squares fits to the points. The dashed curve fits the ($\times$) points. The solid-dashed curve fits the ($+$) points. The solid curve fits all the points.

Second, the clusters appear to shrink with increased color excess. This is shown by the difference between the fits to reddened and unreddened clusters. However, if we assume $\mathbf{R} \simeq 2.6$ then there is no shrinkage at all. The measurement of the apparent shrinkage with E_{B-V} can not be separated from the problem of finding the best value of $\mathbf{R}$. Before we can understand these effects we must attempt to reduce the size of the diameter residuals. In order to do this we will search for dependences of the apparent diameters on other variables, then remove these dependences.

We should expect to find a dependence of apparent diameter on cluster type (Lynds, 1967). To test for such a dependence I have computed residuals in the log-arithm of the apparent angular diameter relative to the least squares fits to all clusters regardless of color excess. These residuals show that Trumpler's diameters depend significantly on both concentration class and richness class. Wallenquist's diameters in contrast show no significant dependence on concentration. In order to correct for the type dependence in Trumpler's diameters I have computed mean type residuals. Residuals for all clusters of a particular concentration class and richness class have been averaged, except for those clusters that markedly deviate from the mean of their type.

Table I presents the preliminary mean type residuals for assumed $\mathbf{R}$ values of 2.8,

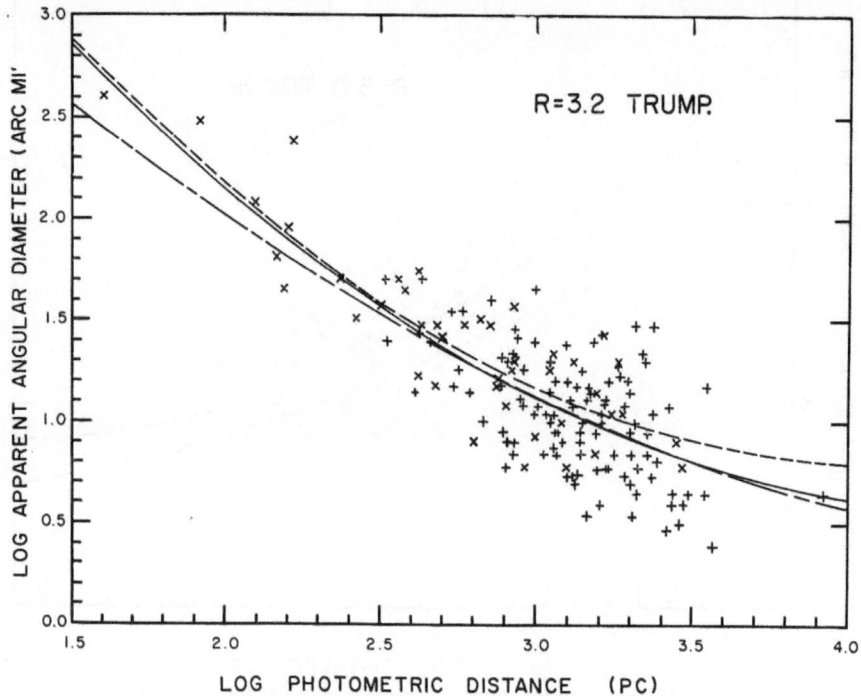

Fig. 2.   The apparent diameter-distance relations for $R = 3.2$.

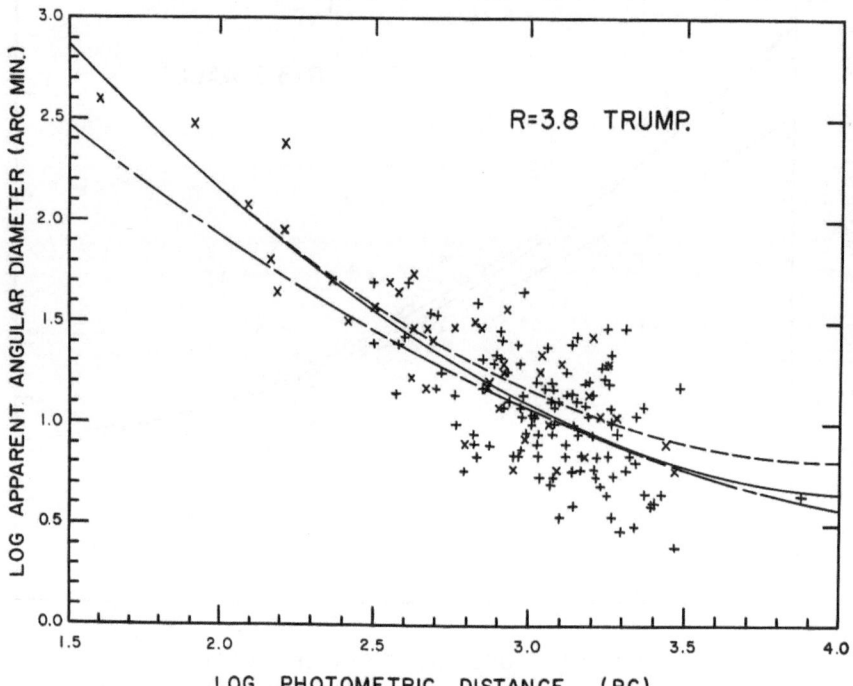

Fig. 3.   The apparent diameter-distance relations for $R = 3.8$.

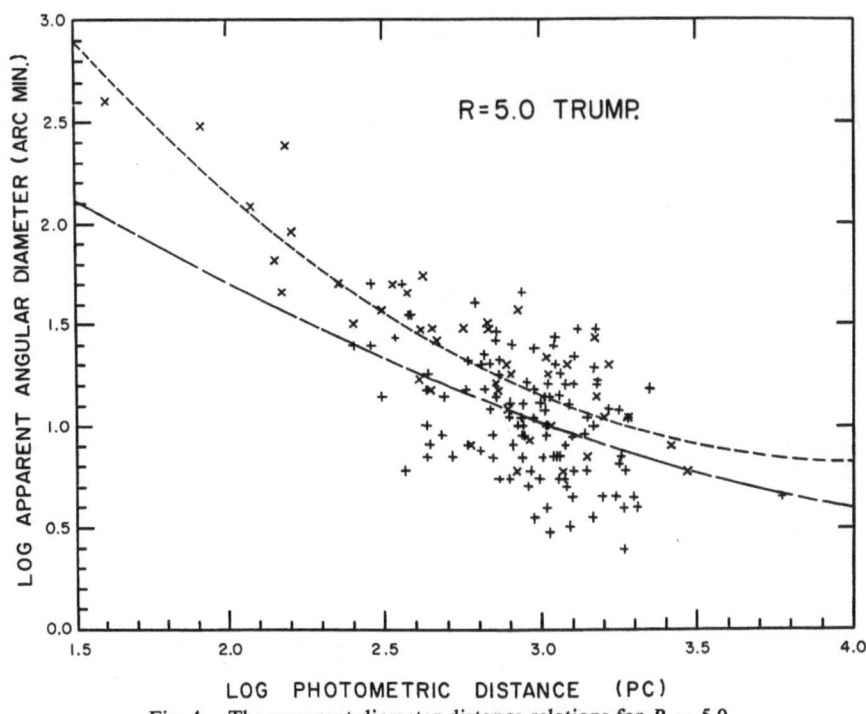

Fig. 4.   The apparent diameter-distance relations for $R = 5.0$.

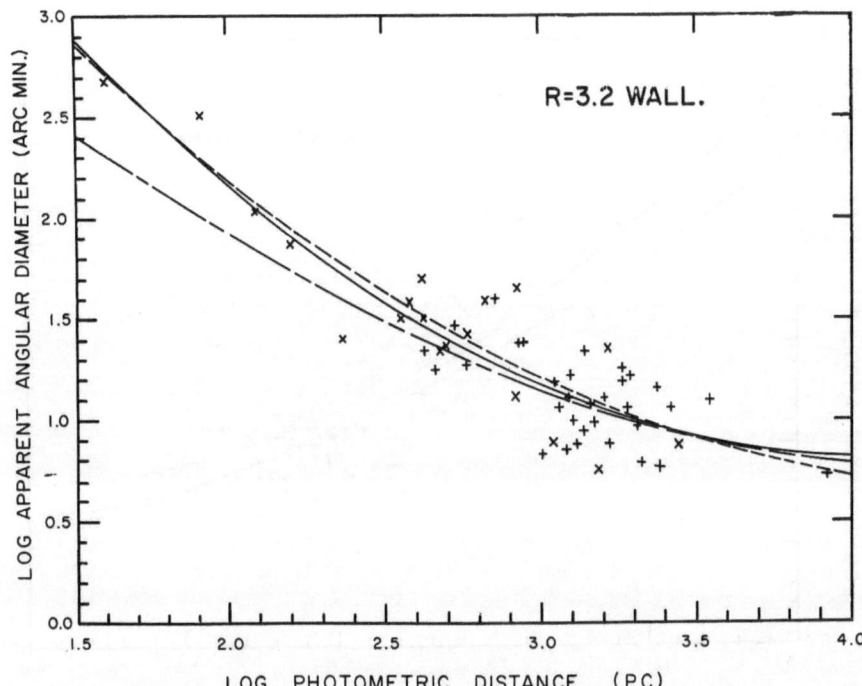

Fig. 5.   The relation between apparent angular diameter and photometric distance for Wallenquist diameters and $R = 3.2$.

TABLE I

Preliminary mean type residuals to the logarithm of the angular diameter

	R = 2.8		
	I	II	III	IV
r	+0.130	+0.130	+0.265	+0.44
m	−0.07	−0.02	+0.265	+0.29
p	−0.185	−0.185	0.0	+0.15

	R = 3.2		
	I	II	III	IV
r	+0.135	+0.135	+0.275	+0.43
m	−0.07	−0.02	+0.275	+0.27
p	−0.185	−0.185	0.0	+0.15

	R = 3.6		
	I	II	III	IV
r	+0.140	+0.140	+0.285	+0.41
m	−0.07	−0.02	+0.285	+0.25
p	−0.185	−0.185	0.0	+0.15

3.2 and 3.6. Because certain cluster types differ in mean E_{B-V} from the mean E_{B-V} of all the clusters, their mean type residuals contain terms like $C \cdot \Delta E_{B-V}$, where ΔE_{B-V} is this color excess difference. If we assume in turn that each value of **R** is correct then we may estimate the shrinkage coefficient **C**, by either the separation of the least squares fits or by the deviation of the mean slope of the distance-apparent diameter relation from minus one. The first method is inaccurate because of the large scatter in the diameter measures. I have therefore corrected the mean type residuals using the slope deviation method. The corrections included in Table I are generally less than ten percent of the mean type residual.

Figures 6, 7 and 8 present the distance-apparent diameter data corrected for dependence on Trumpler type. New least squares fits are also shown. Note that the curvature in the new fits is slightly diminished while the apparent shrinkage with color excess is increased. The best agreement between the reddened and unreddened clusters will seems to be near **R** = 3.2.

With the type dependence removed the Trumpler diameters have noticably smaller residuals. In fact, the revised Trumpler diameters have smaller residuals than the Wallenquist diameters. The larger residuals of the Wallenquist diameters and their smaller number makes them unsuited for further study here.

The remaining Trumpler diameter residuals are still large compared to the residuals which might come from errors in the distance modulii. Therefore we need to search further for diameter dependences.

Figure 9 presents the residuals in the type corrected Trumpler diameters plotted

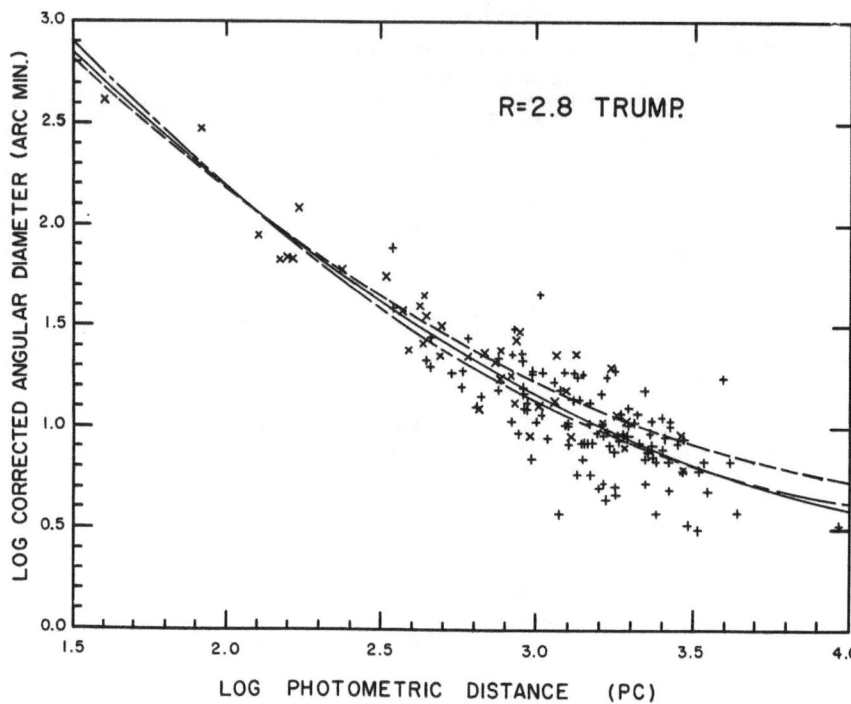

Fig. 6.   The type corrected angular diameter-photometric distance relations for $R = 2.8$.

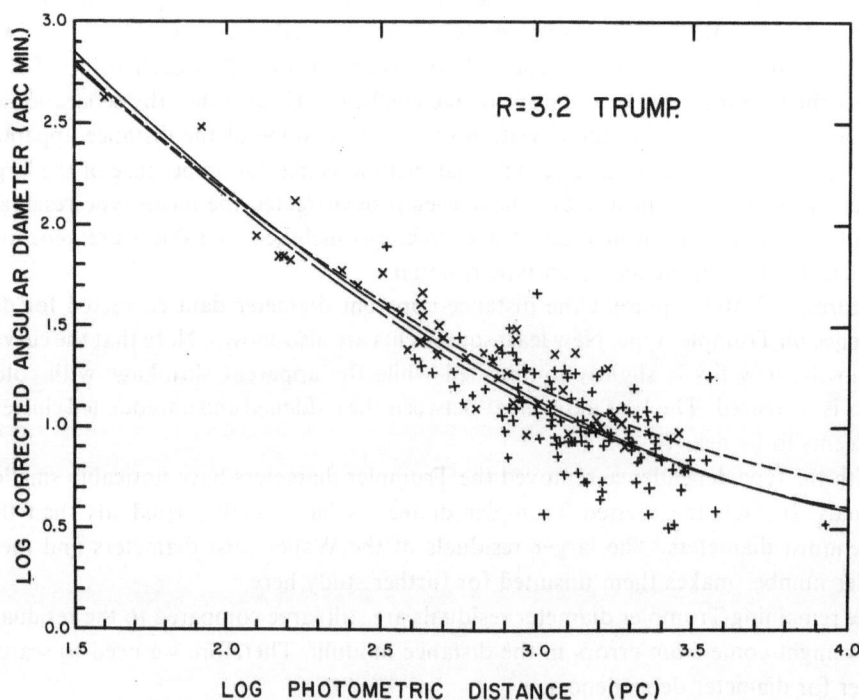

Fig. 7.   The corrected angular diameter-distance relations for $R = 3.2$.

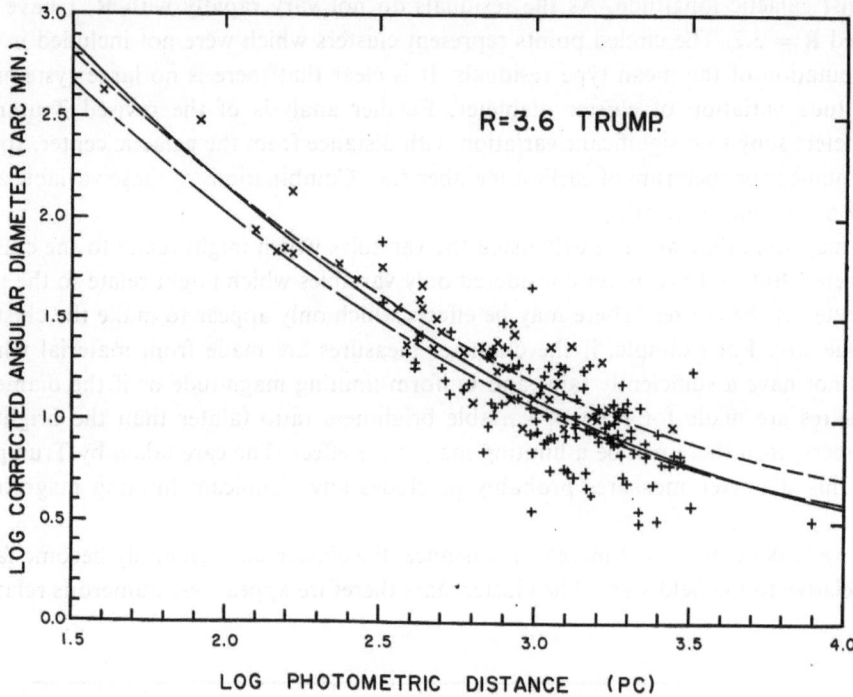

Fig. 8. The corrected angular diameter-distance relations for $R = 3.6$.

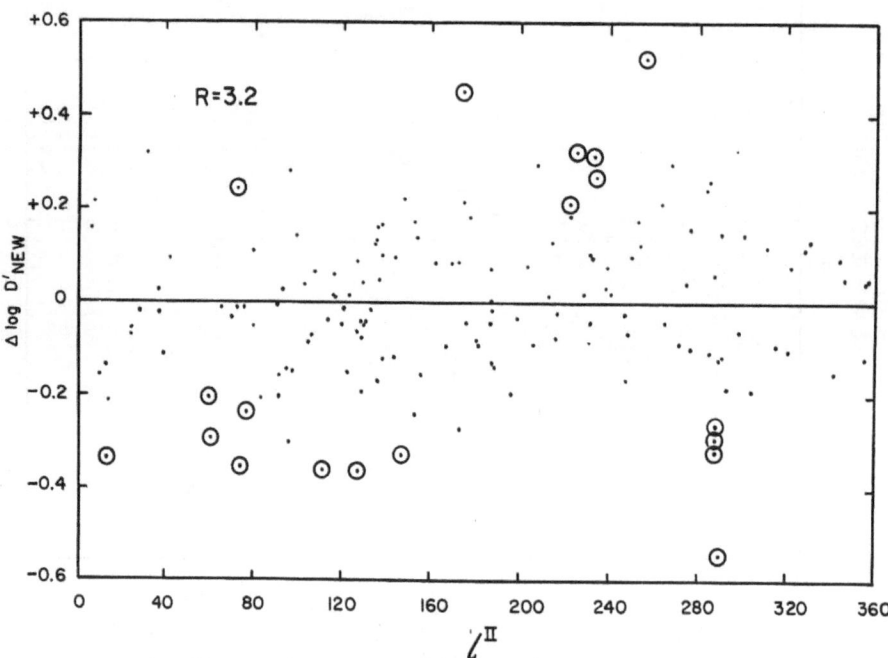

Fig. 9. The residuals in the type corrected angular diameters against longitude for $R = 3.2$. The circled points were not included in the mean type residual computations.

against galactic longitude. As the residuals do not vary rapidly with **R**, I have assumed **R** = 3.2. The circled points represent clusters which were not included in the
computation of the mean type residuals. It is clear that there is no large systematic
longitude variation of cluster diameter. Further analysis of the revised Trumpler
diameters shows no significant variation with distance from the galactic center, spiral
arm number or spectrum of earliest member star. Combinations of these variable also
yield no significant relations.

It may seem that we have exhausted the variables which might relate to the cluster
diameter. But we have as yet considered only variables which might relate to the true
diameter of the cluster. There may be effects which only appear to make the clusters
change size. For example, if the diameter measures are made from material which
does not have a sufficiently faint and uniform limiting magnitude or if the diameter
measures are made for stars of variable brightness ratio fainter than the brightest
members, then there will be a limiting magnitude effect. The care taken by Trumpler
with his diameter measures probably precludes any significant limiting magnitude
effect.

If we look at clusters of increasing distance, the cluster stars generally become fainter relative to the field stars. The cluster stars therefore appear less numerous relative

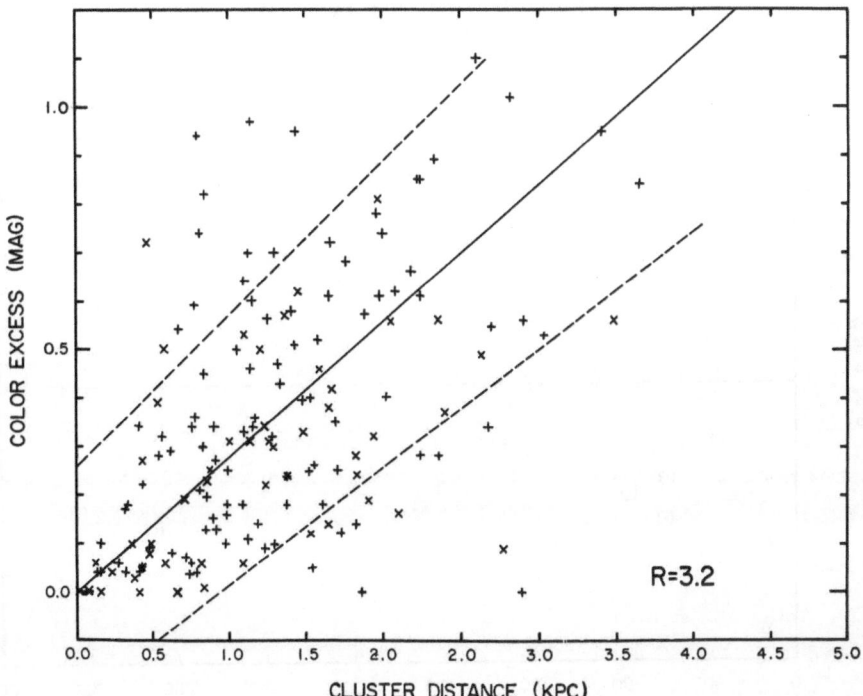

Fig. 10. The color excess, E_{B-V} as a function of distance for **R** = 3.2. The ($\times$) points are clusters
with both Wallenquist and Trumpler diameters. The (+) points are clusters with only Trumpler
diameters.

to the field stars of the same apparent magnitude. As the fainter cluster members are generally spread over a larger area than the brighter ones, the radius at which the cluster stars appear significant decreases with distance (Trumpler, 1922). If the cluster is at a large enough distance then the brighter field stars are ignored. Thus there is a distance beyond which the diameter is not further affected. This effect may be present in Figures 6, 7 and 8 for clusters with log $r < 2.8$.

Consider the influence of variations in the background star density. Near the cluster boundary the number density of stars falls off slowly. As a consequence, slight variations in the background may strongly influence the apparent cluster diameter. Trumpler has given a cluster density profile from an average of several clusters (Trumpler, 1922). The boundary slope is about five to one normalized to radius of one and central density of one. The background density is about one-quarter the central density. With such a boundary slope and background density a fractional change in the background density produces a two to three times larger fractional change in the apparent diameter. This effect may explain the increasing size of the inferred linear diameters of clusters beyond log $r = 2.8$.

These effects on the apparent diameters undoubtedly contribute to the residuals. But, as they are not well understood, nor are there adequate supplementary data available (e.g., background star densities), they can not be removed. Even unknown effects may be present. For expediency I will assume that the mean distance-apparent

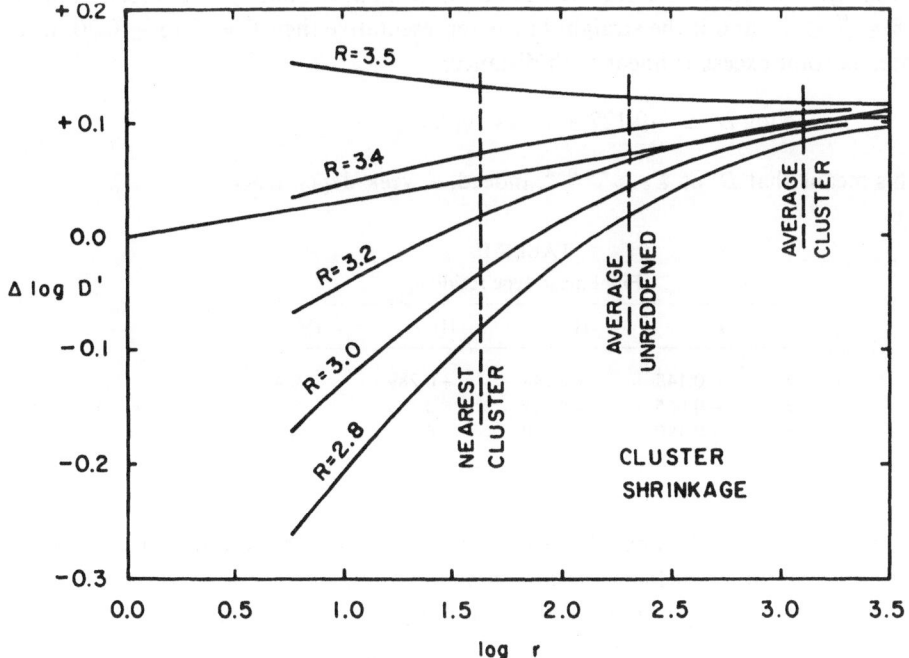

Fig. 11. The clusters shrinkage in going from unreddened to reddened clusters Δ log D' against distance for various **R** values.

diameter relations presented in Figures 6, 7 and 8 are good mean representations of the variation of apparent diameter with cluster distance.

Figure 10 presents the distance-color excess data for all clusters and $R = 3.2$. A series of such plots for various R values shows that E_{B-V} increases more linearly with distance for R values near 3.2. If $R \simeq 3.2$ then $E_{B-V}(r) = (0.28 \pm 0.05P)$ mag./kpe.

Figure 11 presents the distance dependence of the difference between the least squares fits to the reddened and unreddened clusters for various assumed R values. If each R value is in turn presumed correct then its curve represents the cluster shrinkage. The distance shown for the average unreddened cluster was computed using the mean reddening with distance given above. Note that for $R > 3.5$ the cluster shrinkage $\Delta \log D'$ decreases with distance and E_{B-V} while for $R < 2.8$, $\Delta \log D'$ is negative for nearby clusters.

We will presume that clusters shrink slightly with reddening. This may be explained as the effect of extinction on the relatively faint stars near the cluster boundary. The diminished brightness of these stars reduces the apparent cluster diameter. If the cluster shrinkage is a smooth function of color excess only, then clusters can not expand with distance and reddening nor can nearby clusters be significantly larger than the mean unreddened clusters. Consequently we must have $2.8 \le \bar{R} \le 3.5$.

The shrinkage curves are uncertain by about 0.017 in $\Delta \log D'$. Thus only the curves for $R \le 3.0$ have significant curvature. Symmetrically between the limits in $\bar{R}$ is $\bar{R} = 3.15$. A straight line fit to the curve for $\bar{R} = 3.15$ seems the best representation for the shrinkage of all the clusters. If we can be confident at the ninety-five percent level that $2.8 \le \bar{R} \le 3.5$ and if the straight line is representative then $\bar{R} = 3.15 \pm 0.20$. In this case, as color excess is linear with distance,

$$\Delta \log D' \simeq +0.127 + 0.044 \log E_{B-V}.$$

This means that $D' \propto E_{B-V}^{-0.044}$, indeed, a week dependence.

TABLE II

Final mean type residuals

	I	II	III	IV
r	+0.148	+0.148	+0.289	+0.40
m	−0.065	+0.018	+0.289	+0.23
p	−0.190	−0.190	0.0	+0.15

Table II presents the final shrinkage corrected mean type residuals. Note that they are not much different from the preliminary values. Further work on the diameters of open clusters may prove of value to the study of interstellar grains and in the study of galactic structure.

References

Becker, W.: 1963, *Z. Astrophys.* **57**, 117.

Hagen, Gretchen L.: 1970, *Publ. David Dunlap Obs.* **4**.

Johnson, H. L., Hoag, A. A., Iriarte, B., Mitchell, R. I., and Hallam, K. L.: 1961, *Lowell Obs. Bull.* **5**, No. 113, 133.

Lynds, B. T.: 1967, *Publ. Astron. Soc. Pacific* **79**, 448.

Trumpler, Robert J.: 1923, *Publ. Allegheny Obs* **VI**, 45.

Trumpler, Robert J.: 1930, *Lick Obs. Bull.* **14**, No. 420, 154.

Wallenquist, Åke: 1959, *Ann. Upp. Obs.* **17**, No. 7.

THE U, B, V, R, AND I EXTINCTIONS IN FOUR AREAS
OF THE SOUTHERN COALSACK

S. TAPIA

Steward Observatory, University of Arizona, Tucson, Ariz., U.S.A.

Abstract. From star counts the extinctions in five spectral bands were determined for four areas of the Southern Coalsack. The wavelength dependence of the extinction values follows the Whitford curve. Minimum masses for six globules are given.

1. Introduction

In 1971, Dr B. J. Bok decided to undertake a multicolor program of star counts for the region of the Southern Coalsack. He accumulated about thirty U, B, V, and R photographic plates of the region, most of them obtained with the 24–36 in. Curtis-Schmidt telescope at Cerro Tololo Inter-American Observatory. Dr E. W. Miller, during an observing run in Chile in 1972, supplemented the observational material with three infrared plates and photoelectric photometry for two sequences in the same region. At Dr Bok's suggestion, this material has been analyzed in order to study the extinction in the Coalsack in five spectral bands.

The early work of Unsöld (1929) and Lindsay (1941) indicated that the average photographic extinction in the Coalsack is approximately one magnitude. Based on observations of stars behind the Coalsack, Houck (1956) found that A_V can amount to 3.4 mag. From star counts in circular areas of 10' in diameter, Mattila (1970) computed values of A_B as large as 2.4 mag. Rodgers (1959) determined the photographic extinction and approximate distances for seven areas of about one sq deg each. His values of the extinction cover the range 0.7 to 2.4 mag. and his average distance, the best available to date, places the Coalsack at 175 pc from the Sun. The total infrared extinction in the line-of-sight of the Coalsack was studied by Westerlund (1960a, b). In areas of one to two sq deg, Westerlund derived values of A_I between 1.1 and 3.2 mag.

TABLE I

Coordinates and sizes of the four areas 1975

Area	R.A.	DEC.	Size (sq min arc)
I	12^h57^m6	$-61°28'$	1495
II	12 59.7	-62 47	1225
III-a	12 33.0	-63 42	30ʋ
IV	12 29.5	-65 30	809

Greenberg and Van de Hulst (eds.), Interstellar Dust and Related Topics, 43–51.

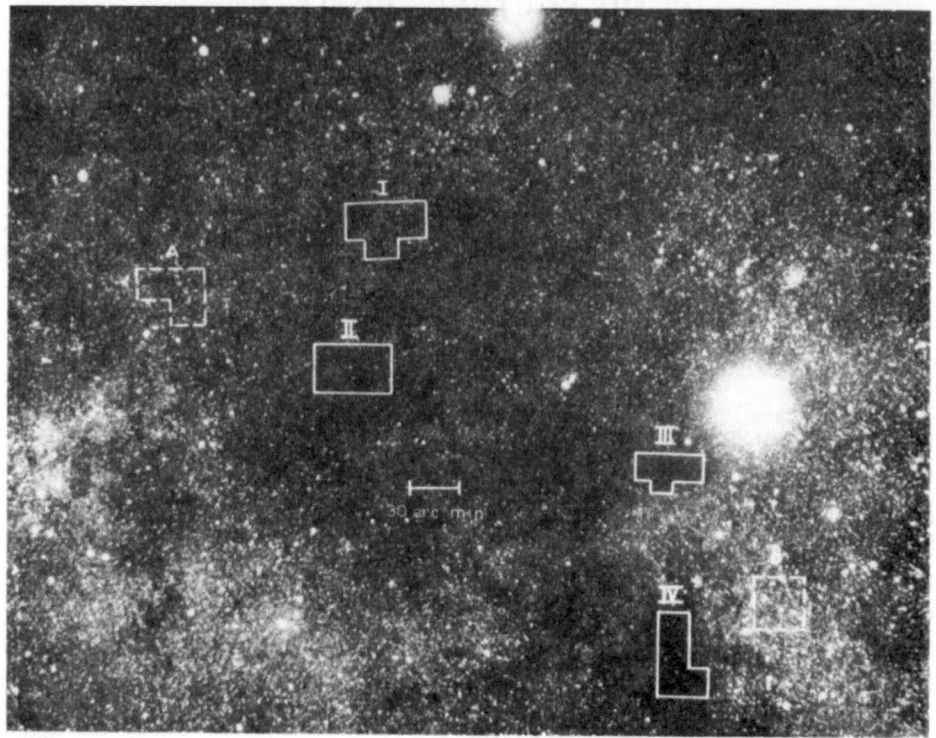

Plate A. *The Southern Coalsack*. A high contrast reproduction of the Franklin-Adams chart. The brightest star is α Crucis, at the top is β Crucis. North is at the top, East is to the left.

For the present study four areas of heavy obscuration have been selected. In general, these four areas show a homogeneous star density on the blue plates; however, inside two of them, a few small dark nebulae are clearly delineated on the red plates. The coordinates of the centers of the four areas and their apparent sizes are given in Table I. In order to determine the extinction by the method of the Wolf diagrams (Bok, 1937), two comparison areas – labeled *A* and *B* – were selected outside of the observable limits of the Coalsack. All six areas are shown in outline on Plate A.

2. Observational Material

The ultraviolet, blue, and visual plates were taken with the standard combination of emulsions and filters, which reproduce approximately the spectral bands of Johnson's (1964) system of magnitudes. The red plates were taken with O98-02 emulsion and an RG-2 filter which define a spectral band between Hα and 7000 Å. For the infrared plates the standard *I-N* emulsion and a Wratten 89-B filter were used.

Two new photoelectric sequences established near the Coalsack were used in the derivation of magnitudes from the photographic material. Finding charts and UBV photoelectric magnitudes for these sequences, and for other sequences in the Southern

Milky Way, are being published by Bok *et al.* (1972). The R and I magnitudes for the stars in both sequences were derived from transformations of the UBV photoelectric data. For the R magnitudes the relation $R = V - 0.5 \times (B - V)$ was used. Bok and Hine (1969) have shown that this transformation gives accurate R magnitudes when the interpolation between UBV photoelectric data is done on O98-02 plates with RG-2 filter. For the I magnitudes the diagram relating the colors $(V - R)$ and $(V - I)$ published by Mendoza (1967) was used.

The star counts in the different magnitude intervals were made with the aid of a Cuffey iris photometer. All plates selected for counting and analysis contain at least one of the four areas and one sequence. The images of the sequence stars were first measured with the iris photometer. The iris readings were then plotted against the magnitudes of the sequence. From this plot, the iris readings corresponding to successive whole magnitudes were then obtained. Next, the image of each star was measured and recorded in its proper interval of magnitude. The same procedure was repeated for the four areas in each of the five spectral bands. The final result of the counts are values of $A(m)$, the number of stars per square degree within magnitudes $(m - \frac{1}{2})$ and $(m + \frac{1}{2})$.

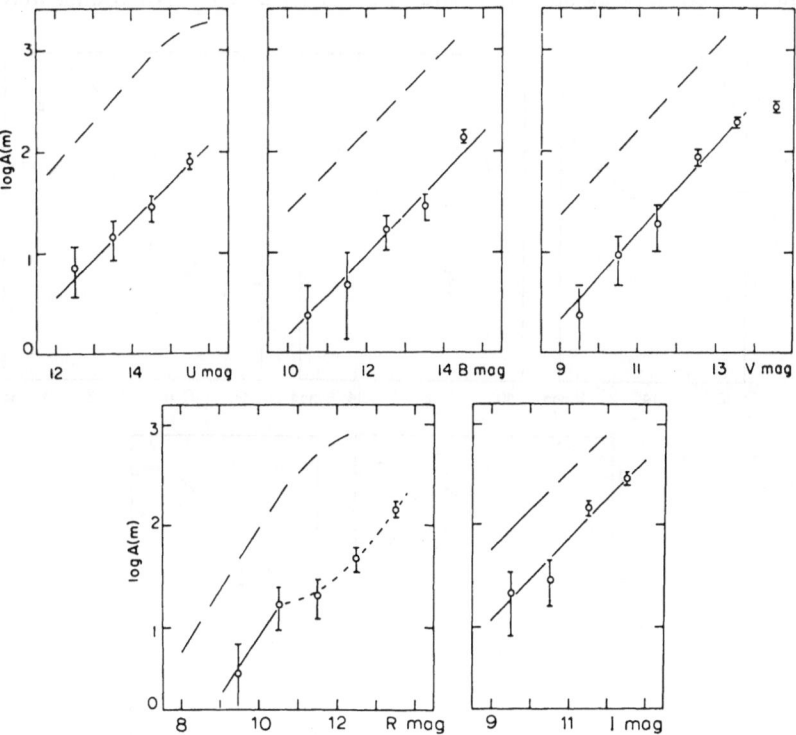

Fig. 1. *The Wolf Diagrams for Area I.* The abscissae are the magnitudes in the UBVRI-system, the ordinates the logarithms of the numbers of stars per square degree. The apparent size of the area is $1495' \times 1495'$.

3. Extinction Determinations

The study of the counts for the comparison areas shows that area *B* is richer in stars than area *A*. Obviously both areas are not free from general extinction and probably they are still affected – to a small degree – by the Coalsack itself. Nevertheless, areas *A* and *B* represent a fair sample of the distribution of stars at the galactic latitude and longitude of the Coalsack. Accordingly, as a satisfactory set of reference counts for the Wolf diagrams, the mean of the counts in both areas has been used. In Figures 1, 2, 3, and 4 the Wolf diagrams for the different spectral bands are shown. For each diagram the dashed line represents the mean of the counts in comparison areas *A* and *B*. The error bars of the open dots correspond to the statistical errors of the counts. Because of the small distance from the Sun to the Coalsack, there are essentially no foreground stars in the counts for small regions and, hence, the difference in magnitudes between the continuous and dashed lines is equal to the extinction. A dotted line is used to indicate that the counts seem to show extinction produced by more than one cloud. The values of the derived extinctions have been collected in Table II. Area IV is not covered very well by any of the red plates and for this reason no value of A_R is listed.

Small isolated dark clouds are found in areas III and IV. No counts were made

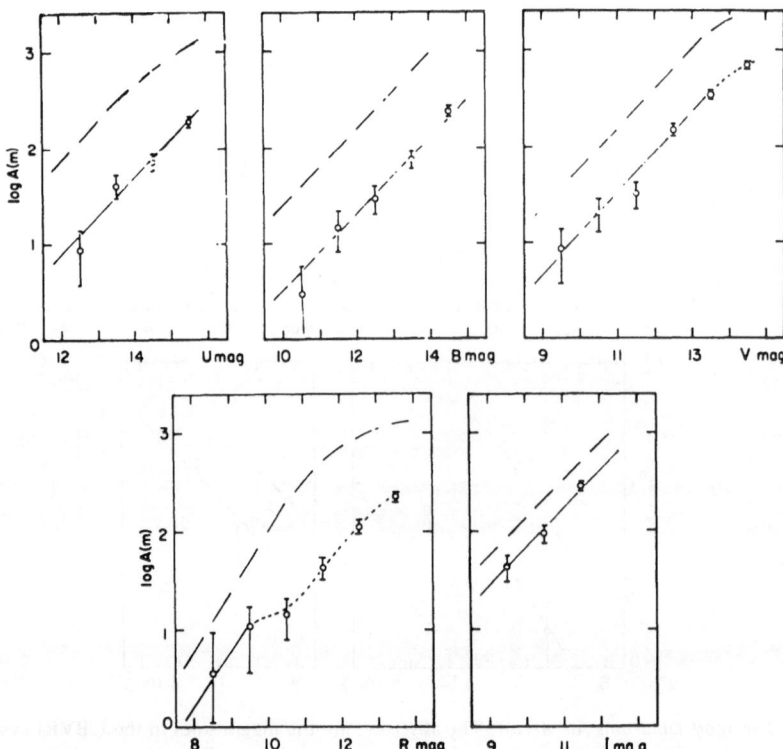

Fig. 2.   *The Wolf Diagrams for Area II*. The apparent size is 1225′ × 1225′.

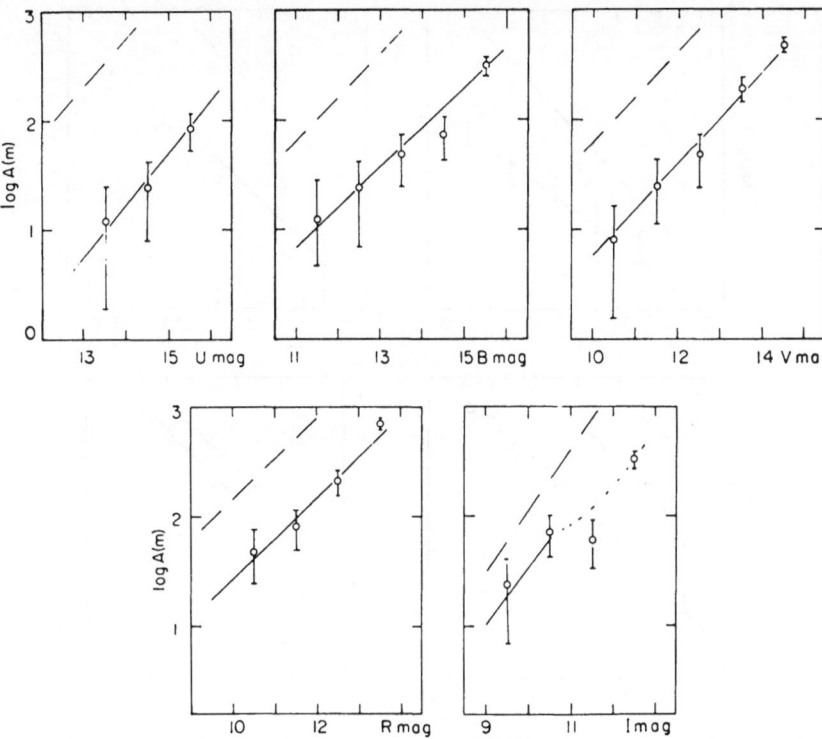

Fig. 3. *The Wolf Diagrams for Area III-a.* The counts refer to the homogeneous smaller area of the original area III. The apparent size is 300′ × 300′. In cases for which the continuous line diverges from the dashed one, the extinction for the faintest apparent magnitude is accepted and listed in Table II.

inside the small areas covered by these clouds since there are no stars – or very few stars – shining through them. Area III contains three very dark clouds and a fairly homogeneous smaller area. Westerlund (1960a) has pointed out that these clouds seem to form the most obscured parts of the Coalsack (see Figure 6 of his paper). Considering the total area III – approx. 777′ × 777′ – the computed values of an average extinction would be meaningless because the small clouds contribute a large percentage of the area but nothing to the number of stars. Therefore, only the counts for the

TABLE II

Values of the derived extinctions in four areas of the southern coalsack

	Areas				Errors
	I	II	III-a	IV	
A_U	3^m6	2^m6	3^m4	2^m8	$\pm 0^m4$
A_B	3.0	2.2	2.8	2.4	± 0.3
A_V	2.2	1.6	2.3	1.8	± 0.2
A_R	1.7	0.9	1.9	—	± 0.1
A_I	1.4	0.7	1.2	0.7	± 0.1

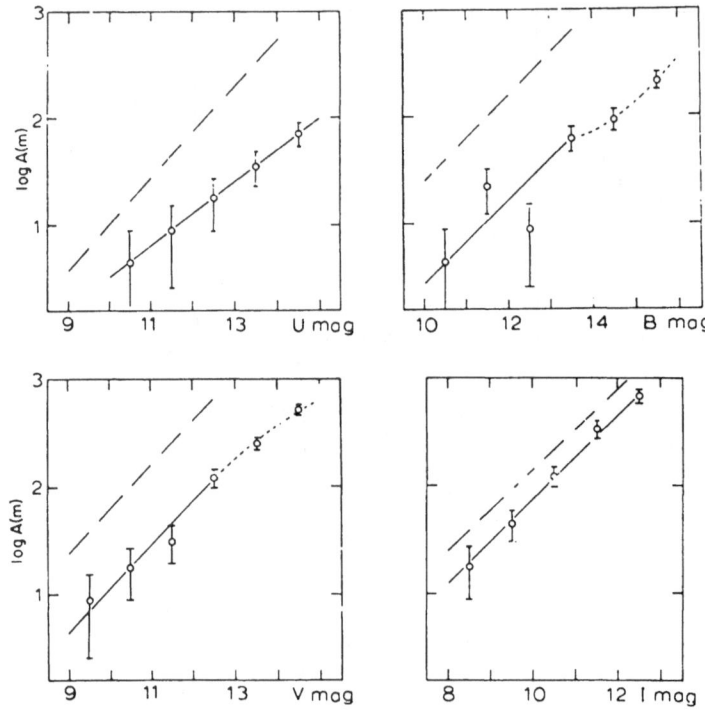

Fig. 4.   *The Wolf Diagrams for Area IV*. The apparent size is 809' × 809'. No counts were made on the red plates. As in Figure 3, the maximum value of the extinction is accepted and listed in Table II.

smaller homogeneous area were used to compute the listed extinction. This smaller area has been named III-a. Inside area IV – covering a total of 920' × 920' – two small dark clouds are easily delineated. To the faintest magnitudes of the sequences no stars are counted inside these clouds either. Again the areas of the dark clouds were subtracted, leaving area IV with 809' × 809', and the extinctions derived for the new area are given in Table II. The errors for the extinction values in the five spectral bands have been estimated from the differences between the reference star counts of the comparison areas *A* and *B*.

The values presented in Table II can be combined to derive an extinction curve. The results are presented in Figure 5, where the values of the extinctions are normalized to $E(B - V) = 1.0$. The continuous curve is Whitford's (1958) extinction-curve applicable to the interstellar medium excluding the region of the Orion Nebula and NGC 6530. To avoid confusion no error bars have been attached to the points in Figure 5. Within the accuracy of these results, the Coalsack seems to follow closely the average properties of extinction known for the obscuring component of the interstellar medium.

4. Masses of the Globules

The observed properties of the small dark clouds found in areas III and IV are consistent with the definition of blobules given by Bok *et al.* (1971). From inspection of the red plates it is possible to see that quite a few globules are contained in the Southern Coalsack.

In order to estimate lower limits for the masses of the globules it may be assumed that their minimum extinctions are at least equal to the extinctions of the adjacent areas. The computations were made using the relation derived by Lynds (1968) between the mass and the extinction of a dark nebula (see Equation (24) of her paper). The distance to the Coalsack is taken to be 175 pc, obtained by Rodgers (1960).

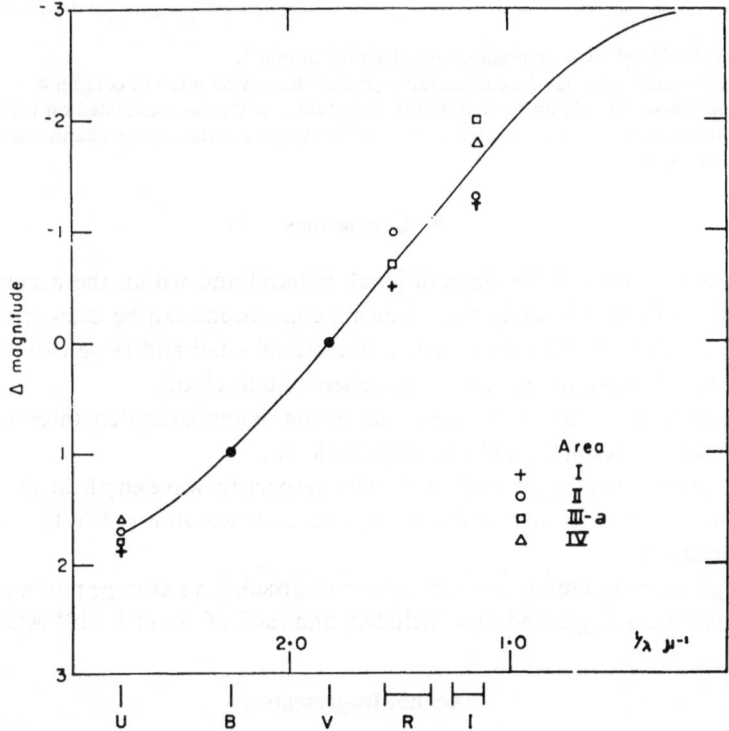

Fig. 5. *The Extinction Curve for the Southern Coalsack*. The derived extinctions are plotted together with Whitford's (1958) extinction curve. The UBVRI extinction values are normalized to $E(B-V) = 1.0$.

For the dust grains forming the globules it is assumed that the efficiency factor for the cross-section of extinction in the visual is 10^{-1}, the mean density is 3 gm cm^{-3}, and the radius is 10^{-6} cm. The computed minimum masses for six representative globules in the Coalsack are given in Table III. Approximate coordinates of the centers and actual sizes are included in the tabulation. These masses correspond to the minimum contributions from the dust grains only. If the ratio of gas to dust in the Coalsack is similar to the general galactic ratio, the total minimum masses would be 100 times larger (Spitzer, 1968) than the values given in Table II.

TABLE III

Computed minimum masses for the globules of areas III and IV

1975 (R.A.)	(DEC.)	Angular area (sq. min of arc)	Radius (10^3 AU)	Visual extinction (minimum value)	$M/M_\odot$ (min. values)
12^h32^m0	$-63°33'$	48	50×20	2^m5	0.035
12 30.5	-63 38	42	38	2.5	0.031
12 28.5	-63 43	19	26	2.5	0.014
12 30.5	-65 10	84	70×20	2.0	0.049
12 27.4	-65 33	27	31	2.0	0.016
12 27.3	-65 44	27	31	2.0	0.016

Notes:
(1) For irregular clouds two dimensions are given in column 3.
(2) The actual extinctions may be considerably greater than those listed in column 4.
(3) The listed masses in column 5 are only the minimum *dust* masses, calculated on the basis of the extinctions in column 4; the actual masses, considering gas and dust, are probably 100 times larger than the listed values.

5. Conclusions

On the basis of the available observational material and within the accuracy of the results derived from this study the following conclusions can be drawn:

(a) The Southern Coalsack is a conglomerate of small and large clouds. It is *not* a sheet of variable extinction, *nor* a nearspherical unit cloud.

(b) The average extinction for the denser regions is approximately three magnitudes in U, two magnitudes in V, and one magnitude in I.

(c) The dependence of the extinction with respect to wavelength in the Southern Coalsack seems to be similar to the average dependence observed in the galactic interstellar medium.

(d) Six globule-like clouds of the Southern Coalsack have average minimum masses of three solar masses (gas and dust included) and radii of about 3×10^4 AU.

Acknowledgements

Grateful acknowledgement is given to Dr V. M. Blanco and Dr J. A. Graham for enabling the presentation of this Report. Thanks are also given to Carolyn Cordwell McCarthy for help during the early stages of the work. Dr Bok and Dr Miller express their fratitude to CTIO for the opportunities given them to gather the observational material. The author is supported by the Joint University of Chile and Cerro Tololo Inter-American Observatory Scholarships Program.

Note added in proof. The discovery of a small dense cloud of formaldehyde in the Southern Coalsack has recently been reported by M. W. Sinclair and J. W. Brooks (*Astrophys. Letters* **11**, 207, 1972). The position of the cloud coincides with the coordinates of the second globule listed in Table III of this report. The diameter of the formaldehyde cloud has been estimated to be close to 0.3 pc. The diameter of the globule, measured on red plates, is approximately 0.37 pc. Further radio astronomical molecular studies of the Southern Coalsack are obviously of great interest.

References

Bok, B. J.: 1937, *The Distribution of the Stars in Space*, University of Chicago Press, copy from University Microfilms, Inc., Michigan, p. 40.

Bok, B. J. and Hine, A.: 1969, *Am. Astron. Soc. Photo-Bull.*, No. 1, 8.

Bok, B. J., Cordwell, C. S., and Cromwell, R. H.: 1971, in B. T. Lynds (ed.), *Dark Nebulae, Globules and Protostars*, Univ. of Arizona Press, p. 33.

Bok, P. F., Bok, B. J., and Miller, E. W.: 1972, *Astron. J.*, in press.

Houck, T. E.: 1956, Ph.D. Thesis, Univ. of Wisconsin. (Copy from University Microfilms, Inc., Michigan.)

Johnson, H. L.: 1964, *Astrophys. J.* **141**, 923.

Lindsay, E. M.: 1941, *Proc. Roy. Irish Acad.* **46A**, 11.

Lynds, B. T.: 1968, *Stars and Stellar Systems* **7**, 117.

Mattila, K.: 1970, *Astron. Astrophys.* **8**, 272.

Mendoza, E. E.: 1967, *Bol. Obs. Tonantzintla Tucubaya*, **4**, 142.

Rodgers, A. W.: 1959, *Monthly Notices Roy. Astron. Soc.*, **120**, 163.

Spitzer, L., Jr.: 1968, *Stars and Stellar Systems* **7**, 1.

Unsöld, A.: 1929, *Harvard Bull.* **13**, 870.

Westerlund, B.: 1960a, *Arkiv. Astron.* **2**, 429.

Westerlund, B.: 1960b, *Arkiv. Astron.* **2**, 451.

Whitford, A. E.: 1958, *Astron. J.* **63**, 201.

INTERSTELLAR DUST: OBSERVATIONS IN THE ULTRAVIOLET AND THEIR INTERPRETATIONS

ADOLF N. WITT

Ritter Astrophysical Research Center, The University of Toledo, U.S.A.

Abstract. Recent observations of interstellar extinction and diffuse galactic light in the ultraviolet are interpreted in terms of a bi-modal size distribution of interstellar grains.

The successful operation of the Orbiting Astronomical Observatory OAO-2 has provided abundant ultraviolet observations from which detailed information on the extinction and scattering properties of interstellar dust grains has been derived. Figure 1 summarizes some of these results. In the upper part the average wavelength dependence of interstellar extinction as determined by Bless and Savage (1972) is shown. In the lower part–using the same inverse wavelength scale–we plot the wavelength dependence of the albedo of the interstellar grains as determined by Witt and Lillie (1973) from an analysis of the diffuse galactic light observed in approximately 30 of Kapteyn's selected areas.

There is good evidence from the diffuse galactic light observations that the scattering at visible wavelengths is governed by a much more strongly forward-directed

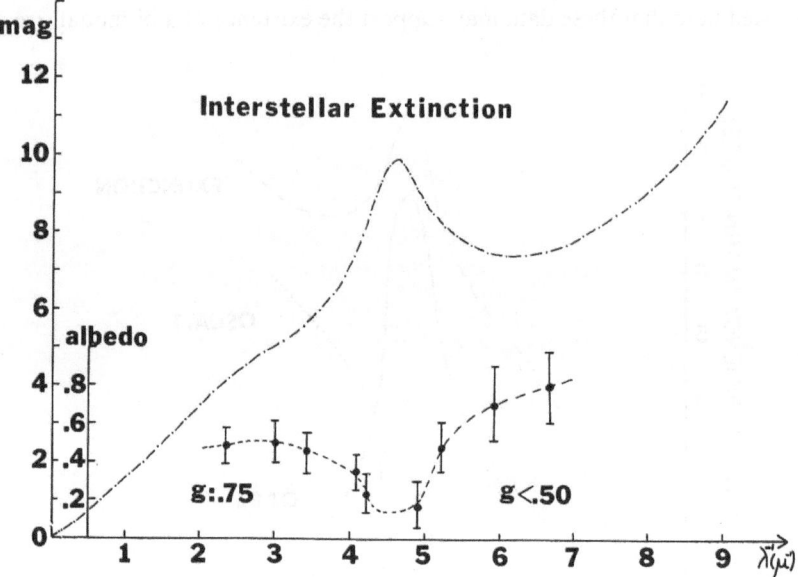

Fig. 1. The wavelength dependence of interstellar extinction as found by Bless and Savage (1972) and the wavelength dependence of the albedo of interstellar grains as determined by Witt and Lillie (1973).

Greenberg and Van de Hulst (eds.), Interstellar Dust and Related Topics, 53–57.

phase function as compared to the far-ultraviolet region, where the scattering appears to be more nearly isotropic. This result is indicated by the estimated values of the phase-function asymmetry of $g = 0.75$ at around 4000 Å wavelength and $g < 0.5$ at wavelengths below 2000 Å.

The pronounced decrease of the albedo near 2200 Å wavelength supports the idea of an absorption nature of the corresponding bump in the extinction curve. The subsequent rapid rise of the albedo shortward of 2000 Å wavelength indicates that the far-ultraviolet extinction is caused by a scattering process. These results do not support the suggestion of Stecher and Williams (1969) that the far-ultraviolet rise in the extinction might be due to photo-dissociation of the H_2^+ molecule.

Witt and Lillie (1973) have also derived the relative efficiencies for scattering and absorption of interstellar grains on the basis of the results shown in Figure 1. In Figure 2 these efficiencies are presented in comparison with the extinction curve.

How can these results be interpreted? It has been concluded by several authors in the past that the high ultraviolet extinction indicates the existence of a much larger *proportion* of small particles in interstellar space. Based on the result that the asymmetry factor of the scattering phase-function is considerably smaller in the far-ultraviolet than in the visible, a conclusion may be suggested that goes somewhat farther.

It appears quite impossible to construct a simple *continuous* size distribution of Mie-scattering particles of any given composition or index of refraction such that the scattering should become less forward – directed as one goes to shorter wavelengths, while it would still approximately reproduce the extinction curve. Therefore, it is suggested here that these data may support the existence of a bi-modal size distri-

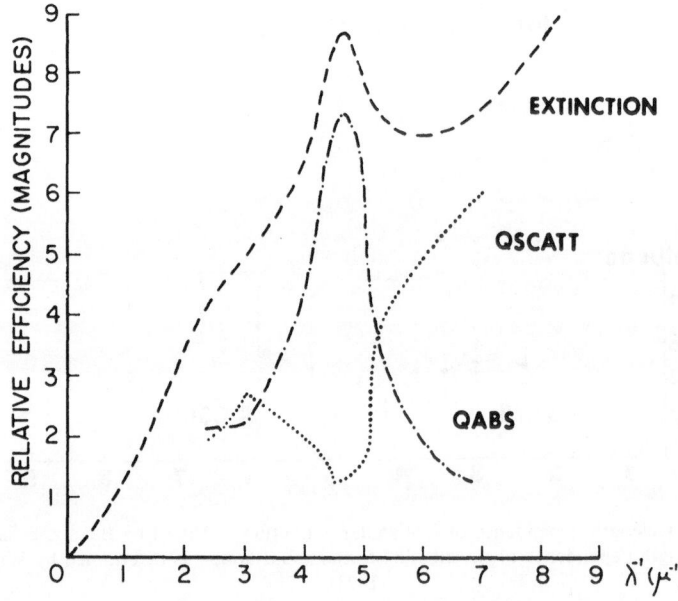

Fig. 2.   The relative efficiencies for extinction, absorption, and scattering of interstellar grains.

bution of interstellar dust particles. This distribution would contain *large* particles with diameters in the range of 5000 Å and larger, and *small* particles with diameters of 500 Å and smaller. The large particles would provide extinction in the visible region including scattering which is strongly forward directed, whereas the small particles – needed in large numbers due to their small intrinsic efficiencies – would dominate the ultraviolet region and contribute nearly isotropic scattering. It would also be very reasonable to assume that particles existing in these two size ranges would be chemically different and would have resulted from quite different processes of formation or sources of origin.

A very simple model along these lines has been examined in order to study the conditions under which the observed properties of interstellar grains may be approximated. Mie calculations for the following ingredients were carried out: large, icy particles with a diameter of 5000 Å and an index of refraction $n = 1.33 - 0.1\ i$, small graphite particles with a diameter of 400 Å, and small dielectric particles with a diameter of 400 Å and $n = 1.66$. There are only two variable parameters in the model: the respective ratios of the number of graphite particles or dielectric particles to the number of large icy particles.

While the chemical composition selected for the grains is in no way original, the size spectrum proposed here appears to be required by the scattered light data. The fact that small particles of *two compositions* are suggested is based on the observations by Bless and Savage (1972) that the magnitude of the absorption bump at 2200 Å appears to be quite unrelated to the amount of far-ultraviolet extinction. The bump is thought to be due to graphite grains in interstellar space.

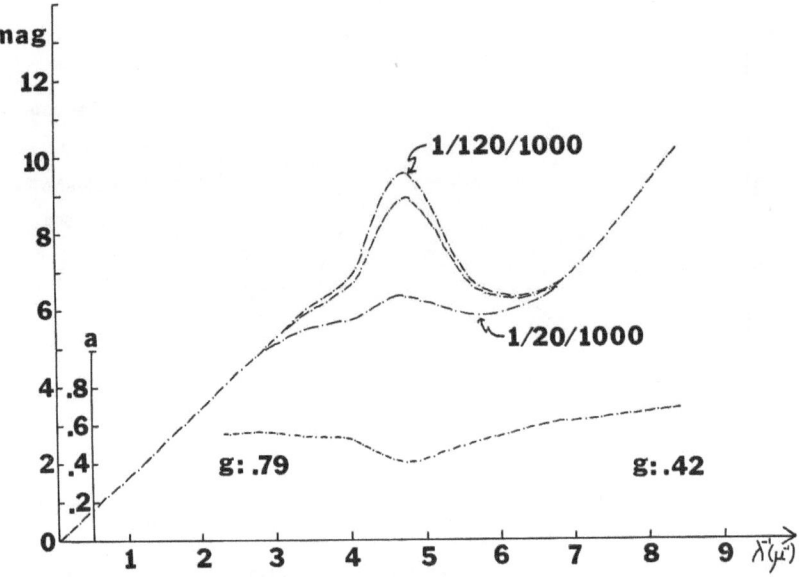

Fig. 3.   Theoretical extinction and albedo curves for a bi-modal size distribution of interstellar grains. The effect of varying the relative number of graphite particles is shown.

What number ratios of particles do we require in order to produce the observed extinction curve and how does the variation of these ratios affect the appearance of the resulting wavelength dependence of extinction. Figure 3 indicates that a mixture of one large particle to every 100 small graphite grains and every 1000 small dielectric grains appears to yield an extinction curve closely resembling the average observed one. The variation of the number of the graphite grains alone changes the magnitude of the absorption peak at λ 2200 without affecting the remainder of the curve. In the lower part of Figure 3 the theoretical albedo curve is plotted for this model. Qualitatively, it has the observed features with a minimum at 2200 Å wavelength and a rise in the ultraviolet, but, quantitatively, the agreement with observations is not perfect. Better data on the indices of refraction of different materials are obviously needed for the ultraviolet, in particular, the question of the existence of materials which remain dielectric into the far-ultraviolet requires further study.

The principal result of the bi-modal size distribution is that the phase-function of scattering, in agreement with the observations, is strongly forward directed in the visible and tends to be more isotropic in the far-ultraviolet, as indicated by the values of the asymmetry factor $g = 0.79$ and $g = 0.42$, respectively.

Figure 4 illustrates the effect of varying the relative number of small dielectric particles in the size distribution. The amount of far-ultraviolet extinction varies accordingly, without producing much change in the other parts of the spectrum. Also, in accordance with the observations of Bless and Savage (1972), the position of the far-ultraviolet extinction minimum moves to shorter and shorter wavelengths as the

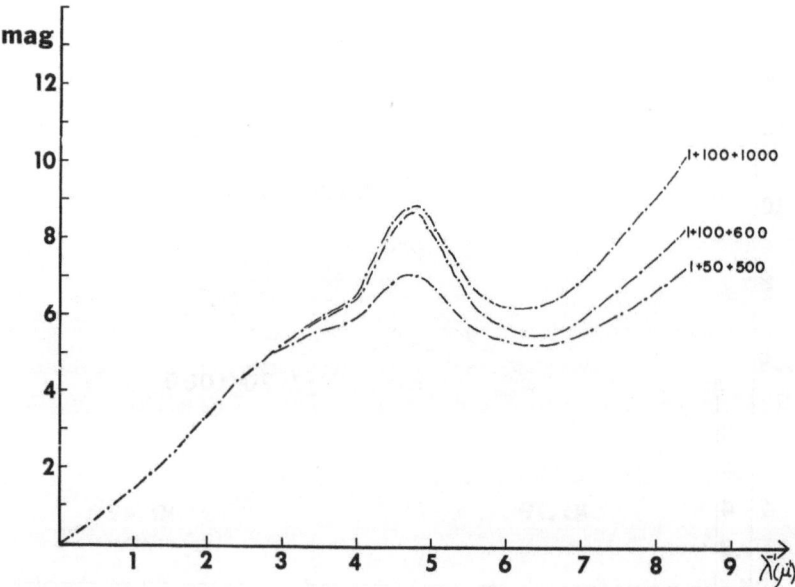

Fig. 4. Theoretical extinction curves for a bi-modal size distribution of interstellar grains. The effects of changing the contribution of small dielectric particles as well as all small particles is demonstrated.

extinction in the extreme UV diminishes. Finally, when both small particle components are reduced in number, the extinction can become nearly grey in the UV, which is similar to a case observed in Orion.

In summary, as a result of observations of the diffuse galactic light the existence of a bi-modal size distribution of interstellar dust particles is suggested. It has been shown that a very simple model which incorporates particles of the two basic size ranges can indeed explain the observations of interstellar extinction and diffuse galactic light.

References

Bless, R. C. and Savage, B. D.: 1972, *Astrophys. J.* **171**, 293.
Stecher, T. P. and Williams, D. A.: 1969, *Astrophys. Letters* **4**, 99.
Witt, A. N. and Lillie, C. F.: 1973, *Astron. Astrophys.* **25**, 397.

A KRAMERS-KRONIG ANALYSIS OF
INTERSTELLAR EXTINCTION*

R. Y. CHIAO**, M. J. FELDMAN, and P. T. PARRISH

Dept. of Physics, University of California, Berkeley 94720, Calif., U.S.A.

Abstract. Integrals over the observed spectrum of interstellar extinction to $9 \, \mu m^{-1}$ are related to known laboratory properties of specific compounds. Using $L\alpha$ hydrogen densities and cosmic abundance ratios, many suggested grain materials can be eliminated.

Little is known about the identity and density of the interstellar grains. It would be desirable then to develop a rigorous analysis which would lead to strong quantitative conclusions. We discuss an approach involving integrals of the extinction over the infrared, optical and ultraviolet portions of the spectrum. Using cosmic abundance ratios and recently determined $L\alpha$ hydrogen column densities, several suggested grain candidates are evaluated.

We employ the Kramers-Kronig relation connecting the real part of the dielectric susceptibility $\chi'(\omega)$ to an integral over frequencies of the imaginary part $\chi''(\omega)$

$$\chi'(\omega_0) = \frac{2}{\pi} \int_0^\infty \frac{\omega \chi''(\omega)}{\omega^2 - \omega_0^2} \, d\omega . \tag{1}$$

This is applied to the grain component of the interstellar medium as a whole.[†] $\chi''(\omega)$ is given by

$$0.921 \, A(\omega) = \kappa(\omega) = \frac{4\pi}{c} \omega \chi''(\omega),$$

where A is the extinction measured in units of (magnitude cm^{-1}) and κ in units of (cm^{-1}). Assuming a specific grain species, $\chi'(\omega)$ is simply related to its column density and optical and solid state properties.

We take two limits of Equation (1), $\omega_0 = 0$ and $\omega_0 \to \infty$, which yield two independent sums $\sum_0$ and $\sum_\infty$. These two sums (explicitly, integrated extinction) are then used to evaluate various grain candidates. For each sum a quantity Δ is derived which is the ratio of the amount of a given material expected on the basis of $L\alpha$ hydrogen column density, $N(L\alpha)$, and cosmic abundances (CARs), to the amount of material required to account for the integrated extinction.

* Work partially supported by the National Aeronautics and Space Administration Grant NGL 05-003-272 and the National Science Foundation Grant GP30424X.
** Alfred P. Sloan Foundation Fellow.
[†] For the first application of the Kramers-Kronig relations to the interstellar medium, see Purcell (1969).

$$\Delta_0 = 3.21 \times 10^5 \left[\frac{N(L\alpha)}{\sum_0}\right] \frac{CAR \times F}{n_g} \tag{2}$$

$$\Delta_\infty = 9.68 \times 10^{-17} \left[\frac{N(L\alpha)}{\sum_\infty}\right] CAR \times Z_{eff}^{lab}. \tag{3}$$

F is a function of the assumed grain shape and dielectric function (Purcell, 1969), n_g is the number density of molecules in the grain, and Z_{eff}^{lab} is essentially the number of electrons per atom participating in the extinction. (An extention of this work, to be published later, will include a detailed discussion of the assumptions made in deriving these equations).

The data sample consists of those stars with published ultraviolet extinction measurements by Bless and Savage (1971), supplemented by optical and infrared extinction information. The infinite sum is taken out to two frequencies: a full sum involving extinction out to 9.0 μm^{-1} and a partial sum out to 5.4 μm^{-1} which serves to isolate the effect of the ubiquitous 4.6 μm^{-1} extinction peak.

Table I is an evaluation of Equations (2) and (3) for individual compounds. Each entry represents an upper bound to the proportional contribution of a specific compound to the integrated extinction. The grains are assumed to be extreme prolate spheroids aligned parallel to the electric acis. This geometry maximizes the extinction efficiency of the grain. Each compound is ascribed a CAR on the basis of the elemental CARs of Aller (1961). The second iron entries use a recent value determined by Foy (1972). These should represent lower and upper limits to the actual CAR for iron. CARs are the major source of uncertainty in this analysis (the entries in Table I scale with the actual CARs). Although Lα densities are at best measured to $\pm 20\%$ (Savage and Jenkins, 1972), by measuring the ratio $N(L\alpha)/\sum$ for each star and averaging over our sample a best fit value gives one sigma uncertainties of 10% for these ratios.

Little information is provided by Δ_∞. The integrated extinction in the spectral region 5.4 μm^{-1} to 9 μm^{-1} is strongly dependent upon particle size and shape. We expect then that our generous assumptions regarding geometry might be a gross overestimate in this case. In view of the strong results provided by $\Delta_{\alpha/p}$, we emphasize

TABLE I

Ratios of actual to required material abundances

Compound	Δ_0	$\Delta_{\alpha/p}$	Δ_∞
SiO$_2$	0.13	0.06	0.40
Fe	0.02	0.09	0.06
	0.19	1.02	0.73
(MgFe)Sio$_4$	0.04	0.01	0.13
	0.12	0.02	0.41
C (Diamond)	0.46	< 0.01	1.47
C (Graphite)	0.61	5.02	3.22
SiC	0.11	0.43	1.01
CH$_4$	0.94	< 0.01	3.29
NH$_3$	0.44	< 0.01	1.17
H$_2$O	1.87	< 0.01	3.04

that these numbers are strict upper limits for the respective compounds, assuming only that ascribed CARs hold.

In specific we point out that (1) compounds involving iron or silicon (often suggested for the low frequency extinction (Wickramasinghe and Nandy, 1971; Gilra, 1971)) seem underabundant to account for the zero-frequency sum; (2) consequently, more attention should be paid to the lighter compounds such as graphite and the frozen gases, CH_4, NH_3, H_2O; and (3) only graphite, SiC, and iron may substantially contribute to the infinite-frequency partial sum.

References

Aller, L. H.: 1961, in *The Abundances of the Elements*, Interscience Publishers, New York, p. 178.
Bless, R. C. and Savage, B. D.: 1971, *Astrophys. J.* **171**, 293.
Foy, R.: 1972, *Astron. Astrophys.* **18**, 26.
Gilra, D. P.: 1971, *Nature* **229**, 237.
Purcell, E. M.: 1969, *Astrophys. J.* **158**, 433.
Savage, B. D. and Jenkins, E. P.: 1972, *Astrophys. J.* **172**, 491.
Wickramasinghe, N. C. and Nandy, K.: 1971, *Monthly Notices Roy. Astron. Soc.* **153**, 205.

PART II

DIFFUSE FEATURES

A SEARCH FOR DIFFUSE INTERSTELLAR FEATURES
IN STARS WITH CIRCUMSTELLAR DUST SHELLS

THEODORE P. SNOW, JR. and GEORGE WALLERSTEIN

Astronomy Dept., University of Washington, Seattle, Wash. 98195, U.S.A.

Abstract. We have searched for diffuse features at λ 4430, 5780 and 6613 in the spectra of 17 stars which show infra-red excesses or other grounds for believing that they have dusty envelopes. In no case could we identify diffuse bands that may be ascribed to the circumstellar envelope. We conclude that the grains in circumstellar envelopes are not identical with the grains in the general interstellar medium.

1. Introduction

Although the diffuse interstellar absorption bands have been studied for over 35 yr no certain identification of the source of the absorption has been made. The general correlation between interstellar reddening and the strength of the diffuse feature at λ 4430 (Duke, 1951) has been used to associate the diffuse absorption features with interstellar dust. Furthermore, the width of the bands indicate that they are not due to the absorption by discrete transitions of atoms or molecules in the gaseous state; more likely they are caused by a substance that is adsorbed on the grains.

While the increase in band strength with distance indicates that the diffuse bands are in most cases interstellar, rather than circumstellar; the possibility is not ruled out that stars with a sufficient concentration of circumstellar dust should show anomalously strong diffuse bands. Recent observations in the infra-red have shown that many stars have an excess of infrared radiation that can be explained most readily by the hypothesis that circumstellar dust is absorbing radiation from the photosphere and re-radiating it in the infrared. The question naturally arises as to whether the circumstellar dust also results in absorption in the diffuse bands. We have investigated the possibility that the diffuse bands may be present in stars with circumstellar extinction by examining the spectra of stars with infrared excesses or other grounds for suspicion that circumstellar dust may be present. We have investigated stars which we believe to be evolved objects rather than contracting stars*; but can never be certain of the evolutionary state of each object.

2. Spectroscopic Observations

We have searched for the diffuse bands at λ 4430, 5780, 5796 and 6613 on spectra loaned from the Hale Observatory plate collection as well as some taken at the Hale Observatories in connection with other investigations of infrared stars by one of us

---

* Except for AB Aur (Herbig, 1960).

Greenberg and Van de Hulst (eds.), Interstellar Dust and Related Topics, 65–70.
All Rights Reserved. Copyright © 1973 by the IAU.

(G.W.). In stars later than about type F0 the blue region of the spectrum is so crowded with lines that the recognition of the λ 4430 band becomes nearly impossible, so we have searched for the features in the visual and red in the cooler stars. Microdensitometer tracings were prepared of the blue spectra, since the 4430 feature is too broad to identify readily by visual inspection. Continuum points 50 Å on either side of 4430 Å were used to establish a baseline against which the absorption in the band could be measured. The red spectra were examined with a visual projection measuring engine, to distinguish the λ 5780 and 6613 features from stellar lines.

3. Photometric Data

Since we are comparing the correlation of diffuse band strength with visual absorption in the circumstellar environment and the interstellar environment, we must find the absorption due to circumstellar material for each star. Published photometric data were used to estimate the visual absorption in stars with infra-red excesses. The absolute energy distribution was plotted for each star and the emergent stellar flux was separated from the infra-red excess. We can fairly easily establish the total reradiated energy this way except when the infra-red flux is still rising at the longest observed wavelength. In that case we have assumed a characteristic temperature and integrated under the black body distribution at that temperature from the longest observed wavelength to infinity.

While the total absorbed flux may be set equal to the reradiated infra-red flux, the energy distribution of the absorption is not easily determined. For interstellar particles the albedo in the visual region is about 0.7 (Van de Hulst and de Jong, 1969); i.e. only about $\frac{1}{3}$ of the energy subtracted from the beam is absorbed by the particles. For circumstellar particles there is some evidence that the total extinction curve differs from the interstellar case and that the ratio of extinction to reddening also differs (Hyland et al., 1969; Strom et al., 1972). Hyland et al. found that the extinction curve for VY CMa is flat from 3500 Å to 10000 Å and then decreases rapidly to nearly zero at 3.5 μ. However, the absorption could not be absolutely zero in the infra-red, since the circumstellar dust radiates profusely out to at least 20 μ. In the second paper cited above the authors found a very large ratio of extinction to reddening indicating that the circumstellar dust absorbs rather nonselectively. In calculating the visual absorption we have assumed that the absorption is equal at all wavelengths. For the purpose of calculating the visual absorption we do not need to know the ratio of absorption to scattering, but will comment on that later.

In view of the difficulties presented by the lack of data in the far infra-red, the ambiguity in determining the cut-off point between emergent stellar flux and reradiated infra-red flux, and the unknown spectral distribution of the absorption in optical wavelengths, we have estimated crude bolometric circumstellar extinctions from

$$A = -2.5 \log \left(\frac{F_v}{F_{\text{bol}}} \right), \tag{1}$$

where F_v is the estimated stellar flux which escapes the dust shell directly without loss, F_{bol} is the total flux of the star, and A is the extinction in magnitudes. Since the bolometric flux is given by the sum of the emergent stellar flux plus the infrared excess flux, we have

$$A = -2.5 \log\left(\frac{F_v}{F_v + F_{IR}}\right), \tag{2}$$

where F_{IR} is the estimated infra-red excess flux. We can also estimate crude bolometric optical depths of the dust shell for optical wavelengths from

$$\tau = -\ln\left(\frac{F_v}{F_v + F_{IR}}\right). \tag{3}$$

In the case of VY CMa, where we had enough information to estimate a true visual extinction as well as a bolometric extinction from Equation (2), we found them to be approximately equal.

In some cases the stars under consideration were variables with large amplitudes. For these it was necessary to attempt to match phases by using available light curve data. This was difficult and at best quite inexact, because very little information can be found concerning the IR light curves of variable stars (Lockwood and Wing, 1971; Lockwood [preprint] have done some work on Mira variables). In a few cases satisfactory phase matching was not achieved and we could only place upper limits on the circumstellar extinction.

4. Results

Table I lists the stars whose spectra we have examined. After the star name or number we list the region searched and the circumstellar absorption calculated as described above. With the exception of RS Ophiuchi and ε Aur we could find no stars with diffuse band absorption. At λ 4430 we can set an upper limit of 5% absorption. No quantitative limit can be set for the bands in the visual region since most of our stars have atomic or molecular lines near λ 5780 and 6613. We can only state that there is no noticeable enhancement or widening of these features.

The only star to show evidence of absorption at λ 4430 is ε Aur. It is an eclipsing system with a period of 27 yr and an eclipse duration of about two years. The eclipse lasts too long for the eclipsing object to be a normal star and it has been suggested that a small central object surrounded by a cloud of gas or dust may be responsible for the eclipse. For a recent discussion of ε Aur see Wilson (1971) which contains many further references. We are concerned here not with the nature of the central object, but with its large surrounding envelope. We find that the λ 4430 band is of equal strength both in and out of eclipse, showing that the λ 4430 absorption is not formed in the object that is in front during the eclipse. Furthermore, the central depth of the λ 4430 band in ε Aur is 11%, which is just about what may be expected from an interstellar reddening of 0.34 mag. and its distance of about 1000 pc. We found from

T. P. SNOW, JR. AND G. WALLERSTEIN

TABLE I

Stars investigated for diffuse bands

Object	Spectral type	Region searched	Circumstellar absorption (mag.)	Reference	Remarks
ε Aur	A8Ia	a b c	0.0	13	*
AB Aur	B9	a	0.4	7, 12, 15	
R Aql	M5–M8e	b c	0.4	4, 12, 17	
W Cep	K0 Ia	b c	≤0.05	4, 17	
VY CMa	M3–M5 Ib	b c	2.1	11, 12	
T CrB	M3III+sdB	a	≤0.05	8, 13, 15	▽
RW Cyg	M3 Ia	b c	≤0.30	6, 17	
U Her	M6–8e	b c	0.50	2, 3, 16	
AC Her	F2–K4Ib	a	0.20	4, 18	△
89 Her	F2 Ia	a	0.50	5	
17 Lep	B8+M2III	a b c	0.50	1, 13, 15, 17	
48 Lib	B pec	a	0.15	13, 14	
T Mon	F8–K0Ib	a b c			+
U Ori	M6–8e	b c	≤0.15	6, 17	
W Ser	F5e	a		19	
VX Sgr	M4	b c	1.0	9, 10, 12	
SV Vul	F8–K0Ib	a b c		4	

a = 4430   * at max. extinction at min. is 0.8 mag.
b = 5780   + not known but presently under study
c = 6613   ▽ spectra taken during outburst of 1946
            △ spectra taken near maximum light.

1. Blanco, V. M., Demers, S., Douglas, G. G., and Fitzgerald, M. P.: 1968, *Publ. U.S. Naval Obs.* **XXI**.
2. Dyck, H. M. and Lockwood, G. W.: private communication.
3. Forrest, W.: private communication.
4. Gehrz, R. D. and Woolf, N. J.: 1970, *Astrophys. J. Letters* **161**, L 213.
5. Gillett, F. C., Hyland, A. R., and Stein, W. A.: 1970, *Astrophys. J. Letters* **162**, L21.
6. Gillett, F. C., Merrill, K. M., and Stein, W. A.: 1971, *Astrophys. J.* **164**, 83.
7. Gillett, F. C. and Stein, W. A.: 1971, *Astrophys. J.* **164**, 77.
8. Hofleit, D.: 1964, *Yale Catalogue of Bright Stars*.
9. Humphreys, R. M. and Lockwood, G. W.: 1972, *Astrophys. J. Letters* **172**, L59.
10. Humphreys, R. M., Strecker, D. W., and Ney, E. P.: 1972, *Astrophys. J.* **172**, 75.
11. Hyland, A. R., Becklin, E. E., Neugebauer, G., and Wallerstein, G.: 1969, *Astrophys. J.* **158**, 619.
12. Johnson, H. L.: 1966, *Ann. Rev. Astron. Astrophys.* **4**, 193.
13. Johnson, H. L., Mitchell, R. I., Iriarte, B., and Wisniewski, W. Z.: 1966, *Comm. Lunar Planet. Rec. Lab.* **4**, 99.
14. Lee, T.: (private communication).
15. Low, F. J.: 1970, Air Force Cambridge Research Laboratories Semi-Annual Technical Report AFCRL-70-0179.
16. Mayall, M. W.: private communication.
17. Neugebauer, G. and Leighton, R. B.: 1969, *Two Micron Survey*, NASA P-3047.
18. Preston, G. W., Krzeminsky, W., Smak, J., and Williams, J. A.: 1963, *Astrophys. J.* **137**, 401.
19. Geisel, S. L.: 1970, *Astrophys. J.* **161**, L105.

a plot of color excess vs distance for stars in the U.S. Naval Observatory Photoelectric Catalogue (Blanco *et al.*, 1968) that the distance to ε Aur is approximately 500 pc, while spectral classification yields the 1000 pc estimate.

The nova RS Oph shows strong absorption at λ 6613, 5780, and 5796 Å (Wallerstein, 1958). We find these bands in the nearby cepheid, Y Oph, which suffers about the same amount of interstellar extinction as RS Oph (Kraft, 1961; Sandage and Tammann, 1971) and is thought to be at roughly the same distance. Therefore in this case also we believe that circumstellar diffuse bands are absent. A detailed comparison of the two stars would be very difficult because of blends with atomic lines in Y Oph and uncertainty in the distance to RS Oph. Photometry of RS Oph shows no infrared excess in the V-K color assuming a K0 star to be present (Wallerstein, 1969) that is reddened by 0.85 in $B-V$ (Svolopolous, 1966) but a considerable excess in the K-N color (Geisel et al., 1970). Observations out to 3.5 μ by Harvey (1972) in August and September 1970 confirm the absence of an excess in the near infra-red.

We wish to compare the diffuse band strength or upper limit on it for circumstellar material with the strength to be expected had the reddening material been of the same nature as interstellar dust. The absorption listed in Table I for each star is only a lower limit of the total visual extinction because the scattering term has been omitted. If the ratio of scattering to absorption is similar to that for interstellar material the absorptions must be increased by a factor of three to find the total visual extinction. The proposed factor of three is an upper limit on the range of possible corrections since the particles may be close to the star and have a phase function that favors forward scattering. In the limiting case the scattering will not add to the absorption, since as much light will be scattered into the beam as out of it.

By inspection of Table I we see that a number of stars have sufficient circumstellar absorption that we would expect to see λ 4430 if it followed the interstellar relationship between λ 4430 and extinction. If we recognize that the absorption of Table I may be only $\frac{1}{3}$ of the total extinction many of the stars should show λ 4430, but they do not.

Gahm and Hultquist (1972) report an anomalous feature at λ 6613 in the spectra of several late type supergiants, and suggest that it may be the λ 6613 diffuse band. In our stars we have used a spectra comparator to compare the lines from different stars, and we conclude that the line we see at 6613 Å is the same in each star, rather than being more broad and diffuse for later type stars. We have no stars in common with Gahm and Hultquist.

5. Discussion

The lack of circumstellar diffuse bands in stars which have IR excesses may be due to some intrinsic difference between the grains in circumstellar envelopes and those in the general interstellar medium, or to the physical conditions in the near environs of stars. There is some evidence for the reduction of λ 4430 strength in regions of strong heating (Greenberg, 1968; Stoeckly and Dressler, 1963). On the other hand, A'Hearn (1971) has found evidence of λ 4430 absorption in reflection nebulae, near hot stars. Although the temperatures in circumstellar dust envelopes are much lower than those in H II regions they are much higher than in dense interstellar clouds.

If the lack of circumstellar diffuse band absorption is due to some intrinsic property of circumstellar dust grains which is different from that of the interstellar grains, then the question arises what that difference may be. Two possibilities are that the composition of the grains is different, or that the size of the grains is different. The size distribution in the circumstellar environment may differ from the interstellar environment because stellar radiation pressure may selectively deplete its circumstellar cloud of the smaller grains. Perhaps our results indicate that circumstellar envelopes around evolved stars are not the main source of the general interstellar dust.

Acknowledgements

We thank Dr C. R. O'Dell for a valuable discussion of the ratio of absorption to scattering by interstellar grains. This research was supported by NSF grant GP-28882.

References

A'Hearn, M. F.: 1971, *Astron. J.* **76**, 264.
Blanco, V. M., Demers, S., Douglas, G. G., and Fitzgerald, M. P.: 1968, *Publ. U.S. Naval Obs.* **XXI**.
Duke, D.: 1951, *Astrophys. J.* **113**, 100.
Gahm, G. F. and Hultquist, L.: 1972, *Astron. Astrophys.* **16**, 329.
Geisel, S. L., Kleinmann, D. E., and Low, F. J.: 1970, *Astrophys. J. Letters* **161**, L101.
Greenberg, J. M.: 1968, in B. M. Middlehurst and L. H. Aller (eds.), *Nebulae and Interstellar Matter*, University of Chicago Press, p. 221.
Harvey, P.: 1972, private communication.
Herbig, G. H.: 1960, *Astrophys. J. Suppl.* **4**, 337.
Hyland, A. R., Becklin, E. E., Neugebauer, G., and Wallerstein, G.: 1969, *Astrophys. J.* **158**, 619.
Kraft, R. P.: 1961, *Astrophys. J.* **134**, 616.
Lockwood, G. W.: preprint.
Lockwood, G. W. and Wing, R. F.: 1971, *Astrophys. J.* **169**, 63.
Sandage, A. and Tamman, G. A.: 1971, *Astrophys. J.* **167**, 293.
Stoeckly, R. and Dressler, K.: 1963, *Astrophys. J.* **139**, 240.
Strom, S. E., Strom, K. M., Brooke, A. L., Bregman, J., and Yost, T.: 1972, *Astrophys. J.* **171**, 267.
Svolopolous, S. N.: 1966, *Publ. Astron. Soc. Pacific* **78**, 157.
Van de Hulst, H. C. and De Jong, T.: 1966, *Physics* **41**, 151.
Wallerstein, G.: 1958, *Publ. Astron. Soc. Pacific* **70**, 537.
Wallerstein, G.: 1969, *Publ. Astron. Soc. Pacific* **81**, 672.
Wilson, R. E.: 1971, *Astrophys. J.* **170**, 529.

DIFFUSE INTERSTELLAR FEATURES IN THE SPECTRA
OF DUST-EMBEDDED STARS

G. E. BROMAGE* and K. NANDY

Royal Observatory, Edinburgh, U.K.

Abstract. Eleven diffuse interstellar features have been studied in the spectra of dust-embedded stars in Cyg OB2 (VI Cyg) (including a Wolf-Rayet star) and field stars in the same direction. The equivalent widths of the lines are all strongly correlated with each other and with E_{B-V}, but there is virtually no correlation with D line strength for these stars. The wavelength positions and half-widths of the diffuse lines appear to be the same for all stars studied.

A programme of extending observations to stars as faint as $B = 14^{m}.0$ for the study of diffuse interstellar features is being continued. The object of this programme is to determine whether any systematic differences exist between the relative strengths and positions of the features produced in circumstellar clouds, and those produced in the general interstellar medium.

Observations of eleven diffuse features in the wavelength range 5500–7000 Å are presented here for dust-embedded stars in the association Cyg OB2 (VI Cyg) and for field stars in the same direction.

TABLE I

Stars observed

Star	Assn.	Sp. Type	V	E_{B-V}
HDE 229059	Cyg OB1	B1.5Ia	8.70	1.71
HD 194279	Cyg OB9	B1.5Ia	7.02	1.20
HD 194839		B0.5Ia	7.49	1.19
HD 195229		B0.5III	7.66	0.44
HD 195592		O9.5Ia	7.08	1.14
BD+40°4219	Cyg OB2 4	O8	10.22	1.50
BD+40°4220	Cyg OB2 5	O7f	9.2 var	2.02
BD+40°4227	Cyg OB2 8A	O6f	9.03	1.59
BD+41°3804	Cyg OB2 10	O9.5Ia	9.88	1.80
	Cyg OB2 12	B5Ia–O	11.51	3.44
	Cyg OB2 WR2	WN5	11.8:	2.4:
HD 199478		B8Ia	5.68	0.47
HD 202850	Cyg OB4	B9Iab	4.23	0.11

The spectra were obtained at a dispersion of 60 Å mm^{-1} and a resolution of 2 Å with the Cassegrain spectrograph of the 98″ Isaac Newton telescope. The Cygnus stars observed are listed in Table I.

* Now at Department of Structural Chemistry, University of Bradford, Bradford, 7, Yorkshire, England.

Greenberg and Van de Hulst (eds.), Interstellar Dust and Related Topics, 71–77.

The spectra are measured in terms of Baker density

$$\Delta = \log_{10}\left(\frac{1}{T} - 1\right)$$

and a calibration curve is derived for the wavelength of each feature. The profile of the broad feature at 6176 Å is obtained by the comparison of stars of similar spectral type. Equivalent widths are derived directly from the digitised spectra: spectral densities are reduced to intensities and then rectified by linearly interpolating the continuum over the integration region for each feature. The same integration regions are used for all stars: they are listed in Table II. The integration steplength is $\frac{1}{2}$ Å.

For the narrow features, the error on an equivalent width estimate arises mainly from the emulsion granularity; but for the two broad ones, the uncertainty in the continuum position is more important. Errors from other sources (such as calibration uncertainties and corrections for blends with stellar and telluric lines) rarely amount to more than 3% of the total equivalent width. Standard errors for all stars have been computed from the data for those with 3 or more spectra.

The features appear very strong in the spectra of the heavily reddened stars of Cyg OB2, which range in spectral type from Wolf-Rayet and Of to B5Ia. A microdensitometer trace of the spectrum of the Wolf-Rayet star No. WR2 (Herbig and Mendoza, 1960), which is believed to be a member of the association, is shown in Figure 1a. From the slope of the continuum, the $(B-V)$ colour is estimated to be 2.1 ± 0.2, which agrees with the value obtained by Reddish (1968) from UBV photometry. This star was previously classified as WN7, but from the observed ratio of C IV to He I emission line strengths, it must be WN5.

The O7f star Cyg OB2 No. 5, which is an eclipsing binary, shows Hα and He I (λ 6678) in emission and He I (λ 5875) in absorption, all shifted to the blue by approx-

TABLE II

Spectrum lines measured

Line name	Integration region (nm)
5705 (+N II 571.1)	570.1–571.5
N II 571.1 only	570.8–571.5
5780+5776	575.5–579.1
5780 only	577.3–578.7
5797 (+C IV 580.2)	579.1–580.5
C IV 581.2	580.8–581.5
D lines (Na I)	588.3–590.3
6010	600.4–601.6
6176+6203	613.5–622.0
6203 only	619.7–620.9
6270	626.5–627.4
6284 (+O₂ 627.8)	627.4–629.5
6379	637.5–638.3
6614	660.8–662.0

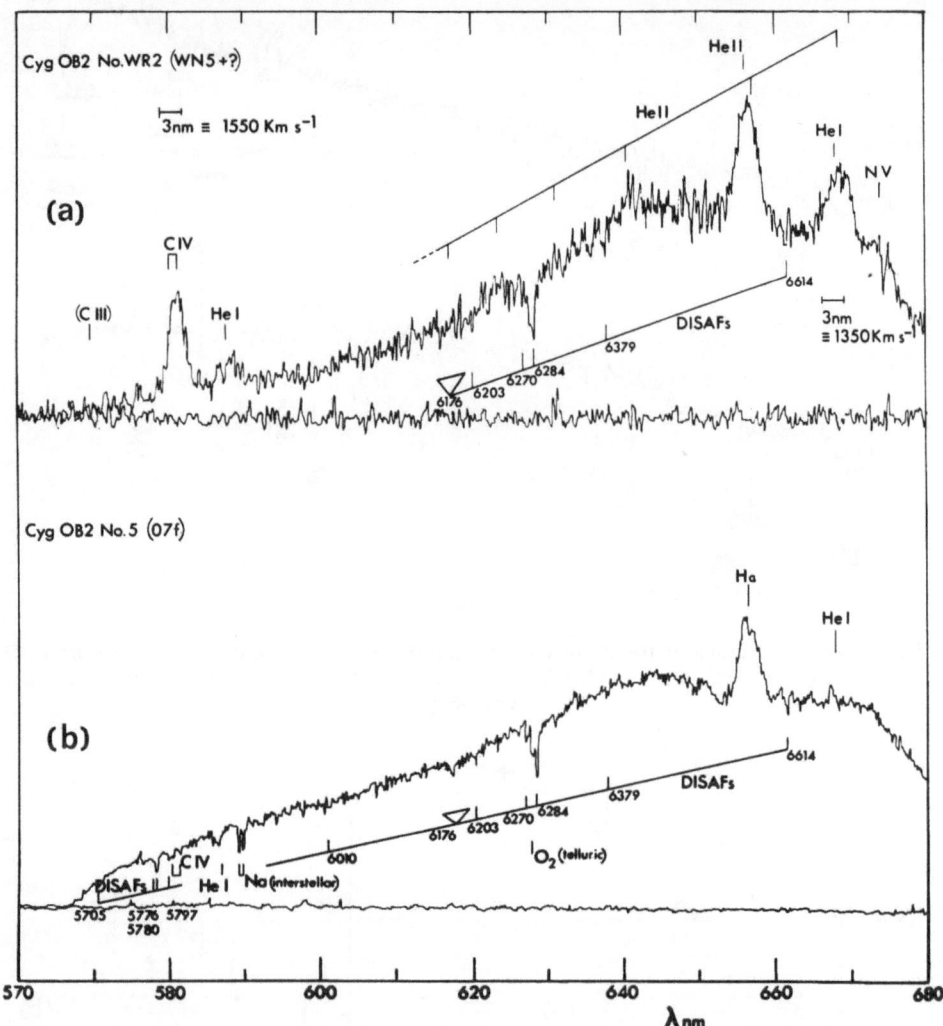

Fig. 1. Microdensitometer tracings of the spectra of two members of Cyg OB2: (a) No. WR2, type WN5, $V = 11.8$, $E_{B-V} = 2.4$, $M_V \simeq -6$. (b) No. 5, type 07f, $V = 9.2$, $E_{B-V} = 2.0$, $M_V \simeq -8$.

imately 150 km s^{-1} (see Figure 1b). It is interesting to note that the C IV absorption lines (5802 Å and 5712 Å), on the contrary, do *not* seem to show the same shift.

The most heavily reddened star Cyg OB2 No. 12 shows the strongest diffuse features in the red so far recorded (Bromage, 1971), and also has Hα in emission. A microdensitometer tracing of the spectrum of this star is reproduced in Figure 2. The '6284' line has a central depth of almost 50%.

A result which is of considerable interest is that for all the stars studied here, whether dust embedded or field stars, the wavelength positions of the observed diffuse features are the same, within the spectral resolution.

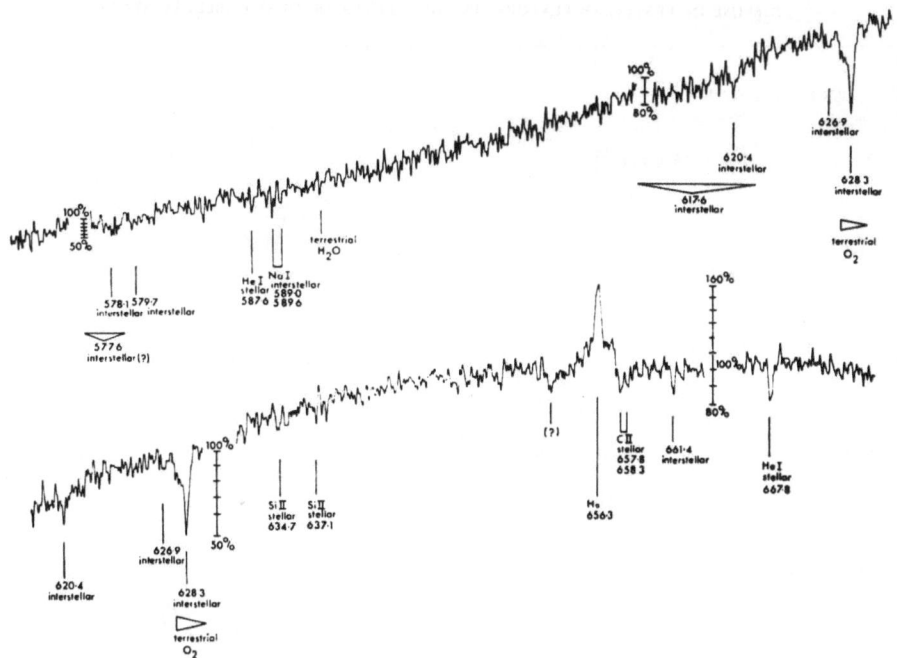

Fig. 2.   Microdensitometer tracing of the spectrum of the star Cyg OB2 No. 12, type B5 Ia–O, $V = 11.5$, $E_{B-V} = 3.5$, $M_V \simeq -12$. Upper trace, wavelength range $\lambda\lambda$ 575–630 nm; lower trace, $\lambda\lambda$ 620–675 nm.

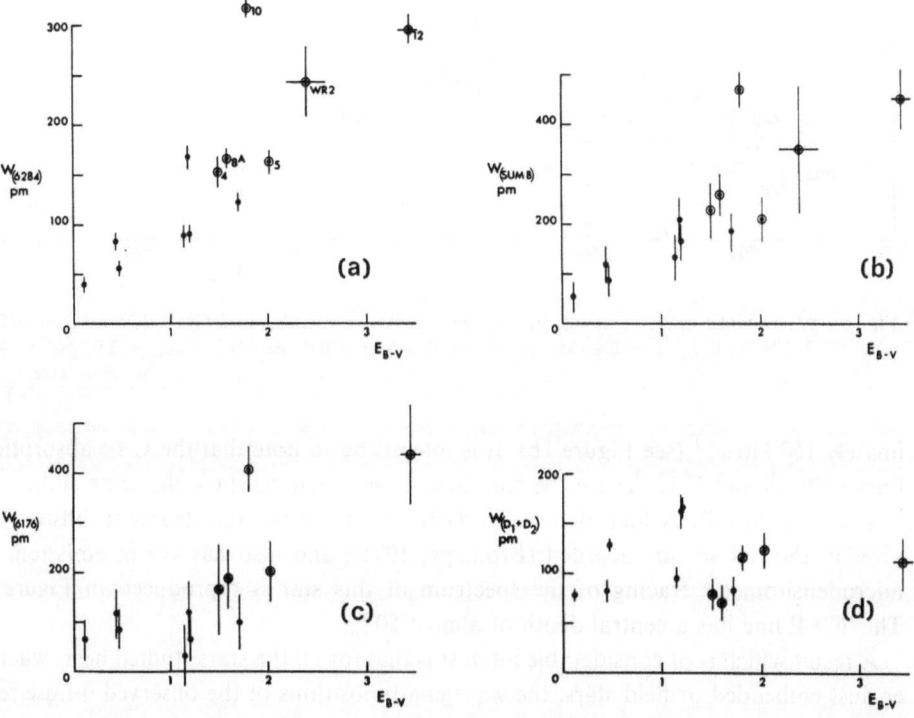

Fig. 3.   Graphs of equivalent widths of interstellar lines (in picometres $\equiv$ 0.01 Å) against E_{B-V}, for Cygnus stars. Points referring to members of Cyg OB2 are shown as open circles. The error bars refer to standard errors. (a) The typical narrow diffuse line '6284' (half-width $\simeq$ 0.4 nm); (b) The total equivalent width of 5 narrow lines in the range $\lambda\lambda$ 620–665 (SUM B); (c) The typical broad diffuse feature '6176' (half-width $\simeq$ 2.5 nm); (d) The total equivalent width of the interstellar sodium D lines.

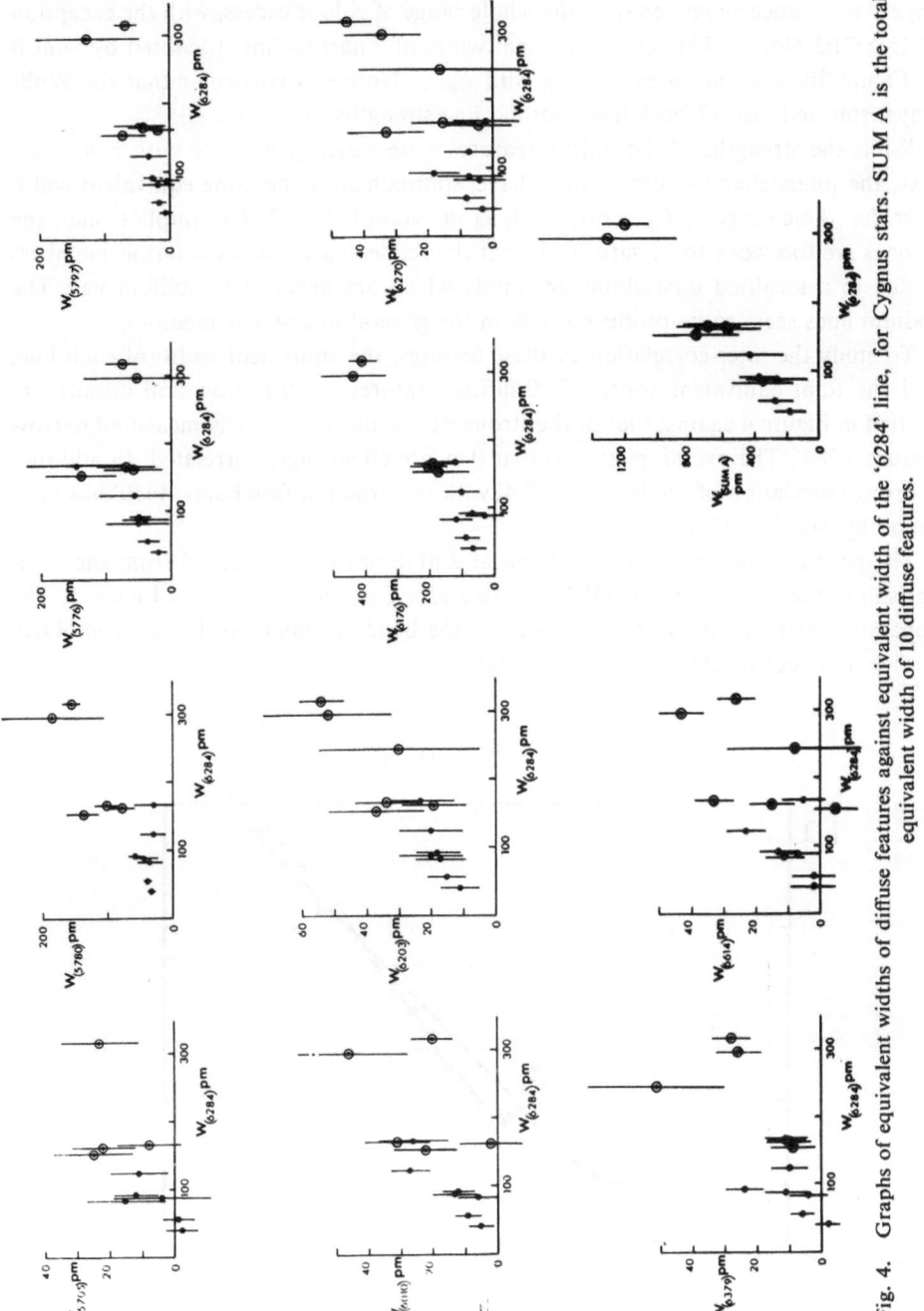

Fig. 4. Graphs of equivalent widths of diffuse features against equivalent width of the '6284' line, for Cygnus stars. SUM A is the total equivalent width of 10 diffuse features.

The relations between the equivalent widths of the typical narrow line '6284' and the broad one '6176', and colour excess E_{B-V}, are shown in Figure 3. In general, they appear to be linearly related over the whole range of colour excess, with the exception of Cyg OB2 No. 10. The total equivalent width of 5 narrow lines (denoted by Sum B in Figure 3b) also increases linearly with E_{B-V}. Notice in particular that the Wolf-Rayet star and No. 12 both have normal line strengths.

While the strengths of the diffuse features systematically increase with colour excess, the interstellar sodium D lines have approximately the same equivalent width over the whole range of E_{B-V} observed, as shown in Figure 3. This implies, since the D lines are too weak to be saturated, that the reddening of the association members is due to a localised dust cloud or clouds which are deficient in sodium gas. The sodium lines seem to be produced only in the general interstellar medium.

To study the inter-correlation of these features, the equivalent width of each line, and the total equivalent width of 10 diffuse features (both narrow and broad), are plotted in Figure 4 against that of the strongest and most accurately measured narrow feature '6284'. The overall picture is that they are all strongly correlated. In addition a strong correlation of the feature '6284' with the broad diffuse band '4430' has been found by Murdin (1972).

The profile of the broad diffuse band at λ 6176 has been obtained from the comparison of the spectra of Cyg OB2 No. 10 and κ Cas, and is shown in Figure 5. The apparent emission wing at the blue side of the band is similar to that seen in '4430' (Brück and Nandy, 1968; Bromage, 1972).

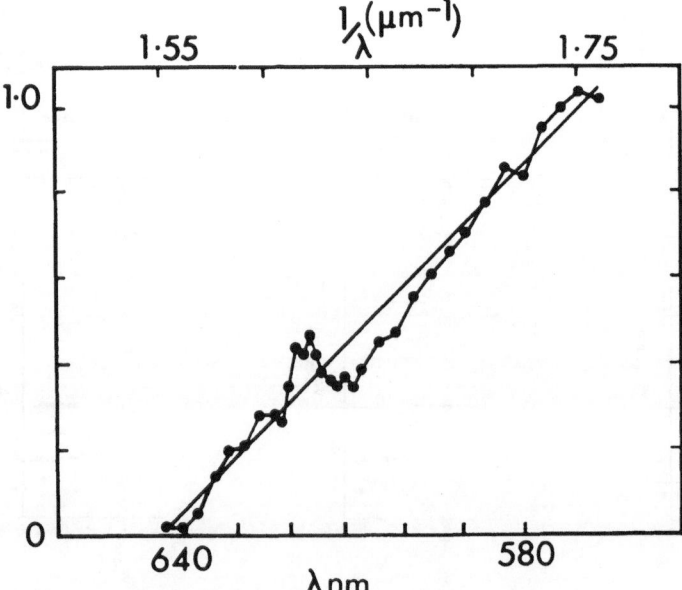

Fig. 5.   The profile of the broad '6176' feature for Cyg OB2 No. 10 compared with κ Cas ($\Delta E_{B-V} =$ 1.47).

Our interpretation of these observations is that the strong correlation of strength of diffuse features with colour excess, the profiles of the broad features, and the inter-correlation of the strength of the features, together with no correlation with D line strength, all support the view that these features are produced by the grains rather than the gas. However, the invariance of the wavelength positions and of the half-widths of these features requires a high degree of uniformity of physical conditions of the grains, including temperature, effective size and nature of the impurities.

References

Bromage, G. E.: 1971, *Nature* **230**, 172.
Bromage, G. E.: 1972, *Astrophys. Space Sci.* **15**, 426.
Brück, M. T. and Nandy, K.: 1968, *Nature* **220**, 46.
Herbig, G. M. and Mendoza, E. E.: 1960, *Bol. Obs. Tonatzintla y Tacubaya*, No. 19, p. 21.
Murdin, P.: 1972, *Monthly Notices Roy. Astron. Soc.* **157**, 461.
Reddish, V. C.: 1968, *Observatory* **88**, 139.

POLARIZATION PROFILE NEAR THE 4430 BAND

K. NANDY and H. SEDDON

Royal Observatory, Edinburgh, U.K.

It has been reported by several authors that the extinction profile near the wavelengths of unidentified diffuse features shows fine structure (Brück and Nandy, 1970; Walker, 1971; York, 1971; Bromage, 1972); the question arises whether these features also exhibit fine structure in wavelength dependence of polarisation. It was first suggested by Greenberg *et al.* (1970, 1971) that interstellar bands produced by grain impurities may be polarized differently from the neighbouring continuum. Similar conclusions were also derived by Wickramasinghe and Nandy (1971) and Kelly (1971). Our first attempt to derive a profile of polarization against wavelength across the 4430 Å band was reported earlier (Nandy and Seddon, 1970). In this method we have used a Wollaston prism at the Cassegrain focus of the 36″ Telescope of the Royal Observatory, Edinburgh to produce orthogonally polarised spectra on the photographic plate. The results for the star 55 Cyg (Figure 1) indicated a definite profile. It was felt than an extension of the programme to acquire further profiles and to establish instrumental polarisation to a greater accuracy was necessary.

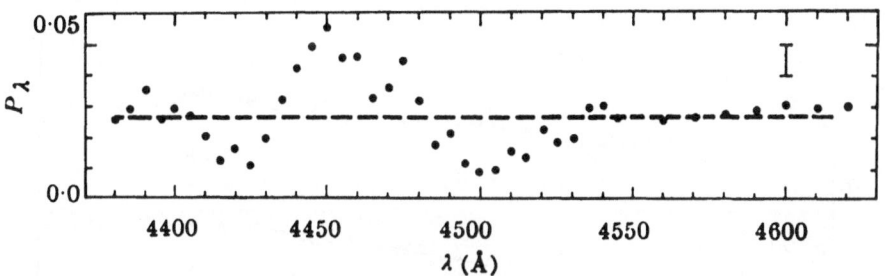

Fig. 1. Wavelength dependance of polarization across in 4430 Å feature for 55 Cyg.

For partially plane polarized star light, the monochromatic magnitude difference Δm_λ between the pair of orthogonally polarized spectra is given by

$$\Delta m_\lambda(\alpha) = p_\lambda \cos 2(\alpha - \theta_\lambda),$$

where

p_λ = the amount of polarization expressed in magnitudes

α = the angle between the sky oriented reference axis and one of the two directions of the prism axis.

θ_λ = the position angle of the starlight measured from the same reference axis as α.

The position of the Wollaston prism is adjusted so that initially $\alpha = 0$. Stokes

Greenberg and Van de Hulst (eds.), Interstellar Dust and Related Topics, 79-81.

parameters $p_x(\lambda)$ and $p_y(\lambda)$ expressed in magnitudes are given by (see Hall and Serkowski, 1963)

$$p_x(\lambda) = \Delta m_\lambda(0°)$$
$$p_y(\lambda) = \Delta m_\lambda(45°).$$

Stokes parameters p_x and p_y have been determined for the unpolarized star α Cyg
(Deneb) and for two other stars with differing degrees of polarization and 4430 Å
absorption. The spectra are measured and reduced digitally producing a complete
plot of p_x and p_y against wavelength, as shown in Figure 2. The profiles derived from
α Cyg establish the instrumental polarization profile, which must then be compared
with the profiles of the other stars to ascertain the significance of the difference inside
and outside the band.

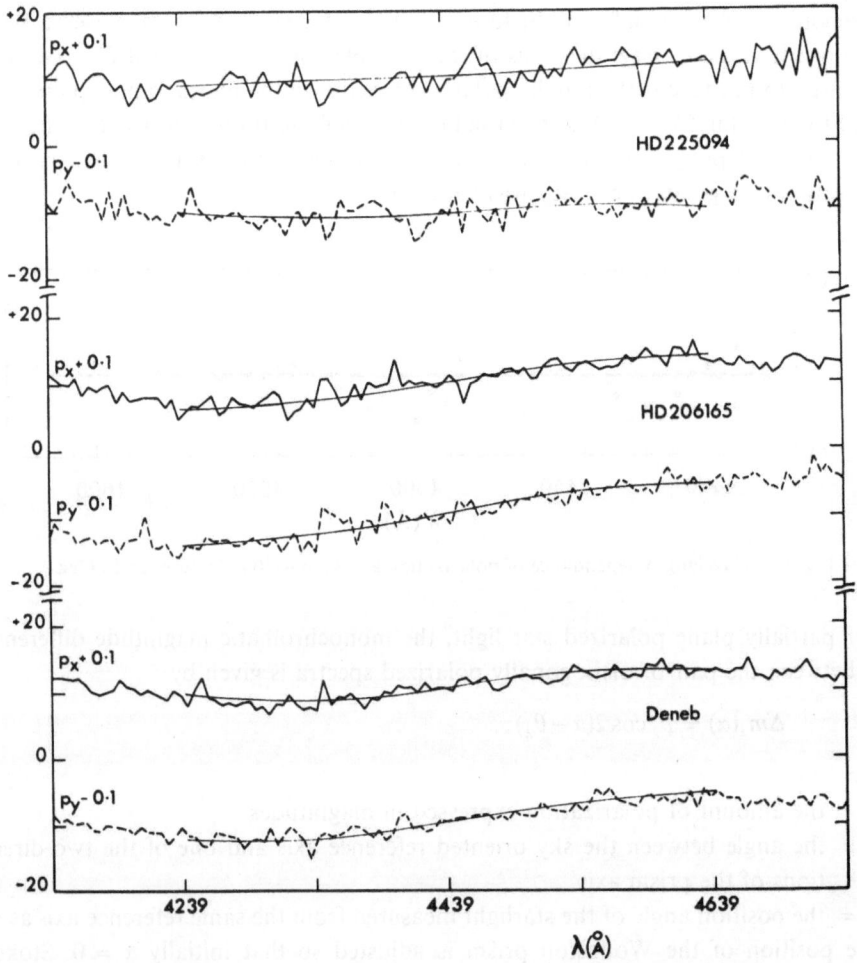

Fig. 2. Wavelength dependance of p_x and p_y for Deneb (α Cyg), HD 206165 and HD 225094.

The noise level of the α Cyg traces is 2.5 times smaller than that of the other two stars, because of the difference in number of spectra used. The 4430 absorption in HD 225094 is much greater than that in HD 206165 and there is indeed an indication of the difference of p_x and p_y values inside and outside the bands.

We now intend to extend these observations to other stars with large 4430 absorption in different galactic longitudes. We shall then combine these observations appropriately to give weighted mean profiles in these longitudes. We have already obtained spectra of good quality for several stars by using the Wollaston prism at the Cassegrain focus of the 98″ Isaac Newton Telescope and their reduction is in progress.

References

Brück, M. T. and Nandy, K.: 1968, *Nature* **220**, 46.

Bromage, G.: 1972, *Astrophys. Space Sci.* in press.

Greenberg, J. M. and Stoeckly, R.: 1970, in L. Houziaux and H. E. Butler (eds.), 'Ultraviolet Stellar Spectra and Related Ground-Based Observations', *IAU Symp.* **36**, 36.

Greenberg, J. M. and Stoeckly, R.: 1971, *Nature Phys. Sci.* **230**, 15.

Hall, J. S. and Serkowski, K.: 1963, *Stars and Stellar Systems* **3**, 293.

Kelly, A.: 1971, *Astrophys. Space Sci.* **13**, 211.

Walker, G. A. H., Hutchings, J. B., and Younger, P. F.: 1970, in L. Houziaux and H. E. Butler (eds.), 'Ultraviolet Stellar Spectra and Related Ground-Based Observations', *IAU Symp.* **36**, 52.

Wickramasinghe, N. C. and Nandy, K.: 1971, *Nature Phys. Sci.* **229**, 234.

York, D. G.: 1971, *Astrophys. J.* **166**, 65.

BROADBAND STRUCTURE IN THE INTERSTELLAR
EXTINCTION CURVE*

D. S. HAYES, G. E. MAVKO, R. R. RADICK, and K. H. REX

Rensselaer Polytechnic Institute, U.S.A.

and

J. M. GREENBERG

State University of New York at Albany, N.Y., U.S.A.

Abstract. We have analysed observations of the interstellar extinction in the range 3400 Å–11 000 Å. The observations have high photometric accuracy and wavelength resolution, and allow a detailed examination of broadband structure as well as the general shape of the wavelength dependence of the extinction curve. The broadband structure has a characteristic size of several hundred ångströms, and may be as important as the diffuse bands in indicating the physical nature of the grains.

We describe our preliminary theoretical interpretation of this broadband structure, and the new observation which it predicts.

1. Introduction

In this paper we report a preliminary observational and theoretical analysis of broadband structure in the interstellar reddening curve in the range 3500–11000 Å. By 'broadband', we mean observations with a wavelength resolution on the order of 100 Å. There are two 'features' in the interstellar extinction curve which we will discuss: (a) The 'knee' in the interstellar extinction curve, taken to be at $1/\lambda = 2.25\ \mu^{-1}$ or $2.30\ \mu^{-1}$ in various published accounts (Whitford, 1958; Nandy, 1964; Underhill and Walker, 1966, 1967; Harris, 1969) and (b) a broad, shallow feature centered on $1/\lambda = 1.8\ \mu^{-1}$.

The 'knee' in the interstellar extinction curve is an apparent sharp break in the slope of the interstellar extinction curve in the blue region of the spectrum. The interstellar extinction curve has been described as having the appearance of two intersecting straight lines by some observers (Whitford, 1958; Nandy, 1964; Harris 1969; Underhill and Walker, 1966, 1967). This 'knee' has been attributed to graphite (Wickramasinghe, 1967) because the index of refraction of graphite has a change in slope a short distance to the blue of the position of the knee, as determined by Nandy (1964).

The broad shallow feature centered on $1/\lambda = 1.8\ \mu^{-1}$ was noticed by Whiteoak (1966) and Walker (1967), and was discovered independently by one of us (DSH) in 1967. Seen in 'emission', it has not been noticed by other observers because of its extreme width (~ 2000 Å) and shallowness ($0.^m05$ for 1^m total reddening between $1/\lambda = 2.09$ and $1.136\ \mu^{-1}$). It is most probably a feature connected with the grains, because of its great width. One would not expect such a broad feature to be caused by an independent constituent of the interstellar medium, especially in 'emission'.

* Work supported in part by grants from the Research Committee of Rensselaer Polytechnic Institute.

Greenberg and Van de Hulst (eds.), Interstellar Dust and Related Topics, 83–90.

In this paper we will show that the 'knee' in the interstellar extinction curve is explained in a straightforward way by a change in the curvature of the theoretical extinction curve in this region. The wavelength of this curvature change is related only to the size distribution of the grains. Thus, there will be no need to invoke graphite to explain the 'knee'.

The feature at $1/\lambda = 1.8 \; \mu^{-1}$, however, must be explained by a change in the index of refraction of some component of the interstellar medium. A plausible model of this feature is that it is part of a double dispersion curve resulting from two broad absorptions centered on 4170 and 6670 Å. The most important consequence of this model is that it predicts that structure should also be seen in the wavelength dependence of the interstellar polarization.

2. The Observed Features

Because of the relative shallowness of these features compared to the general slope of the extinction curve, they are difficult to see in an ordinary graph of the interstellar extinction vs λ or $1/\lambda$. We have chosen the graph of the residuals after a straight line is subtracted from the measured extinction curves. In Figure 1 are shown data for a mean extinction curve derived from spectrophotometric measurements of 28 O-type stars. The data are those of Whiteoak (1966). The extinction curves for the individual stars have been derived by subtracting off the mean energy distribution of seven stars

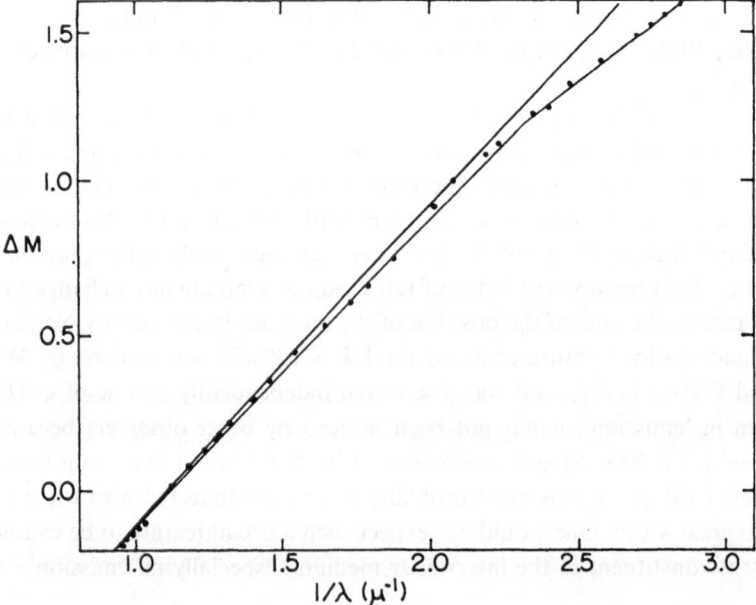

Fig. 1.　The mean reddening curve for twenty-eight O-type stars. Also shown are two straight lines intersecting at 2.3 μ^{-1}, exhibiting a common interpretation of the observations, and the reference straight line, passing through 2.19 μ^{-1} and 1.136 μ^{-1}.

of small reddening. Whiteoak compared the reddened stars with model atmospheres, a procedure which can be successful only if the absolute calibration of stellar energy distributions is correct, within the error of the observations. The Oke (1964) calibration, even with the modifications introduced by Whiteoak, was sufficiently in

TABLE I
The 'unreddened' stars

		B	$m_{2.09} - m_{1.136}$
ι Ori	O9 III	0.12	-1.01
υ Ori	B0 V	0.18	-1.07
ζ Ori	O9.5 Ib	0.13	-0.95
δ Ori	O9.5 II	0.15	-0.97
σ Ori	O9.5 V	0.18	-0.99
15 Mon	O7	0.15	-1.03
10 Lac	O9 V	0.13	-1.00
mean		0.15	-1.00

TABLE II
The 'reddened' stars

		B	$m_{2.09} - m_{1.136}$	$E(m_{2.09} - m_{1.136})$	D
HD	13268	0.09	-0.31	$+0.69$	0.09
	17520	0.13	0.11	1.11	0.07
	17505	0.09	0.30	1.30	0.05
	14947	0.06	0.42	1.42	0.05
	16429	0.11	0.83	1.83	0.04
	15570	0.05	1.00	2.00	0.04
	46966	0.14	-0.62	0.38	—
	47129	0.11	-0.37	0.63	0.03
	46150	0.11	-0.21	0.79	0.04
	46149	0.13	-0.14	0.86	0.04
	46223	0.07	0.00	1.00	0.04
	46573	0.07	0.21	1.21	0.05
	188209	0.11	-0.72	0.28	—
	193322	0.16	-0.36	0.64	0.04
	191978	0.15	-0.27	0.73	0.06
	190864	0.06	-0.13	0.87	0.04
	192281	0.06	0.22	1.22	0.06
	193443	0.12	0.21	1.21	0.07
HDE	228368	0.08	0.49	1.49	0.04
	228854	0.08	0.98	1.98	0.06
HD	194334	0.05	1.20	2.20	0.05
BD+40°4227		0.09	2.23	3.23	0.04
HD	218915	0.11	-0.52	0.48	—
	209975	0.10	-0.37	0.63	0.06
	210839	0.08	-0.04	0.96	0.08
BD+60°2522		0.06	0.43	1.43	0.03
HD	216898	0.15	0.57	1.57	0.05
	217086	0.13	0.82	1.82	0.05

error (Hayes, 1970; Oke and Schild, 1970) that Whiteoak's comparison of the wavelength dependence of extinction for different regions (his Figure 3) only shows the error in the absolute calibration scaled by the relative mean reddening for each region. The same calibration error partially vitiates the work of Underhill and Walker (1966, 1967) which also referred the reddened stars to model atmospheres.

For our 'unreddened' star, we have used the mean energy distribution of seven stars with little reddening. They are listed in Table I, along with their spectral types, Balmer discontinuities, and color $m(\lambda_1) - m(\lambda_2)$ where $1/\lambda_1 = 2.09 \mu^{-1}$ (4785 Å) and $1/\lambda_2 = 1.136 \mu^{-1}$ (8800 Å). Also given are the data for the 'mean unreddened star'. In this range of spectral types (O5–B0), there is little change in the slope of the continuum. In the wavelengths used by Whiteoak for this observations, only the Balmer discontinuity will show a significant change with spectral type or luminosity class, and we have applied corrections to our reddening curves for the difference in Balmer discontinuity between the mean unreddened star and each reddened star. Table II gives a list of all the reddened stars used, with their Balmer discontinuities, colors and color excesses for the wavelength points corresponding to $1/\lambda = 2.09 \mu^{-1}$ and $1.136 \mu^{-1}$.

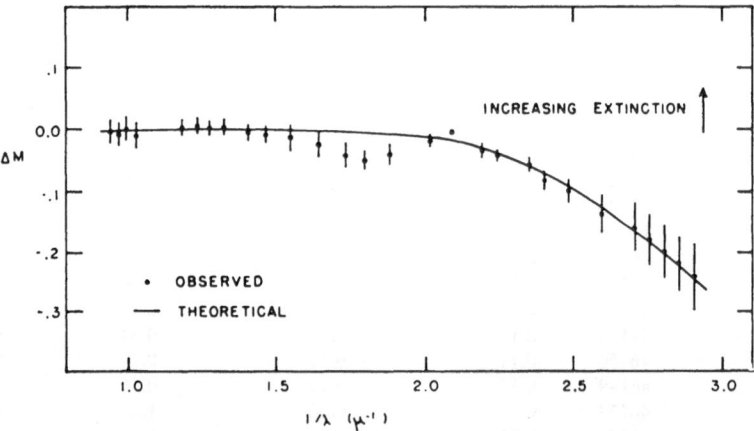

Fig. 2.   The extinction residuals, after the reference straight line is subtracted from the mean extinction curve. Also shown is a theoretical curve.

The individual extinction curves are normalized to 1 mag. reddening between $1/\lambda = 2.09$ and $1.136 \mu^{-1}$, and then averaged and plotted against $1/\lambda$. The average curve is shown in Figure 1, along with our reference straight line, passing through the normalization points. Also shown are the two straight lines which best describe the interpretation often found in the literature. They show the 'knee' at their intersection at about $2.30 \mu^{-1}$. The residuals produced when the reference straight line is subtracted from the extinction curve are shown in Figure 2. The residuals allow a closer look at the 'knee' in the interstellar extinction law. It should be evident, from Figure 2, that since Nandy drew his straight lines in such a way as to average over the feature at $1.8 \mu^{-1}$, he has arrived at a value of the wavelength of the 'knee' which is too far to

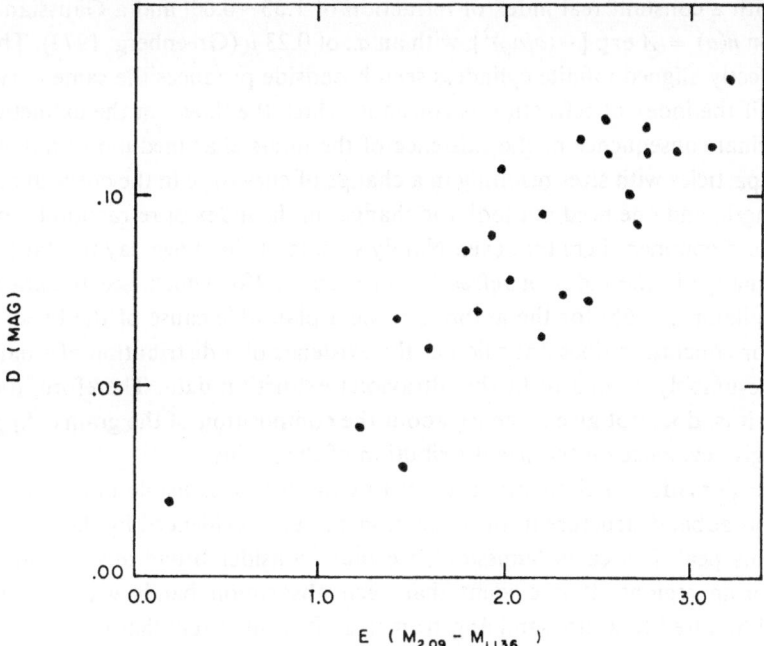

Fig. 3.   The central depth of the 1.8 μ^{-1} 'feature', plotted against $E(m_{2.09} - m_{1.136})$.

the blue. If one must assign a value to the position of the 'knee', it must be placed, not at 2.30 μ^{-1}, but at about 2.09 μ^{-1}:

Figure 2 also clearly shows the broad, shallow feature peaking at about $1/\lambda = 1.8$ μ^{-1}.

That the 1.8 μ^{-1} feature is definitely interstellar is shown by its close correlation with the total reddening, as shown in Figure 3. We have defined the strength of the feature by the central depth, D, in magnitudes, of a smooth curve drawn through the residuals. The central depth is measured at 1.8 μ^{-1} where the peak is almost always found. The zero level is at 2.09 μ^{-1} and 1.732 μ^{-1}. The total reddening is measured by the color excess defined by the points at 2.09 μ^{-1} and 1.136 μ^{-1} as given in Table II. The value of D, for each star is also given in Table II. Although the central depth shows large scatter (as would be expected, since the largest value is only 0.^{m}07, and the observational accuracy is about 0.^{m}02), the linear correlation is obvious. This means that this feature is closely associated with whatever causes the reddening, although it does not prove that it is caused by some characteristic of the interstellar grains, themselves.

3. Theoretical Interpretation

Also shown in Figure 2 is a theoretical curve. It has been derived from a simple analytic approximation to the wavelength dependence of scattering from spherical par-

ticles with a constant real index of refraction of $1.33 - 0.00i$ and a Gaussian size distribution $n(a) = A \exp[-(a/a_0)^2]$, with an a_0 of $0.23\,\mu$ (Greenberg, 1973). The theory
for perfectly aligned infinite cylinders seen broadside produces the same curve in this
region, if the index of refraction is constant. Thus, the 'knee' in the extinction curve
is a normal consequence of the existence of the interstellar medium of a distribution
of dust particles with sizes resulting in a change of curvature in the curve at about this
wavelength, and one need not look for changes in the index of refraction to explain it.
We should comment here that even Nandy's value of the 'knee' lay too far to the red
of the change in the index of refraction of graphite (for which, see Wickramasinghe
and Guillaume, 1965) for the graphite to be a plausible cause of the knee. Incidentally, our conclusion does not rule out the existence of a distribution of smaller sizes,
as is presumably needed to fit the ultraviolet extinction data. Therefore, the 'knee',
such as it is, does not give evidence about the composition of the grains. Apparently,
it *does* give evidence on the size distribution of the grains.

We have made a preliminary attempt to construct a plausible model of the cause
of the broadband structure in the extinction curve, as evidenced by the $1.8\,\mu^{-1}$ peak.
Since this peak is seen in 'emission', we must consider broad absorption bands in
neighboring regions. It is evident that such absorption bands will be very broad
(several hundred to a thousand ångströms), so it seems likely that they are caused by
an impurity in the grains which cause the interstellar extinction, itself. Following
Greenberg and Stoeckly (1970), we predict that structure will also be produced in the
wavelength dependence of the interstellar polarization, and that the observed existence or absence of such broadband polarization structure will be evidence for or
against this model

In order to better define the structure to be fitted in the extinction law, we must
find the residuals from a smooth 'continuum' extinction law, instead of from a straight
line, as in Figure 2. We have defined the smooth continuum by a theoretical curve
similar to that shown in Figure 2. It is also based upon a model of perfectly aligned
infinite cylinders, but with a size distribution of the radii of a the cylinders given by
$n(a) = \exp[-5(a/a_0)^3]$ where a_0 is taken to be $0.225\,\mu$. The index of refraction is here
taken to be $1.66 - 0.01i$. The residuals which result when this curve is subtracted from
the observed extinction curve are shown in Figure 4. It is clear that the peak at 1.8
μ^{-1} is not an isolated feature, but part of a complex structure extending over the entire range of observation. One should therefore include observations extending into
the UV and IR. The examination of the UV and IR extinction data which already
exists has been started, but is not complete at this date.

We will assume that the impurities in the grains responsible for this structure result
in a limited number of broad absorptions. The frequency dependence of the complex
index of refraction m is therefore given by the Clausius-Mossotti relation (in mks
units):

$$\frac{m^2 - 1}{m^2 + 2} = \frac{m_0^2 - 1}{m_0^2 + 2} + \sum_{j=1}^{2} \left\{ \frac{N_j e^2}{12\pi^2 \varepsilon_0 m_e} \left[\frac{1}{v_{0,j}^2 - v^2 + i v \gamma_{1,j}} \right] \right\},$$

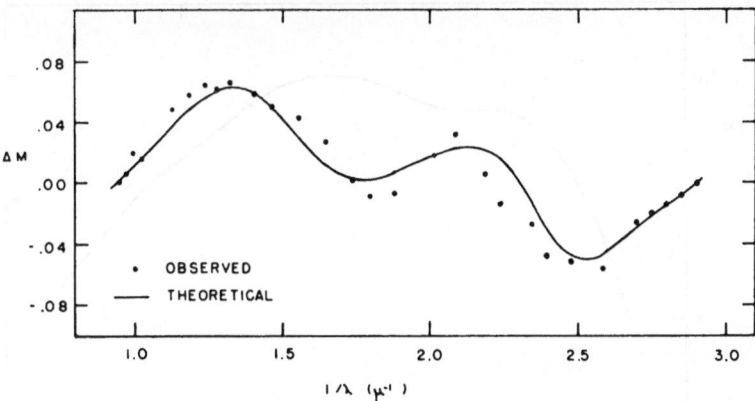

Fig. 4. The extinction residuals, after the reference 'continuum' curve is subtracted from the mean extinction curve. Also shown is a theoretical curve.

where we will attempt to find values of the natural frequencies v_0, damping constants γ_1 and effective densities N of impurity atoms which will result in extinction residuals which fit the observations shown in Figure 4. This wavelength dependent index of refraction is then used to calculate the theoretical extinction curve for perfectly aligned infinite cylinders with the cubic size distribution given above.

TABLE III

Absorption band parameters used in Clausius-Mossotti relation

λ_0	v_0	γ_1	$\Delta\lambda$	N
4170 Å	$7.2 \times 10^{14}\ \text{s}^{-1}$	$1.8 \times 10^{14}\ \text{s}^{-1}$	1000 Å	$5.04 \times 10^{25}\ \text{m}^{-3}$
6670	4.5	2.1	2600	5.04

The theoretical curve is also shown in Figure 4. Table III lists the parameters of the Clausius-Mossotti relation for this curve, including the halfwidth, $\Delta\lambda$, for each absorption.

For this theoretical curve, we have taken $a_0 = 0.225\ \mu$, and the 'continuum' index of refraction m_0 to be $1.66 - 0.01i$.

All calculations of the scattering and absorption efficiencies were made with the computer code developed by Shah (1967).

The agreement between the theoretical residuals and the observed residuals shown in Figure 4 is on the order of the size of the measurement errors. Thus, until better observations are available, we will consider this physical model to be at least a plausible one, and examine its implications with respect to the wavelength dependence of the interstellar polarization. The theoretical normalized polarization is shown in Figure 5. There is clearly a measureable structure present. Work is now underway to try to confirm this feature observationally.

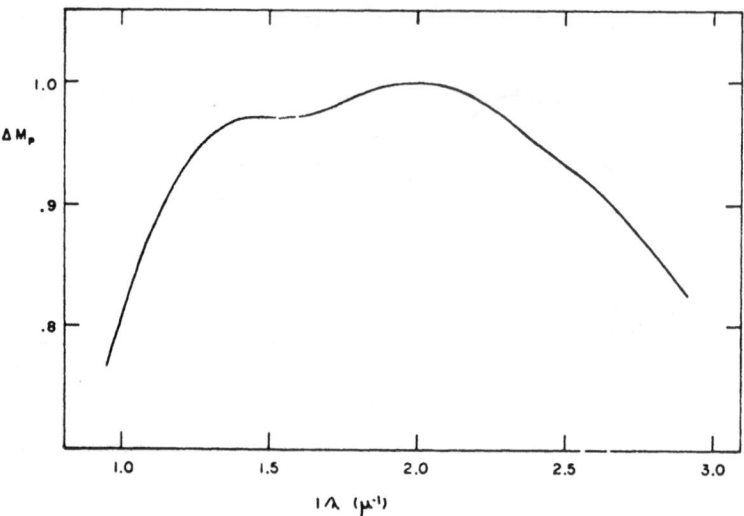

Fig. 5.   The normalized polarization predicted for the theoretical model shown in Figure 4.

References

Greenberg, J. M.: 1973, in M. A. Gordon and L. E. Snyder (eds.), *Molecules in the Galactic Environ-ments* (John Wiley and Sons), Proceedings of a *Symposium on Interstellar Molecules*, held at Char-lottesville, Virginia, October 1971, p. 93.

Greenberg, J. M. and Stoeckly, R.: 1971, *Nature Phys. Sci.* **230**, 15.

Harris, J. W.: 1969, *Nature* **223**, 1046.

Hayes, D. S.: 1970, *Astrophys. J.* **159**, 165.

Nandy, K.: 1964, *Publ. Roy. Obs. Edinburgh* **3**, 142.

Oke, J. B.: 1964, *Astrophys. J.* **140**, 689.

Oke, J. B. and Schild, R. E.: 1970, *Astrophys. J.* **161**, 1015.

Shah, G.: 1967, Unpublished dissertation, Rensselaer Polytechnic Institute.

Underhill, A. B. and Walker, G. A. H.: 1966, *Monthly Notices Roy. Astron. Soc.* **131**, 475.

Walker, G. S. H.: 1967, *Astron. J.* **72**, 37.

Whiteoak, J. B.: 1966, *Astrophys. J.* **144**, 305.

Whitford, A. E.: 1958, *Astron. J.* **63**, 201.

Wickramasinghe, N. C.: 1967, in J. M. Greenberg and T. P. Roack (eds.), *Interstellar Grains*, NASA SP-140, p. 37.

Wickramasinghe, N. C. and Guillaume, C.: 1965, *Nature* **207**, 366.

A STUDY OF THE
INTERSTELLAR DIFFUSE ABSORPTION FEATURES

CHI-CHAO WU

Space Astronomy Laboratory, Washburn Observatory, University of Wisconsin, Madison, Wis., U.S.A.

Abstract. At a resolution of 0.17 Å, the profile of $\lambda\lambda$ 5780, 5797 were found to be asymmetric and steeper to the blue. The FWHM of λ 5780 and λ 5797 are respectively 2.7 Å and 1.1 Å. There is also a broad, shallow feature shortward of λ 5780. Equivalent widths of $\lambda\lambda$ 5780, 5797 were measured at a resolution of 0.26 Å, and they present a very tight correlation. Therefore, it is highly likely that these two diffuse lines have a common origin. The correlation between the equivalent width of λ 5780, $EW(\lambda$ 5780), and $E(B-V)$ is good for normal field stars. But there is a systematic weakening of λ 5780 in the spectra of Be stars, and the reverse is true for stars in the Scorpius-Ophiucus region. Excluding Be stars, a least square fit gives $EW(\lambda$ 5780$) = 0.75 \, E(B-V)$ Å mag.$^{-1}$. From the profiles and structures of the diffuse features, and the correlations with various parameters of the interstellar medium, it is suggested that: (1) the diffuse band λ 4430 may be produced by the preionization of H^{-} and, (2) $\lambda\lambda$ 5780, 5797 may be due to the pure electronic transitions of the impurity centers in solid grains.

Here I wish to report a recent study of the interstellar diffuse absorption features $\lambda\lambda$ 4430, 5780, 5797. Earlier work concerning the diffuse features was discussed by Wilson (1964), Johnson (1967), and Herbig (1967). Photoelectric measurements for the central absorption depth of λ 4430 were made by Stoeckly and Dressler (1964) and Wampler (1966). Equivalent widths of $\lambda\lambda$ 5780, 5797 were obtained for 66 early type stars by the author with a high dispersion echelle spectrograph. The results of the study are summarized here and a more detailed discussion will be published elsewhere.

The profiles and structures of $\lambda\lambda$ 5780, 5797 are shown in Figure 1 for the heavily reddened supergiant HD 183143. At a resolution of 0.17 Å, the profiles are asymmetric with the steeper side in the blue. There is also a broad, shallow feature shortward

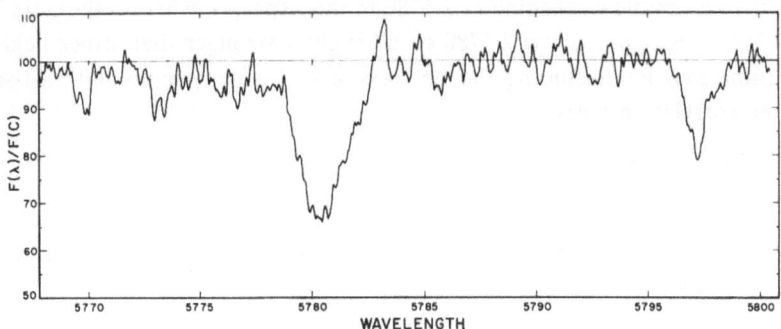

Fig. 1. The profile of $\lambda\lambda$ 5780, 5797 for HD 183143 at a resolution of 0.17 Å. The continuum is obtained by least square fitting a 3rd order polynomial. Note that the profiles are asymmetric and steeper to the blue, and there is a broad shallow absorption feature shortward of λ 5780.

Greenberg and Van de Hulst (eds.), Interstellar Dust and Related Topics, 91-95.

C.-C. WU

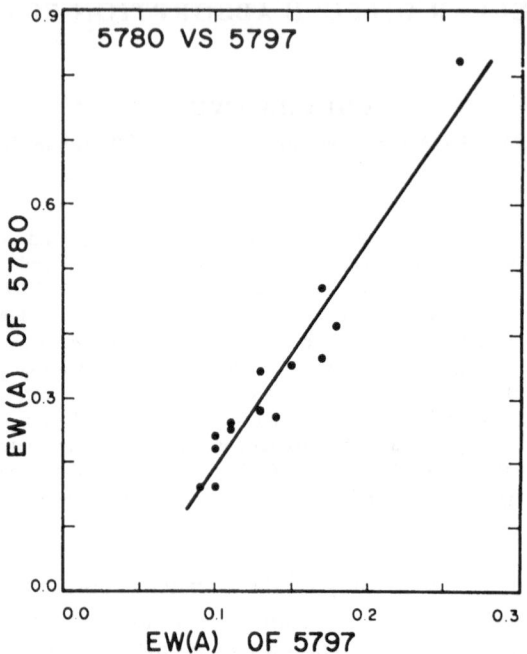

Fig. 2.   Correlation between the equivalent widths of λ 5780 and λ 5797.

of λ 5780. If there is also such a broad feature associated with λ 5797, it is not expected
to be detectable in the spectrum of HD 183143. The FWHM of λ 5780 and λ 5797 are
respectively 2.7 Å and 1.1 Å. Equivalent widths were measured at a resolution of
0.26 Å. As shown in Figure 2, the correlation between the two diffuse lines is tight, indicating that they are very likely carried by the same agent. Henceforth, λ 5797 will
be omitted from further discussion. Figure 3 shows the correlation between the equivalent width of λ 5780 and $E(B-V)$. The Be stars and the close binary system Omicron
Per show the systematic weakening of λ 5780 in their spectra, whereas the stars in the
Scorpius-Ophiucus region show λ 5780 to be slightly stronger than other field stars
with the same $E(B-V)$. Excluding the Be stars, a first order polynomial least-square
fitted to the correlation gives

$$EW(\lambda 5780) = 0.75\, E(B-V)\, \text{Å mag.}^{-1} \tag{1}$$

where $EW(\lambda 5780)$ is the equivalent width of λ 5780. Since λ 5780 is not saturated
and correlates well with $E(B-V)$, for normal stars, Equation (1) should give a good
estimate of the $E(B-V)$ excess produced by interstellar grains.

This investigation further proposes that λλ 5780, 5797 are produced by the pure
electronic transitions of the impurity centers in solid grains. Pure electronic transitions can give rise to very sharp optical lines with FWHM ranging from less than 1 Å
to a few angstroms and sometimes maybe even broader (Rebane, 1970). Therefore
the FWHM of λ 5780 (2.7 Å) and λ 5797 (1.1 Å) are well within the FWHM range of

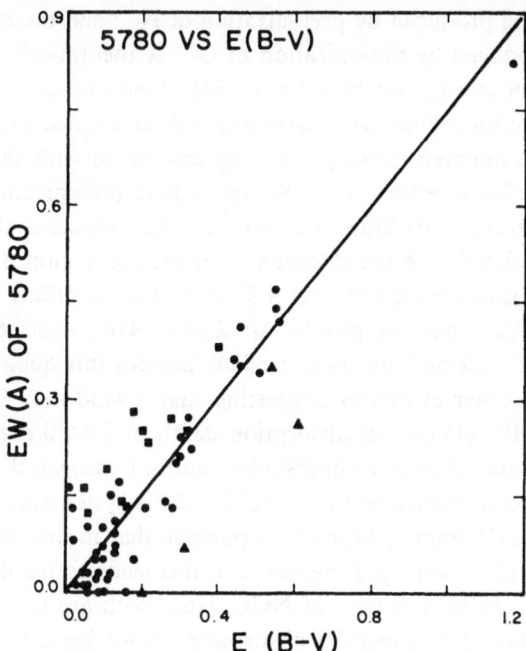

Fig. 3. Correlation between the equivalent width of λ 5780 and E(B−V). λ 5780 is relatively weak in the spectra of Be stars and the close binary system Omicron Per. λ 5780 is systematically stronger for Scorpius-Ophiucus stars. ● fields stars, ▲ Be stars, ■ Sco-Oph stars.

the pure electronic transitions. In the bulk solid, the pure electronic lines are predicted to have a Voight profile (Rebane, 1970). But a symmetric spectral feature in the bulk material will become asymmetric in the extinction spectra of small grains, with the degree of asymmetry depending on the size of the grains (Van de Hulst, 1957, Greenberg, 1968, 1971). Calculations by Wickramasinghe and Nandy (1970) with the observed central wavelength and width as input parameters show that the asymmetric absorption profiles are steeper to the blue. As shown by Figure 1, the profiles of λλ 5780, 5797 are indeed asymmetric and steeper to the blue. This pure electronic transition model can also explain the broad shallow absorption feature shortward of λ 5780. Associated with each parent pure electronic transition, there are weak absorption sidebands shortward of the parent line (e.g. see Yen *et al.*, 1964). As for the systematic weakening of λ 5780 in the spectra of Be stars (Figure 3), it is possible that circumstellar grains are efficient in visual extinction but not carriers of λ 5780 or that circumstellar grains being freshly formed, have not accumulated appreciable concentration of impurity centers, whereas interstellar grains have been subjected to bombardment by cosmic rays and high energy photons for long periods of time, thus contain appreciable concentration of lattice defects.

Rudkjøbing (1969a, b; 1970) and Ingemann-Hilberg and Rudkjøbing (1970), by extrapolating the wavenumbers of resonance lines along isoelectronic sequences and also quantum mechanical calculations, found that the diffuse bands λλ 4430, 4760,

4890, 6180 might be produced by preionization of H⁻; and the diffuse lines $\lambda\lambda$ 5780, 5797 might be produced by preionization of O⁻. A theoretical investigation of pre-ionization has been carried out by Fano (1961). Preionization processes give rise to asymmetric absorption profiles with the steeper side associated with an apparent emis-sion wing, and an apparent absorption wing associated with the side which drops more slowly to the band center. If λ 5780 were due to preionization process, then the broad, shallow feature to its blue corresponds to the apparent absorption wing and should not be associated with the steeper side of the absorption profile. This suggests that preionization is not responsible for λ 5780. As for the diffuse bands, the observa-tion of Wilson (1958) gives the profiles of $\lambda\lambda$ 4430, 4760, 4890, 6180. These profiles agree qualitatively with preionization profiles. Besides this qualitative agreement of profiles, there are other evidences suggesting that λ 4430 may be produced by the preionization of H⁻: (1) central absorption depth of λ 4430 correlates significantly better with the column density of interstellar neutral hydrogen than with $E(B-V)$ and the 2200 Å ultraviolet extinction bump, (2) λ 4430 is systematically weaker for stars behind dust clouds (Wampler, 1966). It is possible that in dust clouds H_2 formation is more efficient and H density decreases, and this leads to the decrease of H⁻ con-centration, (3) two close members of NGC 7762, both are O7 stars with the same $E(B-V)$, but the one with stronger CN absorption has weaker λ 4430. This is consistent with the above argument, because if λ 4430 does originate from H⁻, one will expect the region which is more efficient for molecular formation to give rise to weaker λ 4430. This model has its difficulties: (1) H⁻ has to be formed at high enough rate to satisfy the high destruction rate (Wilson, 1964), (2) observed λ 4430 profile changes from star to star as reported by Wampler (1966) and Brück *et al.* (1969), and (3) polarization changes in the λ 4430 spectral region as reported by Nandy and Seddon (1970). However, the most recent observations by A'Hearn (1972) showed that the feature is unpolarized.

Acknowledgements

I wish to thank Drs B. D. Savage and A. D. Code for helpful discussions and also acknowledge the financial support provided by the Space Astronomy Laboratory of the University of Wisconsin.

References

A'Hearn, M. F.: 1972, *Astron. J.* **77**, 302.
Brück, M. T., Nandy, K., and Seddon, H.: 1969, *Physica* **41**, 128.
Fano, U.: 1961, *Phys. Rev.* **124**, 1866.
Greenberg, J. M.: 1968, in B. M. Middlehurst and L. H. Aller (eds.), *Nebulae and Interstellar Matter*, U. of Chicago Press, Chicago, Chapter 6.
Greenberg, J. M.: 1971, *Nature Phys. Sci.* **230**, 15.
Herbig, G. H.: 1967, in H. van Woerden (ed.), 'Radio Astronomy and the Galactic System', *IAU Symp.* **31**, 85.
Hulst, H. C. van de: 1957, *Light Scattering by Small Particles*, J. Wiley & Sons Inc., New York.

Ingemann-Hilberg, C. and Rudkjøbing, M.: 1970, *Astrophys. Space Sci.* **6**, 101.
Johnson, F. M.: 1967, in J. M. Greenberg and T. P. Roark (eds.), *Colloquium on Interstellar Grains*, Rensselaer Polytechnic Institute, 1965, NASA, Washington, NASA SP-140, p. 241.
Nandy, K. and Seddon, H.: 1970, *Nature* **227**, 264.
Rebane, K. K.: 1970, *Impurity Spectra of Solids*, Plenum Press, New York-London.
Rudkjøbing, M.: 1969a, *Astrophys. Space Sci.* **3**, 102.
Rudkjøbing, M.: 1969b, *Astrophys. Space Sci.* **5**, 68.
Rudkjøbing, M.: 1970, *Astrophys. Space Sci.* **6**, 157.
Stoeckly, R. and Dressler, K.: 1964, *Astrophys. J.* **139**, 240.
Wampler, E. J.: 1966, *Astrophys. J.* **144**, 921.
Wickramasinghe, N. C. and Nandy, K.: 1970, *Astrophys. Space Sci.* **6**, 154.
Wilson, R.: 1958, *Astrophys. J.* **128**, 57.
Wilson, R.: 1964, *Publ. Roy. Obs. Edinburgh* **4**, 67.
Yen, W. M., Scott, W. C., and Schawlow, A. L.: 1964, *Phys. Rev.* **136**, A271.

PART III

REFLECTION NEBULAE AND
DIFFUSE GALACTIC LIGHT

SURFACE POLARIMETRY OF THE MILKY WAY

RAMON D. WOLSTENCROFT

Institute for Astronomy, University of Hawaii, H.I., U.S.A.

Abstract. Two methods of determining the state of polarization of the integrated light from the Milky Way are discussed. In the symmetry method it is assumed that the polarized foreground zodiacal light possesses north/south symmetry relative to the ecliptic: an average polarized intensity of at least $10\ S_{10}(R)$ is derived for the Milky Way at $l \simeq 90^\circ$ and 270°, which is consistent with the result of Sparrow and Ney (1972) who obtained $18\ S_{10}\ (R)$ at $l = 35^\circ$ and 210°. In the method of two epochs the above assumption is not needed but instead time variations of the zodiacal light are assumed to be insignificant: the results obtained by this method (in $S_{10}\ (V)$ units), although compatible with the values found by the symmetry method, do however show that time variations may have influenced this result. Although the evidence for a definite polarization of the Milky Way is persuasive it cannot yet be considered conclusive.

1. Introduction

The integrated light from the Milky Way is the sum of integrated starlight and diffuse galactic light, both of which are expected to be polarized. If the polarization is large enough to be measured reliably, surface polarimetry of the Milky Way could very probably provide valuable new information on the properties of the interstellar grains. In this paper we confine our attention to the question: 'is the Milky Way appreciably polarized?'

To determine the polarization of the Milky Way from observations of the night sky, corrections must be made to allow for the strongly polarized foreground zodiacal light and also care must be exercized in observing only at those optical depths where the influence of tropospheric scattering may be neglected. In order to make an intelligent estimate of the foreground zodiacal light it is necessary to know how the zodiacal light polarization varies with time and angular coordinates. Unfortunately our knowledge of this is quite poor and we are forced to make assumptions which although plausible may conceivably be greatly in error. Two such assumptions are: (1) the time variations of the zodiacal light are insignificant; and (2) the zodiacal light at a given instant has the same properties at mirror points in solar ecliptic coordinates, i.e., at $(\beta, \lambda - \lambda_\odot)$, $(-\beta, \lambda - \lambda_\odot)$, $(\beta, \lambda_\odot - \lambda)$ and $(-\beta, \lambda_\odot - \lambda)$, where $|\lambda - \lambda_\odot|$ runs from 0° to 180°. Schmidt and Leinert (1966), using both assumptions, obtained a degree of polarization, p, in the range $0,3\%$ to 2.3% at six longitudes between $l = 26^\circ$ and 122° ($|b| < 10^\circ$). Sparrow and Ney (1972), using assumption (1) only, determined values of the polarized intensity at $l = 35^\circ$ and 210° of $4\ S_{10}$ (blue) and $18\ S_{10}$ (red): the former value corresponds to $p \simeq 2\%$. The observations of Schmidt and Leinert had to be corrected for scattered artificial light and thus should be considered less reliable than those of Sparrow and Ney which were obtained from the OSO-5 satellite.

Greenberg and Van de Hulst (eds.), Interstellar Dust and Related Topics, 99–108.

2. The Symmetry Method

In an ideal approach to this problem the distribution of polarization over the entire sky should be determined at monthly intervals throughout the year: this would permit the assumptions (1) and (2) to be checked and if valid the polarization of the Milky Way could be deduced at all longitudes. In a more practical approach we have determined the distribution of polarization over the sky at a single epoch and have applied assumption (2); we shall call this the symmetry method. The measurements we have used were obtained during a three-week period at Chacaltaya, Bolivia, in

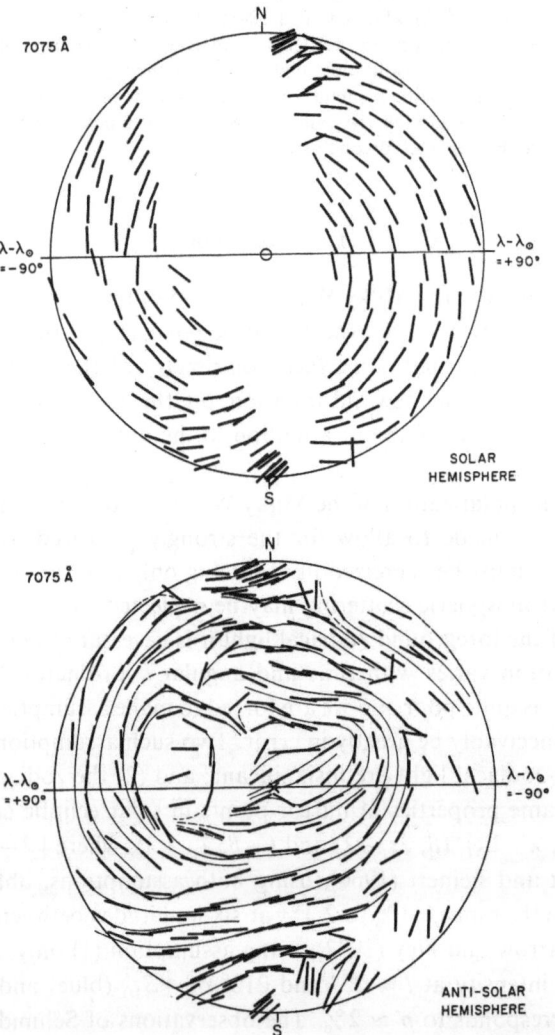

Fig. 1.   Orientation of the polarization plane of the night sky at 7075 Å during August, 1964. In the anti-solar hemisphere streamlines are drawn parallel to the polarization vectors. The Sun and anti-solar point are indicated by an open circle and a cross respectively.

August 1964. The equipment has been described elsewhere (Wolstencroft and Brandt, 1967). The orientation of the polarization plane of the observed night sky radiation at 7075 Å is shown in Figure 1. In the solar hemisphere the polarization orientations define a circular pattern centered on the Sun supporting the widely held view that the zodiacal light is the principal source of the polarization in most of this hemisphere. In the anti-solar hemisphere this circular pattern is seen in the outer parts of the distribution, although it is somewhat distorted: closer to the anti-solar point the distribution is markedly complex and the origin of the polarization is less certain. The average polarized intensity, pI, at 7075 Å is given in Figure 2 and Table I for $10° \times 10°$ areas in $(\beta, \lambda - \lambda_\odot)$ for $| \lambda - \lambda_\odot | > 60°$. A minimum of 50 and an average of 200 observations per area were used: this is also true for Figure 1. The mean $(\beta, \lambda - \lambda_\odot)$ for each $10° \times 10°$ area is not in general at the center of the area and no attempt has been made to correct for this. Although these measurements were made at a high altitude site ($h = 5.2$ km) in the red ($\tau = 0.08$ at the zenith) the influence of tropospheric scattering cannot be ignored. The dependence of pI on air mass has been examined for each $10° \times 10°$ area: in most cases pI is independent of air mass for $z < 70°$ to within $\pm 2 S_{10}(R)$ for regions with $pI < 20 S_{10}(R)$ (the unit used throughout this paper is $S_{10}(R, A0V)$, abbreviated to $S_{10}(R)$; it is the equivalent number of A0V stars of magnitude $R = 10$ per sq deg). The values of pI in Figure 2 and Table I are free of the influence of tropospheric scattering to these limits. The influence of tropospheric scattering on the polarization orientation is more severe, particularly for regions of

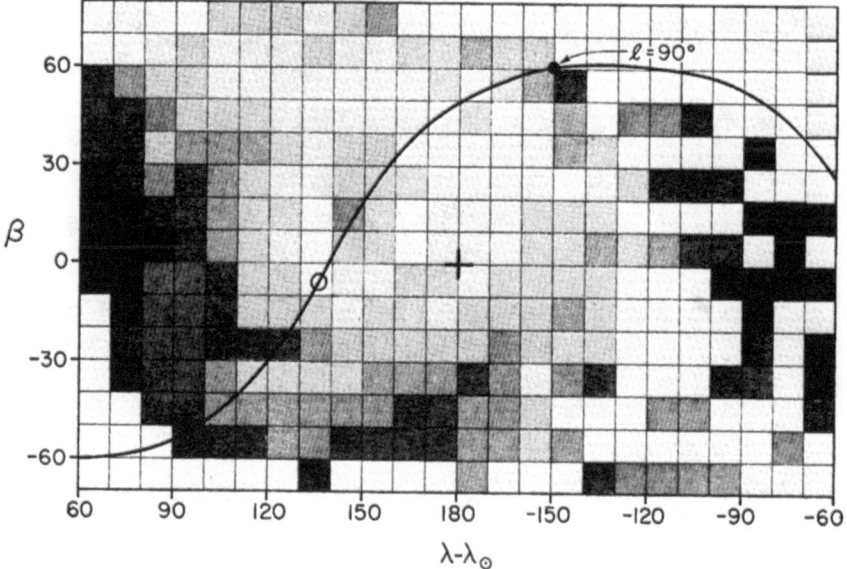

Fig. 2.   The average polarized intensity in $10° \times 10°$ areas of the night sky at 7075 Å in $S_{10}(R, A0V)$ units during August 1964. The polarized intensity increases as the shading goes from lightest to darkest in the sequence the sequence 1 to 7, 8 to 14, 15 to 21, 22 to 43 and 44 and greater in $S_{10}(R, A0V)$ units. There is no data for areas without shading. The continuous line inclined to the ecliptic is the galactic equator; the open circle represents the galactic center.

TABLE I

Values of pl in S_{10} (R, A0V) units

| $\beta°$ | | | | | | | | | | | | | | | |
(λ−λ☉)°	−70 −60	−60 −50	−50 −40	−40 −30	−30 −20	−20 −10	−10 0	0 10	10 20	20 30	30 40	40 50	50 60	60 70	70 80
60 to 70		37	41	48	39	64	105	83	71	55	42	33	36		
70 80		42	29	27	34	35	65	61	51	37	30	25	16	4	
80 90		24	18	23	25	37	26	44	37	20	12	20	14	9	8
90 100	24	19	16	18	26	27	43	43	33]	29	15	14	10	3	11
100 110		21	20	14	27	12	23	17	16	20	20	12	7	3	10
110 120		25	18	13	23	11	11	14	10	10	22	14	4	6	7
120 130		28	20	13	18	13	11	7	8	7	11	6	4	9	12
130 140	17	25	20	12	14	3	5	10	18	3	5	5	1	4	19
140 150		23	23	16	11	13	7	8	8	14	8	5	7	3	
150 160		15	24	21	10	12	10	3	3	8	13	5	5	8	
160 170		17	18	21	13	6	10	6	5	9	1	5	7	14	
170 180		12	15	23	10	13	3	4	1	1	3	3	8	16	
180 −170		17	14	20	18	8	8	8	9	7	7	10	4	14	
−170 −160				5	10	13	8	8	12	6	5	4	11	13	
−160 −150				19	14	19	14	18	10	4	2	3	18		
−150 −140			17	22	10	9	11	9		2	19	17	27		
−140 −130	22		18							14	8	16			
−130 −120	15									26		16			
−120 −110	16							19		37		34			
−110 −100	17						48	25		31	48				
−100 −90	15			22			50	26	56						
−90 −80		19		24	36	45	90		66						
−80 −70			37	46	57		89	71	70						

small pI; no selection of data at small air mass was made in preparing Figure 1 so that some of the true structure is distorted, for example that near the ecliptic poles.

The distribution of polarized intensities shows several unexpected features. The smallest values of pI are not centered on the anti-solar point, as would be expected, but instead define a region mostly in the northern ecliptic hemisphere which is inclined to the ecliptic by about 40° with a centroid at about $\beta = 30°$, $\lambda - \lambda_\odot = 160°$. The galactic plane passes close to the centroid, although because of the great extent of the region of low pI this should probably be considered a coincidence. In the evening hemisphere the values of pI are considerably greater in the south than at mirror points in the north and this may also be true in the morning hemisphere although there is less data there. It is also interesting to note that the polarization orientations in the anti-solar hemisphere are rather more ordered in regions well away from the galactic plane. The possible interpretations of the asymmetry in pI are: (a) the zodiacal light is asymmetric relative to the ecliptic and the anti-solar meridian and the polarization of the Milky Way is negligible; (b) the zodiacal light is symmetric and the Milky Way is polarized; (c) a combination of (a) and (b). We cannot yet distinguish with certainty between possibilities (a) and (b). However, it is worth noting that current analysis of data at $z = 70°$ to 80° (not included in Figure 2) for which a moderate correction for tropospheric scattering is necessary, indicates that the values of pI in the south/morning quadrant are less than in the south/evening quadrant but greater than in the north/evening quadrant: in order to draw conclusions about the Milky Way we shall adopt assumption (b).

In Figure 3 pI is plotted against $|\beta|$ for constant values of $\lambda - \lambda_\odot$. The galactic latitude is indicated only for low latitude points ($b < 30°$). The error bar is the standard deviation based on about 200 observations for each $10° \times 10°$ area. The difference between the curves at high and low galactic latitude represents a lower limit to the polarized intensity of the galactic component since pI is a vector quantity. The observations by Sparrow and Ney (1972) at high galactic latitude are shown for $\lambda - \lambda_\odot = 90°/100°$; they agree very well with our values although this may be fortuitous because it is not certain that their unit, S_{10} (red), coincides with our S_{10} (R, A0V) unit. The high galactic latitude points in the north and south differ on average by 4 $S_{10}(R)$ for the six such points in Figure 3. At $\lambda - \lambda_\odot = 80°$ to 110° and 160° to $-150°$ the points of low b have greater and lesser values respectively than the points of high b; this would be true if the orientations of the galactic and zodiacal polarizations were respectively parallel or perpendicular (approximately) to one another.

The difference between the high and low galactic latitude points is plotted (Figure 4) for various values of the galactic longitude at which the line of constant ($\lambda - \lambda_\odot$) crosses the galactic equator. A few negative values are present because ecliptic south minus north values, rather than the modulus of the difference were used. Average values for the the two ranges of l at 55° to 92° and 278° to 289° are shown; we may use the curves to estimate an approximate value of pI for the galactic component; the average value for the two ranges of l is 9 $S_{10}(R)$ which, since it is strictly a lower limit, is consistent with the value 18 $S_{10}(R)$ obtained by Sparrow and Ney for $l =$

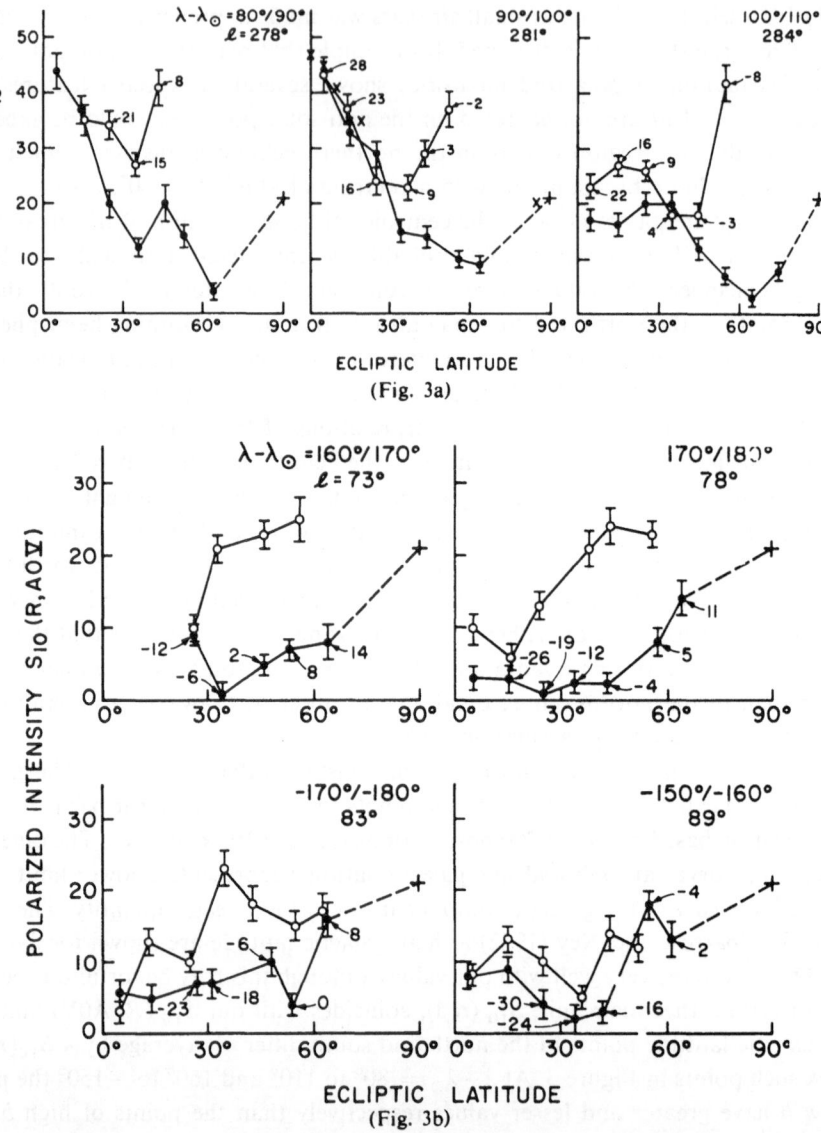

Fig. 3.   Polarized intensity at a fixed $\lambda - \lambda_\odot$ vs $|\beta|$ derived from Table I and Figure 2. Filled and open circles denote respectively points at northern and southern ecliptic latitudes. The number alongside certain points is the galactic latitude for points of low b; for points without numbers $b > 30°$. The cross at $|\beta| = 90°$ represents the average value for the northern ecliptic pole after applying an empirical correction for tropospheric scattering (the error of this point is estimated to be ± 7). At $\lambda - \lambda_\odot = 90°/100°$ X denotes high galactic latitude observations by Sparrow and Ney (1972).

35° and 210°. A more precise estimate of the galactic polarization will be obtained by subtraction of the Stokes parameters (Q, U) in opposite ecliptic hemispheres once the polarization orientations uninfluenced by trophospheric scattering have been derived;

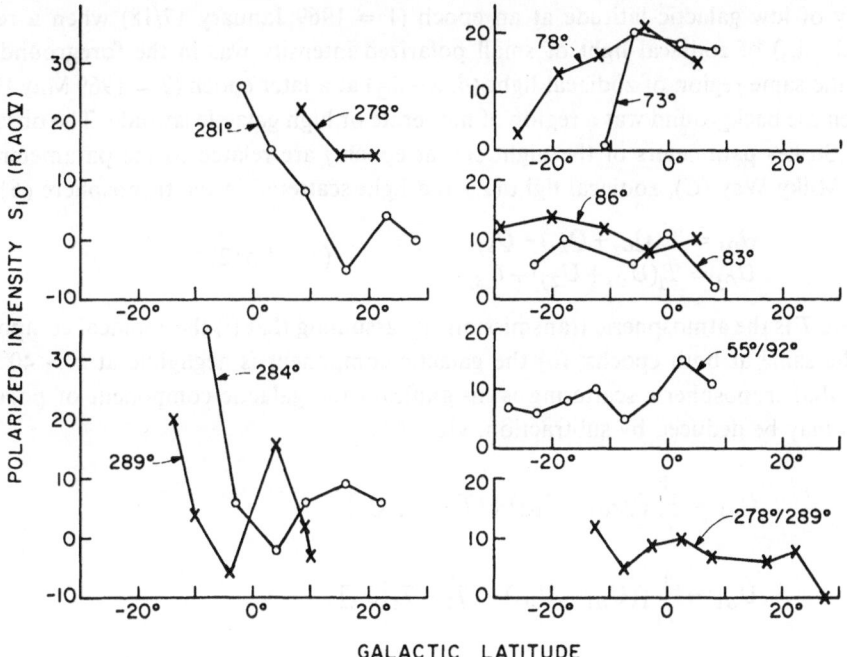

Fig. 4.   Difference of polarized intensity ($S_{10}(R, \text{A0V})$ units) between high and low galactic latitude points of equal 1β and $\lambda - \lambda_\odot$ vs the low galactic latitude coordinate. The number against each curve is the galactic longitude at which the great circle of constant $(\lambda - \lambda_\odot)$ crosses the galactic equator. The curves $55^\circ/92^\circ$ and $278^\circ/298^\circ$ represent averages over their respective longitude ranges.

the necessary assumption in carrying this out is that polarization orientations of the zodiacal light in the two ecliptic hemispheres are identical when the hemispheres are folded into one another about the ecliptic.

The method described above assumes that no significant time variations occurred during the three week observing period; in fact, all nights except one were contained in a one week period and further the observations were made close to the minimum of the solar cycle, so that this assumption is probably valid. The correctness of the assumption that the zodiacal light is symmetric in the four ecliptic quadrants is difficult to asses. It seems likely that the nodes and perihelia of the orbital planes of the dust particles are randomly distributed so that no east/west asymmetry is present; although the mean orbital plane of all particles is quite possibly inclined by a small angle to the ecliptic plane it is an open question whether this could produce a measurable north/south asymmetry.

3. The Method of Two Epochs

We have also carried out a study which makes no assumption about the angular distribution of the zodiacal light: the only assumption is that the time variations of the zodiacal light are insignificant. Observations were made of (a) a region in the Milky

Way of low galactic latitude at an epoch ($1 = 1969$ January 17/18) when a region $(\beta, \lambda - \lambda_\odot)$ of zodiacal light of small polarized intensity was in the foreground and (b) the same region of zodiacal light $(\beta, \lambda - \lambda_\odot)$ at a later epoch ($2 = 1969$ May 19/20) when the background was a region of moderate or high galactic latitude. The observed (O) Stokes parameters of the night sky at epoch j are related to the parameters for the Milky Way (G), zodiacal light (Z) and light scattered in the troposphere (S) by

$$Q_{oj} = T_j(Q_{Gj} + Q_{zj}) + Q_{Sj}$$
$$U_{oj} = T_j(U_{Gj} + U_{zj}) + U_{Sj}, \qquad (j = 1 \text{ or } 2)$$

where T is the atmospheric transmission. By assuming that (i) the zodiacal component is the same at both epochs, (ii) the galactic component is negligible at $b > 40°$ and (iii) that tropospheric scattering is insignificant the galactic component of polarization may be deduced by subtraction, viz.

$$Q_{G1} = \frac{1}{T_1}[(Q_{o1} - Q_{o2}) + (T_2 - T_1)Q_z]$$

$$U_{G1} = \frac{1}{T_1}[(U_{o1} - U_{o2}) + (T_2 - T_1)U_z],$$

where $Q_z = Q_{z1} = Q_{z2}$ and $U_z = U_{z1} = U_{z2}$. The validity of the first two assumptions cannot be directly tested but this may be done for the third assumption. Observations were made at Haleakala, Hawaii at $\lambda = 5080$ Å ($\tau = 0.12$ in the zenith) at air masses less than 1.15: we may place an upper limit on the polarized intensity of the light scattered in the troposphere of about $2 S_{10}(V)$ on the basis of both empirical studies carried out at Chacaltaya and theoretical studies (Wolstencroft and Van Breda, 1967).

The results are shown in Figures 5 and 6. Figure 5 shows the dependence of the polarized intensity $p_{G1} I_{G1} = (Q_{G1}{}^2 + U_{G1}2)^{\frac{1}{2}}$ on b for four $10°$ ranges of longitude between $155°$ and $195°$. When $p_{G1}I_{G1}$ is averaged over all longitudes we obtain the values 6.6, 8.8, 11.4, 9.8, 7.7, 6.9, 6.1, 3.8 and 6.3 $S_{10}(V)$ at $l = 3°$ to $19°$ (in $2°$ steps). The approximately linear trend for $b < 7°$ when extrapolated to $b = 0°$ gives $3 S_{10}(V)$. The orientation χ_{G1} of the polarization plane vs b (Figure 6) is measured from the direction of the north galactic pole in a counter clockwise direction as seen from outside the celestial sphere. The average standard error of a single point is $15°$. For $l = 160°$ the average value of χ_{G1} is close to $90°$, i.e., the polarization plane is parallel to the galactic equator, but with increasing l the average value increases to $120°$ at $l = 190°$. At each longitude χ_{G1} shows a maximum at $b = 11° \pm 2°$: the maximum in $p_{G1}I_{G1}$ occurs at a different latitude, namely $b < 8°$.

While the values of pI are approximately consistent with data in the red discussed earlier in this paper (see also Sparrow and Ney for a discussion of the color of pI) nevertheless, the assumption that the time variations of the zodiacal light may be ignored is a moot point. Fortunately the assumption may be tested. We have sought a correlation between the polarized intensities of the deduced galactic component

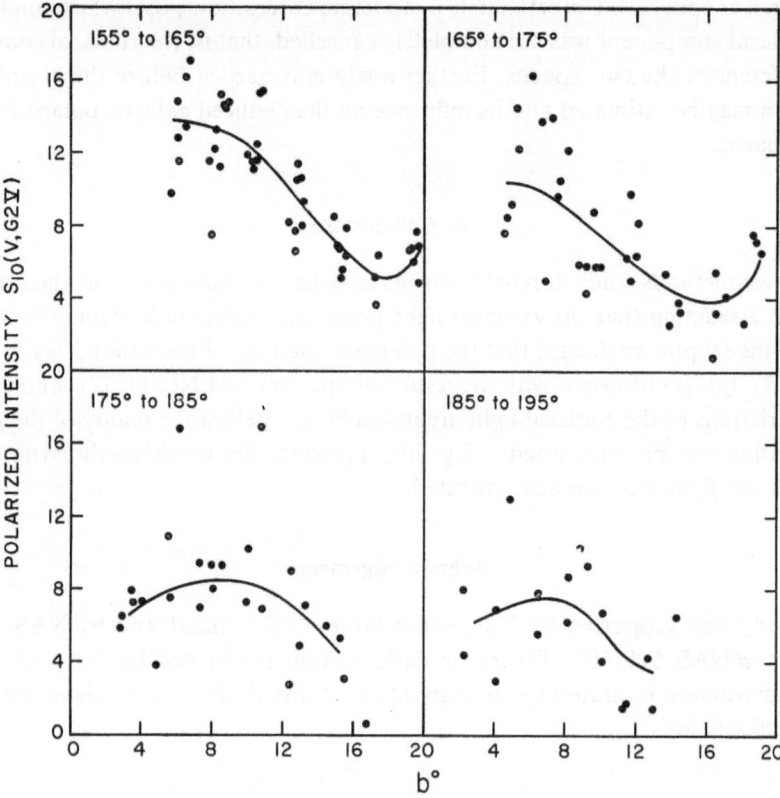

Fig. 5. Polarized intensity of the light from the Milky Way ($S_{10}(V, G2V)$ units) obtained by the method of two epochs for four longitude ranges from 155° to 195°.

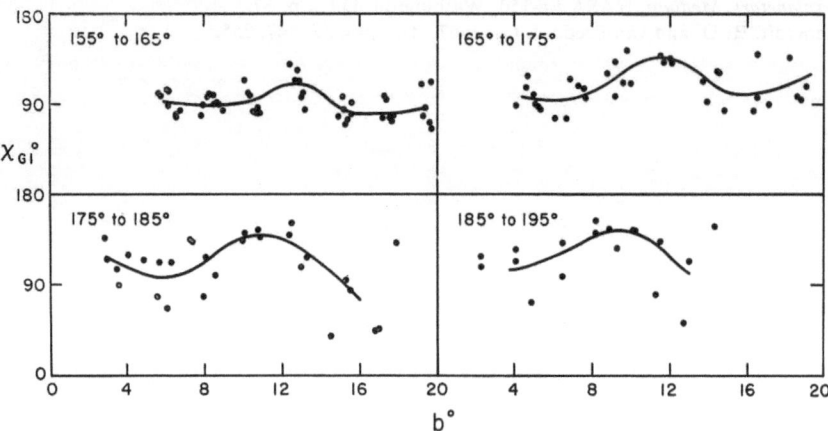

Fig. 6. Orientation of the polarization plane of the light from the Milky Way (method of two epochs): the orientation is measured counter-clockwise from the direction of the north galactic pole as seen from outside the celestial sphere.

(measured at epoch 2). Unfortunately a moderate correlation exists which implies that the zodiacal component was not completely cancelled, that is, the zodiacal component was different at the two epochs. Further analysis is needed before the extent of the variation may be estimated and its influence on the deduced galactic polarization can be evaluated.

4. Conclusions

Of the two methods studied, reliable results have been obtained only by the symmetry method. Assuming that the zodiacal light possesses north/south symmetry with respect to the ecliptic we deduce that the polarized intensity of the Milky Way is at least $10 \ S_{10}(R)$: this is consistent with the results of Sparrow and Ney (1972), who assumed time variations of the zodiacal light are insignificant. Before the reality of the galactic polarization can be established using this approach, the north/south asymmetry of the zodiacal light must be demonstrated.

Acknowledgements

This study was supported by NSF under Grant #GA 28201 and by NASA under contract #NAS 5 11303. The computations were performed by Jerry D. Wilson whose assistance is gratefully acknowledged. I thank Lothar W. Bandermann for helpful discussions.

References

Schmidt, T. and Leinert, C.: 1966, *Z. Astrophys.* **64**, 110.
Sparrow, J. G. and Ney, E. P., 1972, *Astrophys. J.* **174**, 705.
Wolstencroft, R. D. and Brandt, J. C.: 1967, in J. L. Weinberg (ed.), *The Zodiacal Light and the Interplanetary Medium*, NASA Sp-150, Washington, D.C., p. 57.
Wolstencroft, R. D. and van Breda, I. G.: 1967, *Astrophys. J.* **147**, 255.

DUST GRAINS IN REFLECTION NEBULAE

BEN ZELLNER

Lunar and Planetary Laboratory, The University of Arizona, Ariz., U.S.A.

Abstract. Observations and Mie theory calculations of colors and polarization in several reflection nebulae are compared.

A new study of light scattering in reflection nebulae was begun in 1968. Details of the observations and of the theoretical treatment are given elsewhere (Zellner, 1970, 1973). Briefly, observations of color and polarization were made in a few selected regions of NGC 2068, NGC 7023, and IC 5067 with the two-channel digital polarimeter described by Coyne and Gehrels (1967). Most of the data were taken at the 154-cm Catalina reflector of the Lunar and Planetary Laboratory. About 40 h of integration time were invested per region, in six filters ranging in effective wavenumber from 1.21 to 3.03 per micron. The color data were generally in good agreement with previous results, but polarizations in NGC 2068 and NGC 7023 were found to be systematically larger than those reported by Elvius and Hall (1966).

The observations were compared with single-scattering Mie calculations for a wide variety of complex refractive indices $m^* = m' - im''$. Two types of broad size distributions were used, namely the Oort-van de Hulst distribution as modified by Greenberg (OHG distribution) (Greenberg, 1966), $n(a) = n_0 \exp - 5[a/a_0]^3$, and the exponential distribution, $n(a) = n_0 \exp - 5[a/a_0]$. The factor of 5 is introduced in order to make the scale radius a_0 roughly comparable with the equivalent single radius a in the absence of a size distribution. Wavelength-independent refractive indices were assumed, so that a_0 could be left as a free parameter to be determined by the agreement between theory and observation. In most theoretical studies of reflection nebulae, the dust cloud has been assumed to be homogeneous and bounded by a simple geometrical figure such as a sphere or a plane-parallel slab. As illustrated in Figure 1, my models are free of geometrical constraints; in each separate nebular region, the extreme scattering angles θ_1 and θ_2 are independently adjusted to best match the observations. The computed polarization is very sensitive both to the scattering geometry and to the scale radius. The color data, being more perturbed by internal and foreground extinction, are only indicative.

Reflection nebula polarizations usually decrease linearly with wavenumber from the infrared into the ultraviolet. This behavior is characteristic of forward-scattering dielectric grains, but can be produced by particles with a large m'' only under contrived conditions of internal attenuation. Thus pure graphite or metallic grains are immediately ruled out. It has been stated (Hanner, 1971) that silicate grains ($m' = 1.5$ to 1.7) would give negative polarization at short wavelengths, contrary to all observations. Figure 2, however, shows this statement to be a halftruth. With an appropriate shift

Greenberg and Van de Hulst (eds.), Interstellar Dust and Related Topics, 109-113.

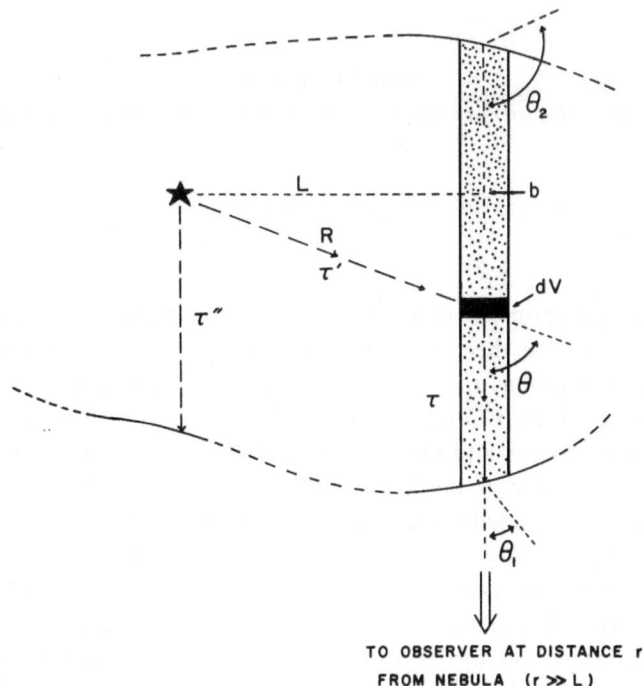

Fig. 1. The scattering plane.

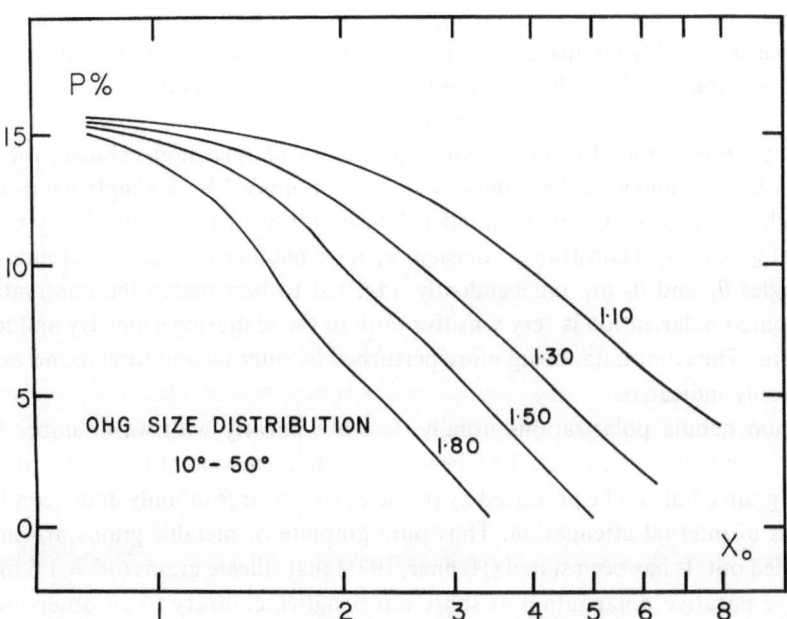

Fig. 2. Polarization vs $x_0 = 2\pi a_0/\lambda$ for single scattering by spherical grains of various refractive indices in the OHG size distribution.

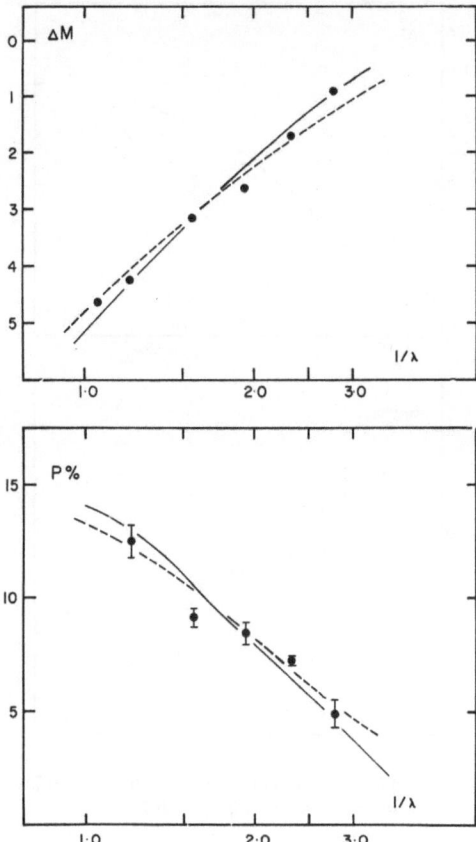

Fig. 3. Observations of color (top; $\Delta M = M_{nebula} - M_{star}$) and polarization (bottom) for a region in NGC 2068. The curves are computed for the OHG distribution with range of scattering angles 10° to 50′. Solid curve: refractive index $m^* = 1.50$, $a_0 = 0.19\,\mu$; dashed curve: $m^* = 1.10$, $a_0 = 0.38\,\mu$.

in radius, ices and silicates have almost identical scattering properties, and I doubt that they can be clearly distinguished by other than spectroscopic methods.

In bright reflection nebulae the optical depths are high*, so that single-scattering models are of doubtful validity. The agreement between observed and computed colors and polarizations, however, is in many cases quite good. Examples are given in Figures 3 and 4 for a region of NGC 2068, and in Figure 5 for a region in IC 5076. The fits become progressively worse as the imaginary component m'' is increased above 0.05 for $m' = 1.30$, or above 0.10 for $m' = 1.65$.

The derived grain properties are summarized in Table I for three fashionable refractive indices. The radii come out to be surprisingly small, smaller by a factor of 2 or 3 than are appropriate for the interstellar extinction curve. The meaning of this result is unclear. While the models are crude, an a_0 as large as the classical value of

* The conclusion by Gehrels (1967) that NGC 7023 is optically thin was based on an incorrect surface brightness, taken from Martel (1958).

B. ZELLNER

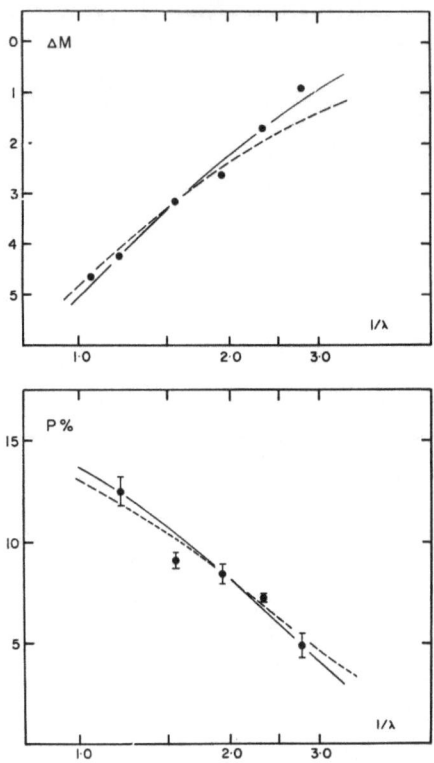

Fig. 4. As in Figure 3, except the solid curve is for $m^* = 1.30$, $a_0 = 0.26\ \mu$, and the dashed curve $m^* = 1.30 - 0.10i$, $a_0 = 0.29\ \mu$.

TABLE I

Derived grain sizes. The listed values are scale radii a_0 in the OHG size distribution

m^*	Nebula	a_0 microns	Mean a_0
1.30	2068 I	0.26	0.20 ± 0.04
	2068 II	0.15	
	5076	0.17	
	7023 NE	0.22	
	7023 N	0.16	
$1.30 - 0.05i$	2068 I	0.27	0.22 ± 0.05
	2068 II	0.20	
	5076	0.16	
	7023 NE	0.26	
	7023 N	0.17	
$1.65 - 0.10i$	2068 I	0.18	0.15 ± 0.02
	2068 II	0.13	
	5076	0.14	
	7023 NE	0.14	
	7023 N	0.12	

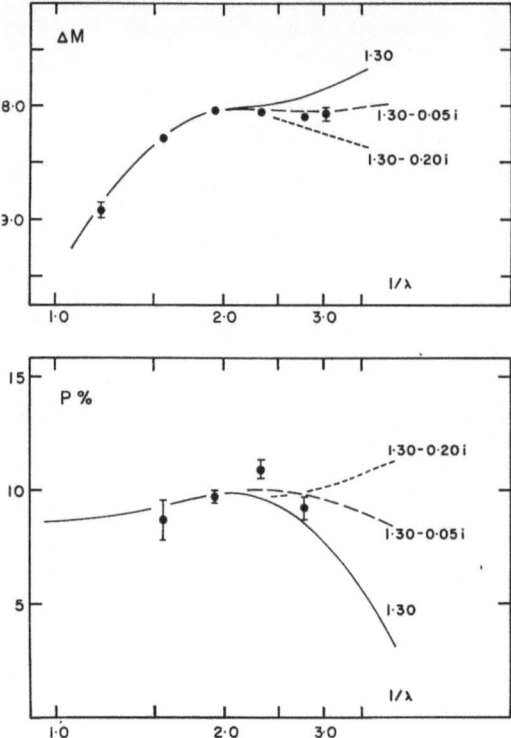

Fig. 5. Observations of color (top) and polarization (bottom) in a region of IC 5076. The scale radii are 0.17 μ for $m^* = 1.30$, 0.16 μ for $m^* = 1.30-0.05i$, and 0.15 μ for $m^* = 1.30-0.20i$. The nebula is backscattering at $150° < \theta < 165°$.

0.6 μ for ices is certainly excluded since, as noted above, negative polarizations would inevitably be produced. Even smaller grains, with radii near 0.1 μ, are implied by the polarizations produced in circumstellar dust shells (Zellner, 1971; Shawl, 1972). It is possible that we are seeing a failure of the Mie theory, in that the extinction and angular-scattering properties of the interstellar grains cannot be modeled by smooth spheres with a unique radius. More extensive polarimetry, together with a theory able to handle arbitrarily high optical depths, is needed to resolve this dilemma.

References

Coyne, G. V. and Gehrels, T.: 1967, *Astron. J.* **72**, 887.
Elvius, A. and Hall, J. S.: 1966, *Lowell Obs. Bull.* **6**, No. 135, 257.
Gehrels, T.: 1967, *Astron. J.* **72**, 631.
Greenberg, J. M.: 1966, in K. Lodén, L. O. Lodén and U. Sinnerstad (eds.), 'Spectral Classification and Multicolour Photometry', *IAU Symp.* **24**, 291.
Hanner, M. S.: 1971, *Astrophys. J.* **164**, 425.
Martel, M.-T.: 1958, *Publs. de l'Obs. de Haute-Provence* **4**, No. 20.
Shawl, S. J.: 1972, Dissertation, University of Texas.
Zellner, B.: 1970, Dissertation, University of Arizona.
Zellner, B.: 1971, *Astron. J.* **76**, 651.
Zellner, B.: 1973, in T. Gehrels (ed.), *Planets, Stars, and Nebulae Studied with Photopolarimetry*, University of Arizona Press.

THE ULTRAVIOLET SURFACE BRIGHTNESS
OF REFLECTION NEBULAE AND THE
ALBEDO OF INTERSTELLAR GRAINS

CHARLES F. LILLIE

Dept. of Physics and Astrophysics, Laboratory for Atmospheric and Space Physics, The University of Colorado, Boulder, Colo., U.S.A.

and

ADOLF N. WITT

Dept. of Physics and Astronomy, Ritter Astrophysical Research Center, The University of Toledo, Toledo, Ohio, U.S.A.

Abstract. Satellite observations of the Merope and Barnard's Loop nebulae in the 1050 to 4250 Å spectral region show that the increase in the albedo of interstellar grains below 2000 Å occurs in individual reflection nebulae as well as in the general interstellar medium. Evidence for differences in the composition and size distribution of grains due to variations in the local radiation field is presented.

1. Introduction

The scattering properties of interstellar dust grains in the ultraviolet are important input parameters for models of interstellar and circumstellar clouds. Recently, we have determined the average albedo of particles in the general interstellar medium in the 1500–4250 Å region (Witt and Lillie, 1971). We find the albedo of interstellar grains (see Figure 4) is fairly constant at $a \approx 0.5$ from 3000–4000 Å, drops rapidly to a minimum ($a \approx 0.1$) at 2200 Å, then rises to $a \approx 1.0$ at 1500 Å. We tentatively attribute this to the presence of three different types of particles in the interstellar medium: a few micron-sized, ice-coated particles which dominate scattering in the visible and near ultraviolet; numerous sub-micron, graphite-like grains which absorb very strongly near 2200 Å; and large numbers of submicron, silicate-like grains which dominate scattering in the far ultraviolet (Witt, 1973).

The rapid increase in particle albedo below 2000 Å may be questioned because we did not correct for the possible contribution of faint, blue, horizontal-branch stars at high galactic latitudes (Hills, 1972); also Henry and Carruthers (1970) were unable to detect Barnard's Loop Nebula in the 1230–2100 Å spectral region with a rocket-borne electronographic camera. To answer these questions, and to search for variations in the scattering properties of grains in different radiation fields, we have begun the analysis of ultraviolet surface brightness measurements of reflection nebulae obtained with the Orbiting Astronomical Observatory OAO-2. In this paper we report the preliminary results of this investigation for the Merope and Barnard's Loop Nebulae.

2. The Observational Material

All of the observations discussed in this paper were obtained with the four stellar

Greenberg and Van de Hulst (eds.), Interstellar Dust and Related Topics, 115–122.

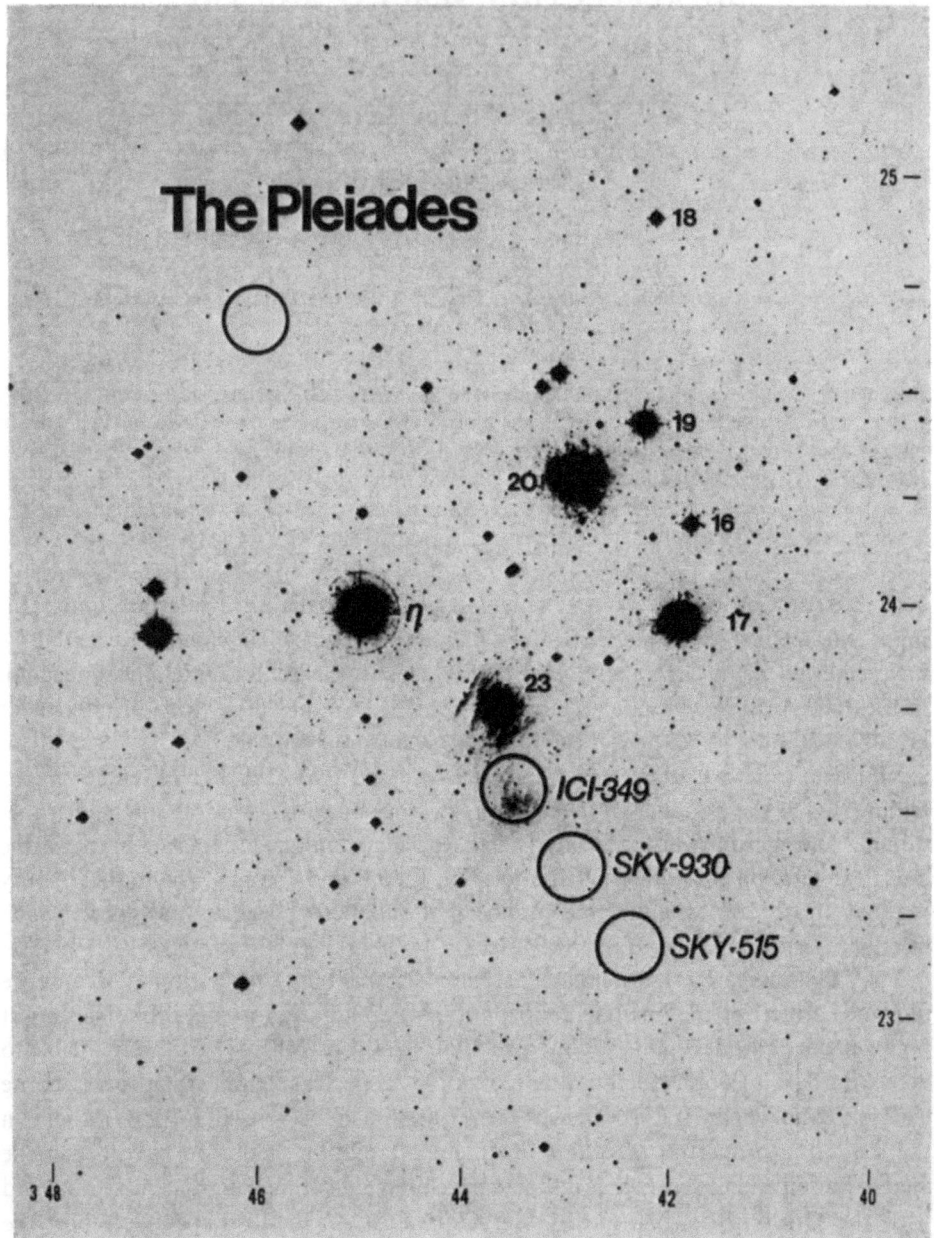

Fig. 1.   Location of areas in the Merope Nebula observed with OAO-2 shown on a reproduction from the red plate of the *Palomar Sky Atlas*. The circles indicate the size of the diaphragm used.

photometers of the University of Wisconsin Experiment Package (WEP) on OAO-2. The instrumentation and operation of this experiment have been described in detail by Code *et al.* (1970). Each of the stellar photometers is located at the focus of an

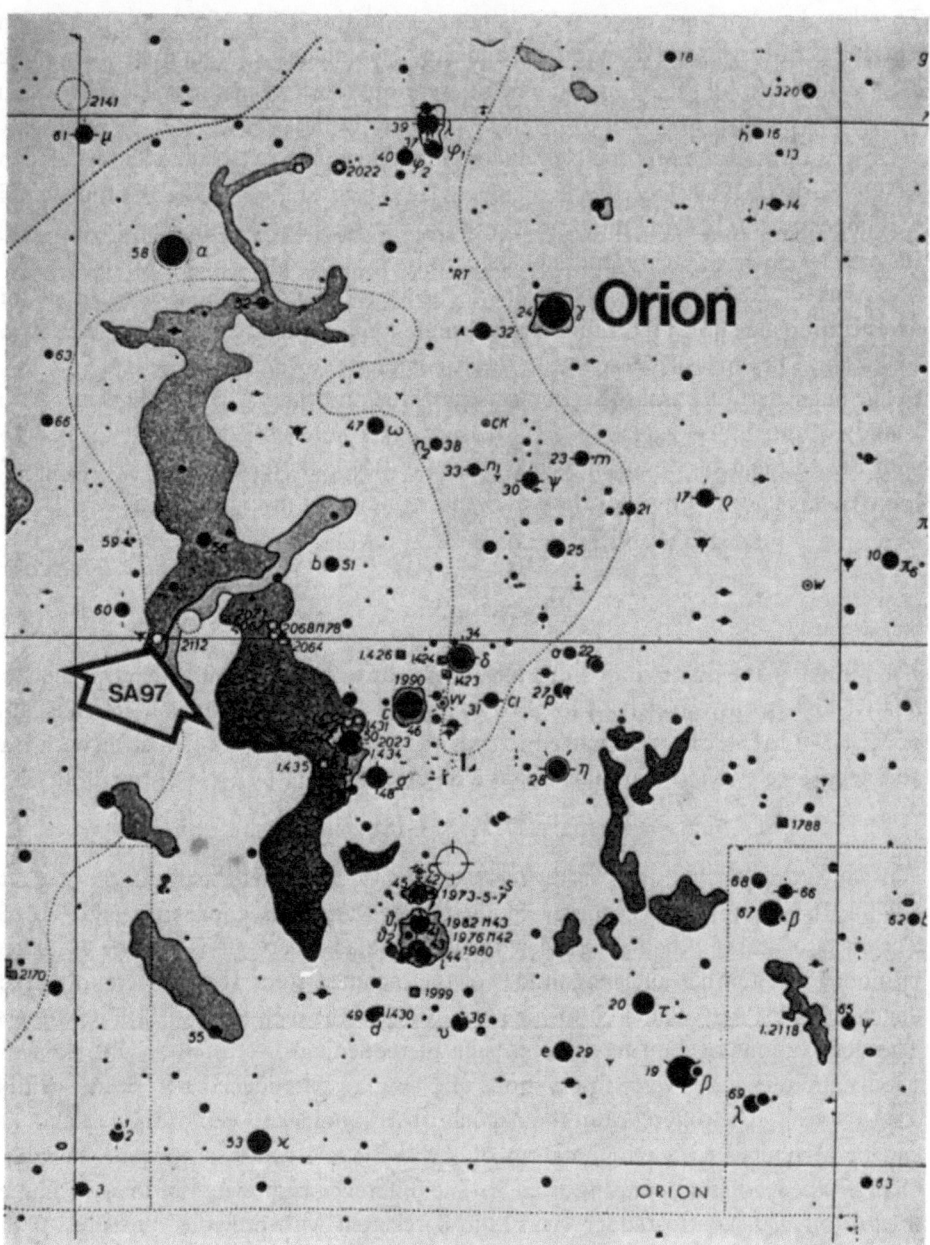

Fig. 2. Location of Kapteyn's Selected Area 97 with respect to Barnard's Loop shown on a reproduction from the *Atlas Coeli*.

eight-inch, off-axis, parabolic mirror and consists of an aperture mechanism; a five-position filter wheel with three medium band interference filters, a dark slide, and a radioactive calibration source; a Fabry lens; and a photomultiplier tube. Both analog and digital (pulse counting) electronics measure the photomultiplier tube output. Together the photometers cover 12 bandpasses in the 1050–4300 Å spectral region. The spacecraft is three-axis stabilized and can be pointed to a target area with an accuracy of $\pm$ 1'. All of the measurements were obtained when the spacecraft was in the Earth's shadow and well away from the inner radiation belt.

Observations of the Merope Nebula were obtained on the 515th and 930th orbits of OAO-2 early in 1969. Due to the relatively large field of view of the photometers, it was impossible to observe the brightest portion of the nebula near Merope (23 Tau). Instead, we observed the bright nebulosity in IC 1-349 about 15' SSW of the star. We also observed 23 Tau and the sky nearby. The location of the regions selected for this investigation are shown in Figure 1, a reproduction from the red print of the *Palomar Sky Atlas*. The circles correspond to the 10' diaphragm which was used.

The outer part of Barnard's Loop was observed on orbits 1267 and 4330 as part of a study of the diffuse galactic light. Measurements were obtained at 12 wavelengths with the 10' field of view centered on BD+0° 1227, the central star of Kapteyn's Selected Area No. 97. The location of SA 97 is indicated in Figure 2.

3. Analysis

3.1. MEROPE

The results of the observations of Merope are shown in Figure 3, where we plot the relative flux versus wavelength for 23 Tau and IC 1-349. We also show the flux ratio of IC 1-349 to 23 Tau, normalized to 5 at 4250 Å. The measurements for both star and nebula were found from the digital data by taking

$$\text{Net Counts} = (\text{Signal} - \text{Dark})_{\text{Star}} - (\text{Signal} - \text{Dark})_{\text{Sky}}$$

and converting to flux units using the preliminary photometer calibration of Code (1970). The photometric accuracy of these observations is approximately $\pm$ 0.^{m}03. The effect of stars in the field of view for the nebular observations may be estimated by noting the number and magnitudes of the included stars. If they were all unreddened B6 stars, they would contribute only 10% of the measured surface brightness. The observations of 23 Tau show no sign of reddening or scattered light from the nebula and they agree with observations of ζ Dra, an unreddened B6 III star, within a few percent. The difference in the nebular observations between orbits 515 and 930 can be attributed to a recollimation of the WEP telescopes in the interim which changed their nominal pointing by 2–3'. The difference is in the sense that the fainter area (orbit 930) is also redder. This is in agreement with previous measures of the Merope Nebula (Collins, 1937; O'Dell, 1965). The $U - B$ colors which we form from our data are in good quantitative agreement with previous measures (O'Dell, 1965).

Because a complete analysis of these data is a complicated problem in radiative transfer involving assumptions about the geometry of the nebula and the reddening curve within it, we confine ourselves at this time to qualitative statements about the nebula.

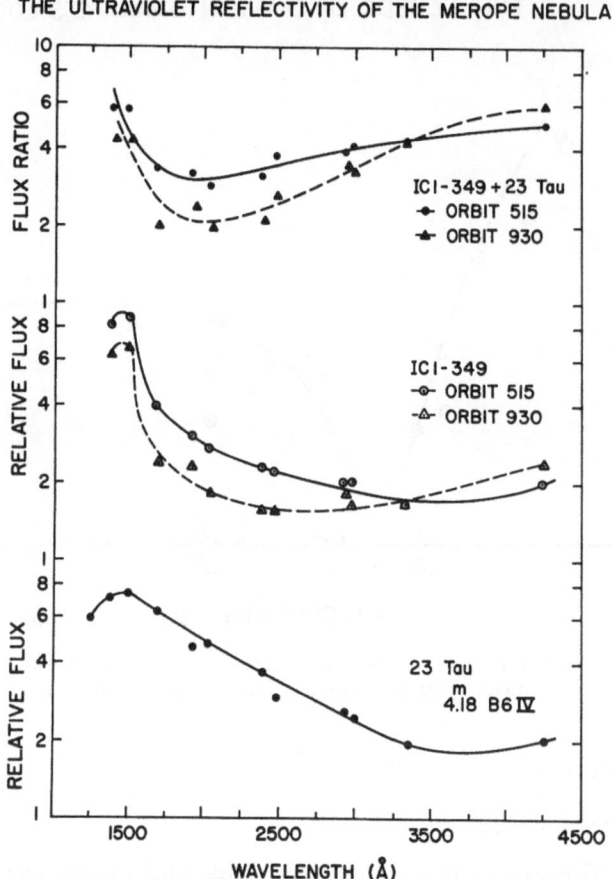

Fig. 3.   The relative flux from 23 Tau and the Merope Nebula and their flux ratio vs wavelength.

Providing the scattering phase function does not change too drastically with wavelength, our flux ratio is proportional to the albedo of the scattering particles at every wavelength. This is the case in the general interstellar medium, and we assume it for the Merope nebula.

We then compare our albedo curve (flux ratio) for Merope and the albedo for the general interstellar medium (see Figure 4) and note that:

(1) The upturn below 2000 Å is present in both cases, and

(2) The absorption feature at 2200 Å is less pronounced or missing in the nebula.

This seems clear evidence that the upturn in the albedo of interstellar particles is real and not due to some unexpected source of far ultraviolet radiation (such as horizontal-branch stars or the extragalactic sky background). The lack of a clear absorption feature at 2200 Å seems to indicate a deficiency in the number of absorbing graphite-like grains in the vicinity of 23 Tau due to evaporation, radiation damage, or particle size segregation in the local radiation field.

THE ALBEDO OF DUST IN BARNARD'S LOOP

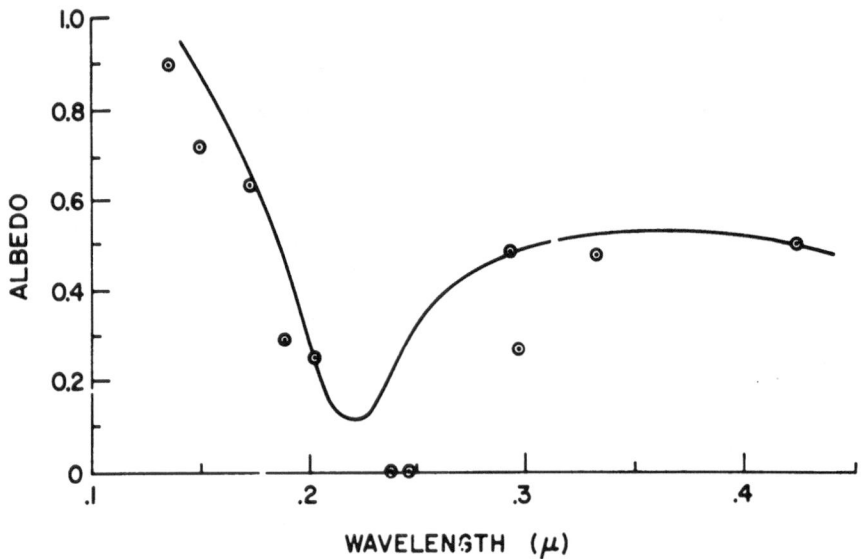

Fig. 4.   The albedo of dust in Barnard's Loop (open circles) compared with the albedo of dust particles in the general interstellar medium (solid line).

3.2. BARNARD'S LOOP

3.2.1. *OAO Measurements*

Selected Area 97 is located in the outer part of Barnard's Loop. Detailed starcounts exist for it in the *Harvard-Groningen Durchmusterung of Selected Areas* and the *Mount Wilson Catalogue of Photographic Magnitudes in Selected Areas 1–139*. These starcounts provide an excellent means of determining the contribution of integrated starlight to the measured surface brightness. In this case, the principal contribution is due to the central star itself, and its ultraviolet brightness can be inferred from its visual magnitude, spectral type, and OAO observations of stars of similar spectral characteristics. The MK spectral type of BD+0° 1227 is B9.5V, as determined from grating spectra obtained with the Cassegrain spectrograph on the 40-in. reflector of the Ritter Astrophysical Research Center. The UBV magnitudes were measured with a photometer on a 16-in. telescope at Kitt Peak National Observatory and found to be: $V = 6.91$, $B - V = -0.01$, $U - B = -0.13$.

The other sources of the measured surface brightness are the zodiacal light and light scattered by dust particles in Barnard's Loop and along the line of sight. The zodiacal light contribution can be determined quite easily from the tables of Smith *et al.* (1965) and an investigation of the ultraviolet spectrum of the zodiacal light by Lillie (1971).

The albedo of the interstellar grains may be found by taking the ratio of the scattered light intensity to that of the illuminating starlight, assumed to be that of a B6V star. The choice of a spectrum was based on an observational study of the interstellar radiation field by Lillie (1968) who found its spectrum to be that of a B7V + G8V combination in the vicinity of the Sun. In the vicinity of the Orion Association, however, we expect the illuminating starlight to have a somewhat earlier spectral type.

In Figure 4 we have plotted the ratio of scattered light to illuminating starlight for SA 97 normalized to 0.5 at 4250 Å at 11 wavelengths (open circles). For comparison, we show the albedo of dust in the general interstellar medium (solid line) from a previous study (Witt and Lillie, 1971). The agreement over the entire wavelength range is quite satisfactory, although there seems to be somewhat more absorption at 2200 Å than we observe for the average field. We consider these observations of Barnard's Loop an independent confirmation of our previous results. We conclude that the dust particles in the Barnard's Loop nebula have approximately the same scattering properties as particles in the general interstellar medium.

3.2.2. *Other Observations*

The Barnard's Loop region has been observed in the ultraviolet by at least two other space experiments: from the Gemini 11 spacecraft with a hand-held camera with an objective prism which viewed the 2200–4900 Å spectral region (O'Dell *et al.*, 1967), and from a rocket with an electronographic camera sensitive in the 1230–2100 Å region (Henry and Carruthers, 1970). On the Gemini photograph the nebula is unexpectedly bright, amorphous in appearance, with clear delineation of the amorphous nebulosity in its western extension. On the rocket photograph, however, there is no trace of the nebulosity.

Why did Henry and Carruthers fail to observe Barnard's Loop? An examination of their data (Weber *et al.*, 1971) indicates the faintest source they could detect was a $V = 8^m2$ B9 star. From OAO photometry data we find an 8^m2 B9V star has an average flux of 5.6×10^{-12} erg cm^{-2} s^{-1} Å^{-1} over the bandpass of their instrument (1230–2100 Å). Since this flux is distributed over a 2′ diameter image, their detection threshold for extended objects is 2.1×10^{-5} erg cm^{-2} s^{-1} Å^{-1} sr^{-1}.

By comparison, O'Dell *et al.* measured a surface brightness of 5.1×10^{-8} erg cm^{-2} s^{-1} Å^{-1} sr^{-1} in one of the brightest parts of the Loop in their bandpass (2200–4900 Å). Using OAO data to find the difference in grain albedo and illuminating flux at the two effective wavelengths, we find the nebula should have a surface brightness of

$$\left(\frac{0.9}{0.4}\right)(2.9)(5.1 \times 10^{-8}) = 3.3 \times 10^{-7} \, \text{erg cm}^{-2} \text{s}^{-1} \text{Å}^{-1} \text{sr}^{-1}$$

at 1500 Å. This is a factor of 64 below the detection threshold of Henry and Carruthers.

Selected Area 97 is located in the same nebulosity 50′ SSE of the point measured by O'Dell *et al.* (1967). Examining our OAO data, we find average scattered light intensities of 7.9×10^{-8} erg cm^{-2} s^{-1} Å^{-1} sr^{-1} in the 2200–4900 Å region, and 3.9×10^{-7} erg cm^{-2} s^{-1} Å^{-1} sr^{-1} in the 1230–2100 Å region. This is quite good agreement with O'Dell *et al.*, well within their estimated error. We conclude the Barnard's Loop nebula has an ultraviolet surface brightness consistent with our albedo curve and O'Dell *et al.*'s measurements, and that Henry and Carruthers' camera was not sensitive enough to detect it.

4. Summary

Our preliminary analysis of OAO observations of the Merope and Barnard's Loop Nebulae shows the increase in the albedo of interstellar grains below 2000 Å is real and occurs in individual reflection nebulae as well as the general interstellar medium. Henry and Carruthers (1970) did not observe Barnard's Loop in the 1230–2100 Å region because its surface brightness is far below the detection threshold of their camera. The OAO measurements of Barnard's Loop are in excellent agreement with those of O'Dell *et all.* (1967).

A comparison of the albedo of grains in Merope and Barnard's Loop Nebulae with the albedo of grains in the general interstellar medium shows evidence of differences in the composition and size distribution of grains due to variations in the local radiation field. This difference is in the sense that the absorbing particles, and then the small scattering particles disappear as the intensity of radiation from nearby stars increases.

Acknowledgements

It is a pleasure to acknowledge the financial support furnished by NASA grant NGR 36-010-016 to Adolf N. Witt and NASA grant NGR 06-003-179 to Charles F. Lillie.

References

Code, A. D.: 1970, private communication.
Code, A. D., Houck, T. E., Bless, R. C., McNall, J. F., and Lillie, C. F.: 1970, *Astrophys. J.* **161**, 377.
Collins, O. C.: 1937, *Astrophys. J.* **86**, 529.
Henry, R. C. and Carruthers, G. R.: 1970, *Science* **170**, 527.
Hills, J. G.: 1972, *Astron. Astrophys.* **17**, 155.
Lillie, C. F.: 1968, *Bull. Am. Astron. Soc.* **1**, 198.
Lillie, C. F.: 1971, in A. D. Code (ed.), 'Scientific Results from the Orbiting Astronomical Observatory (OAO-2)', NASA SP-310.
O'Dell, C. R.: 1965, *Astrophys. J.* **142**, 604.
O'Dell, C. R., York, D. G., and Henize, K. G.: 1967, *Astrophys. J.* **150**, 835.
Smith, L. L., Roach, F. E., and Owen, R. W.: 1965, *Planetary Space Sci.* **13**, 207.
Weber, S. V., Henry, R. C., and Carruthers, G. R.: 1971, *Astrophys. J.* **166**, 543.
Witt, A. N.: 1973, this volume, p. 53.
Witt, A. N. and Lillie, C. F.: 1971, in A. D. Code (ed.), 'Scientific Results from the Orbiting Astronomical Observatory (OAO-2)', NASA SP-310.

SOUTHERN REFLECTION NEBULAE

SIDNEY VAN DEN BERGH and RENÉ RACINE

David Dunlap Observatory, University of Toronto, Canada

Abstract. The Curtis Schmidt telescope at the Cerro Tololo Observatory has been used for a two-color survey of reflection nebulae in the Southern Milky Way. Major concentrations of reflection nebulae are found in Carina and in the zone $250^\circ < l^{II} < 270^\circ$. A conspicuous gap in the distribution of reflection nebulae occurs at $l^{II} \simeq 275^\circ$. A detailed photometric and spectroscopic investigation of the stars imbedded in southern reflection nebulae is now in progress.

1. Introduction

In previous papers (van den Bergh, 1966; Racine, 1968) it has been shown that stars which are imbedded in reflection nebulae occur in associations. Typically the stars in such R associations are ~ 100 times brighter than the unstable main sequence dwarfs that populate T associations. R associations can therefore be studied conveniently out to considerably greater distances than T associations.

Racine (1968) and Racine and van den Bergh (1970) were able to show that R associations outline the Orion spiral arm rather better than do OB associations. The reason for this is that the number of R associations per kpc^2 in the galactic plane is greater than the number of OB associations. This is so because most R associations consist of late B- and early A-type stars which have a much greater space density than do the O- and early B-type stars that outline OB associations.

2. The Southern Survey

The interpretation of the optical spiral arm pattern in the vicinity of the sun is still the subject of considerable controversy (Bok, 1970). The true nature of the spiral pattern is particularly uncertain in the southern sky. It therefore seemed worthwhile to extend the northern reflection nebula survey (that had been carried out on the prints of the Palomar Sky Survey) to the southern hemisphere. A strip with an average width $\sim 12°$ was therefore photographed in two colors with the 24–in. (61-cm) Curtis Schmidt Telescope at the Cerro Tololo Observatory in Chile. The plates for this southern survey were taken in 1969, 1970 and 1971. The blue exposures were mostly on IIa-O emulsion behind a GG13 filter; the red exposures on 098–02 emulsion behind and RG2 filter. Most of the plates were developed in MWP 2.

Compared to the 48–in. (126-cm) Palomar Schmidt the Curtis telescope has two disadvantages:

(1) Faint nebulosity is more easily detected on plates taken with the $f/2.5$ Palomar instrument than on plates taken with the $f/3.5$ Curtis telescope.

Greenberg and Van de Hulst (eds.), Interstellar Dust and Related Topics, 123–125.

(2) The large size and complex structure of the brighter stellar images produced by the Curtis Schmidt makes it much more difficult to detect small reflection nebulae around bright stars on Tololo plates than on Palomar plates.

3. Discussion

Approximately 100 stars embedded in reflection nebulae were found during the course of the present survey (to guard against plate defects only those nebulae that could be seen on both blue *and* red plates were catalogued). The distribution of these objects is plotted in Figure 1. Also plotted in this figure are the reflection nebulae that had previously been found on the Sky Survey prints of the northern sky.

Figure 1 shows that the southern reflection nebulae between $250° < l^{II} < 350°$ are located much closer to the galactic plane than are those that had previously been found on the Palomar Sky Survey prints. The reasons for this are the following:

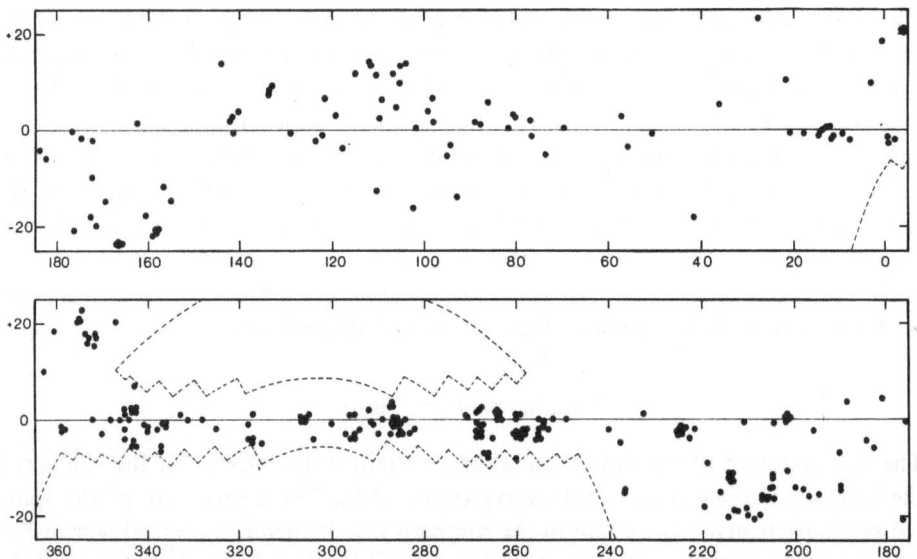

Fig. 1.   Distribution of stars in reflection nebulae in the northern (top) and southern (bottom) Milky Way. Note the very clumpy distribution of reflection nebulae. Dotted lines show the limits of the present survey.

(1) The southern survey only extends to $| b^{II} | \simeq 6°$.

(2) The southern survey includes *all* stars in reflection nebulae whereas the northern survey was confined to BD stars. On the average the southern stars in reflection nebulae are therefore fainter and hence more distant than their northern counterparts.

(3) The southern survey covers a region in which 'Gould's Belt' crosses the galactic equator.

Inspection of Figure 1 shows that the distribution of southern reflection nebulae is quite clumpy. Major concentrations of nebulae are found in Carina and in the zone

$250° < l^{II} < 270°$. A conspicuous gap in the distribution of reflection nebulae occurs at $l^{II} \approx 275°$.

Mr Bill Herbst has recently used the University of Toronto 24-in. (61-cm) reflector on Las Campanas in Chile to obtain classification spectra at a dispersion of 128 Å mm^{-1} of all southern stars with $B < 10$ that are located in reflection nebulae. Spectra of fainter stars will be obtained with an image tube spectrograph. UBV photometry of some of the faint stars in reflection nebulae has been obtained with the Cerro Tololo 60-in. (1.5-m) telescope. Additional UBV observations are planned with the Toronto 24-in. (61-cm) telescope in Chile. Many of the stars in reflection nebulae appear to be members of small clusters. Photometric observations of the stars in these clusters are also planned.

The photometric and spectroscopic observations outlined above will be used to determine the distances to southern associations of reflection nebulae. It is hoped that the combined data on the distances of northern and southern reflection nebulae will lead to an improved picture of nearby spiral structure.

Acknowledgements

We are deeply grateful to Dr Victor Blanco for making large blocks of observing time on the Curtis Schmidt telescope available to us for this project. We also thank Steven van Agt, Tom Barnes, Chris Coutts, and Barry Madore for contributing to the Southern Milky Way Survey. This investigation was supported by the National Research Council of Canada.

References

Bergh, S. van den: 1966, *Astron. J.* **71**, 990.
Bok, B. J.: 1970, *Sky Telescope* **39**, 21.
Racine, R.: 1968, *Astron. J.* **73**, 233.
Racine, R. and Bergh, S. van den: 1970, in W. Becker and G. Contopoulos (eds.), 'The Spiral Structure of Our Galaxy', *IAU Symp.* **38**, 219.

DEPOLARIZATION EFFECT IN REFLECTION NEBULAE

V. VANÝSEK and M. ŠOLC

Dept. of Astronomy and Astrophysics, Charles University, Prague, Czechoslovakia

Abstract. The depolarization effects in reflection nebulae are caused by the size distribution, multiple scattering and irregularity of the grains. In the following abbreviated version of a study of the depolarization effect it is shown that the multiple scattering in reflection nebulae can lead to considerable decrease of the linear polarization. A sample of results obtained by the Monte Carlo method is tabulated.

Several attempts have been made to estimate the nature of dust grains in reflection nebulae by comparison of the observed data with the scattering pattern computed for composite models of polydisperse media using the assumed properties of particles likely to exist in the circumstellar and interstellar space.

Usually some simplifying assumptions for the refractive index and size distribution function for small spherical particles are introduced into the computer program and the Mie theory is used to determine the normalized scattering functions. This method has limits in application as well as in the interpretation of the results, since it gives an exact solution only for homogeneous spherical particles.

On the other hand, the scattering pattern of polydisperse media determined experimentally in the laboratory cannot provide any better guarantee for representing reliable models. The use of high-speed computers made the somewhat idealized polydisperse hypothetical media a relatively efficient tool for this kind of investigation. An excellent discussion of this problem is given in Deirmendjian's monograph (Deirmendjian, 1969).

The computation of scattering properties of optically thin media for this or similar purpose was carried out by several authors. Hanner (1971) compared extensive computations for silicate-like grains with the observed color and polarization in the Merope Nebula. She found that the complex refractive index $m = 1.65 - 0.05i$ may be more representative of the physical properties of submicron grains than pure dielectrics.

The polarization pattern of the computed models seems to be a suitable tool for estimating the physical properties of real dust grains.

There are, however, large discrepancies between the computed degree of polarization and observed values in reflection nebulae.

Polarimetric data obtained for reflection nebulae show that the degree of polarization is generally less than 20% in the visual spectral region and slightly increases with wavelength (see Martel, 1958; Elvius and Hall, 1967; Zellner, 1973). There is only one special case when the linear polarization found by Herbig (1972) in the circumstellar nebula of VY CMa is 60 or 70%.

Greenberg and Van de Hulst (eds.), Interstellar Dust and Related Topics, 127–130.

The computed models usually show large polarization, especially for models where the star is in front of the nebula (relative to the observer). It is obvious that some depolarization effect must be taken into consideration.

The degradation of the degree of polarization is due to the size distribution, multiple scattering or irregularities in the grain shape. The heterogeneity in the chemical composition of the grains (which cannot be excluded in the vicinity of a star's radiation field) can also lead to considerable depolarization of the scattered light.

The behavior of the polarization pattern for media containing absorbing particles differs significantly from those containing dielectric material. A positive polarization is to be found for absorbing clouds (even with moderate absorbers) with the maximum of polarization degree between phase angles $\vartheta = 60°$ to $90°$ (Greenberg and Hanner, 1970). The shift and increase of the maximum value towards 'forward scattering' (i.e. to lower phase angles) with an increase of the imaginary part of the refractive index is very typical.

The negative polarization (i.e. electric vector parallel to the polarization plane) near phase angles $150°$–$170°$ is typical of models containing dielectric particles (Greenberg and Hanner, 1970; Vanýsek, 1970). (These differences, of course, disappear, when the dominant size approaches the Rayleigh scattering domain where the polarization is always positive.)

From analysis of the available models it is evident that the large change of polarization degree with the phase angle together with the unknown geometry of the nebula may cause significant errors in the interpretation because the range of scattering angles along the line of sight is one of the unknown parameters. For instance, only positive polarization was detected in the scattered light of reflection nebulae. This may be due to the simple fact that the forward scattering is dominant in every studied case, and not to the nature of grains.

The depolarization effect caused by multiple scattering must be taken into consideration for dense clouds and the depolarization factor is a function of the optical thickness of the nebula.

In our recent study, the depolarization effect depending on the optical thickness was estimated by the Monte Carlo method. The best approach to the real conditions requires a method which is somewhat heterogeneous in the computing procedure. The phase function which determines the probability of photon propagation after the scattering event was computed for a polydisperse medium with a polynomial size distribution function. Because the determination of linear (and further elliptical) polarization caused by the subsequent scattering involves the difference in the dimensionless components of the Mie scattering complex amplitudes S_1 and S_2 (in van de Hulst's notation), the size distribution is then limited to the Γ function. Therefore the cross-sections and polarization pattern are computed for a *single* particle of some dominant parameter while the phase function follows the polydisperse pattern. This has led to results representing the depolarization effect which is due to the multiple scattering only. However, the photons' trajectory remains the same as in the polydisperse medium.

Table I presents one sample from the set of our results obtained for slightly absorbing silicate particles 0.15 μm in diameter and $m = 1.55 - 0.02i$ (albedo 0.84 for $\lambda = 5500$ Å) in a spherical nebula which may be typical of a nebula where the multiple scattering plays a significant role. The Monte Carlo procedure involving 10^3 photons shows that the multiple scattering may be neglected up to the total optical thickness of the nebula 0.4, when it contributed 15% of the total photon output from the nebula.

TABLE I

Polarization of light in a spherical nebula[a] (Monte Carlo results)

$\rho \backslash \tau$	0.37	0.74	1.10	1.47	1.84	2.21
0.1	9.2	6.4	5.2	4.7	4.3	4.0
0.2	18.7	12.8	10.6	9.4	8.5	7.9
0.3	27.1	19.2	15.7	14.1	12.8	11.9
0.4	38.2	26.0	21.1	18.6	16.8	15.7
0.5	47.0	34.4	26.4	23.4	20.9	19.5
0.6	53.9	38.4	31.7	27.7	25.1	23.3
0.7	66.5	44.0	37.1	31.6	28.4	26.2
0.8	72.8	50.8	39.6	35.8	33.5	30.0
0.9	72.8	56.3	46.7	39.6	35.2	34.0
1.0	91.5	63.1	51.1	45.6	40.5	37.0

[a] Computed for grains with diameter $a = 0.15 \mu$, refractive index $m = 1.55 - 0.02i$, $\lambda = 5500$ Å, size parameter $x_0 = 2\pi a/\lambda = 1.8$, albedo $= 0.84$. The phase function was assumed to be given by a polydisperse model with the size distribution $(x/x_0)^{-4}$ and limits $0.5 < x < 10$.
ρ = projected distance from the star relative to the radius of the nebula.
τ = optical thickness.

The polarization decreases considerably with increasing optical thickness. For instance, when τ increases from 0.4 to 1.5 the value of the polarization is approximately one half of the original value at the same distance from the star. The depolarization effect following from Table I is not combined with the depolarization caused by the size distribution and therefore for optically thin nebulae the values are high. The integration processes over particle sizes lead to the depolarization effect. The probability that the linear polarization in a polydisperse cloud at a given phase angle remains the same as the polarization of dominant size particles is very low. Even a relatively narrow size distribution can lead to a considerable depolarization. However, if both effects are combined then the resulting linear polarization for a moderately thick nebula might be less than 15% in the visual spectral region.

Conclusions

The above-mentioned results for one typical sample of various sets of models indicate that multiple scattering cannot be neglected in the interpretation of the observed linear polarization in reflection nebulae. Consequently, the interpretation of measured values of optically thick nebulae (as for instance NGC 7023) may be seriously misleading if based only on the polarization by single scattering.

Acknowledgements

One of us (V.V.) carried out some computations used here at the University of Massachusetts, Amherst, in 1968–69, and was supported by NSF Grant GP 7793.

References

Deirmendjian, D.: 1969, *Electromagnetic Scattering on Spherical Polydispersions*, Elsevier, New York.
Elvius, A. and Hall, J. S.: 1967, *Lowell Obs. Bull.* **7**, 17.
Greenberg, J. M. and Hanner, M. S.: 1970, *Astrophys. J.* **161**, 446.
Hanner, M. S.: 1971, *Astrophys. J.* **164**, 425.
Herbig, G. H.: 1972, *Astrophys. J.* **172**, 375.
Martel, M.-T.: 1958, *Ann. Astrophys. Suppl.*, No. 7.
Vanýsek, V.: 1970, *Acta Univ. Carol., Ser. Math. et Phys.* **11**, 3.
Zellner, B.: 1973, this volume, p. 109.

ELLIPTICAL POLARIZATION BY LIGHT SCATTERING
BY SUBMICRON SPHEROIDS

TH. SCHMIDT

Max-Planck-Institut für Astronomie, Heidelberg, Germany

Abstract. Calculations have been carried out of single light scattering by Rayleigh type spheroidal dust particles. It can be shown that elliptical polarized radiation with $e = v/p \approx 0.15$ can be produced from unpolarized incident radiation, in the case of dirty quartz particles with a refractive index $m = 1.54 \pm 0.4i$ and a geometrical axial ratio $c : a \approx 1:2$.

1. Introduction

An elliptical polarimetry of the 30 Doradus and η Carinae nebulae, undertaken with the Heidelberg telescope at the Boyden Observatory (South Africa) in the blue spectral region, has shown a circular polarized component greater than 1.5 times the rms error of measurement in 3 cases:

$$30 \text{ Dor: } (v/p)_1 = 0.15 \pm 0.05; \ (v/p)_2 = 0.08 \pm 0.05$$

$$\eta \text{ Car: } (v/p)_3 = 0.09 \pm 0.045,$$

v and p being the circular and linear amount of polarization respectively (see Schmidt, 1971).

The question arises whether it may be possible to explain these ellipticities $e = v/p$ by dust particle scattering, even in the case of unpolarized incident radiation and single scattering (i.e. considering optically thin dust clouds). A direct interpretation of the amount of circular polarization $v = V/I$ does not seem meaningful because the intensity I is confused by strong gaseous emissions in both nebulae.

2. Theory

Concerning these questions calculations are presented of scattering properties of ellipsoidal submicron (i.e. Rayleigh like) particles.

Let P be the vector of the 4 Stokes' parameters

$$P = \begin{pmatrix} I \\ Q \\ U \\ V \end{pmatrix}$$

and F the transformation matrix for an arbitrary optical process

$$P = F \cdot P_0$$

Greenberg and Van de Hulst (eds.), Interstellar Dust and Related Topics, 131–137.

after van de Hulst (1957, p. 43 f.). For unpolarized incident light $I_0 \neq 0$, $Q_0 = U_0 = V_0 = 0$. In this case from a circular polarization parameter $V \neq 0$ follows the condition $F_{41} \neq 0$, while the ellipticity $e = v/p$ is identical to

$$e = F_{41}/\sqrt{F_{21}^2 + F_{31}^2}\,.$$

The transformation F has to be computed for the light scattering process by ellipsoidal Rayleigh particles (i.e. in the blue spectral region for particle diameters $d < 0.13\ \mu\mathrm{m}$).

In Figure 1 the co-ordinate notation of an arbitrarily oriented spheroid is illustrated. The X, Y, Z-system describes the geometry of scattering, Z pointing towards the observer and Z, Y being the plane of scattering. The co-ordinate system 1, 2, 3 is connected with the particle geometry, 3 being the axis of symmetry and 1 the intersection between the X, Y- and 1, 2-plane; ϑ is the nutation and ψ the precession angle of the particle in the X, Y, Z-system.

From the equations of van de Hulst (1957, pp. 44 and 78 f.) follows

$$F_{41} \propto \mathrm{Im}\,(\alpha_3 \alpha_1^*)\sin\psi\sin\vartheta\sin\theta\,\{\cos\psi\sin\vartheta\sin\theta + \cos\vartheta\cos\theta\} \qquad (1)$$

θ being the angle of scattering; α_j is the polarizability tensor of the particle (α_j^* is the conjugate complex):

$$\alpha_j \propto (1/(m_j^2 - 1) + L_j)^{-1} \qquad (2)$$

m_j being the complex refractivity tensor and L_j a particle shape tensor. Some cases of L_j are tabulated by van de Hulst (1957, p. 72); for spheres is $L_1 = L_2 = L_3 = \frac{1}{3}$, for flat rotationally symmetric disks $L_1 = 1$, $L_2 = L_3 = 0$ and for needles $L_1 = 0$, $L_2 = L_3 = 0.5$. F_{21} and F_{31} are more complicated functions of α_j, ϑ, ψ and θ than F_{41}.

From Equation (1) and (2) it can be concluded that $F_{41} \neq 0$ is possible only if

(1) $L_j \neq L_k$ or $m_j \neq m_k$ for any $j \neq k$, and

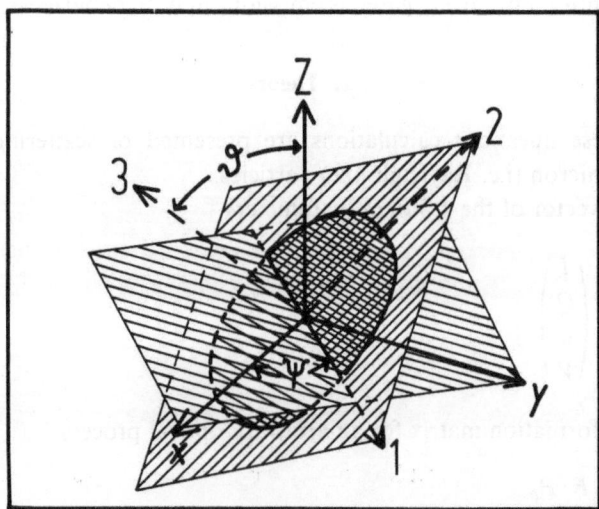

Fig. 1.   Orientation of dust spheroids. Z: direction to the observer, 3: particle axis of symmetry.

(2) $m_j = n_j - \varkappa_j i$ with a nonzero imaginary compenont $\varkappa_j$.
The maximum of F_{41} occurs at $\vartheta = \theta = 90°$, $\psi = 45°$ and is equal to

$$F_{41,max} \propto \tfrac{1}{2} \text{Im}(\alpha_3 \alpha_1^*),$$

but these angles would not produce generally the maximum of ellipticity

$$F_{41}/\sqrt{F_{21}^2 + F_{31}^2} \text{ as well.}$$

3. General Results

In the following calculations will be discussed in some detail for 'dirty quartz' type spheroids, i.e. a scalar refractive index

$$m = 1.54 - \varkappa i.$$

Figure 2 shows the computed ellipticities in case of $F_{41, max}$ for different values of $\varkappa$ and axial ratios c/a (c being the geometrical axis of symmetry). For values $c/a = 1:2$ or $2:1$ $\varkappa \approx 0.4$ is necessary to explain the maximum observed ellipticity $e = 0.15$ (somewhat smaller for the water ice case, shown for comparison).

In Figure 3 ellipticies are drawn for different angles θ, ϑ and ψ, taking $m = 1.54 - 0.4i$ and $c/a = 0.5$ (i.e. oblate spheroids); some more detailed calculations of the same type are shown in Figure 4. It turns out that ellipticities $e \approx 0.5$ (and even of more than 1) are possible for angles, where v itself has no maximum but the linear polarization is small.

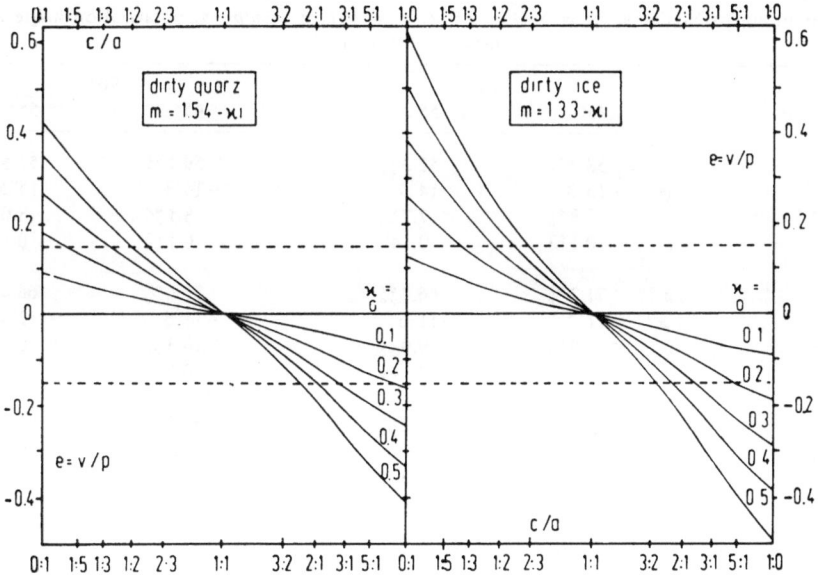

Fig. 2.   Ellipticities $e = v/p = V/\sqrt{Q^2 + U^2}$ by Rayleigh scattering at dirty quartz and ice spheroids in case of $V_{max} = V(\psi = 45°, \vartheta = 90°, \theta = 90°)$ and different absorbing indices $\varkappa$ and axial ratios of the particle c/a (c: axis of symmetry).

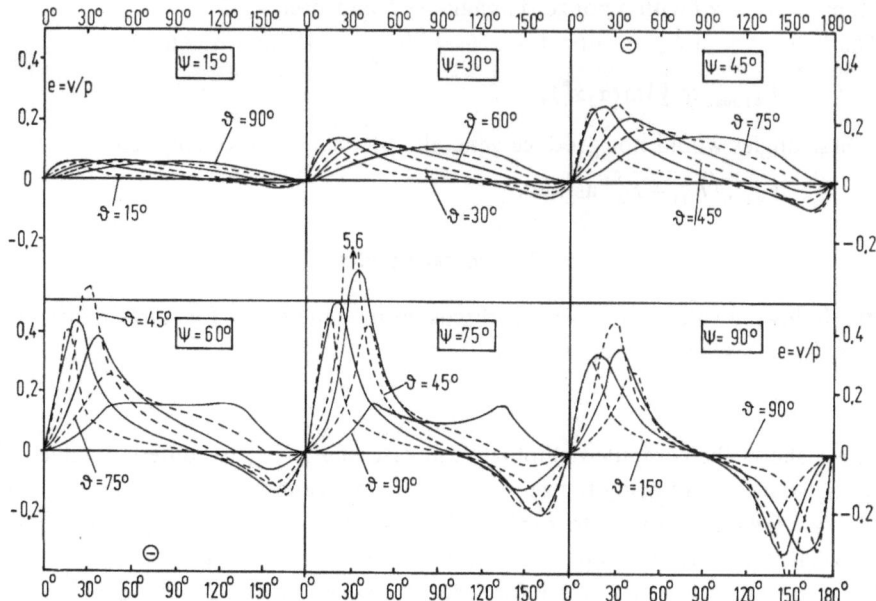

Fig. 3.   Ellipticities $e = v/p$ by Rayleigh scattering at dirty quartz spheroids ($m = 1.54 - 0.4i$ and $c/a = 0.5$) for different angles of scattering (θ), nutation (ϑ) and precession (ψ). (For angles $\psi' = -\psi$ take: $e' = -e$; for $\psi' = 180° - \psi$ take: $e' = -e$ and $\theta' = 180° - \theta$; for $\vartheta' = 180° - \vartheta$ take: $e' = e$ and $\theta' = 180° - \theta$.)

TABLE I

Polarization of scattered light from an optical thin spheric symmetrical cloud of dust particles corresponding to Figure 4. ($p = \sqrt{Q^2 + U^2} / I$; $v = V/I$; $e = v/p$; φ: polarization angle against tangential direction)

| ϑ | | 80° | | 90° | |
ψ		45°	55°	45°	55°
	p	58.7%	52.3%	59.1%	52.5%
	φ	$-13°.8$	$-14°.9$	$-14°.3$	$-15°.5$
$r_0 = 0.5\,r_1$	v	7.8%	7.7%	8.1%	8.0%
	e	0.133	0.148	0.137	0.153
	p	71.2%	66.3%	71.6%	66.4%
	φ	$-11°.2$	$-11°.6$	$-11°.6$	$-12°.1$
$r_0 = r_1$	v	9.9%	9.9%	10.3%	10.3%
	e	0.138	0.149	0.143	0.156
	p	84.9%	82.2%	85.3%	82.3%
	φ	$-9°.3$	$-9°.3$	$-9°.7$	$-9°.7$
$r_0 = 2\,r_1$	v	12.1%	12.3%	12.5%	12.8%
	e	0.142	0.150	0.147	0.156
	p	98.6%	98.6%	98.9%	98.8%
	φ	$-7°.9$	$-7°.7$	$-8°.3$	$-8°.1$
$\theta = 90°$	v	14.2%	14.7%	14.8%	15.4%
	e	0.144	0.149	0.149	0.156

4. A Spherically Symmetrical Dust Cloud

The data of Figure 4 have been taken for calculating the properties of integrated polarisation of light scattered by a spheric symmetrical, optical thin dust cloud with a central illuminating star. In this case the integral

$$\bar{P}(\vartheta, \psi) \propto \int_{-\infty}^{+\infty} P(\vartheta, \psi, \theta(Z)) \varrho(r) \frac{dZ}{r^2} \tag{3}$$

has to be calculated, Z being the co-ordinate along the light beam, θ the angle of scattering and r the distance from the central star; ϱ is the dust density and P represents the Stokes' parameters I, Q, U, V or $F_{11}, F_{21}, F_{31}, F_{41}$ of the transformation matrix F respectively. Assuming a Gaussian density distribution

$$\varrho(r) = \varrho_0 \exp(-r^2/r_1^2)$$

and remembering that

$$\tan \theta = r_0/Z, \quad dZ/d\theta = -r_0/\sin^2 \theta \quad \text{and} \quad \sin \theta = r_0/r$$

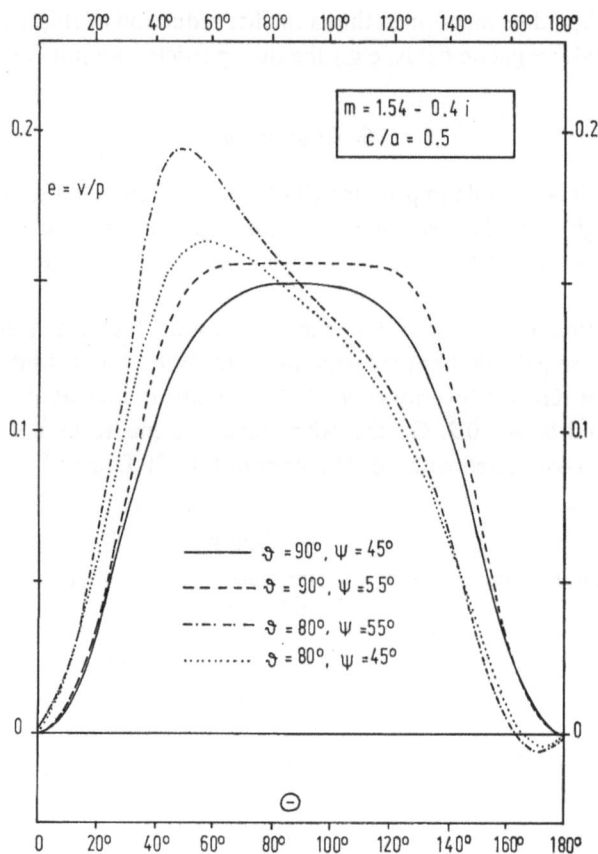

Fig. 4.   Ellipticities $e = v/p$ for 4 special particle orientations (notations as in Figure 3).

(r_0 being the minimum distance between light beam and central star), then

$$\bar{P}(\vartheta, \psi) \propto \int_0^\pi P(\vartheta, \psi, \theta) \exp\left\{-(r_0/r_1)^2/\sin^2\theta\right\} d\theta. \tag{4}$$

Table I shows the results for all cases drawn in Figure 4, together with the non integrated values of $\theta(v_{max}) = 90°$ for comparison. Integrated ellipticities up to nearly $e = 0.16$ ocour, while the deviation of polarization angles from tangential directions remains rather small.

5. Spinning Particles

According to the alignment by the mechanism of Davis and Greenstein (1951) the dust particles spin around their main axis of inertia. Therefore some supplementary test calculations have been carried out for a sample of prolate spheroids spinning around one of their small axes of geometry. As shown by Table II particles of that kind are between 35% and 50% less effective in producing circular polarization than oblate spheroids (the ratio of particle axes being 1:2 both).

Moreover it should be mentioned that a further reduction of ellipticity would occur, if (in case of weak magnetic fields, e.g.) the dust particles are not completely aligned.

6. Conclusion

It turns out that it is possible to produce elliptical polarized light up to $e = v/p \approx 0.15$ by single Rayleigh scattering processes, assuming aligned dirty quartz spheroids with a geometrical axial ratio of 1:2 and an absorbing part of the refractive index $\varkappa = 0.4$ to 0.5.

This value seems to be greater than suggested by the high interstellar dust albedo found from diffuse galactic light measurements by Witt (1968), Mattila (1970, 1971) and others. After Krishna Swamy et al. (1971) the albedo limited absorption of interstellar dust would be $\varkappa \approx 0.2$. On the other hand it seems to be not impossible that the dust medium connected with hot H II regions (like 30 Dor and η Car) shows some

TABLE II

Ellipticity reduction factor e_p/e_0 for replacing oblate by spinning prolate spheroids (axial ratio = 1:2 both)

ϑ	ψ, $180° - \psi$				
	15°	30°	45°	60°	75°
15°	0.66	0.65	0.64	0.64	0.63
30°	0.64	0.63	0.62	0.61	0.59
45°	0.63	0.62	0.60	0.58	0.56
60°	0.63	0.60	0.58	0.55	0.53
75°	0.62	0.60	0.56	0.53	0.50
90°	0.62	0.60	0.56	0.52	0.49

pecularities *or* that the *actual* ellipticity in 30 Dor is somewhat less than 0.15, due to the rms error of ± 0.5.

Another open question is, how the situation turns out in case of Mie particles, although it seems not improbable that the general situation will not be more disadvantageous, especially as the elliptical polarization is due to individual particle properties being normally more marked in the Mie than in the Rayleigh particle region.

Acknowledgements

The calculations have been undertaken by the computer Siemens 2002 of the Heidelberg University.

I am indebted to Drs R. D. Wolstencroft and L. Bandermann for a valuable discussion which gave me occasion to find an error in calculating the linear polarization of the scattered radiation.

References

Davis, L. and Greenstein, J. L.: 1951, *Astrophys. J.* **114**, 206.
Krishna Swamy, K. S., Jackson, W. M., and Donn, B. D.: 1971, *Astron. Nachr.* **293**, 43.
Mattila, K.: 1970, *Astron. Astrophys.* **9**, 53.
Mattila, K.: 1971, *Astron. Astrophys.* **15**, 292.
Schmidt, Th.: 1971, *Astron. Astrophys.* **12**, 456.
Van de Hulst, H. C.: 1957, *Light Scattering by Small Particles*, New York, London.
Witt, A. N.: 1968, *Astrophys. J.* **152**, 59.

A SURFACE POLARIZATION SURVEY OF THE
MILKY WAY AND THE ZODIACAL LIGHT

J. STAUDE, K. WOLF, and TH. SCHMIDT

Max-Planck-Institut für Astronomie, Heidelberg, Germany

Abstract. An elliptical Surface polarimetry of the Night Sky has been carried out at the Boyden Observatory, South Africa. The Milky Way is linearly polarized by $p = 1$ to 2% (Figure 1); 11 regions also show a circular polarized light contribution (Table I).

The amounts of circular polarization near $v \approx 0.1\%$ or less, observed within the Zodiacal Light, are generally *not* exceeding twice the error of measurement: see Staude and Schmidt (1972). ($v > 0$ is defined by clockwise rotation, looking against the light source).

An elliptical (i.e. linear and circular) polarization survey of the Milky Way has been undertaken, using the Heidelberg 50 cm Cassegrain reflector at the Boyden Observatory, South Africa, during 13 nights between 1968 May 31 and 1968 December 19. In six out of these nights the Zodiacal Light has also been observed.

The telescope was equipped with a digital on channel polarimeter and a photomultiplier EMI 6256 S/A. A circular focal diaphragm of 20' was used. The observations have been carried out in two spectral regions, centered at 4200 Å (during seven nights) and 5260 Å (during six nights), the bandwidth being 900 Å and 225 Å respectively (for details see Leinert *et al.*, 1967 and Schmidt, 1971).

Polarizing effects of the instrument including the filters, (the 5620 Å one being an interference filter) were carefully checked for all Stokes' parameters during every night. The applied corrections for instrumental polarization were in the range from 0.1% to 0.4% for the relative Stokes' parameters $q = Q/I$ and $u = U/I$, and from 0.01% to 0.08% for the Stokes' parameter $v = V/I$.

The measurements of the Milky Way, covering the region $320° \leqslant l \leqslant 40°$, $|b| \leqslant 30°$, consisted of sequences of scans in right ascension α with fixed declination δ, giving a reading every 4.5 deg in α for integrations times of 30 s.

The declination interval between neighbouring scans was $\Delta\delta = 5°$. By means of standard star measurements the readings of the Stokes' parameters were corrected for atmospheric extinction and converted into units S^{10} (stars 10^m sq deg^{-1}).

The foreground luminosity is linearly polarized because of the contributions of the Zodiacal Light and the nearby town of Bloemfontein, and it is time dependent mainly because of the air-glow fluctuations. To make a reliable separation of these foreground effects on all Stokes' parameters (between the previously mentioned sequences of scans) accurate registrations of the night sky brightness and polarization beside the Milky Way were carried out every three hours consisting of a grid of point measurements every 22°.5 in declination δ and hour angle t, covering the whole sky.

To obtain the foreground components at the coordinates and times of the Milky Way measurements, the foreground Stokes' parameters of each night were expanded

Greenberg and Van de Hulst (eds.), Interstellar Dust and Related Topics, 139–141.
All Rights Reserved. Copyright © 1973 by the IAU.

into spherical harmonics in the azimutal coordinate system, including terms linear and quadratic in time. With the expansion coefficients determined from the data by a least squares fit procedure, the Stokes' parameters of the foreground luminosity corresponding to the Milky Way measurements were computed and substracted from those observed in the Milky Way.

Since the surface polarization shows a much weaker color dependence then the brightness, the data obtained in the visual region were normalized to those in the blue by the averaged quotient of the intensities in the blue and visual region. Then, the Stokes' parameters of all nights were added together.

Figure 1 shows the results in general. Table I shows the 11 values of circular polarization with $|v| > 2 \Delta v$ (Δv being the rms standard error).

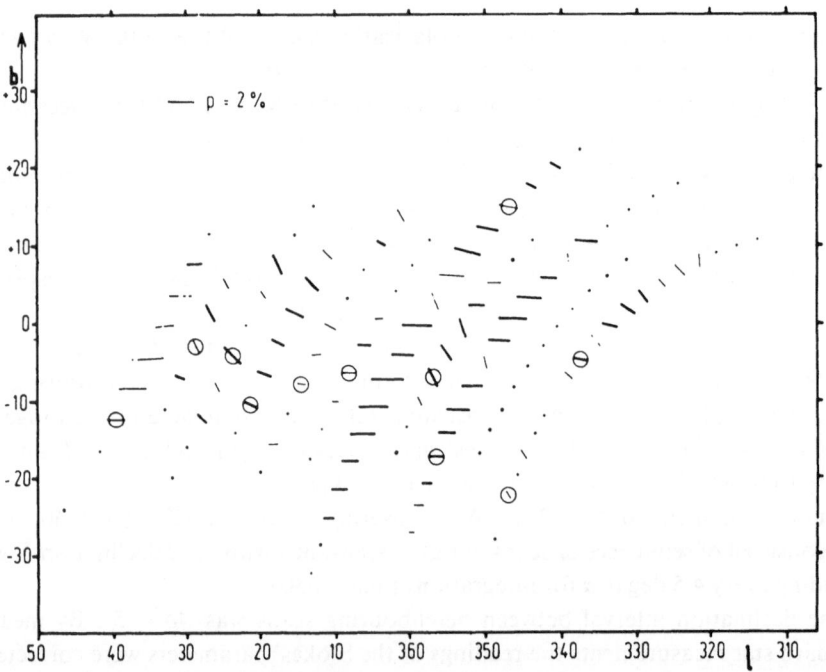

Fig. 1. Surface polarization of the Milky Way. ————: $p > 2\Delta p$, —— ——: $p > \Delta p$, · : $p < \Delta p$ or number of nights of observation $n < 4$, ○: $v > 2\Delta v$.

TABLE I

Circular surface polarization of the Milky Way (n: number of nights)

l	b	$v\,[\%]$	n	l	b	$v\,[\%]$	n
337°	− 4°	0.35 ± 0.11	5	14°	− 8°	−0.25 ± 0.12	6
347°	−22°	−0.78 ± 0.39	5	21°	−10°	−0.94 ± 0.11	3
356°	−17°	−0.64 ± 0.19	7	24°	− 4°	−0.84 ± 0.37	4
357°	− 7°	0.18 ± 0.07	6	29°	− 3°	−0.72 ± 0.07	3
347°	15°	0.72 ± 0.21	5	40°	−12°	−0.82 ± 0.40	3
8°	− 6°	0.32 ± 0.09	7				

The linear polarization is generally in agreement with the results of Wolstencroft (1973), but before interpreting these measurements in more detail, it is necessary to calculate whether an application of corrections is necessary, by including the reffect of scattering by atmospheric Mie particles on the surface polarization of the Milky Way.

Concerning the Zodiacal Light, the main interest was put on the parameter V, because many observations of the linearly polarized component have already been carried out under much better conditions.

A nonzero circular polarization component of the zodiacal light has been reported by Wolstencroft and Rose (1967), using observations at 7030 Å from a sounding rocket.

The Zodiacal Light has been observed during the six nights with the visual filter. The measurements, covering the region $28° \leqslant \varepsilon \leqslant 114°$ and $287° \leqslant \varepsilon \leqslant 332°$, $-13° \leqslant \beta \leqslant 12°$, were arranged in the same manner as those in the Milky Way. After correction of the data for atmospheric extinction and conversion into units S^{10}, V-values of fields $10° \times 10°$ in elongation ε and latitude β were obtained averaging $n = 3$ to 21 single measurements.

Altogether, the degree of circular polarization of the observed Zodiacal Light region is $|v| \leqslant 0.1\%$. No firmly established nonzero result has been found (for details see Staude and Schmidt, 1972).

Acknowledgement

This research has been supported by the Deutsche Forschungsgemeinschaft under contract El 13/14.

References

Leinert, C., Schmidt, Th., and Schwarze, B.: 1967, *Veröff. Landessternwarte Heidelberg-Königstuhl* **20**, 23.
Schmidt, Th.: 1971, *Astron. Astrophys.* **12**, 456.
Staude, J. and Schmidt, Th.: 1972, *Astron. Astrophys.* **20**, 163.
Weinberg, J. L.: 1964, *Ann. Astron.* **27**, 718.
Wolstencroft, R. D. and Rose, L. J.: 1967, *Astrophys. J.* **147**, 271.
Wolstencroft, R. D.: 1973, this volume, p. 99.

PART IV

INTERSTELLAR POLARIZATION

INTERSTELLAR POLARIZATION

K. SERKOWSKI

Lunar and Planetary Laboratory, University of Arizona, Tucson, Ariz., U.S.A.

Abstract. A review of observational data on interstellar polarization is given. The degree of fluctuation in the direction of the galactic magnetic field is discussed. An empirical formula is given which fits all the observations of the wavelength dependence of interstellar polarization when they are normalized with the wavelength λ_{max} of maximum polarization. Well-defined regions on the sky with values of λ_{max} deviating from an average are found. Values of λ_{max} are correlated with the ratio of total to selective interstellar extinction. A unified wavelength dependence of interstellar extinction is obtained by normalizing the observations with the wavelength λ_{max} of maximum interstellar polarization.

In this review I shall try to indicate the principal *observational* facts concerning the interstellar polarization of starlight; these facts may serve as a basis for further discussion on the theoretical explanation of the observed polarization. The problems of interstellar polarization are closely related to those of circumstellar, intrinsic polarization of starlight, which were recently reviewed elsewhere (Serkowski, 1971; Zellner and Serkowski, 1972).

Fruitful discussion of the distribution of the orientations of polarization on the sky became possible after Mathewson and Ford (1970) extended to the southern sky the high galactic latitude polarimetric surveys of Behr (1959) and Appenzeller (1968). Figure 1, prepared by Mathewson and Ford, shows the distribution of the planes of vibration of the electric vector for the interstellar polarization of over 7000 stars*.

Mathewson (1968) explained the elliptical 'flow patterns' seen in Figure 1 by assuming that the magnetic lines of force, along which the dust grains are aligned by the Davis and Greenstein (1951) mechanism, form tightly wound righthanded helices with a pitch angle of about 7°. The helices lie on the surface of tubes having elliptical cross-section of axial ratio 3, with semimajor axes parallel to the galactic plane. The helices have been sheared through an angle of 40° on the galactic plane in an anticlockwise sense. The axis of the helices is directed towards galactic longitudes 90° and 270°, which is the direction of the local spiral arm as determined from H I measurements. Mathewson's helical model of the galactic magnetic field explains not only the optical polarization but also the spurs, ridges, and strongly polarized regions of the galactic radio emission (cf. Verschuur, 1970; Martin, 1971). The regions of strong radio emission are probably regions of magnetic field compression which take the form of elongated tubes, with the magnetic field lines parallel to their length. These tubes have the local directivity of the helical model derived from the optical polarization measurements.

* Polarization of another 1600 stars in the southern Milky Way has been measured by Klare *et al.* (1972).

Greenberg and Van de Hulst (eds.), Interstellar Dust and Related Topics, 145–152.

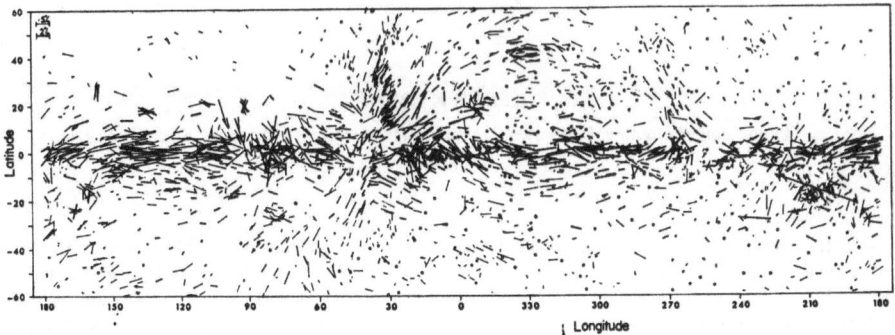

Fig. 1. The interstellar polarization of starlight plotted in galactic coordinates by Mathewson and Ford (1970). The length and position angle of each line indicate the degree of polarization, P, and the plane of vibration of the electric vector, respectively. Small circles are drawn about stars with $P < 0.08\%$. There are two scales for P which are shown in the top left-hand corner of the figure. The first scale operates for $P < 0.6\%$ and these vectors are drawn thin. The second scale operates for $P \geq 0.6\%$ and these vectors are drawn thick.

Mathewson's model suggests that most of the interstellar polarization is produced within about 250 pc of the axis of the local spiral arm, which passes about 100 pc from the Sun in the direction of the galactic anticenter. This is confirmed by Verschuur's (1970) discussion of the dependence of the degree of linear polarization on distance (Figure 2a). To reduce the 'noise' in polarization data and in distance determinations

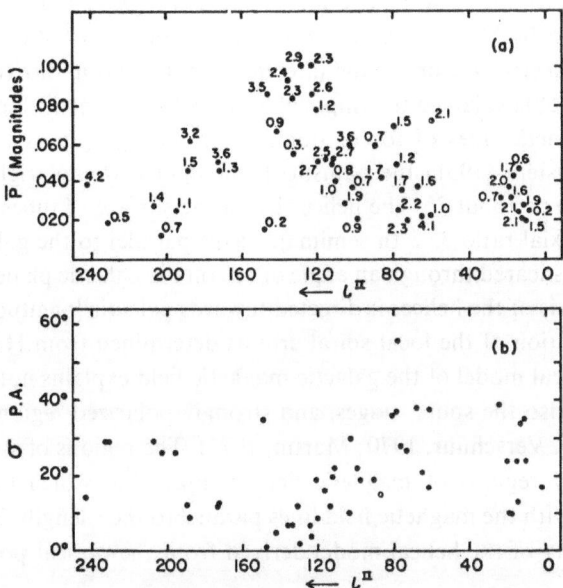

Fig. 2. Polarization properties of star clusters and associations according to Verschuur (1970). (a) Mean polarization as a function of galactic longitude. Numbers indicate the distance to the cluster or association in kpc. (b) The scatter in position angle of member stars as a function of longitude.

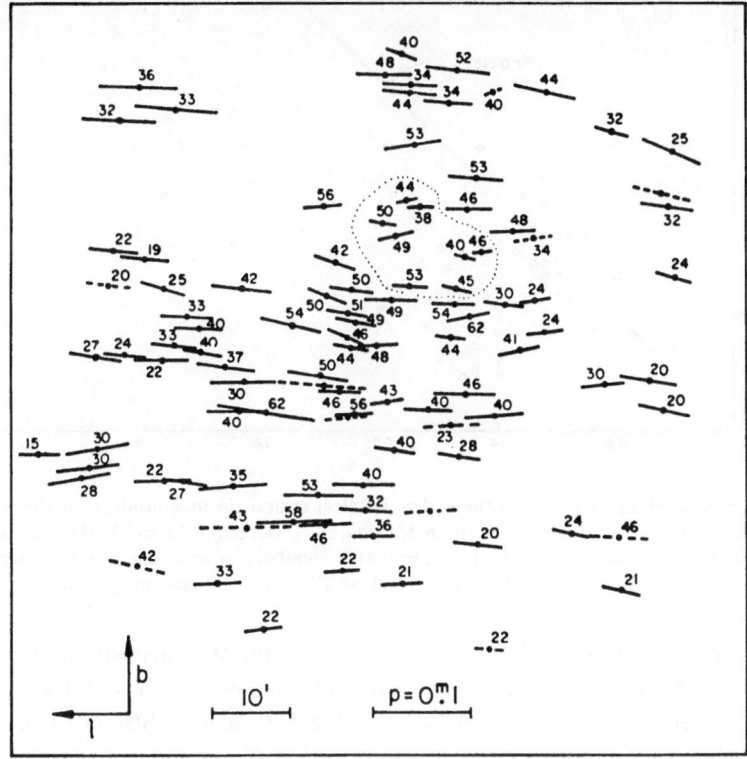

Fig. 3. Polarization of star cluster Stock 2, 320 pc distant, according to Krzemiński and Serkowski (1967). The length and position angle of each line indicate the amount of polarization in the yellow spectral region and the plane of vibration, respectively. The solid lines denote cluster members, dashed lines probable members. Numbers accompanying the lines give the reddening E_{B-V} expressed in hundredths of magnitude. The dotted line delineates the region of small polarization.

he considers only star clusters and associations instead of incividual stars*. Correlation between polarization and distance is very slight.

Figure 2b indicates that the scatter in the polarization position angles is minimum around galactic longitude 140°, which may be explained with the helical model. Of particular interest is cluster Stock 2 at $l = 133°$ (Krzemiński and Serkowski, 1967) which is the only cluster not far from the Sun (320 pc) which shows a strong linear polarization, 2.3% on the average. This cluster is by far the most suitable for studying small scale fluctuations in the direction of magnetic field; the rms deviation of the position angle of the electric vector (Figure 3) from its mean value equals $\pm$ 8°. The fluctuations in the direction of interstellar magnetic field indicated by the polarimetry of Stock 2, as well as several other open clusters (Serkowski, 1968) and field stars (Schmidt, 1968), are quite large: the rms deviation of the direction of magnetic field lines from their mean direction is about $\pm$ 25°.

* For new polarimetric observations of star clusters see Grigoryan (1970) and Breger and Dyck (1972).

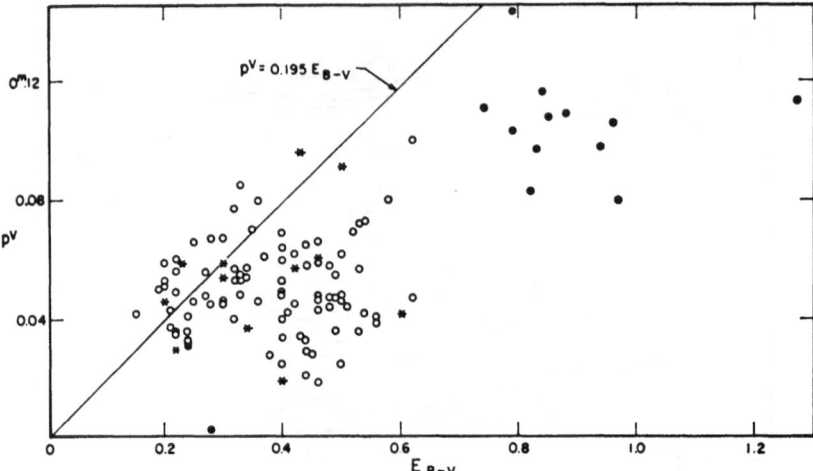

Fig. 4. Amounts of polarization in the yellow spectral region (in magnitudes) for the stars in the region of cluster Stock 2 plotted against the reddening E_{B-V} (Krzemiński and Serkowski, 1967). The cluster members are denoted by open circles, probable members by asterisks, non-members by filled circles. The straight line corresponds to $P/E_{B-V} = 9\%$ per magnitude.

The correlation between the degree of polarization and position angle of neighbouring stars in Stock 2 drops to $1/e$ for an angular distance of about 15'. If we assume that polarization is produced along all the path to the cluster, a microscale of 0.3 pc is obtained for fluctuations in the direction of the magnetic field. A similar value has been obtained by Davies (1968) from a study of depolarization of extragalactic radio sources. The distribution of reddening E_{B-V} in the cluster Stock 2 indicates a microscale of 1 pc for fluctuations in the spatial density of interstellar dust – about 3 times larger than for the magnetic field fluctuations.

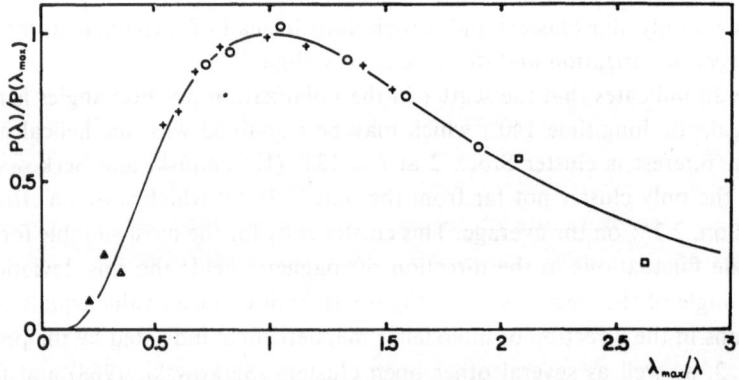

Fig. 5. Normalized wavelength dependence of interstellar polarization averaged for 6 stars in Perseus-Cepheus (crosses) for which $\lambda_{max} \cong 0.52\ \mu$ and for 5 stars in Scorpius (open circles) for which $\lambda_{max} \cong 0.70\ \mu$. Ultraviolet balloon observations for ζ Ophiuchi at $\lambda = 0.225\ \mu$ and $0.286\ \mu$ by Gehrels (1973; squares) and the infrared observations of HD 183143 at $\lambda = 1.6\ \mu$ and VI Cyg *12 at $\lambda = 1.6\ \mu$ and $\lambda = 2.2\ \mu$ by Dyck (1973; triangles) are also plotted. The solid line is calculated from Equation (1).

There is no correlation between polarization and reddening within the cluster Stock 2 (Figure 4). Such lack of correlation is expected if the ratio of space densities of interstellar dust (n_g) and gas (n_H) is constant. For incomplete Davis-Greenstein orientation, the theory of which has been revised recently by Cugnon (1971) and by Purcell and Spitzer (1971), the polarization per unit path length is proportional to n_g/n_H while the color excess is proportional to n_g. For the ratio of percentage polarization P to color excess E_{B-V} an upper limit of 9% per magnitude, derived by Schmidt-Kaler (1958) is usually assumed. Figure 4 suggests, however, an upper limit which may be as high as 12% per magnitude.

The shape of the wavelength dependence of interstellar polarization seems to be the same for all stars. This became evident only when I first normalized this wavelength dependence by plotting $P(\lambda)/P(\lambda_{max})$ against λ_{max}/λ, where λ_{max} is the wavelength of maximum polarization for a given star. The mean wavelength dependence of polarization for several stars with λ_{max} either close to 0.52μ (Perseus-Cepheus) or close to 0.70μ (Scorpius) is plotted in that way in Figure 5, using mainly the observations made at the University of Arizona (Coyne *et al.*, 1973). Ultraviolet balloon observations of ζ Ophiuchi ($\lambda_{max} = 0.60 \mu$) by Gehrels (1973) and infrared observations of HD 183143 ($\lambda_{max} = 0.56 \mu$) and of star No. 12 in the association VI Cyg ($\lambda_{max} = 0.45 \mu$) by Dyck (1973) are also plotted. The observed wavelength dependence of interstellar polarization is well described by an empirical formula (cf. Serkowski, 1971, where it is incorrectly printed)

$$P(\lambda)/P(\lambda_{max}) = \exp\left[-1.15 \ln^2 (\lambda_{max}/\lambda)\right] \tag{1}$$

used for calculating the curve shown in Figure 5. This curve is very close to theoretical curves calculated by Greenberg (1968, Figure 95) for an Oort-van de Hulst size distribution of cylindrical dielectric grains imperfectly oriented by the Davis-Greenstein mechanism. It was not possible to obtain a fit to the observed curve assuming other orientation mechanisms than that of Davis and Greenstein.

There is no observational evidence for any feature in the wavelength dependence of polarization at the λ 4430 absorption band (Wampler, 1966; A'Hearn, 1972). The observations indicate that if such a feature is present it must be limited to a spectral region not wider than 10 Å.

Whenever the orientation of aligned dust grains changes considerably along the light path from a star the presence of a circularly polarized component is expected (cf. Serkowski, 1962). The announcement of the discovery of such interstellar circular polarization was a highlight of this Symposium. The circular polarization was found to change its sign around the wavelength of maximum linear polarization (Kemp, 1973; Kemp and Wolstencroft, 1972). The largest amount of circular polarization, 0.04% for *o* Scorpii, is observed around the wavelength 0.5μ.

The wavelength of maximum linear polarization is expected to be proportional to the mean value of the parameter $(m-1)a$, where m is the refractive index of dust grains and a their radius. Since interstellar extinction considered as a function of $(m-1)a/\lambda$ should have the same shape all over the sky, a unified wavelength dependence of interstellar extinction can be obtained by normalizing it with the wavelength λ_{max} of maximum interstellar polarization. Large differences in the shape of conventional extinction curves for Scorpius, as compared to the Perseus-Cepheus regions (Figure 6a) vanish when $[A(\lambda) - A(\lambda_{max})]/[A(\frac{3}{4}\lambda_{max}) - A(\frac{4}{3}\lambda_{max})]$ is plotted against λ_{max}/λ (Figure 6b). Figure 6 is based on the same stars which were used in Figure 5. Of course, the normalization of the wavelength dependence of interstellar extinction with λ_{max} makes sense only for those spectral regions where the refractive index is

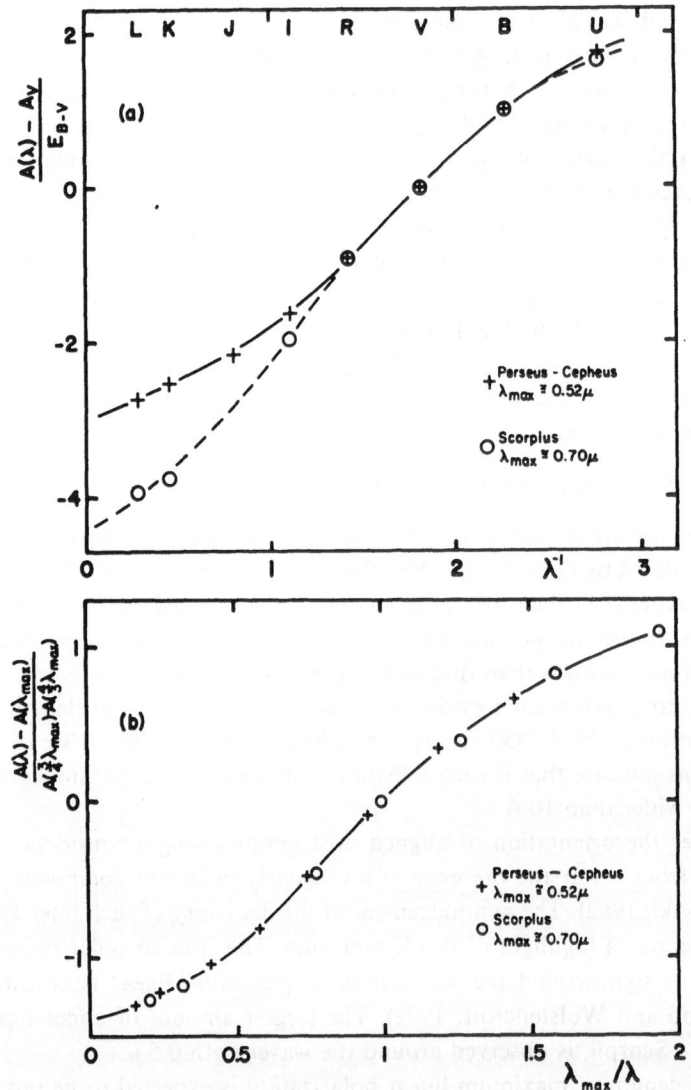

Fig. 6. Wavelength dependence of interstellar extinction for the same stars in Perseus-Cepheus (crosses) and in Scorpius (open circles) for which polarization is plotted in Figure 5. (a) Conventional extinction curve. (b) Wavelength dependence of extinction normalized with λ_{max}, the wavelength of maximum polarization. Infrared photometry in Scorpius by Carrasco *et al.* (1973) is used.

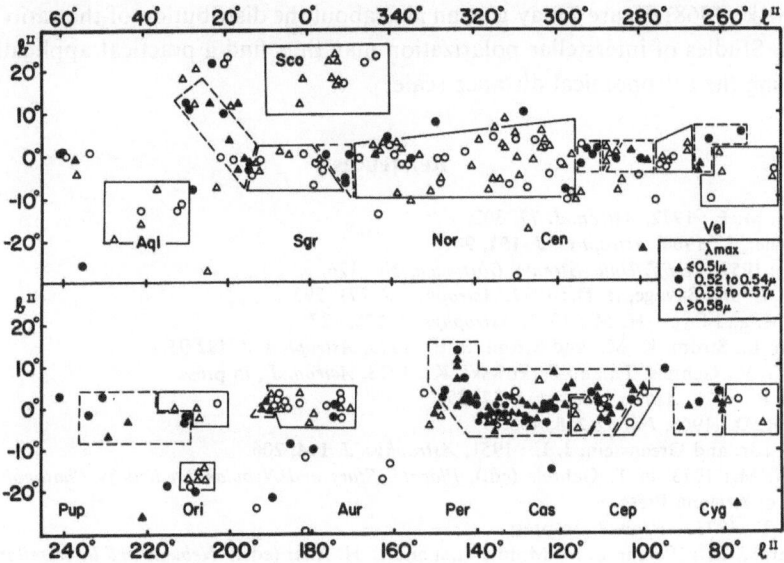

Fig. 7. Wavelength of maximum interstellar polarization, λ_{max}, plotted in galactic coordinates by Serkowski *et al.* (1973). Filled symbols denote the stars with λ_{max} smaller than the median value, open symbols those with λ_{max} larger than the median value. The regions in which the stars of any of these groups predominate are surrounded by dashed or solix lines, respectively.

constant. Therefore it cannot be applied in the far ultraviolet where the wavelength dependence of extinction is dominated by a strong absorption feature around 0.22 μ (Bless and Savage, 1972).

The number of stars with known wavelength of maximum interstellar polarization (λ_{max}) has been nearly doubled by a survey with a multi-channel polarimeter at Siding Spring Observatory recently completed by Serkowski, Mathewson, and Ford (1973); values of λ_{max} are presently known for about 350 stars. The distribution of λ_{max} on the sky is shown in Figure 7. Filled symbols represent the stars with λ_{max} less than the median value of 0.545 μ, open symbols those with $\lambda_{max} > 0.545\ \mu$. An obvious conclusion from inspecting this Figure is that there are some well defined regions on the sky within which λ_{max} is smaller than the median value and regions with λ_{max} larger than the median value. These two types of regions are surrounded in Figure 7 by dotted and solid lines, respectively. The largest and best defined region of low λ_{max} lies along the galactic equator at longitudes 115° to 150°, one of high λ_{max} at longitudes 295° to 350°. The largest values of λ_{max} are observed for these relatively nearby stars in Upper Scorpius and in Orion which are characterized by a large ratio R of total to selective interstellar extinction. In general, the nearby stars have considerably larger λ_{max} than the stars more distant than 0.4 kpc.

The linear size of regions with similar wavelength dependence of interstellar polarization, i.e. with similar average size of dust grains, is about 150 pc, which is approximately the size of large 'clouds' of interstellar medium derived by Scheffler (1967). In view of the correlation between λ_{max} and the ratio R of total to selective extinction

(Serkowski, 1968) Figure 7 may give an idea about the distribution of the ratio R over the sky. Studies of interstellar polarization may thus find a practical application for improving the astronomical distance scale.

References

A'Hearn, M. F.: 1972, *Astron. J.* 77, 302.
Appenzeller, I.: 1968, *Astrophys. J.* 151, 907.
Behr, A.: 1959, *Veröff. Univ. Sternw. Göttingen*, No. 126.
Bless, R. C. and Savage, B. D.: 1972, *Astrophys. J.* 171, 293.
Breger, M. and Dyck, H. M.: 1972, *Astrophys. J.* 175, 127.
Carrasco, L., Strom, K. M., and Strom, S. E.: 1973, *Astrophys. J.* 182 95.
Coyne, G. V., Gehrels, T., and Serkowski, K.: 1973, *Astron. J.*, in press.
Cugnon, P.: 1971, *Astron. Astrophys.* 12, 398.
Davies, R. D.: 1968, *Nature* 218, 435.
Davis, L., Jr. and Greenstein, J. L.: 1951, *Astrophys. J.* 114, 206.
Dyck, H. M.: 1973, in T. Gehrels (ed.), *Planets, Stars and Nebulae Studied by Photopolarimetry*, Univ. of Arizona Press.
Gehrels, T.: 1973, *Astron. J.*, in press.
Greenberg, J. M.: 1968, in B. M. Middlehurst and L. H. Aller (eds.), *Nebulae and Interstellar Matter*, Univ. of Chicago Press, 221.
Grigoryan, K. H.: 1970, *Soob. Byurakan Obs.* 41, 20 and 27.
Klare, G., Neckel, Th., and Schnur, G.: 1972, *Astron. Astrophys. Suppl.* 5, 239.
Kemp, J. C.: 1973, this volume, p. 181.
Kemp, J. C. and Wolstencroft, R. D.: 1972, *Astrophys. J. Letters* 176, L111.
Krzemiński, W. and Serkowski, K.: 1967, *Astrophys. J.* 147, 988.
Martin, P. G.: 1971, *Monthly Notices Roy. Astron. Soc.* 153, 279.
Mathewson, D. S.: 1968, *Astrophys. J. Letters* 153, L47.
Mathewson, D. S. and Ford, V. L.: 1970, *Mem. Roy. Astron. Soc.* 74, 139.
Purcell, E. M. and Spitzer, L., Jr.: 1971, *Astrophys. J.* 167, 31.
Scheffler, H.: 1967, *Z. Astrophys.* 65, 60.
Schmidt, Th.: 1968, *Z. Astrophys.* 68, 380.
Schmidt-Kaler, Th.: 1958, *Z. Astrophys.* 46, 145.
Serkowski, K.: 1962, *Advances Astron. Astrophys.* 1, 289.
Serkowski, K.: 1968, *Astrophys. J.* 154, 115.
Serkowski, K.: 1971, *Proc. IAU Colloq. No.* 15, *Veröff. Remeis-Sternwarte Bamberg*, Bd. 9, No. 100, p. 11.
Serkowski, K., Mathewson, D. S., and Ford, V. L.: 1973, *Astrophys. J.*, in press.
Verschuur, G. L.: 1970, in H. J. Habing (ed.), 'Interstellar Gas Dynamics', *IAU Symp.* 39, 150.
Wampler, E. J.: 1966, *Astrophys. J.* 144, 921.
Zellner, B. H. and Serkowski, K.: 1972, *Publ. Astron. Soc. Pacific* 84, 619.

DEVELOPMENT OF PHOTOGRAPHIC TECHNIQUES
FOR PHOTOMETRY AND POLARIMETRY OF FAINT
STARS IN THE $R-IR$ REGION

P. J. TREANOR, S. J.

Vatican Observatory, Vatican City State

Abstract. Photometric and Polarization measures of fainter stars ($m > 11$) are still very few, especially in the red to infra-red region. In order to exploit large Schmidt cameras to remedy this shortage, improved methods of photographic calibration are needed. This note reviews some of the problems associated with photographic calibration of direct images, and presents a progress report on the development of the ring polarimeter.

The study of the galactic interstellar medium at distances in excess of 1 kpc requires measures of magnitudes, reddening, extinction and polarisation of large numbers of stars fainter than the 10th photographic magnitude. Data of this kind are at present very scarce. It is evident, for example, from the review of polarisation measurements given by Serkowski (1973), that though the polarisation of several thousand stars is now known with considerable accuracy, nearly all these stars are brighter than $B = 10$. Moreover, measures at longer wavelengths, for example in the $R-IR$ photographic region are almost non-existent for fainter stars. The purpose of this note is to analyse the source of this imbalance, and to assess the possibilities of remedying it by the improvement of photographic methods of photometry and polarimetry.

1. Photographic Photometry

The advent of photographic techniques in astronomy at the end of the last century gave early promise of eliminating the large random and systematic errors of visual methods, and of extending photometric methods to fainter magnitudes. For example the early *Carte du Ciel* plates reached easily to $B = 12$ and frequently to much fainter limits. Modern large Schmidt telescopes obtain star images to $B = 19$ or fainter, so that one might hope for reasonably reliable photometry down to at least $B = 17$. However, even with the notable developments in stellar photometers, which ensure highly reproduceable plate measurements (with an internal consistency of three hundredths of a magnitude), the results of photographic photometry were very disappointing.

The non-linear and variable photometric characteristics of photographic plates, combined with the optical photometric errors of the astrographic refractors to create obstacles to accuracy which seemed almost insuperable.

After World War II the development of photomultipliers offered a solution to many of these difficulties. The linearity and the relatively low accidental errors of the photoelectric methods, combined with the possibility of immediate intercomparison of stars all over the sky, made possible the determination of accurate extinction correc-

Greenberg and Van de Hulst (eds.), Interstellar Dust and Related Topics, 153–160.

tions and the relation of the observed magnitudes to the network of stars which constitute the international UBV system. In polarimetry in particular, where the measures are largely differential, the development of multichannel polarimeters offered the possibility of measurements with an accuracy of the order of 0.001 mag. The development of photocathodes sensitive to wavelengths in the red and infra-red region made possible the extension of accurate wavelength dependence measurements to well beyond the limits of the photographic infra-red region. The break-through in accuracy thus achieved led to a concentration of research effort on this technique. The photoelectric method itself made an important contribution to improving the accuracy of photographic determinations, by providing numerous standard sequences which can be used to calibrate photographic plates and to determine field errors. The advantages of the photoelectric method were however so real that the last quarter of a century has seen a progressive lack of interest in photographic techniques of photometry, except for provisional extension of photoelectric studies to fainter magnitudes.

The photoelectric method itself is not without serious limitations. The method is relatively slow. Up to the end of 1966, UBV measures had been published for some 20000 stars, that is, less than one star per two sq deg, and of these the vast majority are brighter than $B = 10$. As a result it is very rare that an adequate photoelectric sequence, even to $B = 12$ is already available for the calibration of photographic plates. Outside the UBV wavelength range, the situation is very much worse. Photomultipliers sensitive to the longer wavelengths have considerably inferior magnitude limits, and have been much less generally used in extensive programmes. This situation does not seem likely to improve radically in the forseeable future. For the study of faint stars and in particular of the usually highly reddened stars at great distances near the galactic plane, and for intrinsically red stars, in statistically useful numbers for galactic survey work, the photoeletric method alone does not seem to offer much hope of providing a solution

It seem therefore appropriate to reconsider the possibilities of photographic methods in this context. In fact, much has happened in the field of photographic techniques since the advent of photoelectric photometry. Automatic measuring machinery, coupled with electronic data-processing methods, now makes it possible to measure up to 900 stellar images per hour. New emulsions, such as the Kodak 3aJ, and other emulsions of this family now under development for longer wavelengths, in virtue of their information storage capacity, reach fainter magnitude limits and promise higher accuracy than the emulsions hitherto available. Large Schmidt telescopes, in addition to their wide field coverage and faint limiting magnitudes, can be used over the whole of the photographic wavelength range. They are also, in principle, free from the troublesome systematic distance corrections, sometimes amounting to several tenths of a magnitude, typical of triplet and quadruplet photographic refractors, which also had the limitation of a narrow useful wavelength range. Schmidt telescopes do however have special difficulties, associated with the small size and low growth rate of their stellar images. This results, first, in an increase in the accidental error in magnitude of the photometer measures of an individual image, and secondly in the difficulty of obtaining an exactly equal focus over the entire plate, which gives rise to a quasi-random distance correction. Methods of alleviating or even removing image-deformation effects have been suggested and tested. Some of these, such as the schraffier cassette and the Fabry lens method achieve this aim, but only at the sacrifice of magnitude limit and convenience of measurement. Others, such as the deliberate introduction of focussing or aberration errors, introduce systematic errors of their own.

A method at present under experiment by the writer aims at avoiding both these difficulties. It is based on an observational technique which is related to the ring-polarimeter technique. A plane parallel glass plate, slightly inclined to the optical axis of the telescope is placed just in front of the photographic plate. In contra-distinction to the ring polarimeter, the angle of the plate is kept so small that the ring images produced have a diameter only two or three times that of the star images themselves. The available light is thus distributed in a symmetrical way over many more grains than in the direct images. Early saturation effects are thus avoided, and the ring image so obtained can be measured and reduced by normal iris diaphragm photometry. There is a small loss of limiting magnitude, but much less than in the schraffier cassette and Fabry lens methods.

The great shortage or absence of faint photoelectric standards, particularly outside the *UBV* region, create a special problem. The limitations of the objective-grating calibration method (Reddish, 1968) are well known. Recently a crossed calcite filter (Bruck *et al.*, 1969) has been used for calibration the random and systematic errors of this system (Pratt, 1968) are of the order of 0.1 mag. Other new methods have been suggested. Reddish (1968) has proposed the use of a mosaic objective prism which should have notable advantages over the objective grating. This would be a very costly expedient. The author (Treanor, 1969) has experimented with a split rotating conical

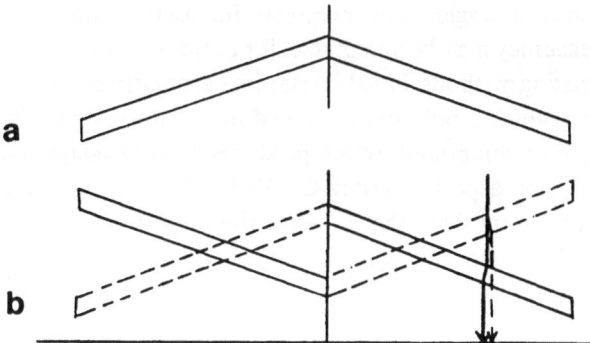

Fig. 1. *Double image formation by means of a split conical lens of zero power.* The circular conical lens shown at a is split through a diameter passing through the cone axis, and one semicircular section is reversed as shown in b. The pair are rotated during exposure about their common axis. Rays passing through the section P are always diverted towards the axis by a constant amount. Rays passing through Q are equally diverted in the opposite direction. The cone angle has been much exaggerated in the figure.

lens of zero power (see Figure 1). One half of the split lens is inverted and covered by a neutral filter. The system produces double images with a known magnitude difference. However it is difficult to avoid image distortion with a system of this kind. The safest method seems still to be the half-filter method, for example in the 'reference' plate version described by Stock and Williams (1960), which is designed to eliminate most of the main systematic errors of successive exposures.

It should be emphasised that strict observance of photometric norms of exposure and processing are essential for this type of work. Apart from more obvious requirements, such as the exclusive use of photometrically good nights, and accurate control of the solutions, temperatures, and mode of plate agitation during processing, one should avoid working within 2 cm of 'original edges' of plates.

It is also advisable, with Schmidt telescopes, to restrict the measured field considerably, say to four sq deg, if greatest accuracy is sought. This applies perhaps even more strongly to lens astrographs, where one should work as far as possible reasonably near to the centre of symmetry of the distance correction which may or may not be the physical centre of the plate. (For the Vatican 4-lens Zeiss astrograph it is actually several cm distant from the plate centre.) Even in the absence of optical errors producing distance corrections, variations in the photographic emulsion lead to increasing errors the further one departs from the region of the calibrating sequence.

2. Photographic Polarimetry

The problems of photographic polarimetry are rather different from those of photographic photometry. Since one is here concerned with highly differential measures, the errors of the photographic process as such can be rendered small by a suitable choice of method. The main problem arises from the greater accuracy required for useful polarimetry. Even for the rather well polarised stars, to which photographic methods are applicable, on accuracy approaching 0.01 mag. in the amount of polarisation and 10° in position angle seems necessary for useful statistical work, although somewhat lower accuracy may be acceptable, for example in the search for very highly polarised stars. Existing methods, based on star image polarimetry use either successive exposures, between which a polaroid is rotated by a known angle (Pratt, 1968), or a calcite plate, simple or compound, which produces pairs of images corresponding to orthogonal polarisation aspects (Serkowski, 1960). The ring polarimeter method of the author is distinct from these that it uses a ringshaped image, the intensity of the circumference of which is modulated continuously by polarisation components at all angles (Treanor, 1968).

3. Errors in Calcite Plate Polarimetry

In this method a single plate gives pairs of images polarised at 90°. The measured intensity-difference between these image, expressed as a fraction of the unpolarised stellar magnitude, is related to the polarisation parameters by an equation of the form

$$f_1 = 2p_0(\theta_1 - \phi), \tag{1}$$

where p_0 is the fractional polarisation, ϕ the position angle of the polarisation, and θ is the position angle of the calcite axis plus an instrumental constant (Treanor, 1960). We need at least two plates with different values of θ_1 to determine p_0 and ϕ. Let us suppose the second observation is made (as is common practice) after rotating the analyser by 45°. We thus obtain

$$f_2 = 2 p_0 \sin 2(\theta_1 - \phi). \tag{2}$$

From (1) and (2) we obtain easily

$$p_0^2 = (f_1^2 + f_2^2)/8 \tag{3}$$

$$\tan 2(\theta_1 - \phi) = f_1/f_2. \tag{4}$$

Introducing the relation between polarisation in magnitudes and percent polarisation

$$p = 2.1717(p_0 + \frac{p_0^2}{3} + ...) \tag{5}$$

we obtain

$$p^2 = 0.54(f_1^2 + f_2^2). \tag{6}$$

A mean magnitude error in a single iris reading corresponding to 0.03 mag. will produce an error in f of about 0.0425. The related error in f^2 is about 0.085, or $0.085\sqrt{2}$ in the sum, considered statistically. The corresponding error in p is therefore 0.033.

In practice the rotation of the analyser will be continued at intervals of 45° through 360°. These observations will reproduce three more pairs of equations, formally identical with (1) and (2), the observed values of f reflecting the effect of random and systematic errors. At best these additional measures can reduce the errors to 0.016. Some twenty plates (40 iris readings) will be needed to reach an accuracy of 0.01 in p. Similar considerations can be applied to the errors in ϕ determined from Equation (4).

Accidental errors of this kind apply also to all point image methods involving successive exposures with single images. In this case, the differences of sky transparency and the effect of successive exposures on the same plate will produce systematic errors which have to be determined by some form of comparison field. Systematic errors due to the deformation of the images in calcite plates also need careful study. These are not wholly eliminated in compound plates (Brand, 1971).

The ring polarimeter method goes some way, in principle, to diminishing these errors, though at considerable sacrifice of limiting magnitude. This last disadvantage is not too serious, since with a 38″ Schmidt, the method is still applicable to moderately reddened stars of $B = 13$. I note here some recent developments of this method, still in course of test, which aim to supply a deficiency in its original formulation. This development allows measurements of the amount of polarisation as well as of position angle to be made with rapidity from a microphtometric analysis of the rings. As will

be recalled, the method as originally proposed, consisted in rotating a polaroid combined with a parallel-sided glass plate, slightly inclined to the axis of the telescope, just
in front of the photographic plate. The star image was thus converted into a small ring,
whose circumference has a varying photographic blackening. When this ring is rotated
in front of a microphotometer slit, an approximately sinusoidal waveform is obtained,
the phase of which is readily related to the position angle of polarisation, and the
amplitude depends on the amount of polarisation. The problem not conveniently
solved in the original exposition of the method was the interpretation of the microphotometer record in terms of the amount of polarisation. This has been solved as
follows. One sector of the rotating polaroid and plate is covered with a filter of known
transmission. The effect of this is to subtract from one sector of the star ring an intensity equivalent to a known magnitude step (say 0.2 mag.). A corresponding step-
displacements occurs in the microphotometer record (see Figure 2). On the reasonable
assumption of near linearity of the photographic characteristic curve over this small
interval, the amount of polarisation can easily be calculated from the amplitude of
the microphotometer record.

In order to test this method, it was found advisable to construct a specialised micro-

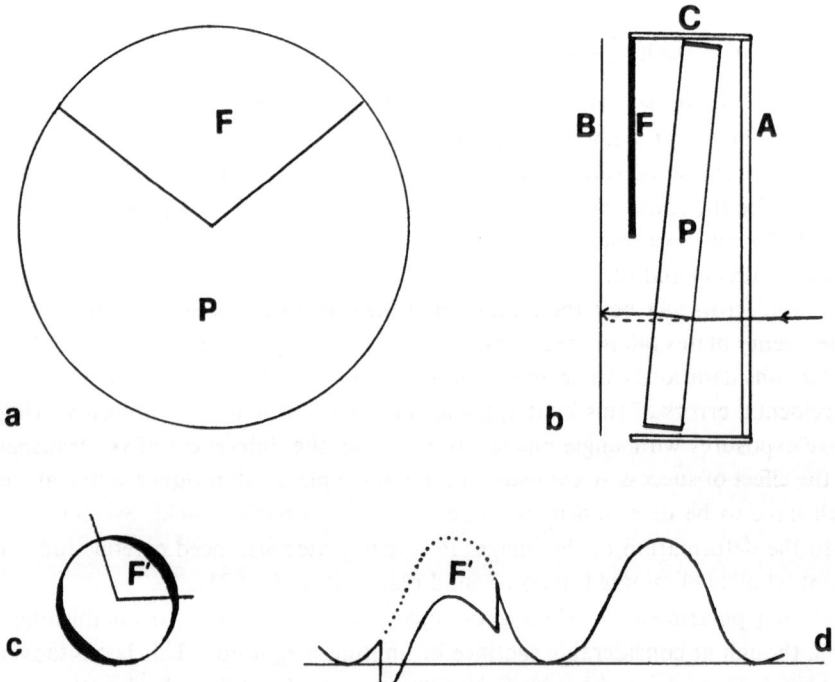

Fig. 2. *Calibration of ring polarimeter.* (a) Filter section F on inclined plate P. (b) Sectional view
of polarimeter, showing polaroid analyser A, inclined plate P, Filter section F, rotating cell C and
photographic plate B. (c) Idealised ring of highly polarised star showing the section F′ affected by the
sector filter. (d) Idealised microphotometer tracing of ring circumference, showing the region F′
depressed by the sector filter of known transmission.

photometer, the essential feature of which is that is that the optics are so designed that the rotation of the slit does not of itself produce any systematic effects on the illumination of the microphotometer photomultiplier. In this way it is possible to leave the plate stationary and to rotate the slit synchronously, rather than to rotate the plate, as was done in my earlier experiments. This is achieved by imaging the source of illumination of the plate on an end-window photomultiplier by means of a Fabry lens placed as nearly as possible in the plane of the rotating slit. This instrument is described in detail elsewhere (Treanor, 1973).

In the first tests, the calibrating step corresponded to about 0.08 mag. and took the form of a double 30° sector on opposite sides of the polaroid. Under these conditions, it was possible to allow the slit to cross the ring completely, so that the step appears on each repetition of the wave form. It is clearly identifiable on the registration (Figure 3). However, two defects are clearly present. First, the step is rather too small to define the scale well in relation to the noise level. Secondly, adjacency influence the height of the peak of the step, and also disturb the wave-form of the curve as a whole. By using a single sector, extending over 90° or more, and scanning with a slit terminating at the ring centre, and using a somewhat denser filter one may hope to eliminate the flaws of this initial experiment.

While it is still too early to predict the final accuracy obtainable by this method, one notes that the main source of random error is due to the rather prominent noise level due to grain emulsion, in this case 103aO. This can certainly be considerably reduced by using emulsions such as 3aJ. Other possible sources of error lie in the non-uniformity of the polaroid plate. This has recently been changed to accomodate a

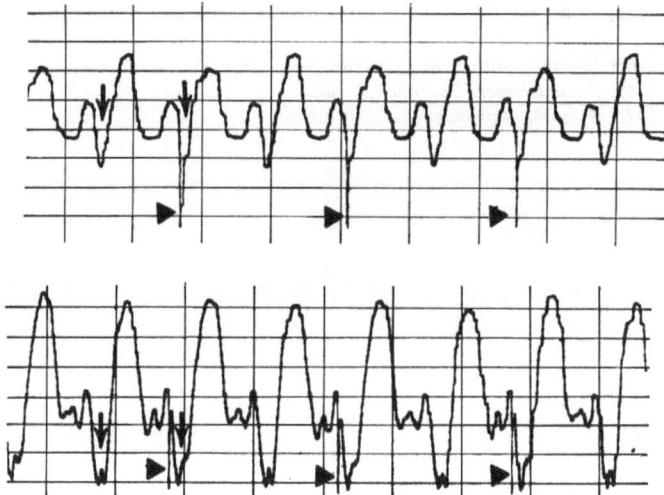

Fig. 3. *Registration of 2 highly polarised stars, showing calibration step.* The step corresponding to an intensity difference of 8 % was obtained using a double sectional filter of 30°, so that the step appears in each cycle of the registration, the single filter, which leaves alternate cycles unaltered, is preferable (see text). The filter step is indicated by a vertical arrow. The pulses which occur in alternate cycles, and which are indicated by the triangular spots are phase references signals.

much larger field (about 8 sq deg). Now that a reference standard is available by means of the sector, these errors can be investigated quantitatively using a fine grained emulsion, in a field of stars with low polarisation. Since earlier experiments, even with 103aE, indicated that polarisations of 0.03 mag. were detectable easily from measures on four plates, the prospect of an ultimate accuracy of 0.01 mag. on finer grained plates does not seem to be excluded.

It must be acknowledged in conclusion that photographic methods remain difficult and of relatively low accuracy. The purpose of this note is to highlight the fact that for the photometric study of most of the Galaxy they are at present almost the only way of achieving adequate coverage. The difficulties of the past should not prejudice unduly our present assessment of their possibilities. With improved emulsions, and eventually with the development of practicable electronic area-scanning techniques capable of effectively replacing the photographic plates, the picture may change radically, and techniques which now struggle with the limitations of conventional photography may come into their own.

References

Brand, P. W. J. L.: 1971, *Roy. Obs. Edinburgh* **7**, 35.
Bruck, M. T., Nandy, K., Caprioli, G., and Smeriglio, F.: 1969, *Astrophys. Space Sci.* **4**, 313.
Pratt, N. M.: 1968, *Publ. Roy. Obs. Edinburgh* **6**, No. 39.
Reddish, V. C.: 1968, *Observatory* **88**, 146.
Serkowski, K.: 1960, *Acta Astronomica* **10**, 227.
Serkowski, K.: 1973, this volume, p. 145.
Stock, J. and Williams, A. D.: 1960, in W. A. Hillner (ed.), *Astronomical Techniques*, University of Chicago Press, p. 400.
Treanor, P. J.: 1960, in W. A. Hiltner (ed.), *Astronomical Techniques*, University of Chicago Press, p. 249.
Treanor, P. J.: 1968, *Monthly Notices Roy. Astron. Soc.* **138**, 325.
Treanor, P. J.: 1969, *Annual Report Specola Vaticana*.
Treanor, P. J.: 1973, *Ricerche Astronomiche* **8**, in press.

INTERSTELLAR CIRCULAR POLARIZATION AND THE COMPOSITION OF INTERSTELLAR DUST

P. G. MARTIN

Institute of Theoretical Astronomy, Madingley Road, Cambridge, England*

Abstract. This paper shows that optical observations of circular polarization produced by aligned interstellar grains could yield valuable information about the grain material. The interstellar medium is known to be linearly dichroic from observations of interstellar linear polarization; many different grain models using a large variety of compositions can be found to reproduce these observations. Since the same aligned grains make the medium linearly birefringent, a small component of circular polarization can result from incident linearly polarized light if the position angle of the linear polarization does not coincide with either principal axis of the medium. Here calculations are presented to demonstrate that the wavelength of the circular polarization is sensitive to the imaginary part of the complex refractive index of the grain material. This provides an opportunity of investigating whether the grains are characteristically dielectric or metallic. Some possible observations are suggested.

1. Introduction

It is proposed that observations of circular (elliptical**) polarization from aligned interstellar grains could be useful in placing restrictions on the type of grain material involved. At the present many grain models based on a large variety of dielectric through metallic composition can be found to produce a reasonable fit to the observed wavelength dependence of extinction and linear polarization in the optical region of the spectrum. However over this same range of models widely different circular polarization is predicted; thus circular polarization measurements offer a new method of resolving the longstanding ambiguity in the interpretation of existing observations. This investigation was prompted by a statement near the end of van de Hulst's monograph (1957) forseeing interstellar circular polarization; since then this interesting possibility has not received the attention it probably deserves.

In Section 2 the phenomena of dispersion and extinction in a medium containing interstellar grains and their relation to the Stokes parameters describing circular and linear polarization are reviewed. Next the wavelenght dependence of circular polarization for different materials is investigated using the Mie theory for circular cylinders. Finally several observations are suggested.

2. Dispersion, Entinction and Polarization

The action of interstellar grains on radiation passing through interstellar space can be shown to be equivalent to that of a medium with complex refractive index

---

* Present address: Dept. of Astronomy, University of Toronto, Toronto, Ontaria, Canada.
** The term circular polarization is used to denote the circular component (Stokes parameter V) of what is generally elliptical polarization.

Greenberg and Van de Hulst (eds.), Interstellar Dust and Related Topics, 161–167.

$$\tilde{m} = \tilde{n} - i\tilde{k} = 1 - i(2\pi)^{-2}\lambda^3 NS(0), \tag{1}$$

where N is the number density of grains, λ is the wavelength of the light and $S(0)$ is the complex amplitude function of the radiation scattered in the forward direction (van de Hulst, 1957); $\tilde{m}$ is close to once because N is so small. From this formal equivalence the amount of extinction and dispersion can be seen immediately. The linear extinction coefficient γ depends on the imaginary part of $\tilde{m}$;

$$\gamma = 4\pi\lambda^{-1}\tilde{k} = \pi^{-1}\lambda^2 N \operatorname{Re}[S(0)] \equiv NC_e, \tag{2}$$

where C_e is the extinction cross-section per grain. The presence of grains also produces a phase lag (or advance if ε is negative)

$$\varepsilon = 2\pi\lambda^{-1}s(\tilde{n}-1) = (2\pi)^{-1}\lambda^2 Ns \operatorname{Im}[S(0)] \equiv \tfrac{1}{2}NsC_p, \tag{3}$$

where s is the pathlength in the medium and C_p has been defined by analogy to C_e.

Both linear and circular polarization depend on interstellar grains of anisotropic shape being aligned so that the averaged profile along the line of sight is non-circular. Observations of linear polarization are usually taken as evidence that such alignment exists, although what alignment mechanism is actually operating is not clearly understood. Except in the direction of the alignment axis of symmetry this medium has different complex refractive indices for orthogonal orientations of the electric vector of the radiation; it is convenient to choose two perpendicular directions, 1 and 2 say, along the long and short principal axes of the grain profile respectively. The symmetry is such that linear dichroism ($C_{e1} \neq C_{e2}$) and linear birefringence ($C_{p1} \neq C_{p2}$) can occur, while their circular counterparts cannot. Linear birefringence requires a non-zero imaginary part of the scattering amplitude function (a non-zero phase change). Note that linear birefringence does not necessarily imply linear dichroism (e.g. non-absorbing particles small compared to the wavelength) or vice-versa (Figure 1). In addition to alignment of the grains and the resulting linear birefringence the incident radiation must be linearly polarized at a position angle other than $0°$ or $90°$ relative to the principal axes 1 and 2 for the birefringence to produce the desired circular polarization.

Changes in polarization can be described conveniently in terms of the four Stokes parameters I, Q, U, V. Here we adopt the convention that for an observer looking towards the source of radiation position angles are measured counterclockwise from axis 2 in the plane perpendicular to the direction of propagation. Thus, for example, $V > 0$ corresponds to right-hand circular polarization in which the position angle of the electric vector increases with time (counter-clockwise rotation).

Two useful expressions can be written down from the amplitude-phase definitions of Q and V:

$$dQ = E_2^2 - E_1^2 \simeq \tfrac{1}{2}I[(1-\gamma_2\,ds)-(1-\gamma_1\,ds)] = \tfrac{1}{2}IN\,ds(C_{e1}-C_{e2}) \tag{4}$$

$$dV = 2E_2E_1\sin(\varepsilon_2-\varepsilon_1) \simeq U(\varepsilon_1-\varepsilon_2) = \tfrac{1}{2}N\,ds(C_{p1}-C_{p2})U. \tag{5}$$

These equations emphasize the following important points: that within the above

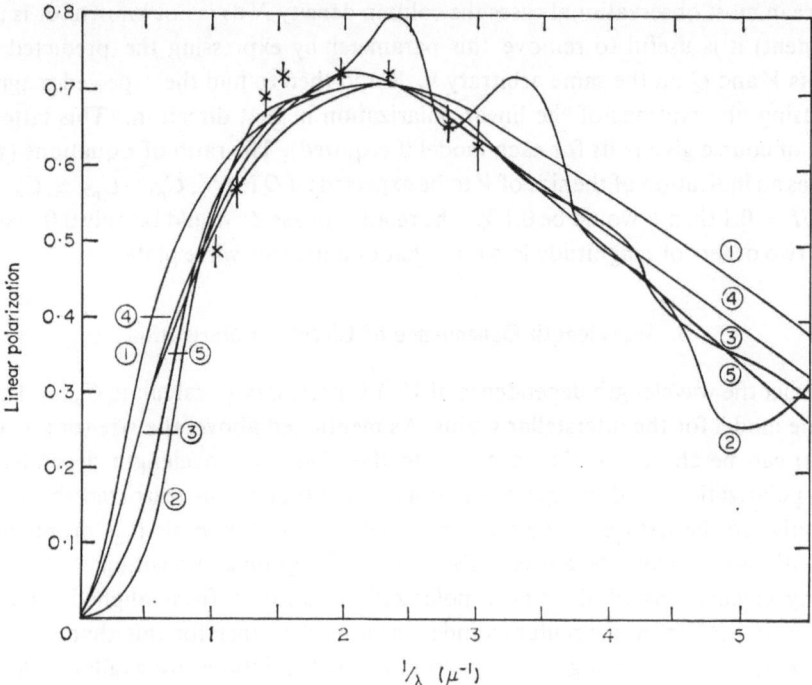

Fig. 1. Wavelength dependence of linear polarization for different complex refractive indices ('picket fence' alignment for long circular cylinders). For each curve the vertical scale and the grain size were chosen to obtain a good approximation to the shape of the mean observed linear polarization (crosses). The vertical scale is arbitrary. **1** $m = 1.1$; **2** $m = 1.5$; **3** $m = 1.5-0.1$; **4** $m = 1.5-0.5\,i$; **5** $m = 1.5-1.5\,i$.

convention interstellar grains produce the Q component of linear polarization, and that a component V results only when the incident radiation has a U component of linear polarization and the medium is linearly birefringent. The latter two requirements must be considered in choosing good observational prospects.

A more general treatment along the lines developed by van de Hulst (1957) and Serkowski (1962) shows that

$$I^{-1}\frac{dI}{ds} = -\tfrac{1}{2}N(C_{e1}-C_{e2}) - \tfrac{1}{2}N(C_{e1}-C_{e2})(Q/I) \tag{6}$$

$$\frac{d(Q/I)}{ds} = \tfrac{1}{2}N(C_{e1}-C_{e2}) - \tfrac{1}{2}N(C_{e1}-C_{e2})(Q/I)^2 \tag{7}$$

$$\frac{d(U/I)}{ds} = \tfrac{1}{2}N(C_{p1}-C_{p2})(V/I) - \tfrac{1}{2}N(C_{e1}-C_{e2})(Q/I)(U/I) \tag{8}$$

$$\frac{d(V/I)}{ds} = \tfrac{1}{2}N(C_{p1}-C_{p2})(U/I) + \tfrac{1}{2}N(C_{e1}-C_{e2})(Q/I)(V/I). \tag{9}$$

Usually only the first term on each right-hand side will be important.

Since in most observational cases the column density $N\,ds$ is not known (or is model dependent) it is useful to remove this parameter by expressing the predicted components V and Q on the same arbitrary scale and then to find the expected magnitude of V using observations of the linear polarization in that direction. (This latter step would of course give $N\,ds$ for each model if erquired). The ratio of Equations (4) and (5) gives an indication of the size of V to be expected; if Q is 1%, $C_{p1} - C_{p2} \simeq C_{e1} - C_{e2}$ and $U/I = 0.1$ then V would be 0.1%. The relative phase $\Delta\varepsilon$ would be only 0.01 radians, about two orders of magnitude less than that of a quarter-wave plate.

3. Wavelength Dependence of Circular Polarization

To predict the wavelength dependence of V, it is necessary to calculate C_p on the basis of some model for the interstellar grains. As mentioned above, the size (or size distribution) can be chosen so that a match to the observed wavelength dependence of linear polarization (and extinction) is obtained. Models thus restricted show that V is sensitive to the nature of the grain material, and on this basis it is proposed that observations of V could be a useful discriminant for grain composition.

Many calculations of the linear polarization expected from aligned interstellar grains have used infinite circular cylinders as a model, since for this shape Mie solutions exist, whereas for a general spheroid no such solutions are available. Actually the Mie solutions are valid for finite cylinders of any radius as long as the conditions $l \gg a$ and $l \gg \lambda$ are satisfied (van de Hulst, 1957). Microwave analogue experiments with spheroids and cylinders (Greenberg, 1961; Greenberg et al, 1968) demonstrate qualitative agreement even when $l \simeq 2a$ (all wavelengths).

The Mie theory for infinite cylinders, including the case of oblique incidence, gives the cross-sections as series expansions (Lind and Greenberg, 1966)

$$Q^E = \frac{C^E}{2a} = \frac{2}{x}\left(b_0^E + 2\sum_{n=1}^{\infty} b_n^E\right) \tag{10}$$

$$Q^H = \frac{C^H}{2a} = \frac{2}{x}\left(a_0^H + 2\sum_{n=1}^{\infty} a_n^H\right), \tag{11}$$

where $x = 2\pi a/\lambda$, $C = C_e + iC_p$ and the superscripts E and H identify the two polarizations with electric vectors respectively in and perpendicular to the plane formed by the cylinder axis and the direction of propagation. Both the real and imaginary parts of the complex coefficients have to be calculated when Q_s, the efficiency factor for scattering, is determined so that usually no new coefficients would have to be calculated for either Q_e or Q_p; only a regrouping of terms is necessary.

A computer program kindly supplied by N. C. Wickramasinghe was appropriately modified to include computations of Q_p. In Figure 1 some results are shown for models chosen to match the mean observed linear polarization curve tabulated by Coyne and Gehrels (1967). No attempt is made at this point to use a size distribution or oblique incidence since we are interested in isolating the effects of changing the complex refractive index m of the grain material. Nevertheless even for this simplified calculation for a single grain size and 'picket fence' alignment the agreement with observation is quite reasonable for each complex refractive index. (Averaging over size will smear out the oscillations in the curve(s) for $m = 1.5$.)

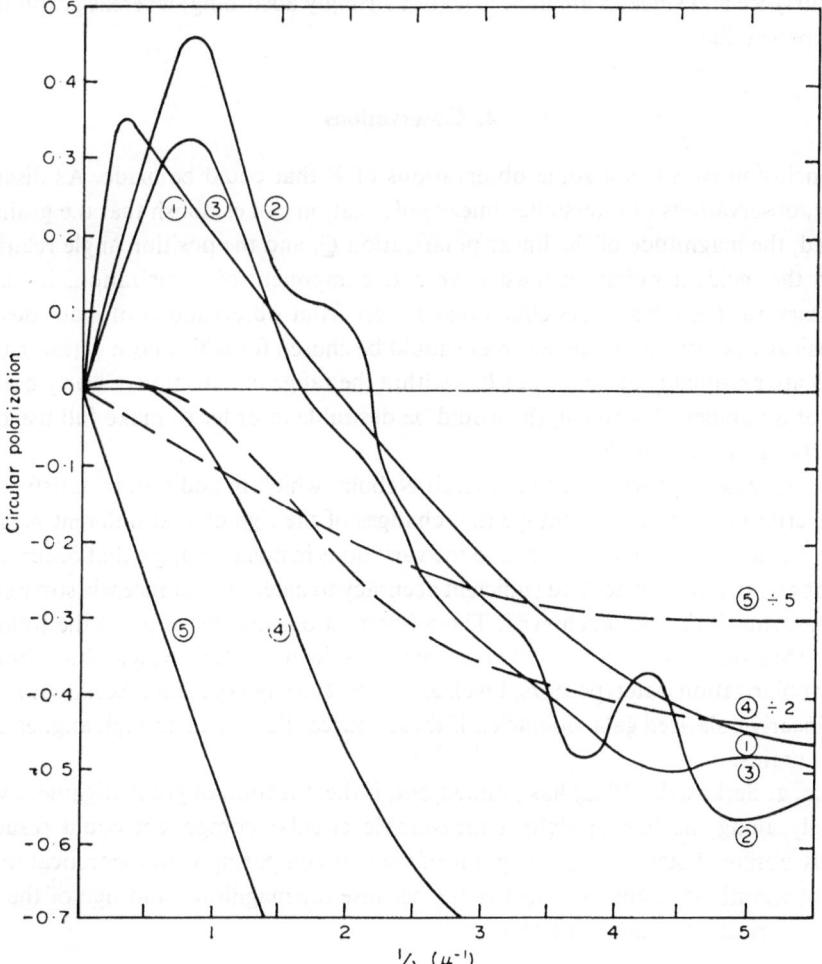

Fig. 2.   Wavelength dependence of circular polarization (for constant intrinsic linear polarization) for the grains producing the linear polarization in Figure 1.

What is important is that these different grains which produce similar linear polarization curves in Figure 1 do not have the same circular polarization, as is clear from Figure 2. Evidently changing n has little influence on the wavelength dependence of V, but it can be seen that variations in k have a large effect. Curves for the two extremes, $k = 0$ and $k \simeq n$, reveal characteristic differences. The former, for dielectric materials, have a relative phase lag a long wavelengths which reverses to an advance in the optical region. On the other hand what we shall call 'metallic' materials produce a relative phase advance which increases towards shorter wavelengths. Some insight into these different behaviours can be obtained through use of approximations to the Mie theory (Martin, 1972).

Several different materials have been suggested for the interstellar grains. Of these, ices, dirty ice and silicates are dielectric (not strongly absorbing) whereas graphite and iron are metallic.

4. Observations

In conclusion we suggest some observations of V that could be made. As discussed above, observations of interstellar linear polarization can establish that the grains are aligned, the magnitude of the linear polarization Q, and the position angle relative to which the incident radiation must have a U component of polarization. It may be necessary to infer this interstellar polarization from observations of stars near the intrinsically polarized source. Sources should be chosen for which both Q (interstellar) and U are reasonably large so that V is within the range of detection. Clearly observations at a number of wavelengths would be desirable in order to make full use of the predictions of the models.

As a specific example we cite the Crab Nebula, which in addition to satisfying the above criteria offers the advantage that changes of the sign of V at different positions in the Nebula can be predicted, due to the variations in position angle that occur. In the past observations have not had sufficient accuracy to detect V, but recently some succes in measuring V has been achieved. These observations are discussed in the following paper (Martin, et al., 1972). Other possibilities include stars which show intrinsic linear polarization (latetype stars, Dyck et al., 1971; early-type stars, Serkowski, 1970) and linearly polarized galactic nuclei, if these sources lie behind enough aligned interstellar grains.

Also, as Serkowski (1962) has pointed out, if the direction of grain alignment varies strongly along the line of sight a measurable circular component could result for heavily obscured stars. From the point of view of comparing with theoretical models such observations might be of less value, because the magnitude and sign of the 'incident' polarization U are not known.*

Acknowledgements

This work was supported through a Commonwealth Scholarship, by the National Science Foundation under grant GP 31356X and by the National Aeronautics and Space Administration under Grant NGR 33-008-102. I thank Prof. R. Novick for his hospitality during my stay at Columbia University. This paper is a brief version of work previously published (Martin, 1972).

* Kemp (1972, 1973) has reported detection of this effect. A detailed discussion has been given by Martin (1972).

References

Coyne, G. and Gehrels, T.: 1967, *Astron. J.* **72**, 892.
Dyck, H. M., Forrest, W. J., Gillett, F. C., Stein, W. A., Gehrz, R. D., and Woolf, N. J.: 1971, *Astrophys. J.* **165**, 57.
Greenberg, J. M.: 1968, *Stars and Stellar Systems* **7**, 221.
Greenberg, J. M., Pedersen, N. E., and Pedersen, J. C.: 1961, *J. Appl. Phys.* **32**, 233.
Kemp, J. C.: 1972, *Astrophys J. Letters* **175**, L35.
Kemp, J. C.: 1973, this volume, p. 181.
Lind, A. C. and Greenberg, J. M.: 1966, *J. Appl. Phys.* **37**, 3195.
Martin, P. G.: 1972, *Monthly Notices Rox Astron. Soc.* **159**, 179.
Martin, P. G.: 1972, Paper presented at *IAU Colloq.* **23**, Nov. 1972, Tucson, Ariz.
Martin, P. G., Illing, R., and Angel, J. R. P.: 1972, *Monthly Notices Roy. Astron. Soc.* **159**, 191.
Serkowski, K.: 1962, *Adv. Astron. Astrophys.* **1**, 290.
Serkowski, K.: 1970, *Astrophys. J.* **160**, 1083.
Van de Hulst, H. C.: 1957, *Light Scattering by Small Particles*, Wiley, New York.

DISCOVERY OF INTERSTELLAR CIRCULAR POLARIZATION
IN THE DIRECTION OF THE CRAB NEBULA

P. G. MARTIN*

Institute of Theoretical Astronomy, Madingley Road, Cambridge, England

and

R. ILLING and J. R. P. ANGEL

Columbia Astrophysics Laboratory, Columbia University, New York, N.Y. 10027, U.S.A.

Abstract. A search in many small regions of the Crab Nebula has resulted in the detection of a small component of circular polarization. The variation of the sign and magnitude with position in the Nebula indicates that the polarization is of interstellar origin. On the basis of the polarity, strength, and colour dependence, it is concluded that the composition of the aligned grains causing this polarization is dielectric. Metallic particles are clearly ruled out.

1. Introduction

This paper reports the detection of circular polarization in the Crab Nebula, interpreted as due to interstellar grains. Since the theoretical discussion which motivated these observations appears in the preceding paper (Martin, 1972, Paper I) it will be summarized only briefly here. The presence of aligned anisotropic (elongated) grains makes the interstellar medium not only linearly dichroic, as observations of the interstellar linear polarization indicate, but also linearly birefringent. Therefore, when linearly polarized radiation passes through the medium, a component of circular polarization is produced if the position angle of this incident polarization does not coincide with either principal axis of the medium. The resulting maximum ellipticity (ratio of ellipse axes) is roughly equal to the fraction of linear polarization P caused by the interstellar medium. Circular polarization from interstellar grains is of interest because, as shown in Paper I, the magnitude and sign vary with wavelength in a manner characteristic of the grain material, offering an opportunity to distinguish clearly between dielectric and metallic grains. Restrictions placed on the composition of the grains by our observations are discussed below.

An almost ideal source for the study of interstellar birefringence is the Crab Nebula. The strongly polarized synchrotron continuum is seen through a foreground interstellar medium whose polarization properties are known independently from stellar observations. The range of position angles of linear polarization in the Nebula allows a check of the angular dependence of the circular component, for it can be assumed that the optical properties of the interstellar medium are virtually constant over the small angular size of the Nebula. The variation of circular polarization across the Nebula is important in establishing that the observed effect is really interstellar. Rees

* Present address: Dept. of Astronomy, University of Toronto, Toronto, Ontario, Canada.

(1971) has predicted that the radiation will be intrinsically circularly polarized if the nebular magnetic field is 30 – Hz electromagnetic radiation rather than a DC field; however, this effect can be distinguished by the predicted variation with position, which is quite different for the two mechanisms.

2. Observations of the Crab Nebula

2.1. EARLY OBSERVATIONS

In the past, observations of the Crab Nebula have failed to show circular polarization above the limits of their observational accuracy. Using a photographic method, Oetken (1966a) found values around 2%, but these were questioned after new photoelectric measurements were taken (Oetken, 1966b). In this latter paper, it is reported that Serkowski also had found no significant polarization using a 96″ diaphragm, the Stokes parameter V being $0.22 \pm 0.5\%$. Although not looking for this particular effect, Wolstencroft (1966) obtained some limits with a 40″ diaphragm, $V = 0.2 \pm 0.65\%$ at visual wavelengths and $0.15 \pm 0.5\%$ in the red. The only measurements of sufficient accuracy to detect this effect were made by Landstreet and Angel (1971) when investigating the predictions of the 'synchro-Compton' emission process (Rees, 1971). For two 7-s spots in the Nebula in the wavelength region 4000–5800 Å, they found $V = -0.024 \pm 0.040$ and $-0.034 \pm 0.047\%$. These latter results are discussed below along with the new measurements.

2.2. INTERSTELLAR AND INTRINSIC LINEAR POLARIZATION

The interstellar linear polarization can be estimated by studying stars in the vicinity of the Nebula. However, there are only three stars in the catalogues of Hiltner (1956) and Hall (1958) within a few degrees and at nearly the same distance. The star HD 36879 shows 2% linear polarization at position angle 152°, HD 36547, 1.4% at 149°, and HDE 245310, 3.9% at 146°. To supplement these data, Cocke et al. (1970) have measured the polarization of 10 stars fainter than twelfth magnitude within a few minutes of arc of the Nebula. However, no distances are known. If these stars are main sequence stars as far away as the Crab, they would have spectral types about B_3 and later. A mean polarization $2.0 \pm 0.2\%$ at $147° \pm 3°$ is given, but there was a larger scatter of about 10° in the position angles (Gehrels, private communication).

Observations of the Crab Nebula pulsar NP0532 could possibly give another determination of the interstellar polarization. It is found that near the intensity peaks of both the main and secondary pulses, the linear polarization decays to nearly zero. Kristian et al. (1970) suggest that the residual of about 2% at 160° is largely of interstellar origin. Observations of the secondary pulse by Cocke et al. (1970) are consistent with this interpretation, but there is a disagreement of about 40° in the position angle at the polarization minimum of the main pulse.

Maps of the linear polarization over the face of the Nebula have been made by Walraven (1957), Woltjer (1957), and Hiltner (1957). No new measurements of the linear polarization were made during this work, and we have adopted Woltjer's measurements for the wavelength range 5200–6400 Å (Plate I). For the regions we have observed, the estimated uncertainty in the polarization is 2–3%, and 4° to 6° in position angle. Removal of the small component of interstellar polarization gives the intrinsic linear polarization used in our analysis.

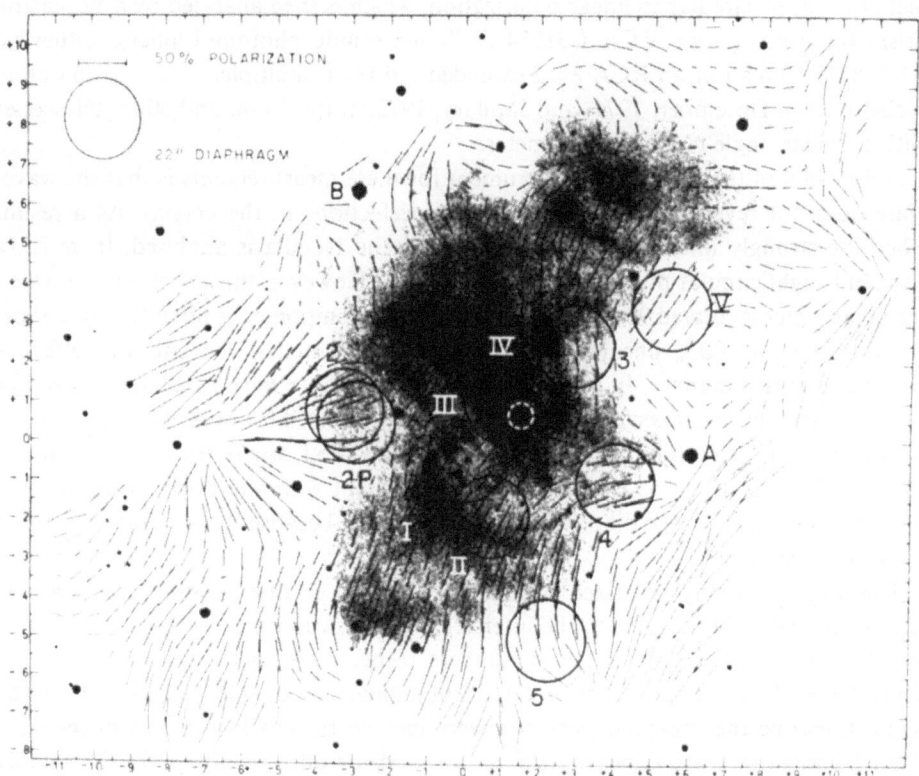

Plate I. A map of linear polarization over the face of the Crab Nebula (from Woltjer, 1957), showing the positions chosen to search for circular polarization. Dashed regions were examined previously by Landstreet and Angel (1971).

It is assumed on the bais of synchrotron emission that the intrinsic polarization is the same at all wavelengths. No Faraday rotation occurs at optical frequencies (Conway, 1971). However, there exists the possibility that in regions with strong filaments the percentage polarization will vary with colour if the contribution to the total radiation from the unpolarized emission lines is important. The dominant lines are from [O II] in the blue, [O III] in the yellow, and Hα + [N II] and [S II] in the red. Of the positions actually observed only two (those labelled 2 and 4) could be affected; with the relevant line and continuum intensities measured by Woltjer (1958), we have calculated that there might be a 20% dilution from the lines. Only position III could be contaminated with starlight. No allowances have been made for these possibilities.

2.3. Instrumentation and Observations

The polarimeter described previously by Angel and Landstreet (1970) and Landstreet and Angel (1972) was used to make observations of circular polarization, mostly at the Cassegrain focus of the 82-in. telescope of McDonald Observatory in February, 1972. This instrument uses a Pockels cell acting as an electrically switched quarter-wave plate to convert circular to linear polarization, which is then analysed by a Wollaston prism followed by two RCA C31034 gallium-arsenide photomultipliers. Either an RCA 8575 bialkali or an RCA 8852 extended-red photomultipler was used to obtain earlier data in December, 1971, and January, 1972, at the 36-in. and 30-in. telescopes with a similar single-channel polarimeter.

A difficulty of the Pockels-cell instrument for these measurements is that the wave-plate does not reverse exactly because of imperfections in the crystal. As a result, when the strongly linearly polarized light from the Nebula is analysed, there is an apparent slight circular polarization of up to 0.3%. However, this effect reverses when the polarimeter is rotated 90°, while the true circular component is not affected. Laboratory tests with 100% linearly polarized light confirm that the spurious effect is eliminated in this manner. In every case the results that we have listed are an average of equal observations taken at 90° orientations. We selected for observation regions of 22″ diameter from Woltjer's (1957) map, shown in Plate I, where the scale is 1 unit = 11.16 s. These areas are as large as possible consistent with a uniform position angle over the aperture. A large diaphragm size is desirable because the accuracy is limited by counting statistics.

The bright, strongly polarized regions were chosen so that position angles nearly parallel to and at both ±45° to the interstellar polarization are represented (labelled with Roman numerals, odd and even numbers, respectively). In this way, the predicted dependence of circular polarization on position angle could be investigated. With the 82-in. telescope the observed positions were located by small offsets from the stars A en B shown in Plate I; nearby stars were used for offset guiding. On the smaller telescopes much larger offsets had to be made relative to a more distant guide star, BD 22°947, because stars A and B were visible only during the best seeing.

2.4. Observational Results and Discussion

The circular polarization measurements, expressed as percentages, are given in Table I for the four different wavelength regions studied (characteristically infrared, broad red, yellow, and blue). The standard deviations quoted are calculated from counting statistics alone. For each colour separate tabulations for the same position in the Nebula represent an observation on a different night. The values are corrected for the night sky background (which was found to be unpolarized), and a small correction has been applied for instrumental efficiency. The sign convention adopted is the same as in Paper I: the Stokes parameter V is positive (right handed) when, to an observer facing the Nebula, the electric vector in a fixed plane rotates counterclockwise. Observations of the circularly polarized white dwarfs Grw + 70°8247 and G99–37 were made to establish the sense of polarization. The linear polarization listed in Table I is the vector-averaged polarization from Woltjer (1957) corrected for interstellar polarization.

The measurements in Table I show small but definite polarization in the broad red and infra-red bands, clearly changing sing from one region to another. It is seen that the measured effect depends on the position angle of the linear polarization and not on position in the Nebula. Thus, positions 2 and 4, which have a common position angle of about 100°, both show negative polarization of about −0.1 % in the broad red band. These two positions are located at diametrically opposite points from the pulsar. Positions 3 and 5, both with a linear position angle of about 10°, show positive circular polarization although they are in quite different directions from the pulsar.

TABLE I

Observations of circular polarization in the Crab Nebula

Position	Linear polarization $\mathscr{P}(\%)$ $\theta(°)$ (p.a.)		Infra-red 8800–7800 Å	Circular polarization (%)		
				Broad red 8800–6000 Å	Yellow 5600–4200 Å	Blue 4700–3500 Å
3	17.3	6.5	+0.235 ± 0.092 +0.147 ± 0.037	+0.110 ± 0.027 +0.142 ± 0.045 +0.109 ± 0.038	+0.09 ± 0.06	+0.05 ± 0.15[b] −0.08 ± 0.06[a]
5	32.2	15.9		+0.280 ± 0.040		
4	32.1	100.1	−0.201 ± 0.062	+0.29 ± 0.11[a] −0.33 ± 0.13[a] −0.110 ± 0.038 −0.061 ± 0.024		−0.14 ± 0.10[a] −0.08 ± 0.07[a] +0.14 ± 0.13[a]
2	38.9	108.8		−0.142 ± 0.033		
2P	30.6	103.3			−0.08 ± 0.06[b]	
II	31.4	151.5	−0.235 ± 0.092 −0.152 ± 0.051	−0.099 ± 0.021 −0.017 ± 0.048 −0.053 ± 0.032		
III	18.9	138.9	−0.105 ± 0.052	−0.016 + 0.023	−0.02 ÷ 0.04	
IV	21.6	141.6	−0.239 ± 0.115	−0.047 + 0.039	−0.024 + 0.040[c]	
V	25.6	145.9		+0.063 ± 0.092	−0.034 ± 0.047[c]	
I	16.9	155.3				

All observations with the 82-in. telescope except (a) 30-in., and (b) 36-in.
(c) From Landstreet and Angel (1971).

The observed dependence of circular polarization on position angle is characteristic of transmission through a wave plate or birefringent medium and is not consistent with an interpretation in terms of intrinsic circular polarization of the Nebula predicted by the 'synchro-Compton' theory. In this theory the polarization would be strongest along the rotation axis of the pulsar, taken to be the axis of the observed linear polarization, and of opposite sign at apposite points from the pulsar. Furthermore, this theory predicts a comparatively large effect (1 %) independent of wavelength which is also in contradiction with the observations. Thus, the new data add weight to the conclusion, already drawn from the absence of circular polarization in visible light (Landstreet and Angel, 1971), that 'synchro-Compton' emission cannot contribute more than a few per cent of the light of the Nebula.

We are also confident that the observed effect is not instrumental for the following reasons. First, the measurements of each position are generally reproducible and show a scatter consistent with the calculated error from counting statistics. This is despite changes of telescope and of polarimeter, and repeated realignments of the optics. The only discrepancy is in the broad red data for position 4 taken with two different telescopes. This might possibly be explained by a slight offset error on the smaller telescope which would mean a different region was being measured. While the rotation cancels all spurious effects in the polarimeter, there is the remote possibility that both telescope mirror systems were acting as wave plates with the same fast and slow axis positions, which would result in the measured dependence of polarization on position angle. Against this we can argue that any wave-plate effect would probably result in a greater polarization at shorter wavelengths, opposite to what is observed. Also, at least for the 82-in. telescope which is symmetrically aluminized for minimum linear polarization (about 0.1 %), there is no reason to expect birefringence.*

Circular polarization from interstellar grains is expected to change for different regions of the Nebula according to

$$V/\mathscr{P} = A(\lambda)\sin 2(\theta - \theta_0), \tag{1}$$

where $\mathscr{P}$ is the intrinsic linear polarization at position angle θ (Table I), θ_0 is the position angle of the interstellar polarization, and $A(\lambda)$ is the retardation which depends on the number of grains and on the grain material. By plotting $V/\mathscr{P}$ (which is very nearly equal to twice the ellipticity) against θ, we should be able to fit a sine curve to determine θ_0 and the sign and magnitude of $A(\lambda)$.

Figure 1 displays our results in this manner. Separate measurements of the same point on different nights are shown individually to indicate the repeatability and have been separated slightly in position angle for clarity. Several symbols are used to distinguish between different parts of the Nebula when the position angles are nearly the same. For each colour the best fitting values of A and θ_0 were determined by a least squares fitting procedure. The corresponding functions $A(\lambda) \sin 2(\theta - \theta_0)$ are plotted.

We make the following observations. First, in the infra-red and yellow the data points from many different regions in the Nebula are very well fitted by a sine curve. The same trend holds in the broad red, but there appear to be some systematic errors in addition to what can be attributed to counting statistics. The fit would be improved if, as mentioned previously, the measured values in positions 2 and 4 were low because of filamentary light. In the blue the poor quality data do not warrant the drawing of the best fit curve; the formal solution has a large amplitude (2), and passes through

---

* The circular polarization measurements have been confirmed by Martin and Angel using a third telescope, the 84-in. at Kitt Peak National Observatory.

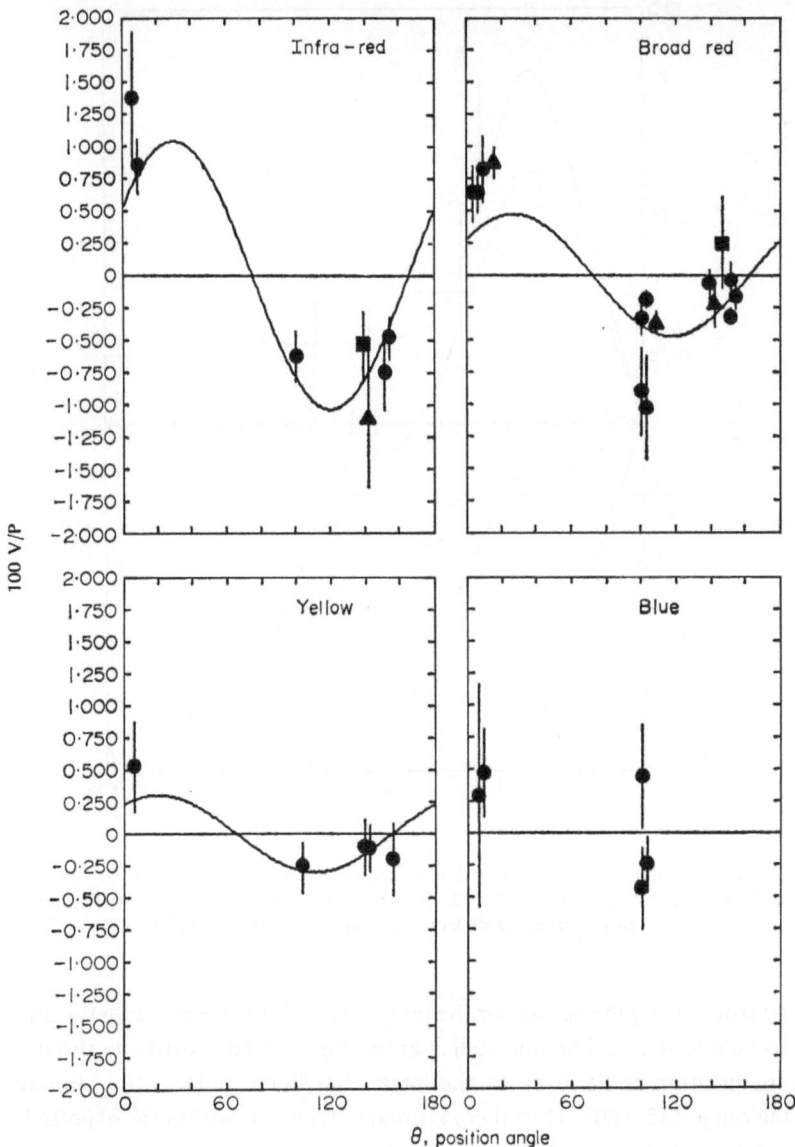

Fig. 1.   Measured values of the normalized circular polarization, $V/\mathscr{P}$, vs the position angle of the intrinsic linear polarization. Curves have been fitted by a least-squares procedure.

zero near the two groups of data points. The linear dependence of V on $\mathscr{P}$ is illustrated by the agreement of the broad red values of $V/\mathscr{P}$ for positions 3 and 5, which have nearly the same position angle but $\mathscr{P}$ different by a factor 2.

Seondly, the values of θ_0 obtained for each colour are consistent with each other, the best fit angles being 165, 162, and 155° for the infra-red, broad red and yellow data respectively, with a weighted average of 163°. The close agreement found is likely

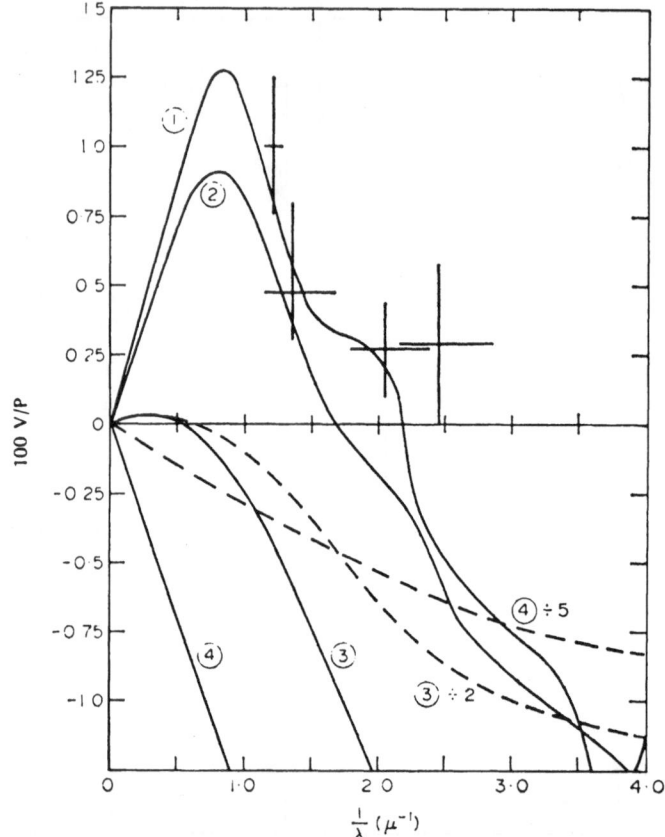

Fig. 2. Comparison of the observed circular polarization with that expected for different grain 'materials': 1 $m = 1.5$, 2 $m = 1.5 - 0.1\,i$, 3 $m = 1.5 - 0.5\,i$, 4 $m = 1.5 - 1.5\,i$. The observations clearly indicate dielectric rather than metallic grains.

fortuitous; from fitting the curves for the individual colours we estimate the uncertainty in θ_0 to be about 10°. The interstellar grain theory predicts that θ_0 should be the same as the position angle given by the interstellar linear polarization, found above to be in the range 145–160°. Thus there is good agreement, within the expected errors, for the direction of the grain orientation deduced independently from the interstellar linear and circular polarization measurements.* Thirdly, $A(\lambda)$ decreases with decreasing wavelength, although the functional dependence is uncertain. This colour dependence holds point by point in the Nebula.**

_____

* If the position angle of the interstellar polarization were not known, there would be an ambiguity in θ_0 of 90°, and in the sign of the amplitudes. Here the good agreement of the two determinations removes the ambiguity.
** It is now clear why interstellar circular polarization was not found in the earlier measurements (Landstreet and Angel, 1971); for the regions studied the position angles were nearly parallel to the interstellar direction, and in any case the effect is small in the yellow wave band.

The measured sign and amplitudes of $A(\lambda)$ can be compared with the theoretical predictions for different complex refractive indices to derive valuable information about the grain material. In Paper I it is shown how the relative amounts of linear and circular polarization for a given grain model can be calculated from the Mie theory for infinite cylinders. Figure 2 in Paper I shows the relative amounts of circular polarization expected from different grain materials when the linear polarization from each grain model has been normalized to the same value. In this calculation the linear polarization observations have been fitted first, because the same grains must be responsible for both the linear and circular polarization. The proper wavelength scale is found by choosing the appropriate grain size for each complex refractive index to obtain a match with the mean observed wavelength dependence of interstellar linear polarization. As shown in Figure 1 of Paper I this fit can be obtained for a large variety of grain materials, leaving a large ambiguity in this respect in the interpretation of the observations. However, the circular polarization is very dependent on composition.

Figure 2 of this paper shows the predicted values of $A(\lambda)$ for the Crab Nebula for different grain compositions;* the scale is determined from Equations (4) and (5) of Paper I assuming the visual interstellar linear polarization to be 2%. This procedure essentially eliminates the number of grains along the line of sight, because it enters equally in the equations describing the degree of circular and linear polarization from the medium. For Figure 2 we have calculated the amplitudes which best fit our observational results, assuming θ_0 to be 163°. The vertical bars show the estimated uncertainty.

The observed values of $A(\lambda)$ agree well with the predicted curve for pure dielectric grains. The sign, strength, and general trend with wavelength all support this interpretation. On the basis of curve 2, drawn for a refractive index $1.5 - 0.1\,i$, it appears that the imaginary part of the refractive index does not exceed 0.1. As discussed in Paper I, some materials which fall in this category are 'ices', including 'dirty ice', silicates, and silicon carbide. Aligned metallic particles, such as graphite or iron, can be ruled out by the difference in sign of the polarization from that predicted and by the conspicuous absence of strong polarization in the blue, which would be expected to be in excess of 0.5%**.

3. Conclusion

The circular polarization detected in the Crab Nebula shows all the characteristic features expected of polarization by aligned interstellar grains. The dependence of the

---

* The curves shown display the general trend as the imaginary part of the complex refractive index increases from zero. Details such as the exact wavelength of the crossing for dielectric grains depend on the size distribution, the degree of alignment, and the variations of the refractive index in the optical region of the spectrum.

** However, we cannot exclude spherical or unaligned metallic particles, or small particles which would not contribute to the polarization (and extinction) at visible wavelengths.

circular component on the linear polarization position angle rather than position in the Nebula indicates an interstellar origin. The sign, strength, and wavelength dependence of the effect all indicate that, at least along a 2-kpc path in the Galactic plane towards the anticentre, the composition of the aligned grains causing the polarization is dielectric rather than metallic. Since the interstellar grain composition may vary in different parts of the Galaxy, it would be desirable to observed more objects in other directions.

In view of the important conclusions to be derived from circular polarization measurements, our observations of the Crab Nebula will be extended to cover more position angles and wavelength bands and will include direct measurement of the linear polarization in each region and colour. Better information about the interstellar linear polarization can be obtained by using narrow-band filters centred on the strongest emission lines of the filamentary structure. This will give the linear polarization directly in front of the Nebula and will also show any changes that might occur across the Nebula. With these improvements it is hoped that accurate curves of the type shown in Figure 1 can be constructed to give detailed information about the wavelength dependence of circular polarization for comparison with theoretical models.

Acknowledgements

It is a pleasure to acknowledge the cooperation of Prof B. Warner and to thank the Director of McDonald Observatory for making the 82-in., 36-in., and 30-in. telescopes available to us. This work was supported by the Research Corporation, the National Science Foundation under grant GP 31356X, and the National Aeronautics and Space Administration under grant NGR 33-008-102. P.G.M. is grateful for financial assistance from Commission 38 of the IAU and from Churchill College, Cambridge. This is Columbia Astrophysics Laboratory Contribution No. 63. J.R.P.A. is an Alfred P. Sloan Research Fellow. This is a part of work published previously (Martin *et al.*, 1972).

References

Angel, J. R. P. and Landstreet, J. D.: 1970, *Astrophys. J. Letters* **162**, L61.
Cocke, W. J., Disney, M. J., Mancaster, G. W., and Gehrels, T., 1970, *Natures* **227**, 1327.
Conway, R. G.: 1971, in R. D. Davies and F. G. Smith (eds.), 'The Crab Nebula', *IAU Symp.* **46**, 292.
Hall, J. S.: 1958, *Publ. U.S. Naval Obs.* **17**, 275.
Hiltner, W. A.: 1956, *Astrophys. J. Suppl.* **2**, 389.
Hiltner, W. A.: 1957, *Astrophys. J.* **125**, 300.
Kristian, J., Visvanathan, N., Westphal, J. A., and Snellen, G. H.: 1970, *Astrophys. J.* **162**, 475.
Landstreet, J. D. and Angel, J. R. P.: 1971, *Nature* **230**, 103.
Landstreet, J. D. and Angel, J. R. P.: 1972, *Astrophys. J. Letters* **174**, L127.
Martin, P. G.: 1972. *Monthly Notices Roy Astron. Soc.* **159**, 179, (Paper I).
Martin, P. G., Illing, R., and Angel, J. R. P.: 1972, *Monthly Notices Roy. Astron. Soc.* **159**, 191.
Oetken, L.: 1966a, *Astron. Nachr.* **289**, 13.

Oetken, L.: 1966b, *Astron. Nachr.* **289**, 189.

Rees, M. J.: 1971, in R. D. Davies and F. G. Smith (eds.), 'The Crab Nebula', *IAU Symp.* **46**, 407.

Walraven, Th.: 1957, *Bull. Astron. Inst. Neth.* **13**, 293.

Wolstencroft, R.: 1966, *Observatory*, **86**, 223.

Woltjer, L.: 1957, *Bull. Astron. Inst. Neth.* **13**, 301.

Woltjer, L.: 1958, *Bull. Sstron. Inst. Neth.* **14**, 39.

INTERSTELLAR CIRCULAR POLARIZATION OF
UPPER SCORPIUS STARS

JAMES C. KEMP*

Institute for Astronomy, University of Hawaii, Honolulu, H.I. 96822, U.S.A.

Abstract. Circular polarization of interstellar or perhaps circumstellar origin has been clearly established in two reddened stars in the Upper Scorpius region, o-Sco and σ-Sco A; and detected preliminarily in a third, HD 154445. In the Corning 5–60 filter band $\lambda\lambda$ 3800–4600, measured values of $10^4 q$ were: o-Sco, $-(2.9 \pm 0.2)$; σ-Sco A, $-(1.7 \pm 0.2)$; and HD 154445, $+(2.8 \pm 0.8)$. Complete wavelength dependences $q(\lambda)$ for o-Sco and σ-Sco A are under study, but so far are known to have the following similarity. Each shows a broad maximum (of negative q) centered in the range $\lambda\lambda$ 3800–4300; and the values are smaller by a factor $\sim$ 4, but still negative, in the range $\lambda\lambda$ 5000–6000. The effect here is most likely due to a twisting grain alignment along the line of sight. In that case a simple model predicts $q(\lambda) \propto (n'_l - n'_r) \cdot (n''_l - n''_r) \theta_0 z_0^2$, where $n = n' + i\,n''$ is a formal, complex refractive index (in van de Hulst's forward-scattering approximation), l and r referring to longitudinal and transverse grain axes, and θ_0 is the total twist angle over the interaction length z_0. An alternate mechanism would be multiple scattering (at large angles, not along the line of sight) within asymmetrical, well-localized clouds around the individual stars. The similar sign and magnitudes of q for o-Sco and σ-Sco A, which are 1.5° apart in the sky, argue for the line-of-sight mechanism – for example through a large cloud in front of the entire local region. Comparative studies of many stars in Upper Scorpius will be of obvious interest.

1. Observations and Results

The theoretical description of interstellar absorption by van de Hulst predicts, or at least permits, the existence of circular as well as linear polarization. The most basic mechanism is that due to a twist in the alignment direction of particles along the line of sight. I report here detection of circular polarization in three reddened stars, which very likely can be attributed to this mechanism. On the first of the three, a separate report is in press (Kemp, 1972).

The stars are o-Sco, σ-Sco A, and HD 154445, A- and B-type stars in the Scorpio-Centaurus association (Garrison, 1967). Measurements on o-Sco were made first on the 88-in. telescope at Mauna Kea Observatory. In view of the special significance of this finding I repeated the observations on another telescope (24-in. at Mauna Kea) to help rule out all possibility of spurious linear-circular effects in the telescope. In reality the ratio of linear to circular polarizations here is quite within manageable limits for our polarimetric system including the telescopes we use (Kemp *et al.*, 1972). In so far as LCC effects in the polarimeter as such are concerned, in all these measurements I applied both of our procedures (modulator and base rotations) to guard against spurious results. In the 5–60 filter band the LCC term Δq that had to be cancelled by such rotations was no more than a third of the measured q for any of the three stars. There was no contamination from sky background, as checked in each case by sampling for circularly polarized flux in the adjacent sky.

* On leave from Dept. of Physics, University of Oregon, Eugene, Ore. 97403.

Greenberg and Van de Hulst (eds.), Interstellar Dust and Related Topics, 181–185.

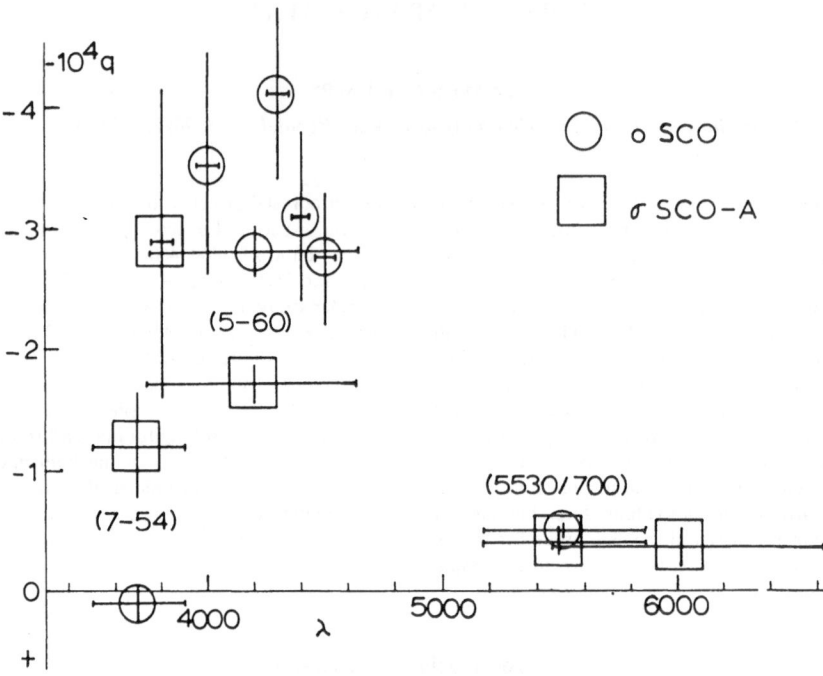

Fig. 1.   Circular polarization data for o-Sco and σ-Sco A.Horizontal bars indicate the half-trans-
mission widths of filters.

The fractional circular polarization values q for these stars are shown in Figure 1
and Table I, variously for certain filter bands. My usual axial-vector sign convention
is used: Positive q means that if a fixed plane is placed across the line of sight between
us and the star, we see the E vector as rotating counterclockwise in that plane. The
o-Sco data are generally those taken as mentioned on the 88-in. Statistics there were
adequate to allow use of some narrow-band (100 Å) filters. These were used to probe
the region $\lambda\lambda$ 4000–4400, of special interest for interstellar effects; of the three spectral
regions examined so far this one showed the largest q. Best signal-to-noise obtained
in a band $\lambda\lambda$ 3800–4600 (Corning 5–60 filter). Measurements at the 24-in. on o-Sco
were made with that filter, with results identical to those obtained on the 88-in. one
month previous, to within $(\Delta q)/q \cong \pm 15\%$ statistical and calibration error. Meas-

TABLE I

Data for three stars, circular polarization q measured in $\lambda\lambda$ 3800–4600 filter band

Star	Spect. type	m_V	E_{B-V}	$10^2 p$ (vis.)	$10^4 q$	$\tilde{\theta}_0$ (radian)
o-Sco HD 147084	A5	4.6	0.70	4.4	$-(2.9 \pm 0.2)$	$+0.4$
σ-Sco A HD 147165	B1	2.9	0.40	1.6	$-(1.7 \pm 0.2)$	$+2.0$
HD 154445	B1	5.6	0.51	3.6	$+(2.8 \pm 0.8)$	-0.6

urements at the 24-in. on the brighter star σ-Sco A were then made, in the filter bands shown in Figure 1; and, with long integration time in the 5–60 band but with adequate statistics in the result, on HD 154445. For comparison, an upper limit $q = -(0.05 \pm 0.10) \times 10^{-4}$ was recorded for α-Sco (Antares); and a result with poor statistics, $q = +(0.3 \pm 0.2) \times 10^{-4}$, was obtained for δ-Sco, both in the 5–60 band.

Results in the broad near UV band (Corning 7–54 filter) were possibly contaminated by the far-red leak in that filter around λ 7200. In o-Sco this was not properly accounted for in Kemp (1972) and the small positive q in that band is suspect, especially in view of the large reddening of the star. In σ-Sco A it was ascertained that the flux from the leak was only 10% of the total flux through the filter and the 7–54 result is probably reliable in this case. (The question is whether a large q, perhaps of positive sign, might obtain at $\lambda \gtrsim 7200$.)

The wavelength dependence is as yet fragmentary. I expect shortly to investigate one or more of the stars using filter widths of 100 to 200 Å throughout $\lambda\lambda$ 3700–8500, at the 88-in. There is a crude indication (Figure 1) that $q(\lambda)$ for o-Sco and σ-Sco A may be similar.

2. Discussion

Using the effective refractive-index formalism of van de Hulst (1957) for line-of-sight propagation, if the alignment angle $\theta(z)$ (assuming elongated particles lying for simplicity in the xy plane normal to the line of sight – Figure 3) is assumed slowly varying, it is easy to construct a pair of coupled differential equations for the E-vector components E_x, E_y governing the propagation. (Apparently equivalent equations in the Stokes-vector components are displayed in Serkowski (1962).) The problem here is that of a general gyrotropic medium and no doubt has been exhaustively studied, though I did not have a convenient reference at hand. For the case in which θ varies linearly from zero (or strictly from some arbitrary initial direction) to a small angle $\theta_0 \ll 1$ over the entire length z_0 of interaction with the particles, I solved these equations by a power-series expansion of $E_x(z)$, $E_y(z)$. If the inital light is completely unpolarized, the resulting circular polarization is then:

$$q \cong \tfrac{1}{3} \left(\frac{2\pi}{\lambda}\right)^2 (n'_l - n'_r)(n''_l - n''_r)\theta_0 z_0^2,$$

where (n'_l, n''_l) are the real and imaginary parts respectively of the formal refractive index $n_l = n'_l + i n''_l$ corresponding to the parallel axis of the grains, while (n'_r, n''_r) pertain to a perpendicular axis. In the same model the linear polarization is:

$$p = \frac{2\pi}{\lambda}(n''_l - n''_r)z_0$$

so that we have the relationship:

$$\theta_0 = \left(\frac{3q}{p^2}\right)\left[\frac{n''_l - n''_r}{n'_l - n'_r}\right].$$

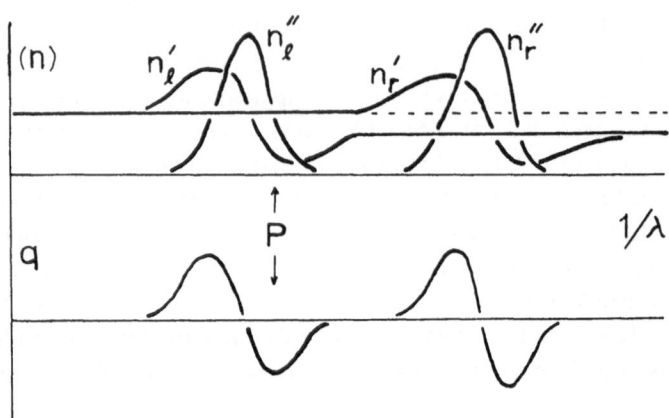

Fig. 2.   Model for parallel and perpendicular refractive indices of hypothetical elongated or anisotropic aligned particles; and $q(\lambda)$ predicted by simple relationships in text.

It is reasonable to assume, on various general considerations including the implications of the Kramers-Kronig relations, that a peak value of q corresponds to a wavelenght such that the dichroic-birefringent ratio in brackets above is of the order of
unity. On the strength of a quite naive model for $n_{l,r}$ (λ) discussed below (Figure 2
etc.), we make the tentative guess that in our case for $\lambda \sim 4200$ this ratio is in fact
negative, of order -1. We therefore define an angle $\tilde{\theta}_0 \equiv -(3q/p^2)$, and list these
values in Table I. This gives a rough measure of the total alignment rotation. If the
sign of the dichroic-birefringent ratio has been guessed correctly, the signs of the $\tilde{\theta}_0$
are the real ones; positive $\tilde{\theta}_0$ means that the alignment pattern along z looks like a
righthand screw. We note that the $\tilde{\theta}_0$ found here (Table I) are of the order of a radian,
and are not really $\ll 1$ as assumed in deriving the above simple relations. Finding
more exact formulae should not be difficult.

What we have here is a probe into the third dimensional aspect of interstellar polarization, thus perhaps of the magnetic alignment fields in space. In the present data, one
question is whether the circular polarization is indeed due to the line-of-sight mecha-

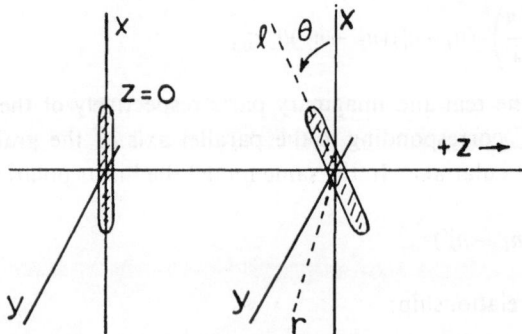

Fig. 3.   Geometry for varying grain alignment along line of sight. Axes l and r lie in the transverse
xy plane.

nism; or alternatively to highly localized asymmetric clouds or dust shells around (or adjacent to) individual stars, which can produce elliptical polarization by reflective effects, i.e. large-angle double scattering. It would seem unlikely that two stars (o-Sco and σ-Sco A), nearby in space, could be chance have similarly disposed, asymmetric dust shells so similar in structure as to produce the observed q values, which have the same sign and approximate magnitudes; this argues convincingly for the line-of-sight mechanism. The question then is whether we are seeing a diffuse galactic absorption, with the indicated variation of grain alignment occuring over a long path; or instead a somewhat localized cloud in front of or encompassing these stars in the upper Scorpius region. Observation of the interstellar q for more stars in this region should bring the answer.

Finally we ask what light is shed by the circular polarization results on the chemistry or morphology of the interstellar grains. In Figure 2 we show the effective birefringence and dichroism (in the sense of the forward-scattering formalism of van de Hulst) of aligned grains of some unspecified character. We assume that the parallel and perpendicular excitations (l, r) have predominant extinctions, respectively, defined by the two peaks n_l'' and n_r'' shown; these have corresponding real refractive indices n_l' and n_r'. Based on the formula above, the resulting $q(1/\lambda)$ is also displayed: It resembles the spectral derivative of the extinction peaks. Since the linear polarization ($p(\lambda)$ supposedly has the shape of the dichroism ($n_l'' - n_r''$), let us suppose that the typical maximum of p around λ 5000 (from the work of Gehrels and others) corresponds to the broad peak of n_l''. The implication is that the peak in q around λ 4300 is in the region P of Figure 2; and we expect to find a reversed sign of q in the near infrared. The prediction will be checked in the near future.

Acknowledgements

This work was by support to Mauna Kea Observatory by NASA, and the NSF.

References

Garrison, R. F.: 1967, *Astrophys. J.* **147**, 1003.
Kemp, J. C.: 1972, *Astrophys. J.* **175**, 35.
Kemp, J. C., Wolstencroft, R. D., and Swedlund, J. B.: 1972, *Astrophys. J.* **177**, 177.
Serkowski, K.: 1962, *Adv. Astron. Astrophys.* **1**, 289.
Van de Hulst, H. C.: 1957, *Light Scattering by Small Particles*, Wiley, New York.

ALIGNMENT OF DUST GRAINS BY MAGNETIC RELAXATION:
A COMPARISON BETWEEN RESULTS OBTAINED BY
TWO DIFFERENT METHODS

P. CUGNON

Observatoire Royal de Belgique, Belgium

Abstract. This paper is devoted to a comparison between results obtained by Purcell and Spitzer (1971) using a Monte-Carlo method and by the author (1971) using a Fokker-Planck equation. It is shown that there is a good agreement between the results within the dispersion expected from the Monte-Carlo method.

In a preceding paper (Cugnon, 1971) I tried to make a comparison between the results obtained by Purcell (1969) using a Monte-Carlo method for solving the problem of grain alignment by magnetic relaxation, and my own results, derived from the solution of a Fokker-Planck equation. For reasons explained in this paper, the comparison, limited to a few cases, could not be considered significant. At the same time, Purcell and Spitzer (1971) published a more detailed paper on the subject, using the same method, but with somewhat different working hypotheses. The following comparison between the Monte-Carlo and the Fokker-Planck methods will refer only to the Purcell and Spitzer paper of 1971 and the author's paper of the same year.

Table I summarizes the main hypotheses made in both methods and the essential limitations in the theories. These arise either from approximations (solution of the Fokker-Planck equation in linear cases – 'sites' approximation in the collision model used in the Monte-Carlo method) or from a lack of precision (increasing dispersion in the results given by the Monte-Carlo method, for decreasing values of δ) (Spitzer, 1972). The definition of δ and some other quantities is shown in Table II. From Table I two important differences appear between the characteristics of both methods:

(1) The use of different shapes.

(2) The limitations of the theories, which indicate that the Monte-Carlo and the Fokker-Planck methods are complementary rather than overlapping. Our aim is to

TABLE I

	Monte-Carlo method (Purcell and Spitzer)	Fokker-Planck method (Cugnon)
Grain shape	square prisms	spheroids
Collision type	'sticking' and evaporation	'sticking' and evaporation or specular reflection
Domains of best accuracy	moderate and high elongation or flatness and $\delta > 1$.	nearly spherical grains any value of δ, or $\delta \ll 1$, any value of the eccentricity.

Greenberg and Van de Hulst (eds.), Interstellar Dust and Related Topics, 187–189.

P. CUGNON

TABLE II

Definition of some parameters

$$\delta = \frac{\text{kinetic relaxation time}}{\text{magnetic relaxation time}}$$

$$\xi = \frac{2\,T_{\text{gra}}}{T_{\text{gra}} + T_{\text{gas}}}$$

T_{gra} = internal temperature of grains
T_{gas} = kinetic temperature of gas
$Q_A = <\tfrac{1}{2}\,[3\cos^2\,(\mathbf{B}, \text{axis of the grain}) - 1]>$

study here the influence of these differences on the results, and, from this discussion, to decide if a comparison remains possible.

Table III shows a comparison between the main shape parameters of spheroids and square prisms; the first column give the aspect ratio which is, for spheroids, the ratio between the transverse axis and the axis of symmetry, and, for prisms, the ratio between the side of the square and the height of the prism. Using this definition, the inertial ratio γ has the same value for both shapes. The third and fourth columns give, respectively for spheroids and prisms, the product $\gamma\varepsilon$, which may be defined as the ratio of the relative increments to the axial and the transverse components of the angular momentum (for a more explicit definition, see Cugnon, 1971). This parameter appears explicitly in the Fokker-Planck equation; it may be considered as a 'hidden' parameter in the Monte-Carlo approach. There is a difference between the values of this parameter, but, fortunately, the effect of this difference may be easily studied by the Fokker-Planck method in the case of infinite needles, i.e. where this difference is the highest, because for this case an analytical solution exists. After some calculation it appears that this effect is negligible. Somewhat arbitrarily, but with a very small risk of error, we assume that this holds true also for finite elongations. It then seems permissible to compare the respective results of spheroids and prisms.

The second difficulty arises from the different domains of values of the parameter δ covered by both methods. This difficulty can be overcome by considering the Fokker-

TABLE III

Shape parameters of spheroids and square prisms

	a/b	γ	$\gamma\varepsilon$	
			spheroids	prisms
needles	0	∞	1	1.667
	0.2	13	1.012	1.432
	0.5	2.5	1.043	1.196
spheres or cubes	1	1	1	1
	4	0.531	0.616	0.739
disks or flat prisms	∞	0.5	0.5	0.5

Planck results as a limiting case, and by checking if there is a good convergence of the Monte-Carlo results to the thus defined limit.

In practice, this may be done by looking at the values of Q_A/δ given by the Monte-Carlo method, Q_A being the degree of alignment of the grains, and by studying the convergence of a sequence of those values for a given aspect ratio, to the parameter

$$\lim_{\delta \to 0} \frac{Q_A}{\delta} = \tfrac{3}{2} F (1 - \xi)$$

which is given by the Fokker-Planck equation method. The results shown in Figure 1 demonstrate that the convergence is quite good within the dispersion in the Monte-Carlo results, and, furthermore, confirms the author's prediction (1971) i.e. that Q_A remains proportional to δ for $\delta \lesssim 1$, for moderate eccentricities.

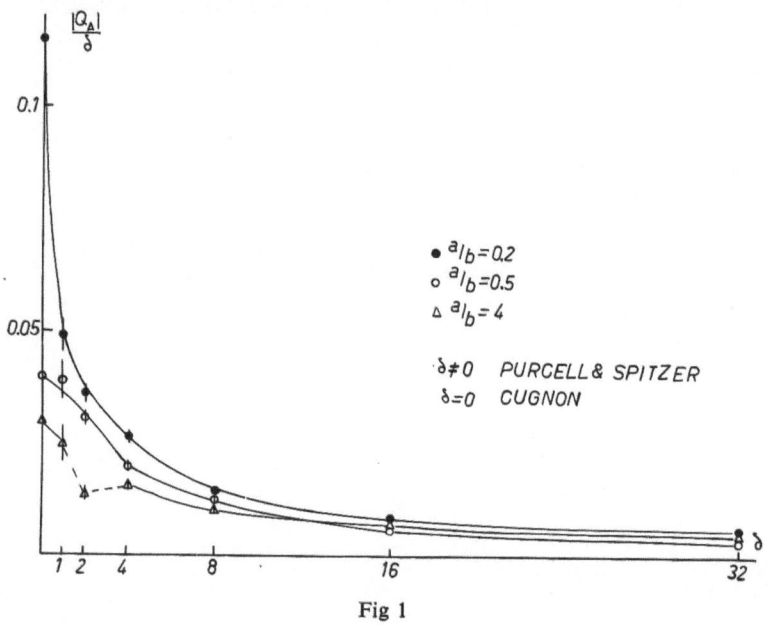

Fig 1

Acknowledgement

I am much indebted to Prof. L. Spitzer Jr., who suggested this work, and gave me many helpful remarks and suggestions.

References

Cugnon, P.: 1971, *Astron. Astrophys.* **12**, 398.
Purcell, E. M.: 1969, *Physica* **41**, 100.
Purcell, E. M. and Spitzer, L., Jr.: 1971, *Astrophys. J.* **167**, 31.

Place

A SEARCH FOR INFRARED POLARIZATION
IN THE GALACTIC CENTER

T. MAIHARA, H. OKUDA, and S. SATO

Dept. of Physics, Kyoto University, Kyoto, Japan

Abstract. Observations of infrared polarization have been tried in the galactic center. A preliminary result is that the polarization at K-band (2.2 μ) is less than 5%, much less than that expected from interstellar polarization.

1. Introduction

A few years ago, Becklin and Neugebauer discovered a strong near infrared source coinciding with the position of the galactic center (Becklin and Neugebauer, 1968). Extension to longer wavelength observations by themselves and by Low and Rieke have revealed some strong middle infrared radiation in more compact regions (Becklin and Neugebauer, 1969; Rieke and Low, 1971). Furthermore, surprisingly strong emission has also been found in the far infrared region by Low's observations from a jet-plane and by Hoffmann and others using balloon borne equipment. (Aumann and Low, 1970; Hoffmann *et al.*, 1971). The emission mechanisms of these radiations have not been well understood.

Measurement of polarization in the galactic center could give some information regarding these emission mechanisms. Low and others have observed it in middle infrared and obtained an upper limit of about 3% for the polarization at 10μ (Low *et al.*, 1969).

Strong infrared emission of the galactic center may also be useful for the investigation of interstellar space especially in the central part of the Galaxy, which is inaccessible by ordinary methods.

Extremely large interstellar extinction of the galactic center would be accompanied by large polarization, provided the interstellar dust particles are partially aligned by a galactic magnetic field or some other mechanism. This is another reason why we tried to measure the infrared polarization in the galactic center.

Since the preliminary report at the Liège symposium (Maihara *et al.*, 1972), we have made observations in 1971 and 1972 by a somewhat different method from the previous one.

2. Observations

The infrared polarimeter which we used is the same as used in the measurements of polarization of the infrared stars (Hashimoto *et al.*, 1970). The details of the polarimeter will be published elsewhere (Maihara *et al.*, 1973). An HR type polaroid put

Greenberg and Van de Hulst (eds.), Interstellar Dust and Related Topics, 191-196.
All Rights Reserved. Copyright © 1973 by the IAU.

between the filter and the detector was rotated with a constant speed of one rotation in about 2 min.

The observations were carried out in the summers of 1970, 1971 and 1972 by attaching the polarimeter to the 36 or 74-in. reflector at Okayama Astrophysical Observatory. The diaphragm was 5 mm in diameter so that fields of 30″ (74-in.) or 1′ (36-in.) were available. The observations were done in so-called K-band or at $2.2.\mu$ by using a liquid nitrogen cooled lead sulphide photoconductor.

Since the signal from the galactic center is so weak observations of many rotations were necessary to improve the statistics. The signal from the detector was recorded on a magnetic tape together with clock pulses which indicate the rotation angle of the polarizer. In the observations of 1970, the data was processed by an analog method, that is, the signals for each rotation of the polarizer were added to preceding data after rewinding the tape and reading it out. From the observations of 1971, the signals were recorded in series for a number of rotations to save the dead time for rewinding the tape. After the observations, they were digitised with a constant sample rate, added and analysed by a computer.

The results thus analysed are given in Table I, which are also illustrated in Figure I. The position angle is measured eastward from the meridian.

Finite degrees of polarization seem to be present statistically. But we must be careful to check for spurious polarizations of instrumental origin. In order to check such instrumental polarizations, we examined the following points.

First, the Fourier components of the first and the fourth harmonics were calculated as well as the second harmonics. Only the second harmonics-which are responsible for the polarization-appear to be significantly larger than the statistical fluctuations.

Second, the polarizations of normal stars were measured. They are much less than the observed values for the galactic center.

TABLE I

Infrared polarization of the galactic center

Date	Rotations	Amplitude	Position angle
1970 Jul.	57	$5.3 \pm 1.3\%$	$19° \pm 7.5°$
1971 May	26	7.4 ± 2.7	-22.6 ± 15
1972 May	34	6.0 ± 3.5	-31 ± 20

Comparison with other harmonics

	1st	2nd	4th
1971 May	$2.3 \pm 2.3\%$	$7.4 \pm 2.7\%$	$4.3 \pm 2.3\%$
1972 May	3.9 ± 3.3	6.0 ± 3.5	4.2 ± 3.3

Calibration by stars
α Sco $1.4 \pm 0.2\%$
γ Leo 0.6 ± 0.2

Polarization of sky background
34 rotations 1.8 ± 1.9

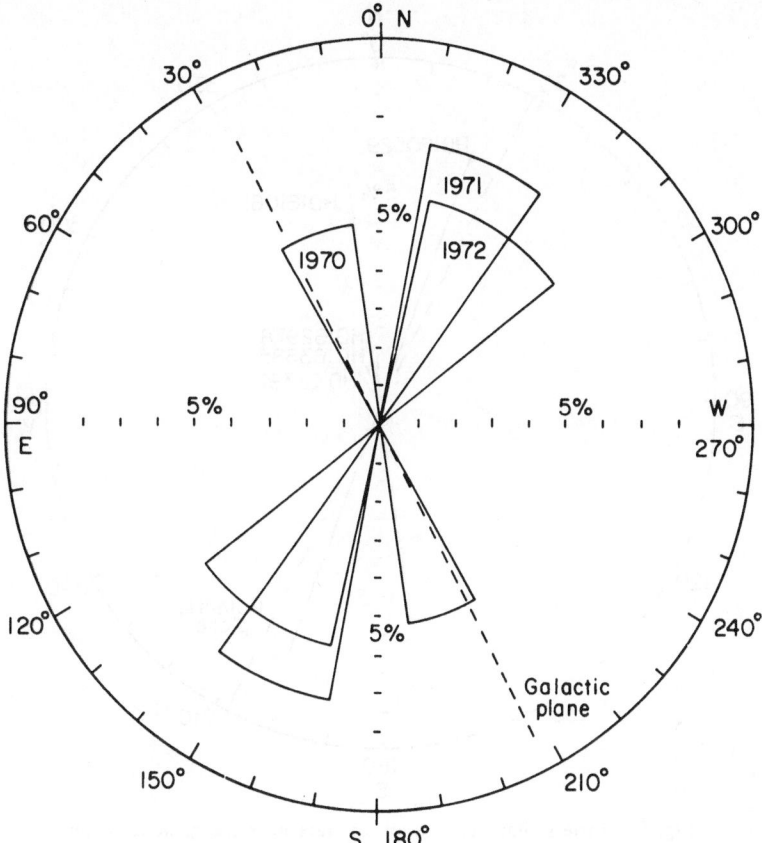

Fig. 1.   Infrared polarizations toward the galactic center observed in 1970, 1971 and 1972.

Third, the signal from the sky background was checked. It did not show any significant polarization larger than the statistical fluctuation.

Results of these checks are also given in Table I.

3. Discussion

So far as we have checked the data, no spurious polarization was definitely seen. But
we must note that the observational conditions were rather poor, that is, only a short
time was available and elevation of the galactic center was low in the southern sky so
that signal to noise ratio was very poor. We therefore reserve our final conclusion to
the more detailed observations in the future and take the observed values of about 5%
as an upper limit. Even in this case, the following discussion may be interesting.

That polarization may appear in the course of propagation through interstellar
space has been well known in the optical region. Mathewson and Ford have made
intensive studies of interstellar polarization of stars (Mathewson and Ford, 1970).
From their data we obtained the degree of polarization and position angles of these

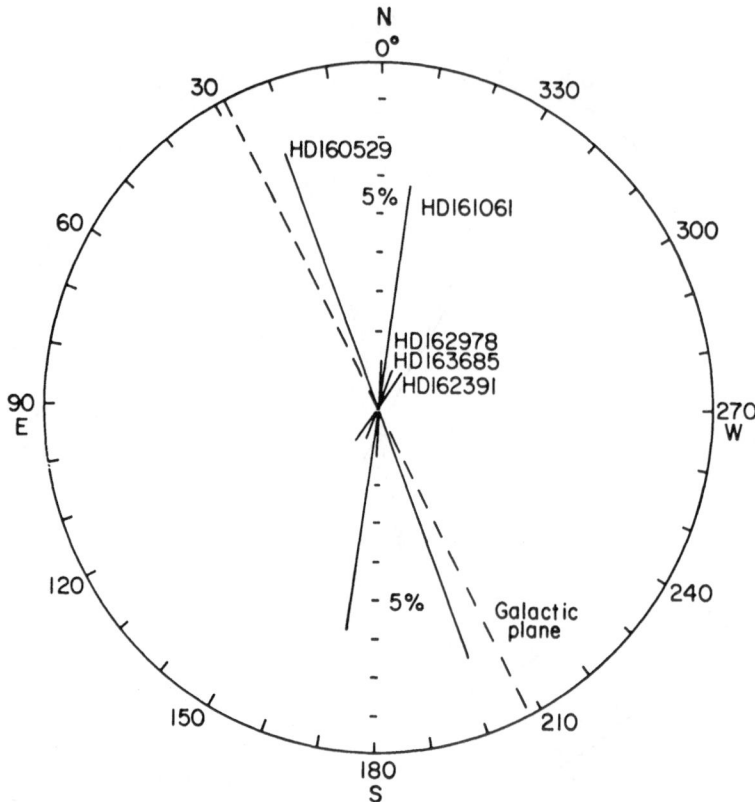

Fig. 2.   Insterstellar polarization of stars near the galactic center.

stars which are located within 5° from the galactic center. These are shown in Figure 2.
HD 161061 is the nearest star to the galactic center (about 1° separation). The degree
of polarization of these stars is relatively large and the position angles are concentrated
in a narrow range of direction, almost coincident with our observations shown in
Figure 1.

Interstellar polarization is correlated with extinction. In Figure 3, the relation between the degree of polarization and the amount of visual extinction is shown by
plotting the data of Mathewson and Ford (Mathewson and Ford, 1970). The stars
near the galactic center are marked by dots with a circle. It is seen that the degree of
polarization generally is proportional to the extinction. On the other hand, interstellar
polarization decreases with wavelength; about one third of visual polarization would
be expected for K-band (Dyck, 1972).

If we adopt 27 mag. of visual extinction for the galactic center as was estimated by
Becklin and Neugebauer (1968) and assume that the above relations of interstellar
polarization hold in the whole path to the galactic center, the polarization at K-band
would be expected as large as 10 to 20%. This is much larger than the observed upper
limit.

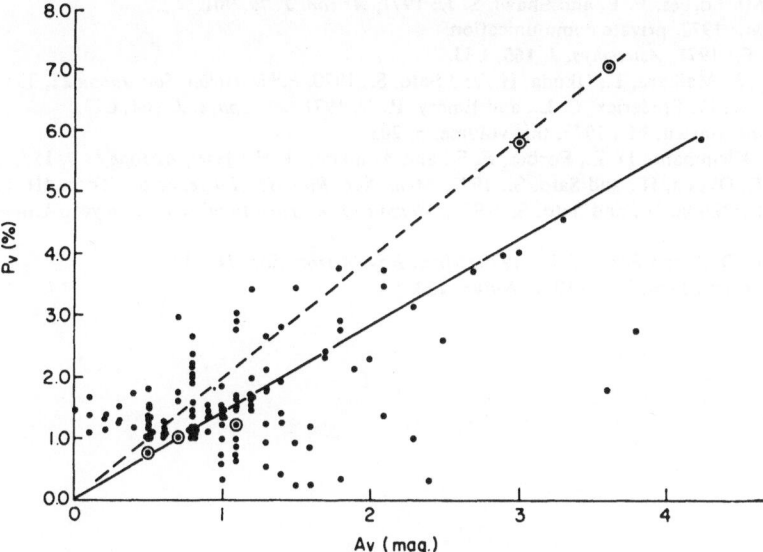

Fig. 3.   Relation between the degree of interstellar polarization and the visual extinction. Stars near the galactic center are indicated by dots with a circle.

As the interstellar polarization is believed to be caused by non spherical dust grains aligned by a regular magnetic field in the Galaxy, the observations might suggest that the magnetic field in the inner part of the Galaxy is not so regular as seen in the solar neiborhood. Stephen Loyd and Harwit have obtained a similar result by their detailed studies of the differential polarization of starlight as a function of distance (Stephen Loyd and Harwit, 1973).

The polarization of the galactic center might also be due in part to scattering of light from the central source by dust grains which are distributed asymmetrically or disk-like around the galactic center, as has been found in infrared stars (Hashimoto *et al.*, 1970; Forbes, 1971; Dyck *et al.*, 1917).

In any case, we must limit further dicussions until more detailed and definite observations are made. We are now preparing to make another observation next year and hope to get more definite results.

Acknowledgement

We want to thank the staff in Okayama Astrophysical Observatory for their kind help in our observations.

References

Aumann, W. F. and Low, F. J.: 1970, *Astrophys. J.* **159**, L159.
Becklin, E. E. and Neugebauer, G.: 1968, *Astrophys. J.* **151**, 145.
Becklin, E. E. and Neugebauer, G.: 1969, *Astrophys. J.* **157**, L31.

Dyck, H. M., Forbes, F. F. and Shawl, S. J.: 1971, *Astron. J.* **79**, 901.
Dyck, H. M.: 1972, private communication.
Forbes, F. F.: 1971, *Astrophys. J.* **165**, L83.
Hashimoto, J., Maihara, T., Okuda, H., and Sato, S.: 1970, *Publ. Astron. Soc. Japan* **22**, 335.
Hoffmann, W. F., Frederick, C. L., and Emery, R. J.: 1971, *Astrophys. J.* **164**, L23.
Lloyd, S. and Harwit, M.: 1973, this volume, p. 203.
Low, F. J., Kleinmann, D. E., Forbes, F. F., and Aumann, H. H.: 1969, *Astrophys. J.* **157**, L97.
Maihara, T., Okuda, H., and Sato, S.: 1972, *Mém. Soc. Roy. Sci. Liège*, 6e sér. Tome III, p. 417.
Maihara, T., Okuda, H., and Sato, S.: 1973, *Memoir of the Faculty of Science*, Kyoto University, in print.
Mathewson, D. S. and Ford, V. L.: 1970, *Mem. Roy. Astron. Soc.* **74**, 139.
Rieke, G. H. and Low, F. J.: 1971, *Nature* **233**, 53.

PHOTON AND MAGNETIC ALIGNMENT
OF INTERSTELLAR DUST GRAINS

LEON W. KING and MARTIN O. HARWIT

Center for Radiophysics and Space Research, Cornell University, Ithaca, N.Y. 14850, U.S.A.

Abstract. We have considered the cooperative effects of photon and magnetic alignment on plausible models for interstellar dust grains. Neither effect, either in cooperation or alone, appears able to produce the observed alignment.

Previous papers (Harwit, 1970a, b; Purcell and Spitzer, 1971; Jones and Spitzer, 1967; Purcell, 1969) have examined various mechanisms to account for the observed alignment of interstellar dust grains. We here examine in more detail the effects of absorption of interstellar starlight photons and re-emission of infrared radiation on the most often mentioned alignment mechanism, paramagnetic relaxation.

Let us take as a model for a grain a cylinder of radius r and length $l = \eta r$. Consider the angular momentum of the grain to be statistically in equilibrium, the various additions and substractions of angular momentum being equal. Take the X-Y plane to be the galactic plane with 95 % of the starlight photon flux lying in the plane; assume that the gas flow around the grain is isotropic and that the magnetic field B is in the Y direction.

Consider now the various contributions to the angular momentum squared of the grain. Tending to add angular momentum in a random walk fashion are ΔL_S^2, the incremental angular momentum squared carried by starlight photons, ΔL_R^2, that carried by isotropically thermally reradiated photons, ΔL_T^2, that carried by anisotropically thermally reradiated photons, the anisotropy being due to the alignment of the grain, and ΔL_G^2, that contributed by isotropic gas collisions with the grain. Tending to remove angular momentum are ΔL_D^2, the damping due to friction with the interstellar gas, ΔL_m^2, the damping due to paramagnetic relaxation, and ΔL_P^2, the damping due to more probable emission of photons which remove angular momentum than those that add. While the first 4 terms are random walk processes, the last 3 are systematic.

Let us take as the alignment the excess angular momentum in the X and Y directions as compared to an isotropically rotating grain. Thus:

$$Q_j = \frac{L_X^2 + L_Y^2}{L_X^2 + L_Y^2 + L_Z^2} - \tfrac{2}{3},$$

where L_i^2 is the rotational angular momentum squared in the ith direction. To find these we must solve the 3 equations:

$$\sum_j \Delta L_{ji}^2 = 0,$$

Greenberg and Van de Hulst (eds.), Interstellar Dust and Related Topics, 197–202.
All Rights Reserved. Copyright © 1973 by the IAU.

where $j = S, R, T, G, D, M, P$ and $i = X, Y, Z$. These must be solved by a self con-sistent calculation as the moment of inertia of the grain in each direction depends on the degree of alignment. We take $\Delta L_{mP} = 0$, as the magnetic field is in the Y-direction.

The various terms are:

Photon Angular Momentum Contributions

Starlight
$$\Delta L^2_{Si} = N_i(r,\eta)\, cq\hbar^2 r^2 \pi\, (\eta+1)(\tfrac{1}{3} + F(\eta,d_i))\, \tau$$
Isotropic Infrared Re-Emission
$$\Delta L^2_{Ri} = (\sum_{i=X}^{Z} \Delta L^2_{Si})\, \beta/3q$$
Anisotropic Infrared Re-Emission
$$\Delta L^2_{Ti} = F(\eta,d_i)\, \Delta L^2_{Ri}$$
Photon Damping
$$\Delta L^2_{Pi} = -I(d_i)\, \omega_i^2\, k'\, \frac{\left(\sum\limits_{i=X}^{Z} N_i(r,\eta)\right)}{3}\, c\beta\hbar^2\, r^2\, \pi\, \frac{(\eta+1)}{3}\, \tau$$
Fractional Excess Area Exposed
$$F(\eta,d_i) = \frac{\eta\sqrt{\eta^2+4}+1 + d_i(\eta\sqrt{\eta^2+4}-1)-(2\eta\sqrt{\eta^2+4}+2)/3}{\dfrac{2\eta\sqrt{\eta^2+4}+2}{3}}$$

Gas and Magnetic Angular Momentum Contributions

Isotropic Gas Collisions
$$\Delta L^2_{Gi} = nv\, \frac{(mvr)^2}{3}\, \frac{r^2}{4}\, (\pi + \frac{\eta^3}{3})\, \tau$$
Gas Frictional Damping
$$\Delta L^2_{Di} = -I(d_i)\, \omega_i^2\, \frac{nvm}{\sqrt{3}}\, \frac{v^{4/3}}{c}\, \tau$$
Paramagnetic Relaxation Damping
$$\Delta L^2_{Mi} = -I(d_i)\, \omega_i^2\, KVB^2\left(\frac{T_{Gas}-T_{Grain}}{T_{Gas}+T_{Grain}}\right)\, \tau$$
Moment of Inertia
$$I(d_i) = V\rho\frac{r^2}{4}(1+\frac{\eta^2}{3})\, d_i + \frac{1-d_i}{2}\left[V\rho\frac{r^2}{4}(1+\frac{\eta^2}{3}) + V\rho\frac{r^2}{2}\right]$$

$$d_i = \frac{\omega_i^2}{\omega_X^2 + \omega_Y^2 + \omega_Z^2}$$

$F(\eta, d_i)$ is a result of the partial alignment of the grain. $N_i(r, \eta)$ is the number of photons absorbed by the grain and is obtained by integrating a suitable efficiency factor over an approximate interstellar radiation field. We take this factor to be $(v/v_0)^2$ below v_0 and 1 above, where $v_0 = c/2\pi r \sqrt{\eta}$ (Purcell and Spitzer, 1971). The radiation field used consists of the superposition of 3 blackbody spectra, corresponding to temperatures of 14 500 K, 7500 K and 4000 K multiplied by dilution factors of 4×10^{-16}, 1.5×10^{-14}, and 1.5×10^{-13} respectively (Werner and Salpeter, 1969). The terms ΔL_{Di}^2 and ΔL_{mi}^2 are taken from Purcell and Spitzer (1971) by integrating the expressions they give for the damping torques on the grain. The factor $3^{-1/2}$ comes from having to use the average angle at which the gas molecules strike the grain; the factor $(T_{gas} - T_{grain})/T_{gas} + T_{grain})$ is inserted to take into account the temperature dependence.

q is the correlation between absorbed and scattered photons, the forward scattered ones having no effect on the angular momentum while the backward scattered ones contribute to the random walk process as do the absorbed ones. We estimate that $q = 1.2$ is reasonable. β is the number of thermalized infrared photons emitted for each starlight photon absorbed. Estimating the grain temperature to be 30 K we divide the average interstellar photon temperature by this and thus obtain $\beta = 143$. k' is a number that determines the excess number of emitted photons which remove angular momentum. From Harwit's paper we have:

$$\frac{\text{Probability}(-)}{\text{Probability}(+)} = \frac{(\omega_p + \omega)^3}{(\omega_p - \omega)^3} \frac{2J+1}{2J} \cong 1 + \frac{3\omega}{\omega_p}.$$

Thus $k' = 3/\omega_p$.

To fix numerical values we choose gas characteristics $v = 1.7 \times 10^5$ cm s^{-1}, $T_{gas} = 120$ K, and $m = 1.6 \times 10^{-24}$ gm: the grain density ρ we take to be 1 gm cm^{-3}; B we take to be 3×10^{-6} G; the frequency of the emitted radiation, ω_p, is 3×10^{12} s^{-1}; V is the volume of the grain and K as defined in Spitzer and Jones (1967) we take to be about 10^{-13}. n is the number of gas molecules per cubic centimeter.

The main results of the calculation are presented in Tables 1 and 2. Table 1 presents a comparison of the amount of alignment obtained with gas, magnetic, and photon effects operating, vs the alignment obtained with just gas and magnetic effects as a function or the radius and the eccentricity of the grain and the number of gas molecules we assume per cubic centimeter. The last line presents the result of averaging over the 5 values of r and eccentricities from 1 to 10. It can be seen that the net photon effects are least for the larger grains and for higher values of the gas density, the effect being one of lessening the alignment for the small grains. As it is the smaller grains that are best aligned by paramagnetic relaxation (Purcell and Spitzer, 1971), the net photon effect is to increase somewhat the magnetic field needed to account for the alignment. It is seen that the net disaligning effect is smaller than that estimated by Purcell and Spitzer (1971) for spherical grains, especially for the larger values of r.

Table 2 presents the ratio of photon and magnetic alignment when they are not working in conjunction. It can be seen that the photon aligning mechanism gives

TABLE 1

Photon and Magnetic Alignment
Magnetic Alignment

EC = Eccentricity = $\eta/2$ R = Radius
Number of Gas Molecules/cm^3

		0.1	1.0	3.0	10.0
R = 3.16 x 10^{-7}cm	EC=2	.273	.337	.457	.678
	EC=5	.166	.331	.550	.810
R = 1.00 x 10^{-6}cm	EC=2	.113	.220	.399	.693
	EC=5	.0886	.328	.607	.911
R = 3.16 x 10^{-6}cm	EC=2	.0725	.252	.508	.858
	EC=5	.110	.527	.862	1.12
R = 1.00 x 10^{-5}cm	EC=2	.128	.542	.854	1.08
	EC=5	.321	.909	1.07	1.16
R = 3.16 x 10^{-5}cm	EC=2	.443	.935	1.03	1.08
	EC=5	.794	1.03	1.06	1.06
Average Over R and EC		.305	.627	.825	1.01

TABLE 2

Photon Alignment
Magnetic Alignment

EC = Eccentricity = $\eta/2$ R = Radius
Number of Gas Molecules/cm^3

		0.1	1.0	3.0	10.0
R = 3.16 x 10^{-7}cm	EC=2	.0101	.0123	.0164	.0237
	EC=5	.0117	.0225	.0354	.0490
R = 1.00 x 10^{-6}cm	EC=2	.0110	.0212	.0385	.0672
	EC=5	.0157	.0562	.100	.144
R = 3.16 x 10^{-6}cm	EC=2	.0138	.0482	.0977	.166
	EC=5	.0286	.131	.207	.262
R = 1.00 x 10^{-5}cm	EC=2	.0228	.0971	.154	.196
	EC=5	.0567	.148	.171	.183
R = 1.00 x 10^{-5}cm	EC=2	.0378	.0797	.0881	.0921
	EC=5	.0533	.0648	.0690	.0685
Average Over R and EC		.0353	.0878	.118	.140
Deviation From Purcell and Spitzer Method		16.9%	8.8%	6.3%	6.8%

alignments far smaller than that given by magnetic effects, which already yields Q_j's which are too small. The difference in the Q_j's obtained here from those of previous papers (Harwit, 1970a, b) comes from taking a larger value of β as indicated from infrared rocket observations (Houck *et al.*, 1971) and considering the systematic disalignment more rigorously. These considerations indicate that photon alignment is not sufficient to produce the observed polarization.

The last line of the table shows a comparison between this method of calculation and that employed by Purcell and Spitzer (1971) for the case in which we have no photon effects. The values given are the average over r and eccentricity of the absolute value of the difference in alignment obtained by the two methods divided by the sum of the alignments. It was found that the agreement is reasonably close in all cases, the best agreement being in the cases of higher gas density and radii between 3.16×10^{-6} cm and 3.16×10^{-5} cm, conditions which are more probable than the other ones treated.

From this calculation we therefore conclude that even taking into account the effect of instantaneous alignment of the grain on the various angular momentum contributions, neither paramagnetic relaxation nor photon alignment are sufficient to produce the observed alignment, either alone or working in conjunction with each other.

Acknowledgements

This work was supported by the Research Corporation. One of us (L.W.K.) was supported by the Ford Foundation through the 6-yr PhD program at Cornell.

TABLE OF SYMBOLS

n = Number of Gas Molecules/cm^3

v = Velocity of Gas Molecules

m = Mass of Gas Molecules

ρ = Density of the Grain

V = Volume of the Grain

r = Radius of Grain

η = Ratio Length of Grain to Radius

ω = 2π Frequency of Rotation of Grain

q = $1 + \dfrac{\text{Number of Backward Scattered Photons}}{\text{Number of Absorbed Photons}}$

β = Number of IR Photons Emitted for Each Photon Absorbed

N_i (r, η) = Integrated Number of Photons Absorbed by the Grain

k' = 3/Frequency of IR Photons Emitted

K = χ''/ω = Imaginary Part of Volume Susceptibility /ω

C = Factor Depending on Shape of the Grain

B = Interstellar Magnetic Field Strength

T = Temperature

References

Harwit, M.: 1970a, *Bull. Astron. Inst. Czech.* **21**, 204.
Harwit, M.: 1970b, *Nature* **226**, 61.
Houck, J. R., Soifer, B. T., Pipher, J., and Harwit, M.: 1971, *Astrophys. J.* **169**, L31.
Jones, R. V. and Spitzer, L.: 1967, *Astrophys. J.* **147**, 943.
Purcell, E. M.: 1969, *Physica* **41**, 100.
Purcell, E. M. and Spitzer, L.: 1971, *Astrophys. J.* **167**, 31.
Werner, M. W. and Salpeter, E. E.: 1969, *Monthly Notices Roy. Astron. Soc.* **145**, 249.

THE DIFFERENTIAL POLARIZATION OF
STARLIGHT AS A FUNCTION OF DISTANCE

STEPHEN LLOYD and MARTIN O. HARWIT

Center for Radiophysics and Space Research, Cornell University, Ithaca, N.Y. 14850, U.S.A.

Abstract. We have plotted the differential polarization as a function of distance and galactic longitude for 1732 stars between galactic latitudes $-5°$ and $+5°$. The data were obtained from several published sources. A fairly clearcut distinction is seen between the effects of spiral arms and inter-arm regions indicating that starlight is systematically polarized only when passing through spiral arms.

One reason why the polarization of starlight is interesting is that it gives us information about two different phenomena, both related to the galactic structure. The first is the distribution of dust; the second is the mechanism that aligns the dust grains.

We have determined the variation of the polarization with distance. The goal was to find what we shall call the *incremental* polarization at a given distance and longitude in the galactic plane. This incremental polarization is the average polarization of the stars in a given area of the galactic plane with the foreground polarization subtracted. It represents the local polarization intrinsic to the area.

In order to do this, we have gathered optical polarization data from a number of sources, principally Hall's (1958) catalog of northern stars and Mathewson and Ford's (1970) catalog of southern stars (see also Appenzeller, 1966 and Behr, 1959). In all, we used 1732 stars between galactic latitudes $-5°$ and $5°$. These stars were divided into 180 sectors in the galactic plane. The plane was divided first into 36 longitude intervals, each $10°$ wide; then each longitude interval was divided into 5 distance intervals: 0–400 pc, 400–800 pc, 800–1500 pc, 1500–2500 pc, and > 2500 pc.

In order to add or subtract polarizations, we made use of the Stokes parameters Q and U for each star. Then, for each sector of space, we averaged Q and U over all the stars in that sector. Finally, from the average values of Q and U in a sector B, we subtracted the average values of Q and U in the adjacent nearer sector A at the same longitude. From these difference values, $Q_B - Q_A$ and $U_B - U_A$, we computed the incremental polarization and polarization angle corresponding to sector B. Actually, this incremental polarization refers to a distance interval that overlaps A and B, rather than just to B. Unfortunately, this does not always work well. There usually is a large dispersion in the polarization and sometimes in the angle, so that often the dispersion in Q and U is as large as the differences $Q_B - Q_A$ and $U_B - U_A$.

In Figure 1 we see some of the results. Each column is one longitude interval. The nearest distance interval is at the top; the farthest at the bottom. As is customary, the length of each line is proportional to the incremental polarization; a vertical line represents a polarization angle of $0°$; a horizontal line – an angle of $90°$. We can see how both the magnitude and the angle of the local incremental polarization vary with distance.

Greenberg and Van de Hulst (eds.), Interstellar Dust and Related Topics, 203–207.
All Rights Reserved. Copyright © 1973 by the IAU.

For example, at $l = 80°$ and $l = 90°$ we see that the incremental polarization is moderately large out to great distances, while at $l = 100$–$120°$ we see the polarization dropping off in the most distant intervals. At $l = 140°$, in Figure 2, we see the polarization drop off and then increase again; the large polarization in the most distant

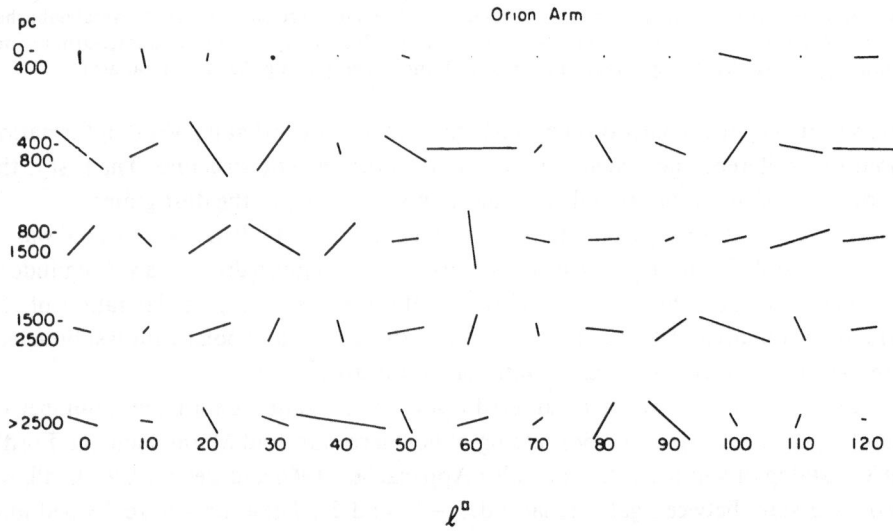

Fig. 1. The differential polarization of starlight in the galactic plane as a function of distance and longitude. The length of the vector at $l = 0°$, 800–1500 pc, represents a polarization of 1%. A vertical vector represents a polarization angle of $0°$.

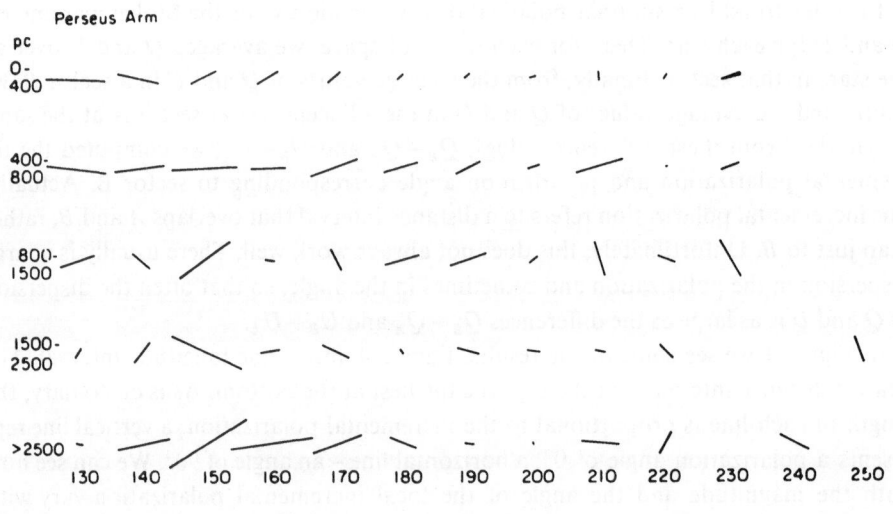

Fig. 2. See caption to Figure 1.

intervals between $l = 140°$ and $l = 170°$ may be due to the presence of the Perseus Arm. Around $l = 130°$, the polarization vectors are most closely aligned to 90°. The polarization here is somewhat larger than in the opposite direction at $l = 310°$. Here we see rather weak polarization, perhaps because the Sun is nearer the inside edge of the local arm. Figure 3 shows a region, in which statistics are somewhat poor.

In Figure 4 we see a new way of presenting the polarization data in the galactic plane. We have taken the polarization vectors from the preceeding figures and put them onto a chart of the local spiral arms. The stars and crosses represent clusters and associations of supergiants; this chart was taken from a recent review paper by Bok (1970).

Each polarization vector was placed halfway between the middle of the distance intervals used in its computation. Together the vectors form rings at distances from the Sun of 400, 900, 1600, and 2500 pc. The rings containing the polarization vectors were placed at distances from the Sun roughly representative of their distance interval – 600 pc, 1200 pc, 2000 pc, and 2500 pc. The innermost distance group was omitted. A vector parallel to the tangent of the circle means a polarization angle of 90°.

The first thing we notice is the familiar strong alignment around $l = 130°$, and, to a lesser extent, $l = 310°$. Also, the incremental polarization seems substantially larger at the 600 pc ring than further out. Along the ends of the local arm, the incremental polarization is generally weak, although the angle is not highly scattered.

In the Orion Sagittarius interarm region the angle is highly scattered, as one would expect. The incremental polarization seems moderately high, although this may be a spurious effect of the large angle scatter.

Around $l = 150°$ we see a large incremental polarization that might be evidence of the Perseus Arm. The polarization in that arm around $l = 120°$, however seems anomalously low. The Sagittarius Arm may be 'hidden' by the large angle scatter in the

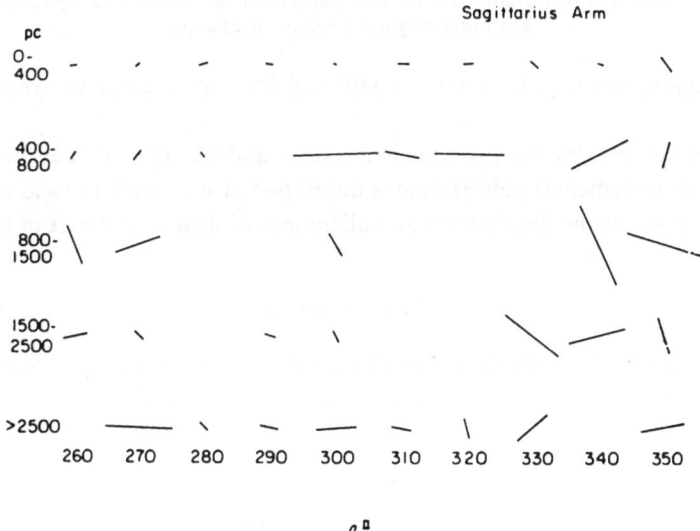

Fig. 3.   See caption to Figure 1.

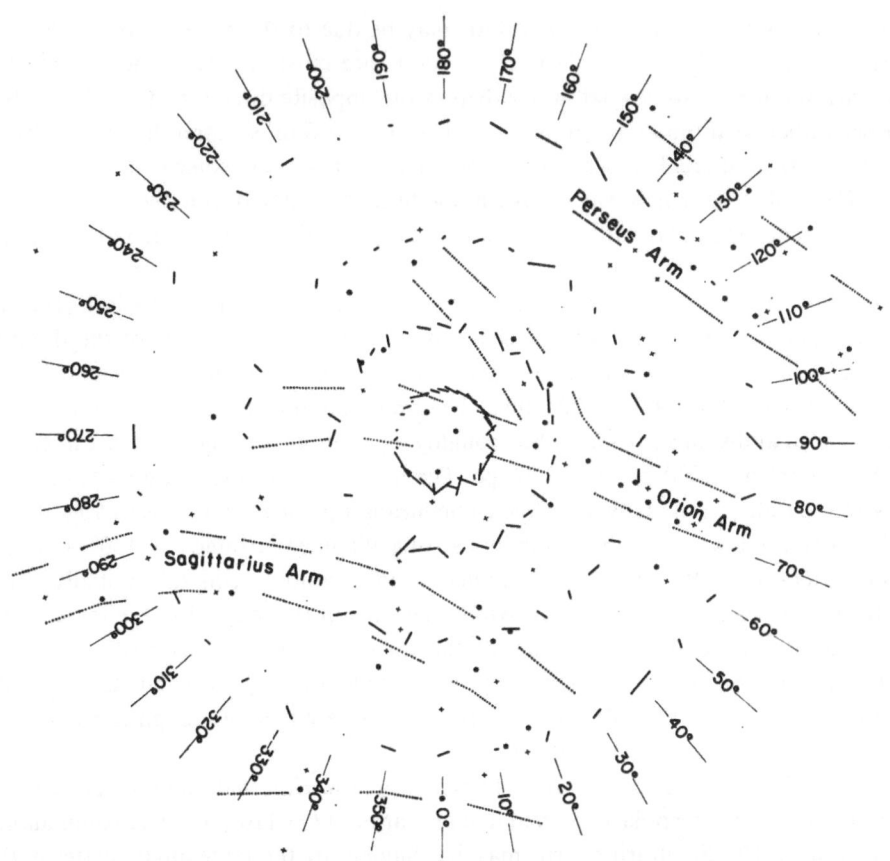

Fig. 4.   The differential polarization vector displayed on a chart of the local spiral arm structure.
The outer ring has a radius of 2500 pc.

interarm region, although between $l = 340°$ and $350°$, at ≈ 2 kpc we see some alignment.

Whatever the case, these are just the first results, and we expect that as this technique
of finding the incremental polarization is developed, it will begin to yield more useful
new evidence about the distribution and alignment of dust in our part of the Galaxy.

Acknowledgements

This work was supported by the Research Corporation. One of us S. L. was supported
by the Ford Foundation through the 6-yr PhD program at Cornell University.

Appendix

The Stokes parameters are characteristics of light beams, not polarizers. The following
argument is intended to justify the use of the Stokes parameters (in this paper to

describe polarizing media. (In this discussion, the fourth Stokes parameter, V, is ignored).

If a beam of unpolarized light with Stokes vector $\bar{S}_i = (I_i, 0, 0)$ is incident on a polarizer, the emerging beam is $\bar{S}_0 = (I_i/(1+P_p))(1, q_p, u_p)$, where P_p is the polarization degree of the emerging beam, $q_p = P_p \cos 2\theta_p$, $u_p = P_p \sin 2\theta_p$, and θ_p is the angle of polarization of the emerging beam.

Consider a weakly polarized beam

$$\bar{S}_i = I_i(1 - P_i, 0, 0) + I_i(P_i, q_i, u_i) \qquad P_i \ll 1$$

incident upon a weak polarizer ($P_p \ll 1$). To the first order in P_i and P_p, the second, polarized component of $\bar{S}_i$ is transmitted unchanged by the polarizer. The emerging beam is then

$$\bar{S}_0 = (I_i(1 - P_i)/(1 + P_p))(1, q_p, u_p) + I_i(P_i, q_i, u_i).$$

Using $(1-P_i)/(1+P_p) \approx (1-P_i)(1-P_p) \approx 1 - P_i - P_p$ and ignoring second order terms, we find

$$\bar{S}_0 = I_i(1 - P_p - P_i, q_p, u_p) + I_i(P_i, q_i, u_i) \sim I_i(1, q_p + q_i, u_p + u_i).$$

References

Appenzeller, I.: 1966, Z. Atrophys. **64**, 296.
Behr, A.: 1959, Veröff. U. Sternw. Göttingen, No. 126.
Bok, Bart J.: (1970) in De Jager (ed.), Highlights of Astronomy, p. 63.
Hall, J. L.: 1958, Publ. U.S. Naval Obs. 2nd Ser. **17**, No. VI.
Mathewson, D. S. and Ford, V. L.: 1970, Monthly Notices Roy. Astron. Soc. **74**, 139.

POLARIZATION OF LIGHT BY DUST IN GALAXIES

AINA ELVIUS

Stockholm Observatory, Saltsjöbaden, Sweden

Abstract. Several galaxies were observed for polarization during April and May 1972. Preliminary results are presented concerning those galaxies where polarization can be expected because of selective extinction or scattering of light by interstellar dust.

Dust clouds in other galaxies seem to be similar to the clouds in the Milky Way. Large-scale structures of the magnetic fields related to the interstellar matter in most types of galaxies may be revealed by observations of the polarization of light from such galaxies. Some observations of this kind have been made earlier.

Most of the observations were made with the 72-in. Perkins reflector of the Ohio Wesleyan and the Ohio State University at the Lowell Observatory. Additional data were obtained with the Lowell Observatory 42-in. telescope at the same site south of Flagstaff, Arizona. Dr J. S. Hall kindly let me use his area-scanning polarimeter (Hall, 1968) and the new computerized data-acquisition system built at the Lowell Observatory (Albrecht *et al.*, 1971).

Table I summarizes some data concerning the observed galaxies. Results of polarization measures in yellow light of NGC 4565 are illustrated in Figure 1. The polarization due to extinction in the dust clouds of the spiral structure projected against the center of the galaxy reaches $2.8 \pm 0.8\%$ in the darkest part of the lane and is about 1% along most of the scan at the same side of the centre. The strongest electric vector is roughly parallel to the symmetry plane of NGC 4565.

In blue light only a few scans were made of NGC 4565 and the results are less precise. The contribution of scattered light also seems to complicate the interpretation.

NGC 4216 is also a galaxy seen almost edge-on. It seems to differ from the type of NGC 4565, however, and the polarization is also different. There is a weak polarization at the bright side of the centre which might be due to scattering of light by particles aligned in a magnetic field along the spiral arms. There is some doubt whether the effect is real, because the degree of polarization is only 2 or 3 times the error expected from pulse count statistics.

NGC 4438 was included in the observing list mainly because of the dust clouds ejected toward the north-east under the influence from a nearby E galaxy. The light from these clouds is very faint and the data uncertain. No remarkable polarization was found in these faint areas. In the bright part of the galaxy a polarization of 1.4% was recorded which is probably to be interpreted as interstellar polarization in the spiral arms of NGC 4438.

The peculiar galaxy NGC 3718 has a rather well defined 'bar' of dark matter through the nucleus and the main body. This dark lane does not look like normal

Greenberg and Van de Hulst (eds.), Interstellar Dust and Related Topics, 209–212.

TABLE I

Galaxy polarizations

Galaxy NGC	Type[a]	Sp	C_0	Interesting feature	Observed polarization effects
3067	Sb⁺III	F2	0.54:	Early type spectrum Similar to M 82?	Possibly polarized. Data uncertain due to bright sky
3718	Snnt	G0	0.54	Dust 'bar' through nucleus	About 4% pol. in dust bar El. vector parallel to bar
4216	Sb II	G3	0.88	Unusual color distribution	Weak polarization at bright side If real, possibly due to scattering
4438	Snntt	G3	0.66	Dust cloud ejected Disturbed by E galaxy	$P = 1.40 \pm 0.23\%$, $\theta = 40° \pm 5°$ at center Weak uncertain pol. in dust cloud
4565	Sb I:	G0	0.68	Edge-on spiral. Dust lane accross nucleus	$P = 2.8 \pm 0.8\%$ along dark band of spiral structure. See Figure 1

[a] Types, spectra and colors mainly after Van den Bergh (1960) who gives references to the original sources.

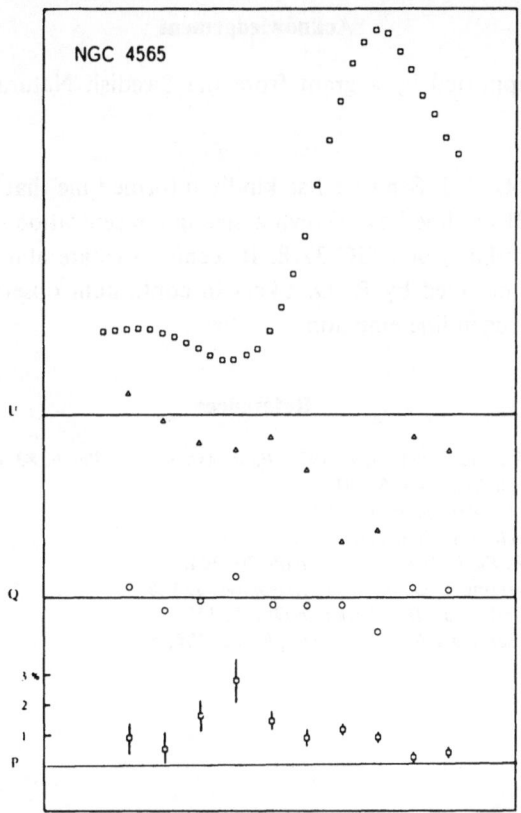

Fig. 1.   Average data from several polarimeter scans with an OG 5 filter along the minor axis of NGC 4565. The top curve shows the distribution of yellow light along the scan. U and Q are Stokes' parameters. The bottom curve P gives the polarization of light in per cent. Statistical errors from the pulse counts are indicated. (Q and U are given to indicate the direction of the polarization. Negative values of U mean that the electric vector is approximately parallel to the major axis of NGC 4565 and thus also parallel to the dark lane.)

spiral structure. My observations show, however, that the light from the dark bar is polarized about 4% with the electric vector along the bar. This indicates that large-scale magnetic fields are associated with the barlike structure. It may be speculated that this unusual dark bar may be the embryo of spiral arms formed by ejection of matter from the nucleus. It is not known to me whether NGC 3718 has been shown to emit radio radiation or to have an active nucleus.

The galaxy NGC 3067 may be a spiral but was classified as a ?I by Morgan (1958) who also published a photograph. This object was observed because of its possible similarity to M 82 which is known to show very high degrees of polarization in its outer faint regions due to scattering of light in dust clouds (Elvius, 1963, 1972; Sanders and Balamore, 1971; Visvanathan and Sandage, 1972). Preliminary reductions indicate the presence of polarization in NGC 3067 but the final conclusion must await a more careful study of the influence of a polarized sky background.

Acknowledgement

This work was supported by a grant from the Swedish Natural Science Research Council.

Note added in proof. U. J. Schwarz has kindly informed me that high resolution observations in the 21 cm line have shown a strong concentration of neutral hydrogen along the bar (dust-lane) of NGC 3718. It seems to rotate almost as a solid body. – The galaxy was detected by R. D. Ekers in continuum observations near 21 cm (including the hydrogen line emission).

References

Albrecht, R., Boyce, P., and Chastain, J.: 1971, *Publ. Astron. Soc. Pacific* **83**, 683.
Elvius, A.: 1963, *Lowell Obs. Bull.* **5**, 281.
Elvius, A.: 1972, *Astron. Astrophys.* **19**, 193.
Hall, J. S.: 1968, *Lowell Obs. Bull.* **7**, 61.
Morgan, W. W.: 1958, *Publ. Astron. Soc. Pacific* **70**, 364.
Sanders, R. H. and Balamore, D. S.: 1971, *Astrophys. J.* **166**, 7.
Van den Bergh, S.: 1960, *Publ. David Dunlop Obs.* **2**, 159.
Visvanathan, N. and Sandage, A.: 1972, *Astrophys. J.* **176**, 57.

PART V

DISTRIBUTION OF DUST AND GAS

DUST NEAR THE NUCLEUS OF M 31

HUGH M. JOHNSON

Lockheed Missiles and Space Company, Palo Alto, Calif., U.S.A.

Abstract. Special photographic observations reveal silhouetted dust patterns as close as 6″ from the nucleus. The spatial distribution and kinematics of the dust are discussed.

1. Introduction

Johnson and Hanna (1972) have copied a plate of the central region of M 31 (Johnson, 1961) with a technique which especially well depicts the abundant dust features in the radii $r = 6″$ to several arc min. We wish to make a series of conclusions about the distribution and motions of the interstellar gas and dust, partly with reference to the kinematics now well known in most of the range within $r = 2' = 400$ pc of center (Rubin and Ford, 1971). The major semi-axis X of the rotationally approaching side of the main body of M 31 is at the position angle $PA = 218°$, and the principal plane of the main body of M 31 is inclined to the line of sight at $i = 13°$ (Rubin and Ford, 1971). Thus the far side of the principal plane of M 31 lies in the range of position angles which increase from $PA = 38°$ by 180°.

2. Discussion and Conclusions

(1) If light and dust were confined to the principal plane of M 31, the dust should be equally silhouetted at all position angles. However, the light of M 31 near center is not at all confined to the principal plane, as is shown by the isophotal ratio of minor axis to major axis, 0.7 at $r = 1'$ with a slow decrease beyond $r = 1'$ (Kinman, 1965), compared with $\sin i = 0.225$ for the principal plane. Nonthermal radio sources centered on the nucleus of M 31 are also nearly spherical to the radius of 3′5 (Pooley and Kenderdine, 1967). Clouds of dust which intercept lines of sight through luminous spheroids will be much better silhouetted in the near hemisphere because of the smaller ratio of foreground to background light. Therefore we interpret the strongest images to be clouds in the near hemisphere. Of course, a given optical depth in the near-hemisphere clouds can mimic the effect of larger optical depth in clouds farther along the same line of sight, so the interpretation if statistical.

(2) Cloud structure becomes more mottled with increasing r. Mottling is fairly equally prevalent in all position angles at $r \geqslant 2'$. We think that this argues against confinement of clouds to the principal plane because of the unequal silhouetting effect in near and far hemispheres as discussed in paragraph 1. We think that the more diffuse clouds are broken into wavy clumps with a typical size of 20″ and they cover about 50% of the area inside $r = 3'$.

Greenberg and Van de Hulst (eds.), Interstellar Dust and Related Topics, 215–217.

(3) An arm-shaped cloud originates (or terminates) very faintly near $r = 6''$, $PA = 120°$, thrusts northeastward, becomes quite strong by $r = 15''$ just before it crosses the major axis at $r = 1'1 - 1'7$ in $PA = 38°$, and continues northward to $r = 2'0$, $PA = 10°$, at least in one branch of several possible branches. Is this a 'spiral arm'? We know of no test in this case to prove spiral structure or, on the contrary, to prove accidental contiguities. If the proposed arm is rectified onto the principal plane, it appears to be overly stretched parallel to the Y-axis, and it cannot then be well approximated by a logarithmic spiral or an Archimedes spiral. We believe that the feature does not lie in the principal plane.

(4) If the feature discussed in paragraph 3 were a spiral arm, we might expect a symmetrically opposite arm. A very faint counterpart indeed appears to exist, and to be identical with Baade's (1958, 1963) S1 arm where it crosses the X-axis at $PA = 218°$. If it is axi-symmetric with part of the paragraph-3 arm, the axis of symmetry may not be close to the rotation axis of the principal plane of M 31 because the arms should then be equally visible where they cross the respective $\pm X$ semi-axes.

(5) We conclude that the dust near the nucleus of M 31 is probably not restricted to the 25-pc-thick disk of gas in the principal plane to which the model of Rubin and Ford (1971) confine the gas. We think that the dust is not even confined to the 100-pc-thick co-planar disk of clouds which Rubin and Ford (1971) postulate. Spatial decoupling implies kinematical decoupling, so we anticipate the radial-velocity jumps which Rubin and Ford (1971) report at the intersections of the spectrographic slit with cloud images. The most pronounced discontinuity on their spectra occurs where the slit crosses the image of a large and dark cloud complex marked 'f' on their Figure 8, at about $r = 70''$ in $PA = 338°$

(6) We conclude from both form and relative contrast of various dust features which are silhouetted on the central bulge of M 31 that their distribution falls into four domains: (i) apparent absence within $r = 6''$, (ii) two arm-like features out of the principal plane within $r = 1'$, (iii) branching and break-up into wavy structure in a nearly isotropic transition zone around $r = 2'$, and (iv) formation into major spiral arms nearer the principal plane and beyond $r = 2'5$ on the minor axis and certainly by $r = 8'-10'$ on the major axis (Baade's N2 and S2 arms). Rubin and Ford (1971) observe velocities of gas in dust to be prevalently redshifted relative to velocities of disk gas. This suggests net infall of dust towards center, possibly with tide-like effects on clouds captured from domain (iii) into domain (ii). The dust clouds contain gas which emits brighter lines than the disk does (Rubin and Ford, 1971). Volume-emissivity must be higher in the clouds than it is in the disk since the cross-sections of the clouds are generally smaller than 100 pc, which is the line-of-sight depth of the gas confined to the disk models of Rubin and Ford.

(7) We take note that the position angle of the major axis of the ellipsoidal nucleus of M 31 is at $PA = 52° \pm 2°$ or 14° greater than the position angle of the major axis of the principal plane of the main body of M 31, as discovered by Johnson (1961). This has important dynamical implications. If the angular-momentum vector of the nucleus is not parallel to the angular-momentum vector of the bulk of the mass of

M 31, the decoupling may be the result of inflow of matter from outside the principal plane of M 31. If 'the nucleus of M 31 is really quite young', as Spinrad (1971) asserts, the input of fresh matter may be required by way of stellar evolution.

(8) Finally we note that S And, the 1885 supernova, appeared at $r = 17''$, $PA = 254°$ (Copeland, 1886), which is in or near a small knot of dust in the counter-arm described in paragraph 4.

Acknowledgements

This work has been done under the Lockheed Independent Research Program. We thank Dr C. R. O'Dell, Director, for forwarding the plate used in this study from the Yerkes Observatory vault, and we thank Dr Vera C. Rubin for helpful comments on an early draft of this paper.

References

Baade, W.: 1958, in D. J. K. O'Connell (ed.), *Stellar Populations*, North-Holland Pub. Co., Amsterdam, p. 3.

Baade, W.: 1963, in C. Payne-Gaposchkin (ed.), *Evolution of Stars and Galaxies*, Harvard Univ. Press, Cambridge, p. 59.

Copeland, R.: 1886, *Monthly Notices Roy. Astron. Soc.* **47**, 49.

Johnson, H. M.: 1961, *Astrophys. J.* **133**, 309.

Johnson, H. M. and Hanna, M. M.: 1972, *Astrophys. J. Letters* **174**, L 71.

Kinman, T. D.: 1965, *Astrophys. J.* **142**, 1376.

Pooley, G. G. and Kenderdine, S.: 1967, *Nature* **214**, 1190.

Rubin, V. C. and Ford, W. K., Jr.: 1971, *Astrophys. J.* **170**, 25.

Spinrad, H.: 1971, in D. J. K. O'Connell (ed.), *Nuclei of Galaxies*, North-Holland Pub. Co., Amsterdam, p. 45.

A STUDY OF THE DISTRIBUTION
OF INTERSTELLAR MATTER

LARS OLOF LODÉN

Astronomical Observatory, Uppsala, Sweden

Abstract. As a first step towards an empirical estimation of the minimum size of interstellar dust formation units (clouds etc.) the correlation between angular separation and difference in colour excess has been studied for a selection of close star couples with coinciding magnitude and spectral type. Principally, the result indicates that there is no palpable correlation between the difference in colour excess and the lateral separation of the stars. This and other properties speak in favour of the conception that obscuring matter in the Milky Way is concentrated in extremely small units of high density.

1. Introduction

The convenient picture of the interstellar medium as being more or less homogeneously distributed over considerably extended parts of the Milky Way is in any respect unallowably oversimplified. Several astrophysical phenomena would be explained in a relatively natural way if the interstellar matter was considered to be concentrated in numerous very small and dense elementary units in the shape of miniclouds, bars, or rods. The most conspicuous indication is apparently the capricious behaviour of the colour excess. It is a well-known fact that in no part of the sky can there be found a good point-to-point correlation between distance and colour excess for the stars. Only a very vague statistical correlation may be established, which will differ appreciably between various parts of the sky. For numerous more or less nearby stars there has been recorded an astonishingly great excess, while for some very distant ones the excess is practically undetectable. This phenomenon is also reflected by the correspondingly great variability in the specific colour excess (units of magnitude per kpars). Firstly there is a general tendency for this parameter to vary strongly with distance in a certain direction, independently of its absolute level for a given distance. In some cases the observed specific excess will diminish with a power of ten when the distance increases from 0.4 kpc to 4 kpc ($m_0 - M$ = 8 to 13), thus indicating that most of the extinction might occur within a relatively short distance from the observer. In general, the variation is more modest. Table I illustrates two different instances. Secondly there is a tremendous amplitude in the individual values of the specific excess. A reasonable average value for 1 kpars in the galactic belt will lie around 0.25, but the individual values range between 0.03 and 5.5 or even more. At least a few per cent of the early-type stars beyond 1 kpars near the galactic plane show values around 0.05 and the same fraction of the more local ones show values greater than 2.5. Certain checks show that these discrepancies cannot be explained only as an effect of differences in altitude above the galactic plane.

Greenberg and Van de Hulst (eds.), Interstellar Dust and Related Topics, 219–225.

TABLE I

Specific colour excess as a function of corrected
distance modulus for the Carina-Centaurus region
according to Lodén (1970) and for the north-
accessible part of the Milky Way according to
Hiltner (1956)

$m_0 - M$	$E\ kpc^{-1}$	
	Car-Cen	NMW[a]
7.0	0.338	1.252
8.0	0.302	1.150
9.0	0.265	0.900
10.0	0.228	0.715
11.0	0.192	0.535
12.0	0.155	0.360
13.0	0.118	0.180

The absolute level for the NMW values is affected
by the fact that most of the stars in question are
particularly selected with respect to high colour
excess.
[a] NMW = Northern Milky Way.

Among other indications of the presence of small and dense units in the interstellar
space we may mention the measured sizes of certain molecular radiation sources. If
these formations may have dimensions of the order of some solar system radii, the
dark interstellar cloud elements may also have it.

2. Tentative Study of Density Gradients in the Interstellar Space by Means of Comparison of Colour Excess for Stars with Small Angular Separation

2.1. COINCIDENCES

By visual inspection of objective-prism plates of Mily Way regions one frequently
detects pairs or higher multiples of stars with very small angular separation and so
closely coinciding spectral type and apparent magnitude that they form a conspicuous
configuration on the plate. It might be reasonable to classify these objects, or at least a
considerable fraction of them, as parts of highly diluted clusters or associations which
at present can only be revealed as such coincidences – in some cases even as binaries
(Stock and Wroblewski, 1972). A statistical discussion of the problem with respect to
the present experience and material supports this idea. In order to establish the real
physical connection between the components in each individual case it is necessary to
secure additional information in the form of dynamical data etc., however.

Independently of the true genetic correlation between the 'coincidence' components,
the objects in question are very useful for a large series of various investigations just
because of the similarity in magnitude and spectral type in conjunction with the narrow
separation. One investigation of this kind concerns the distribution of interstellar
matter. The members may be easily compared with respect to parameters which are

conditioned by the influence of this matter, for instance colour excess, polarization, or interstellar absorption lines, and local variations will be detected even in a superficial analysis. The present study is based upon UBV photometry of 320 'coincidence' groups with a total of 800 stars in the Carina-Centaurus region (Lodén, 1970). The majority of these stars are of spectral types rather close to A0 and hence, their colour excess can be estimated with fair accuracy.

2.2. THE COLOUR EXCESS DIFFERENCES

As a first rough approach to the problem I have computed and studied the mean values of the excess difference for certain intervals in the separation (Table IIA and IIB). If there should be a tendency towards a continuous increase of the difference with the separation over large regions, this may indicate the presence of relatively extended homogeneous clouds or moderate density gradients in interstellar space. If, on the other hand, there is no such tendency detectable one might draw the conclusion that the matter is discontinuously distributed with great density gradients within small regions. According to the table there is practically no correlation at all between mutual distance and colour excess. In other words: the chance of finding a certain difference in colour excess is the same for a very narrow pair as for an appreciably more widely separated one. Consequently, there is already some indication of the suspected distribution of the interstellar matter and the concentration in small dense units. The order of magnitude may be estimated as 10^4 AU or less.

A possible objection that the estimated difference might also be interpreted as a

TABLE IIA

Mean difference in colour excess as a function of lateral separation in angular distance

Minutes of arc	$\Delta(B-V)$	$\Delta(U-B)$	N
0.05 − 1.0	0.076	0.098	162
1.1 − 2.0	0.070	0.100	73
2.1 − 3.0	0.076	0.099	58
3.1 −10.0	0.068	0.100	158
10.1 −26.0	0.086	0.176	33

TABLE IIB

Mean difference in colour excess as a function of lateral separation in linear distance

log AU	$\Delta(B-V)$	$\Delta(U-B)$	N
< 4.0	0.064	0.086	37
4.0 − 4.4	0.084	0.116	56
4.5 − 4.9	0.069	0.098	96
5.0 − 5.4	0.074	0.096	158
5.5 − 5.9	0.063	0.122	114
> 5.9	0.122	0.093	18

TABLE III

Average values of $\Delta(\overline{B-V})$ for certain intervals in $m_0 - M$

$m_0 - M$	$\Delta(\overline{B-V})$	N
6.0– 6.9	0.041	10
7.0– 7.9	0.053	21
8.0– 8.9	0.060	47
9.0– 9.9	0.072	131
10.0–10.9	0.074	154
11.0–11.9	0.094	63
12.0–12.9	0.068	54

difference in intrinsic colour caused by uncertainties in the classification should not be ignored. However, an effect of this type should not be dependent upon the lateral separation. It would merely increase the scatter in the individual values.

Concerning the statistical distribution of the differences in colour excess it is interesting to note that it apparently follows that of a normal distribution with maximum frequency at small values but not at zero. The corresponding distribution of the 'relative' excess differences ($\Delta(B-V)$/separation in parsecs) also gives the impression of a normal frequency distribution that is very similar to the just mentioned one. An investigation of the possible correlation between difference in colour excess and distance from the observer shows a rather clear statistical increase of the difference with the distance, at least up to a certain limit (Table III). This might indicate that the greater the distance, the greater the risk for obscuration of one of the two components, i.e. the result in question also speaks in favour of the discontinuous distribution of the interstellar matter.

2.3. THE ABSOLUTE COLOUR EXCESS

Next we may study the average behaviour of the absolute colour excess itself, represented by the mean value for both components in the same pair of stars. Starting with the frequency we then again find a rather typical normal distribution which does not change very much if we repeat the procedure with the 'specific' colour excess. Thus, there is no visible connection between the distribution of colour excesses and the distribution of stars, which one should have expected if the interstellar matter had been smoothed out along the line of sight.

Another way to study the present material is illustrated in Table IV. The investigated part of the Milky Way is subdivided in three sections, each one covering approximately 15° in galactic longitude, and the average value of the excess difference is computed for each section. For comparison the corresponding average values of the total colour excess are also included. It seems evident from Table IV that there is a very poor correlation between the various parameters. The only striking feature is the apparent minimum of the excess difference in both colours for the central region. This tendency, however, has no visible correspondence in total excess or average distance.

TABLE IV

Average values of colour excess difference for three main sections of the Carina-Centaurus region. Average total colour excess and distance added for comparison

	Galactic longitude intervals l^{II}		
	281°–294°	294°–307°	307°–320°
$\overline{\Delta(B-V)}$	0.078	0.064	0.078
$\overline{\Delta(U-B)}$	0.092	0.088	0.124
N	114	153	213
$\overline{E(B-V)}$	0.218	0.271	0.245
$\overline{E(U-B)}$	0.128	0.144	0.136
$\overline{K}$pc[a]	1.76	2.10	1.56
$\overline{K}$pc[b]	1.30	1.45	1.14

[a] Uncorrected for interstellar extinction.
[b] With tentative correction for interstellar extinction.

Thus there is no indication that large differences in colour excess should be connected with high values of the excess itself and vice versa. Rather there is an indication of anti-correlation, whether spurious or not. This fact might be interpreted as an effect of difference in the cloud element dimensions between different regions so that, for instance, the elements in the Crux Coal Sack region should be somewhat greater than the ones just outside this region.

Although the above observations may be interpreted in more than one manner, they all speak in favour of a very heterogeneous distribution of the interstellar dark matter unless we prefer the application of much more complicated physical procedures in the production of colour excess than the hitherto accepted ones.

In the author's opinion a continued detailed study of narrow stellar pairs with coinciding spectral type will probably be very valuable for our knowledge about the distribution of the interstellar matter.

As a preliminary result of the above discussion we may conclude that the presence of dense distinct interstellar clouds with dimensions of the order of much less than 10000 AU should not be regarded as unreasonable. I do not even find it futile to look for observable time-variation in colour excess for some stars as a result of a combination of proper motion and heterogeneity in the interstellar obscuring medium. A corresponding study of the interstellar polarization might also be conceivable.

Appendix: Hints on a Possible Model for the Distribution of Interstellar Matter

The problem is to find within the framework of hitherto accepted values of physical parameters for the interstellar space, a mechanism that is capable of condensing matter in small units. If this turns out to give a negative result, the next problem will be to reconsider the parameters themselves without getting into conflict with generally accepted physical laws. As a final step, the validity of the laws themselves has to be seriously scrutinized.

For the relative vicinity of the galactic plane some tentative figures of reasonable qualitative characteristics of the *average* physical conditions in the interstellar space have been given and generally accepted. Although these figures are for several reasons subject to great uncertainty, it would at least be encouraging if a modified theory of any kind does not get in too much conflict with them. They do not imply, however, any absurdity in a series of substantial local deviations – even by several orders of magnitude.

In order to obtain a rough picture of one possible organization of a cloud system that is in first approximation non-hierarchical, we regard a huge homogeneous interstellar cloud the diameter of which has the order of magnitude of one kiloparsec. The cloud in question is situated in front of the observer and covers the field of vision in a certain solid angle. Its density is considered to be equal to the average density of the interstellar matter in the Milky Way. For mathematical simplicity, the cloud is regarded as spherical although its real shape is in fact completely arbitrary. Suppose the cloud is divided up in a certain number of equal fragments with the same total surface as the first one and randomly scattered within the same volume. In that case the density of the small clouds will be higher than that of the first one. Table V shows the simple relations between number, radius, and density for this situation.

For stars situated within or beyond the cloud region there is now a certain risk for obscuration by one or occasionally more clouds, but there is also a corresponding possibility that the line of sight should be free from any obstacle. Under highly idealized conditions, the fraction F of the field of vision that will be obscured by k clouds ($k = 0 - n$) may be expressed by the relation

$$F(k, n) = \frac{n!}{k!(n-k)!} \left(\frac{1}{n}\right)^k \left(1 - \frac{1}{n}\right)^{n-k},$$

where n is the total number of clouds. For large n values this relation will converge towards $1/ek!$. We may thus expect that about 1 % of the number of stars beyond the

TABLE V

Correlation between number, radius, and density for constant obscuring surface

log n	log r (cm)	log ϱ (g cm^{-3})	note
0	21.5	−26	kpars
2	20.5	−25	
4	19.5	−24	
6	18.5	−23	pars
8	17.5	−22	
10	16.5	−21	
12	15.5	−20	conceivable region
14	14.5	−19	
16	13.5	−18	o.m. AU

cloud system will be obscured by 4 elementary clouds, about 6% by three, about 20% by two, about 37% by one, and about 37% by none. At a superficial comparison with the observational result these figures are apparently quite reasonable, but this statistical procedure is not realistic because the final result is not sensitive to the number of clouds. Furthermore, we have no empirical indication that the clouds should be equal in size and regular in shape, or that their total surface should be kept constant.

References

Hiltner, W. A.: 1956, *Astrophys. J.* **24**, 11.
Lodén, L. O.: 1970, *Stockholm Observ. S. Milky Way Survey*, Report 1.
Stock, J. and Wroblewski, H.: 1972, *Astron. Astrophys.* **18**, 341.

DUST AND NEUTRAL HYDROGEN
IN THE REGION OF IC 1396

S. CHRISTIAN SIMONSON, III

Astronomy Program, University of Maryland, Md., U.S.A.

Abstract. A 21-cm survey at Dwingeloo of a 5° square containing IC 1396 shows several neutral hydrogen concentrations that correspond in shape and position with dust clouds, particularly the bright-rimmed clouds involved with the H II region.

1. Introduction

IC 1396 is a large H II region of low surface brightness excited by the O6 trapezium system HD 206267 and the stars of the early-type cluster Trumpler 37, whose distance modulus is 9.6 mag. (Simonson, 1968). Since the angular size of the nebula and the bright-rimmed dust clouds with which it is interacting are a few times larger than the beamwidth of the Dwingeloo 25-m radio telescope, the region has some advantages for studying the relation of neutral hydrogen and dust. However, its location at $l =$ 100° presents problems in separating discrete neutral hydrogen features from the general foreground and background neutral hydrogen since the velocity range $v = 0$ to -10 km s^{-1} corresponds to a distance range of 0 to 1.5 kpc. The results of a survey of neutral hydrogen and a comparison with the dust distribution are given briefly here; the details will be presented elsewhere.

2. Observations

A 5° square containing IC 1396, from $l = 97°$ to 102°, $b = +1°$ to $+6°$, was surveyed on a 0°.5 grid with the Dwingeloo 25-m radio telescope using a bandwidth of 8 kHz (1.7 km s^{-1}). Contour maps of brightness temperature were constructed in all three coordinate planes – $l, b; b, v;$ and l, v – at every interval in the orthogonal coordinate. The l, b maps were compared with the Palomar Observatory Sky Survey and with Khavtassi's (1960) and Lynds's (1962) atlases of dark nebulae, all reproduced on the same scale.

3. Results

In order to confirm the reality of supposed correlations of 21-cm and optical features, one must have some idea of the size of fluctuations in the general field. In this instance it is possible to use an unpublished large-scale 21-cm survey by H. van Someren Greve made at Dwingeloo and covering the region $l = 90°$ to 115°, $b = -1°$ to $+15°$, on a 1°.4 grid. The large-scale survey shows a general concentration of 21-cm line emission

Greenberg and Van de Hulst (eds.), Interstellar Dust and Related Topics, 227–229.

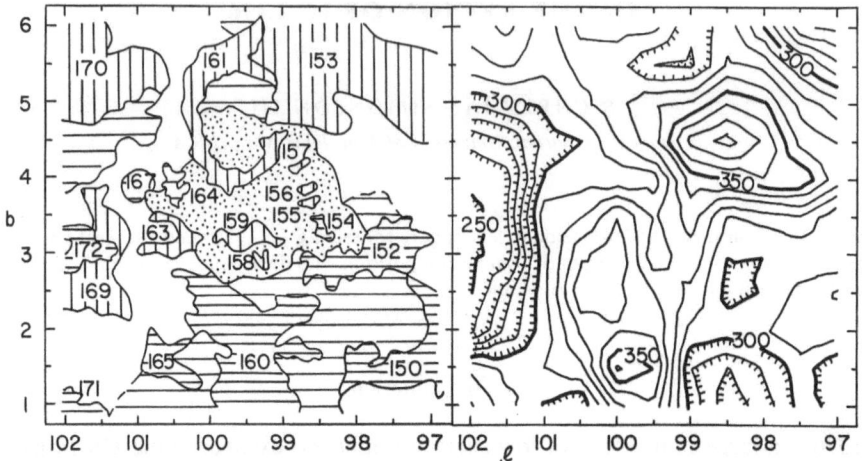

Fig. 1.   Comparison between dust clouds (Khavtassi, 1960), left, and neutral hydrogen brightness temperature, summed over $v = -4$ to 0 km s^{-1}, right. Cross hatches indicate dust clouds of three levels of opacity; stippling indicates Hα emission. The 21-cm contour intervals are 10 K.

at the position of IC 1396 in the velocity range of the cluster stars and the Hα emission, about -5 to $+5$ km s^{-1}. At velocities below -5 km s^{-1} a galactic feature, perhaps a 'spur', extends into the region from the direction of Cygnus.

The present detailed survey resolves the concentration around IC 1396 into several smaller concentrations. Figure 1 shows the comparison with dust clouds in the velocity range containing the strongest features. We note the following correlations between 21-cm features and dust clouds:

(1) Bright-rimmed dust clouds, which are obviously connected with IC 1396, appear to be associated with H I emission features at velocities between -5 and $+5$ km s^{-1}. These include Kh 155, 156, 157, 159, 160, 163, 164, 167, and portions of Kh 153 and 161. There is a tendency for the very densest parts of the dust clouds to be associated with somewhat less 21-cm emission than less opaque parts, but there is no evidence in the line profiles for self-absorption. This would indicate the gas temperature exceeds 80 K, but there may also be some foreground emission.

(2) Dust clouds seen projected against IC 1396 but without bright rims, e.g., Kh 152, 154, and parts of 161, are apparently associated with 21-cm emission of somewhat lower intensity at velocities between 0 and $+10$ km s^{-1}.

(3) Other dust clouds associated with 21-cm emission at $v = -5$ to $+5$ km s^{-1} are Kh 150, 165, 170, 171, and parts of Kh 153 and 161.

(4) A region bright in H I but only moderately obscured occupies about 1 deg^2 at $l = 98°$, $b = +4°.5$, at $v = 0$ to $+5$ km s^{-1}.

(5) Beginning rather abruptly at $l = 101°$ between $b = +1°.5$ and $+5°$ and extending into higher longitudes is a region very low in 21-cm emission and low in optical obscuration except for several sharp-edged opaque dust clouds, e.g., Kh 169 and 172. Van den Bergh (1967) has suggested that this area is a 'fossil H II region'. The ap-

pearance in H I tends to bear this out, as if some general expansion had swept the neutral gas away.

Overall, in the velocity range -4 to $+10$ km s^{-1}, where there appears to be significant correlation of dust and neutral hydrogen, the mean column density of neutral hydrogen, N_H, amounts to 1.5×10^{21} cm^{-2}, uncorrected for the effects of optical depth. For 28 stars in the survey area with a mean distance modulus of 9.6 mag. (Simonson, 1968), the ratio $N_H/3E_{B-V} = 8 \times 10^{20}$ cm^{-2} mag.$^{-1}$.

Judging by van Someren Greve's large-scale survey, about 80 % of N_H may be due to foreground and background hydrogen. (It is unnecessary to allow for absorption against the thermal radiation of IC 1396; at 1400 MHz the peak brightness temperature is less than 3 K (Lynds, 1961).) The mass in the concentrations may therefore be estimated as $2 \times 10^4 \mathcal{M}_\odot$, uncorrected for optical depth and taking the distance as that of IC 1396. For comparison, the mass of ionized hydrogen is about $7 \times 10^3 \mathcal{M}_\odot$ (Pottasch 1965) and the mass of stars earlier than B3 is about $10^3 \mathcal{M}_\odot$ (Simonson, 1968).

4. Conclusions

The dust clouds in the region of IC 1396 are apparently associated with neutral hydrogen concentrations, and the association is clearest for the clouds with bright-rim structures. The major exception is a 'fossil H II region'. Provided the optical depth in the neutral hydrogen does not exceed a value of the order unity, the ratio of neutral hydrogen in dust clouds to ionized hydrogen to early-type stars is 20:7:1 in IC 1396 and its environs. The overall ratio of N_H/A_V amounts to 8×10^{20} cm^{-2} mag.$^{-1}$, quite an ordinary value.

Acknowledgements

The observations were made in collaboration with H. van Someren Greve and with financial support from the Netherlands Organization for the Advancement of Pure Research (Z.W.O.) while the author was on the staff of the Leiden Observatory. Support from the U.S. National Aeronautics and Space Administration is gratefully acknowledged.

References

Khavtassi, J.: 1960, *Atlas of Galactic Dark Nebulae*, Abastumani Obs.
Lynds, B. T.: 1962, *Astrophys. J. Suppl. Ser.* 7, 1.
Lynds, C. R.: 1961, *Publ. Nat. Radio Astron. Obs.* 1, 43.
Pottasch, S. R.: 1965, *Vistas* 6, 149.
Simonson, S. C., III: 1968, *Astrophys. J.* 154, 923.
Van den Bergh, S.: 1967, *J. Roy. Astron. Soc. Can.* 61, 23.

THE ASSOCIATION OF DUST AND H I GAS

E. BRAUNSFURTH

MPI für Radioastronomie, Bonn, Germany

and

K. ROHLFS

MPI für Radioastronomie, Bonn, Germany

and

Astronomy Program, University of Maryland, Md. 20742, U.S.A.

Abstract. The hypothesis that interstellar dust is associated with cold gas is advanced. To test this, such observational features of the gas that depend predominantly on the cold gas have to be isolated. Two such features are discussed, the 21-cm line strength in absorption and the intensity of narrow emission line components in intermediate galactic latitudes. It is shown that both features have a good correlation with the dust extinction. Various measurements of the gas to dust ratio are collected and some implications for the interstellar medium are discussed.

The relative abundance of atomic hydrogen gas vs. dust grains has been a matter of controversy for many years. In most theoretical investigations Lilley's (1955) value of $\varrho_H/\varrho_d = 100$ has been used, but there is little observational evidence that this number is a constant. Already many years ago Bok (1953) remarked that this ratio may well vary from region to region. Indeed several authors have cast severe doubts on the assumption that an average ratio ϱ_H/ϱ_d has much physical meaning. Recently Wesselius and Sancisi (1971) showed quite conclusively that no general relation exists between N_H and A_V that applies to all regions. On the other hand, just as convincing evidence can be found in the investigations of the very same authors that N_H and A_V are correlated, if we restrict the investigations to selected regions. These relations may, however, vary from region to region. Some heuristic arguments may indicate the reasons for this variation and show what to do if one wants to find a relation applicable everywhere.

If a dust grain interacts with the interstellar gas, the growth rate of the grain radius is $da/dt \propto T^{-\frac{1}{2}}$ (Wickramasinghe, 1967, Equation (6.7)) in an isobaric medium. According to Spitzer (1968, p. 181 ff) and Field *et al.* (1969) the interstellar medium does fulfill this condition, and therefore the dust grains will grow predominantly in low temperature (high density) regions. If the sticking probability of the atoms on the grain surface increases with decreasing temperature this tendency will even be stronger. To obtain information pertaining to dust formation, we therefore should compare the gas content of such cool regions with the dust extinction.

Observations of the 21-cm line in emission, however, include a fair amount of hot gas as well, and very little dust, if any, should be expected to be associated with this hot gas, if our working hypothesis is right. Therefore the observed ratio N_H/A_V will be too large by an amount that depends on the ratio of the hot and the cold gas along

Greenberg and Van de Hulst (eds.), Interstellar Dust and Related Topics, 231–235.

the line of sight. Means thus have to be found to distinguish observationally the contribution of the cold gas from that of the hot.

(1) Cold, dense gas has a large optical depth for 21-cm line radiation ($\tau \propto N_H/T$) and thus such regions will be detectable in 21-cm absorption profiles. A comparison of the 21-cm absorption and the dust extinction has been made by Rohlfs (1971). The 21-cm absorption data were taken from Radhakrishnan *et al.* (1972) and mainly extragalactic sources have been used to make sure that all neutral hydrogen gas in the

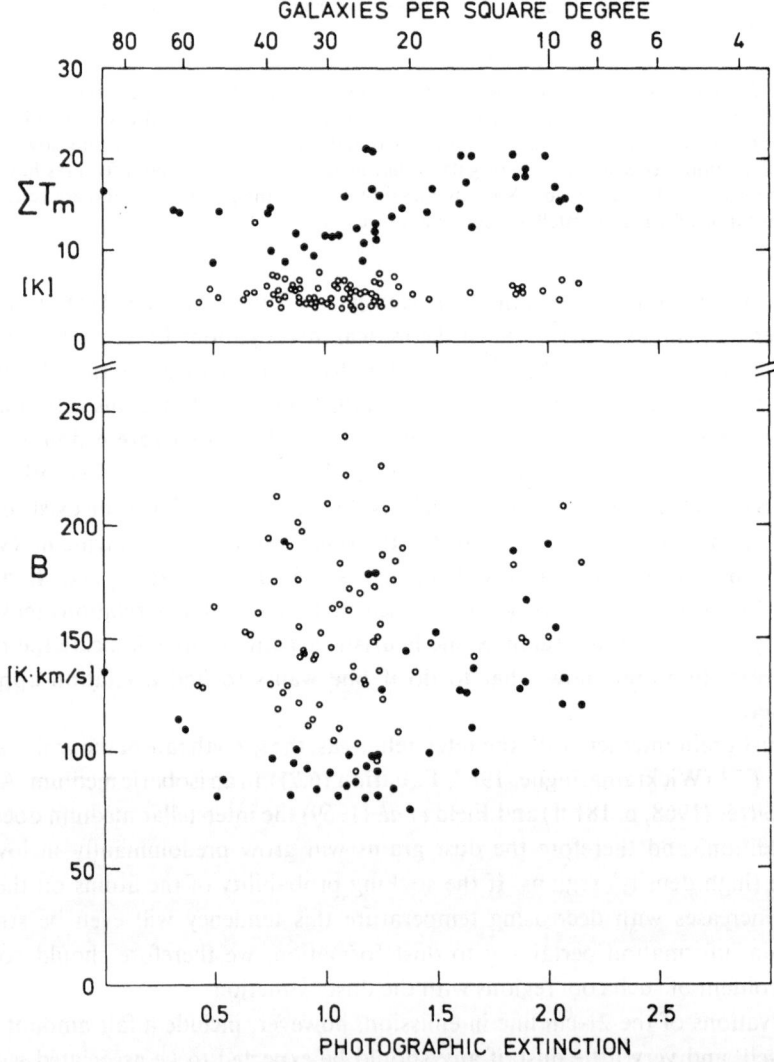

Fig. 1. Comparison of the peak brightness temperature (upper diagram) and the profile area (lower diagram) of emission line profile components with the photographic extinction in the same direction for $b^{II} = 30'$. The extinction is inferred from counts of galaxies averaged over 6 square degrees in latitude, the counts are given at the top of the diagram. Line profile components with dispersions of $10 < \sigma < 15 \, \text{km s}^{-1}$ are marked ○, those with $1 < \sigma < 4 \, \text{km s}^{-1}$ are marked ●.

line of sight will be in front of the source. The optical extinction is taken from observations of OB stars with $z > 500$ pc in directions close to the sources. Although sources from quite different directions in the sky were used, and even several low latitude objects were included, there is a strong correlation between the 21-cm atomic hydrogen absorption and the visual extinction. One relation fits all data, irrespective of which area in the sky they come from. The slope of this mean relation gives $\varrho_H/\varrho_d \approx 50$. This value depends somewhat on the assumed grain model and spin temperature of the gas.

(2) Another attempt to separate the cold gas in the line of sight from the contribution of the hot gas has been made with the help of the intermediate latitude 21-cm emission line survey of Grahl *et al.* (1968). The line profiles have been separated into Gaussian components. Usually a wide component with a dispersion $\sigma \approx 12$ km s^{-1} and a narrow one with $\sigma \approx 3$–4 km s^{-1} were found. Mebold (1972) showed that the wide component is distributed fairly uniformly through space and probably corresponds to the high temperature component of the gas. We therefore have compared the narrow components with extinction data taken from galaxy counts by Shane and Wirtanen (1967) according to the compilation by Kiang (1968). Avoiding the region $-30° < l < 30°$ towards the galactic center, we obtained the results given in Table I for the correlation coefficients. Nonparametric Spearman ranked correlation coefficients were used to avoid problems which stem from the distribution of the data points, but both the common product-moment correlation coefficient and Kendall's ranked correlation coefficients gave essentially identical results. Thus the dust extinction is correlated mainly with narrow emission line profiles and hardly at all with wide ones. Quite similar results concerning the correlation of neutral hydrogen gas and dust extinction have been obtained for the intermediate galactic latitude survey of Takakubo *et al.* (1966). The details of these investigations will be reported elsewhere.

TABLE I

Correlation-coefficients between emission line strength and dust extinction for b = 30°

		Narrow compon. $1 < \sigma < 4$	Wide compon. $10 < \sigma < 15$
B vs A_{p_g}	corr. coef.	0.42 ± 0.12	0.07 ± 0.23
	probabil. P	0.002	0.38
T_{max} vs A_{p_g}	coor. coef.	0.47 ± 0.13	0.10 ± 0.17
	probabil. P	0.0005	0.28

In Table II results for the gas/dust ratio are collected. We have always given three quantities B/A_V, N_H/A_V and ϱ_H/ϱ_d. Each of these contains all the information that the other has. But B/A_V gives the ratio of two observational quantities, and no theoretical assumption enters into its determination ($B = \int T_b (V) dV$ in K·km s^{-1}). The derivation of N_H involves an assumption about the optical depth of the gas, while ϱ_H/ϱ_d in addition requires a dust grain model. Here Lilley's (1955) model ($a = 3 \times 10^{-5}$ cm,

$Q = 2$, $\varrho_d = 1$ g cm^{-3}) was used. The range of values in the last column is due to different assumptions about the optical depth of the line radiation. Although the values given must be regarded as very uncertain and the possibility exist that both the hydrogen self absorption and the narrow emission line component underestimate the hydrogen content somewhat, we believe that the range is due to real differences and not to observational selection. A question of some importance is then, what conclusions concerning the chemical composition of the interstellar medium can be drawn from this. It should be remembered that the observations discussed here refer to atomic hydrogen only; hydrogen in any form different from this remains undetected here. If the usual values $X = 0.63$, $Y = 0.36$, $Z = 0.014$ (Allen, 1963) for the abundance of hydrogen, helium and heavy elements by weight are adopted, and if it is assumed that all heavy elements are locked up in the dust grains, then $\varrho_H/\varrho_d = 45$. Values smaller than this can only be obtained if some of the hydrogen is in a form other than atomic. At most $S = 0.00083$ hydrogen by weight can be locked on the grains as H_2O-ice if the standard oxygen abundance is adopted. This would require that all oxygen is used up in this. The only other alternative is molecular hydrogen.

TABLE II

Gas-to-Dust Ratio according to different methods

Source	General high latitude medium (Sturch 1969)	21-cm abs. (Rohlfs 72)	21-cm self absorption (Knapp 72)	Narrow em. line components (this paper)
B/A_v	600–1100	300	32	40
N_H/A_v	1.1×10^{21}–2×10^{21}	5.5×10^{20}	6×10^{19}	1.3×10^{20}–7.3×10^{19}
ϱ_H/ϱ_d	100–180	50	5.4	11.5–6.6
f	0 0	0	0.88	0.74–0.85

The last line in Table II gives the fractional abundance of molecular hydrogen computed by

$$f = \frac{2n_2}{n_2 + 2n_2} = 1 - \frac{Z}{X} \frac{\varrho_H}{\varrho_d}.$$

If this amount of molecular hydrogen is not observed, either as free gas or as a coating on the grains, then the only alternative is a varying chemical composition of the interstellar medium.

References

Allen, C. W.: 1963, *Astrophysical Quantities*, 2nd ed., Athlone Press, London.
Bok, B. J.: 1953, in J. M. Burgers and H. C. van de Hulst (eds.), 'Gas Dynamics of Cosmic Clouds', *IAU Symp.* 2, 221.
Field, G. B., Goldsmith, D. W., and Habing, H. J.: 1969, *Astrophys. J.* 155, L149.
Grahl, B. H., Hachenberg, O., and Mebold, U.: 1968, *Beitr. Radioastronomie* 1, 3.
Kiang, T.: 1968, *Dunsink Obs. Publ.* 1, No. 5, 109.
Knapp, G. R.: 1972, Thesis, Univ. of Maryland.

Lilley, A. E.: 1955, *Astrophys. J.* **121**, 559.

Mebold, U.: 1972, *Astron. Astrophys.* **19**, 13.

Radhakrishnan, V., Brooks, J. W., Goss, W. M., Lockhart, P., Murray, J. D., and Schwartz, U. J.: 1972, *Astrophys. J. Suppl.* **24**, 1.

Rohlfs, K.: 1971, *Astron. Astrophys.* **13**, 46.

Shane, C. D. and Wirtanen, C. A.: 1967, *Publ. Lick Obs.* **22**, Part 1.

Spitzer, L.: 1968, *Diffuse Matter in Space*, Interscience Pub. N.Y.

Sturch, C.: 1969, *Astron. J.* **74**, 82.

Takakubo, H. and van Woerden, H.: 1966, *Bull. Astron. Inst. Neth.* **18**, 488.

Wesselius, P. R. and Sancisi, R.: 1971, *Astron. Astrophys.* **11**, 246.

Wickramasinghe, N. C.: 1967, *Interstellar Grains*, Chapman and Hall Ltd., London.

GALACTIC DUST DISTRIBUTION IN
THE SOLAR NEIGHBORHOOD

J. H. CAHN and R.D. NOSEK

University of Illinois Observatory, Urbana, Ill. 61801, U.S.A.

Abstract. Using the Shklovsky-Minkowski-Aller constant mass method of distances to planetary nebulae, a model of the Galactic dust distribution in the vicinity of the Sun has been determined. Distances are determined in part from extinctions derived from radio continuum to $H\beta$ flux ratios for an assumed electron temperature of 7000 K. The distance scale is based on the brightness of planetary nebulae in the Magellanic Clouds (Seaton, 1968).

In an extension of the work of Cahn and Kaler (1971) an empirically determined dust distribution in the vicinity of the Sun has been developed. The overall features of the dust distribution are based upon optical color excess measurements and 21 cm hydrogen position determinations. A recently revised set of $H\beta$ extinctions was then used to refine the positions of the dust.

In the Cahn and Kaler paper, hereinafter referred to as CK, a Mills (1959) spiral was picked whose parameters gave the best agreement with the then known $H\beta$ extinctions. It was realized at that time that such an idealization of the local spiral arms was too inflexible to permit accurate predictions of extinctions over the accessible regions of the Galaxy. It was then decided to try to develop an empirical dust map, which would provide the desired flexibility. The first step in constructing the map was to initialize the model with existing measurements. We used the color excess map developed by Fitzegerald (1968) as shown in Figure 1 and the 21 cm maps of Winnberg (1968) shown in Figure 2 and that of Kerr and Weaver as given by Simonson (1970) in Figure 3. In order to refine these positions we then extended and refined the extinctions in CK as given in Table I. These values are based upon an electron temperature of 7000 K rather than 5000 K as suggested both by Kaler and by Peimbert (1971). The radio fluxes include unpublished measurements by Cahn, Rubin and Hermann, and Aller and Milne in addition to those already reported in CK.

The resulting adjusted map is shown in Figure 4. In order to store such a large fraction of the Galaxy on the computer, only two levels of extinction, 0.407 kpc^{-1} and 2.033 kpc^{-1} were used, the latter being the shaded area in Figure 4. To account for the finite thickness of the dust distribution, the values of extinction in the plane were reduced by the Gaussian factor $\exp[-(z/150 \text{ pc})^2]$ where z is height above or below the plane. In homogeneous media of either the low or high specific extinction, the integrals to infinity along the line of sight give extinctions of 0.061cscb_{11} and 0.305cscb_{11} respectively, where b_{11} is the galactic latitude. Figure 4 represents a fit to the measured extinctions resulting in a correlation coefficient of 0.99.

It is to be understood that with such a limited list of calibrators, each of whose distance is only statistically accurate, that the map is still very speculative. In Figure 4,

Greenberg and Van de Hulst (eds.), Interstellar Dust and Related Topics, 237–241.

TABLE I

Galactic coordinates

Nebula	l_{II}	b_{II}	Distance (pc)	Radius (pc)	Extinction
IC 4634	0.38	12.22	3680	0.075	0.61
NGC 6369	2.44	5.85	1040	0.070	2.21
NGC 6620	5.88	−6.15	6360	0.077	1.49
NGC 6445	8.07	3.90	1360	0.109	1.27
NGC 6309	9.66	14.81	2800	0.094	0.93
IC 4593	25.41	40.73	2990	0.093	0.14
NGC 6818	25.86	−17.90	1970	0.087	0.20
NGC 6751	29.23	−5.93	2570	0.131	0.52
NGC 6778	34.61	−6.71	2850	0.109	0.68
NGC 6572	34.62	11.84	1670	0.058	0.44
NGC 7293	36.24	−57.10	150	0.293	0.05
NGC 7009	37.76	−34.58	1270	0.087	0.20
CN3-1	38.26	12.09	4910	0.064	0.59
NGC 6781	41.84	−2.98	670	0.173	1.27
NGC 6210	43.12	37.76	2060	0.081	0.15
NGC 6804	45.75	−4.59	1640	0.095	1.00
NGC 6879	57.23	−8.93	6750	0.082	0.65
NGC 6886	60.14	−7.74	3980	0.073	1.15
NGC 6853	60.83	−3.69	260	0.211	0.10
NGC 6720	63.15	13.98	840	0.141	0.21
NGC 6842	65.91	0.60	1560	0.180	1.01
NGC 6894	69.48	−2.62	1510	0.162	0.92
NGC 7027	84.92	−3.49	1210	0.041	1.54
HU1-2	86.54	−8.83	5350	0.065	0.75
NGC 7354	107.84	2.31	1640	0.080	1.86
NGC 40	120.02	9.87	1180	0.104	0.90
IC 3568	123.66	34.55	2520	0.110	0.34
IC 289	138.82	2.81	1430	0.128	1.49
NGC 1501	144.56	6.55	1150	0.144	1.11
NGC 1514	165.53	−15.29	720	0.120	0.63
NGC 23712	189.16	19.83	1670	0.176	0.22
J 320	190.39	−17.77	5050	0.079	0.72
NGC 2022	196.68	−10.93	2640	0.124	0.47
NGC 2392	197.88	17.40	1190	0.129	0.35
NGC 1535	206.48	−40.57	2230	0.099	0.12
IC 418	215.22	−24.27	1740	0.053	0.33
IC 2165	221.33	−12.40	3520	0.068	0.85
NGC 2440	234.84	2.43	1360	0.109	0.63
NGC 2610	239.64	13.95	1810	0.151	0.77
NGC 3242	261.06	32.06	1040	0.094	0.30
NGC 2818	261.98	8.60	1750	0.170	0.67
NGC 3132	272.11	12.39	1220	0.133	0.28
NGC 4361	294.11	43.62	940	0.186	0.17
NGC 5307	312.38	10.56	2980	0.091	0.76
IC 4406	319.69	15.74	2140	0.104	0.47
NGC 5882	327.84	10.09	2230	0.076	0.44
NGC 6326	338.20	−8.38	2930	0.086	0.81
NGC 6153	341.84	5.46	1390	0.083	1.26
NGC 6072	342.16	10.83	990	0.168	1.15
NGC 6563	358.50	−7.33	1400	0.146	0.62

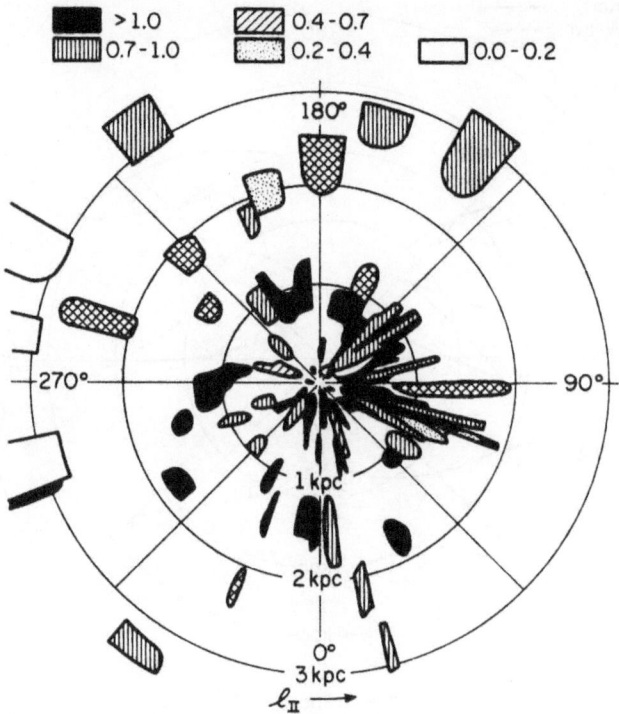

Fig. 1.   Color excesses of Fitzgerald (1968). Darkened areas indicate excesses greater than 1 mag. kpc⁻¹.

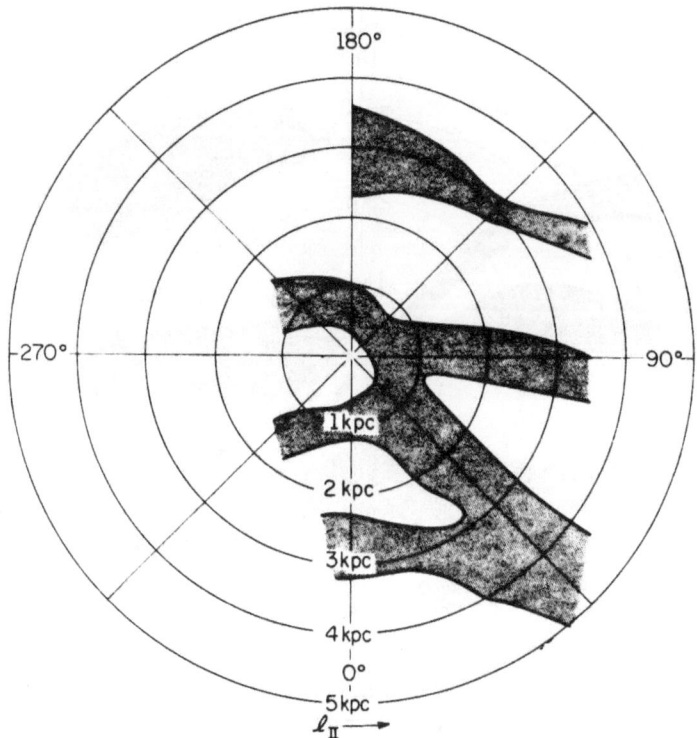

Fig. 2.   Part of 21 cm map of Winnberg (1968)

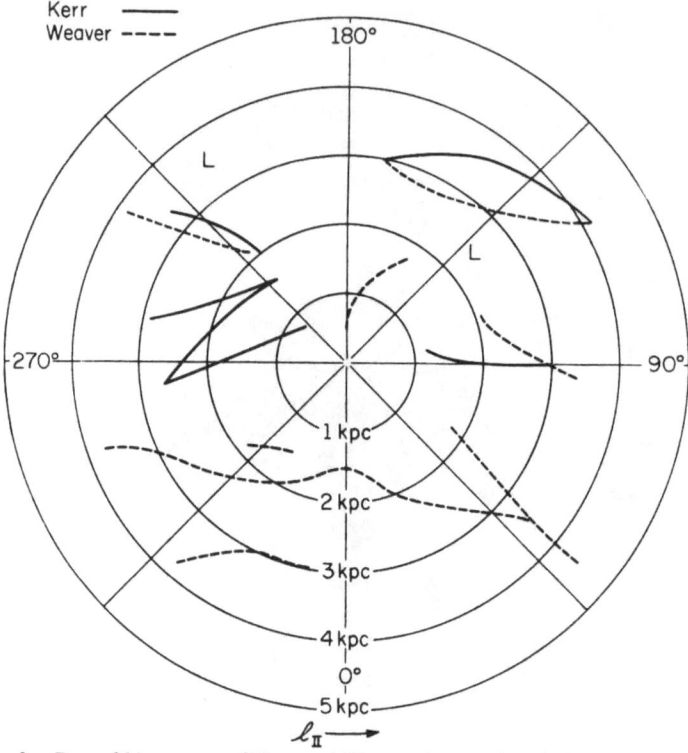

Fig. 3.   Part of 21 cm map of Kerr and Weaver data as given in Simonson (1970).

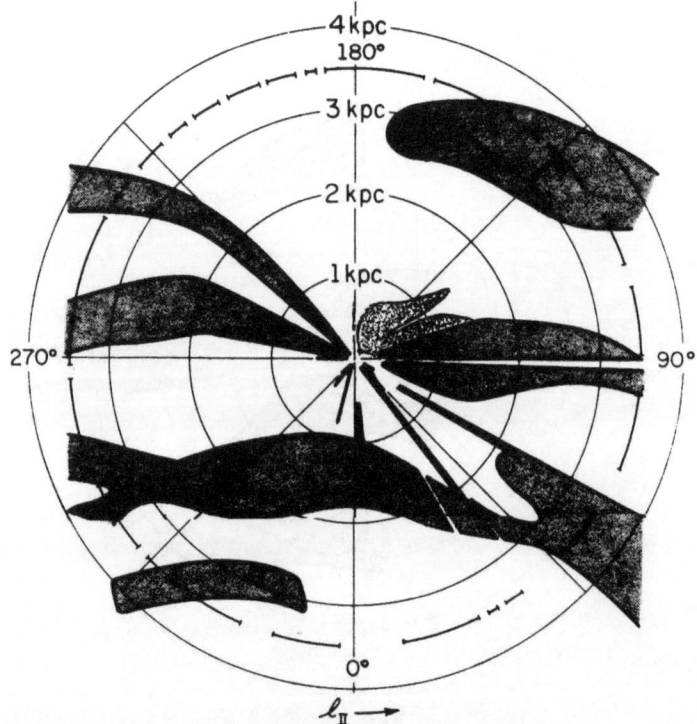

Fig. 4.   Adjusted map of dust distribution in accord with planetary nebular extinctions and distances.
Broken line indicates coverage (c.f. text).

the intermittent line at 3.5 kpc represents the incompleteness of the data. The solid part of the line represents uncalibrated regions of the map. In addition, regions closest to the Sun are most reliable, since the average value of b_{II} is 15°.

Finally, it is to be hoped that the present work can be coordinated with determinations of color excess to help determine R, the ratio of total to selective absorption.

References

Cahn, J. H. and Kaler, J. B.: 1971, *Astrophys. J. Suppl.* **22**, 319.
Fitzgerald, M. Pim: 1968, *Astron. J.* **73**, 983.
Mills, B.: 1959, in R. N. Bracewell (ed.), 'Paris Symposium on Radio Astronomy', *IAU Symp.* **9**, 431.
Peimbert, M.: 1971, *Bol. Obs. Tonantzintla Tacubaya* **6**, 29.
Seaton, M. J.: 1968, *Astrophys. Letters* **2**, 55.
Simonson, S. C.: 1970, *Astron. Astrophys.* **9**, 163.

OBSERVATIONS OF ATOMIC HYDROGEN IN
DENSE INTERSTELLAR DUST CLOUDS

GILLIAN R. KNAPP

University of Maryland, Md., U.S.A.

Abstract. High frequency resolution 21-cm line observations of almost 100 interstellar dust clouds showed that about one-third of them contain significant amounts of cold neutral hydrogen. Simple models for the velocity distribution of hydrogen in the clouds were fitted to the observations, and spin temperatures ranging from 16 to 40 K and column densities from 7×10^{19} to 6×10^{20} atom cm^{-2} were found. The ratios of H I to OH and H I to dust in the clouds were found to have about 1 to 10% of their value in the general interstellar medium.

Most dense interstellar dust clouds have not been detected as excess emission in neutral hydrogen, leading to the speculation that most of the hydrogen in such clouds is molecular. If, however, the H I in these clouds is colder than that in the surrounding interstellar medium, it should be detectable as narrow self-absorption features in H I profiles in the directions of such clouds, assuming that the cold dust cloud is in front of most of the rest of the hydrogen, which then acts as a background to be absorbed. Observations of profiles in the directions of almost 100 dense interstellar dust clouds made with the NRAO 140-ft telescope using high velocity resolution (0.34 and 0.17 km s^{-1}) showed the presence of self-absorption dips for about one-third of the clouds observed (Knapp, 1972). These dips are very narrow (half-widths from 0.9 to 4.5 km s^{-1}), are at the same velocities as the molecular lines observed for the clouds, and are most likely to be observed for fairly dark clouds ($A_v \geqslant 3^m$). A sample profile for the large cloud Kahvtassi 141 is shown in Figure 1.

The observations were analysed to find the neutral hydrogen content of the dust clouds assuming that the gas in each cloud has a gaussian dependence on velocity and a constant spin temperature, and that there is no filling-up of the self-absorption dip by emission from H I in front of the dust cloud. The solutions indicate spin temperatures from 16 to 40 K (which are higher values than those found for the molecules), peak optical depths from about 0.5 to 4.0, and column densities from 7×10^{19} to 6×10^{20} atom cm^{-2}. These values are about 1–10% of those predicted from the observed OH and dust column densities, so that the rest of the hydrogen is presumably in molecular form. Using optical and kinematic distance estimates, values for the neutral hydrogen number densities ranging from 3 to 77 atom cm^{-3} were derived for eleven clouds.

The kinematics of the whole sample of dust clouds was examined using these velocities and distances, and it was found that most of the dust clouds belong to the local Gould Belt expanding system, in which they are embedded more or less at random. The rest of the clouds belong to the other local structural feature, the Orion arm.

Greenberg and Van de Hulst (eds.), Interstellar Dust and Related Topics, 243-245.
All Rights Reserved. Copyright © 1973 by the IAU.

The fact that H_2CO, OH, CO and H I observations for the same cloud often have roughly the same line widths strongly suggests the presence of systematic expansion or contraction motions. The H I line widths are often slightly larger than those for

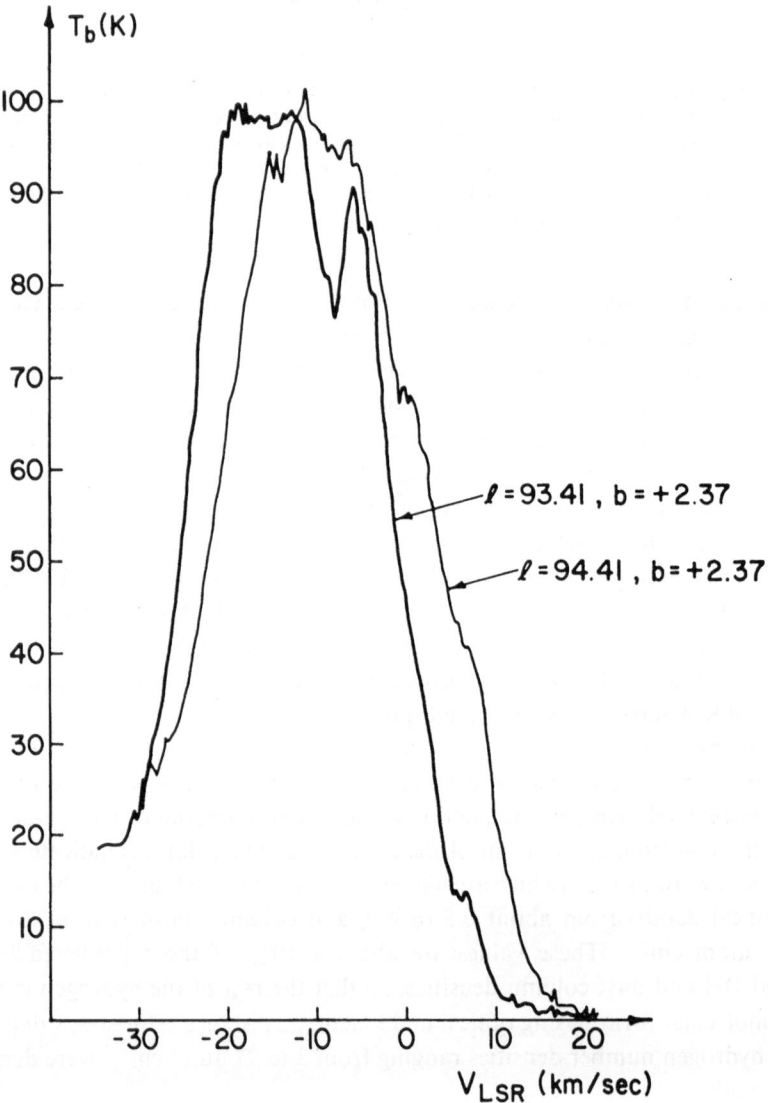

$\ell = 93.41$, b = +2.37

$\ell = 94.41$, b = +2.37

Fig. 1.   Sample H I profiles in the direction of Kh 141. The 'on-cloud' profile is drawn with a heavy line, and the 'off-cloud' (comparison) profile with a lighter line.

the molecules, indicating that the H I distribution extends slightly outside that of the molecules. The H I self-absorption profiles for several clouds show structure suggesting the presence of temperature gradients and expansion/contraction motions (with velocities of about 1 km s^{-1}), and analysis of simple models shows that, if these clouds are contracting, they are slightly hotter in the center than on the outside.

Reference

Knapp, G. R.: 1972, Ph.D. Thesis, University of Maryland.

THE CORRELATION OF INTERSTELLAR
REDDENING AND Lα ABSORPTION

EDWARD B. JENKINS

Princeton University Observatory, Princeton, N. J., U.S.A.

and

BLAIR D. SAVAGE

*Space Astronomy Laboratory, Washburn Observatory, University of Wisconsin,
Madison, Wis., U.S.A.*

Abstract. For 61 stars whose interstellar Lα absorption was observed by OAO-2, the average relation of H I column density to $B-V$ color excess is equal to 5×10^{21} atoms cm^{-2} mag^{-1}. For a recent dust grain model this value implies that the density ratio $\langle \rho_{HI}/\rho_{dust} \rangle \approx 100$. A good correlation was found for deviations in the measurements of N_{HI} and $E(B-V)$ from their mean values expected over the distances to the different stars, indicating an association of H I gas with dust.

The observation of 21-cm line radiation in both emission and absorption has been a traditional and very productive method of learning about the distribution of H I gas in space. The density of interstellar hydrogen may also be found by measuring the width of the Lα absorption feature in ultraviolet stellar spectra (Jenkins, 1970). At present, the amount of space probed by the Lα observations falls far short of the coverage by 21-cm surveys. However the 21-cm data are usually not well suited for comparisons with various absorptions observed toward stars, including reddening by dust, since the radio beam inevitably must sample an indefinitely long cone in space, instead of a volume of limited length and infinitesimal solid angle. In essence, Lα observations provide exactly the same sample volume as the other stellar measurements, and the derived hydrogen column densities are completely insensitive to the temperature of the gas.

At present, we have measured the strength of Lα absorption in the spectra of some 69 stars of spectral type B2 or earlier which were observed by the Wisconsin low-resolution scanner on board OAO-2. The experimental methods and results have already been presented in some detail by Savage and Jenkins (1972), and Figure 1 shows the correlation of hydrogen column densities against $B-V$ color excesses which was plotted in the original article. The determination of color excesses started with the intrinsic colors of Johnson (1963) which were applied to the spectral types assigned either by Lesh (1968) to northern stars or by Hiltner *et al.* (1969) to southern stars. These colors were compared with the photometric data of Iriarte *et al.* (1965) and Cousins and Stoy (1963) for the northern and southern objects, respectively. For all of the stars except those classified as emission line types, whose $E(B-V)$ values were considered to be less reliable, we obtained

$$\sum N_{HI} / \sum E(B-V) = 5 \times 10^{21} \text{ atoms cm}^{-2} \text{ mag}^{-1}.$$

It is of interest to derive a gas to dust mass ratio from the observation of $\sum N_{HI} / \sum E$ $(B-V)$. A recent model for the interstellar grains of Gilra (1971) reproduces reason-

Greenberg and Van de Hulst (eds.), Interstellar Dust and Related Topics, 247–250.

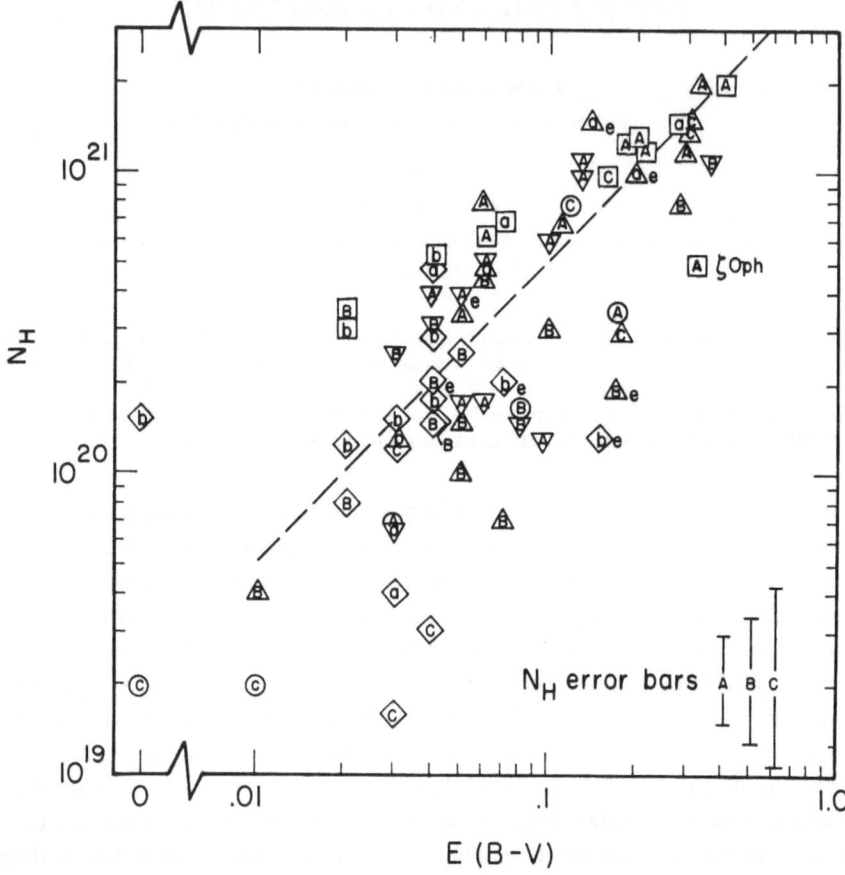

Fig. 1.   A comparison of hydrogen column densities to color excesses for 69 stars. The shapes of the symbols denote different regions of the sky outlined in the original work by Savage and Jenkins (1972). The letters indicate the relative probable error in each of the determinations of N_{HI} with lower case letters indicating those measurements taken from stars of spectral type B2. An 'e' follows symbols for emission line stars, for which the $E(B-V)$ values are probably less accurate. The dashed line represents our average ratio of 5×10^{21} atoms cm^{-2} mag.$^{-1}$ from our observed $\Sigma N_{HI}/\Sigma E(B-V)$. (Reproduced by courtesy of the *Astrophysical Journal*; University of Chicago Press, publisher. © 1972. American Astronomical Society. All Rights Reserved).

ably well the basic observational data on the grains. For this model Gilra finds that 2.5×10^{-5} gm cm^{-2} is needed to produce 1 mag. of extinction at the V-filter. If we assume the ratio of total to selective extinction R_V is 3.0, then the gas to dust mass ratio implied by this model and the OAO-2 result is $< \varrho_{HI}/\varrho_{dust} > \approx 100$. This result can be changed by factors of 2 or 3 depending on the particular grain model selected.

Our figure for the overall ratio of hydrogen column density to reddening is biased toward regions of space having a less than average amount of obscuration. For the 61 stars considered, the average color excess per unit distance

$$\Sigma E(B-V) / \Sigma r = 0.28 \text{ mag. kpc}^{-1}$$

whereas one would expect a value of 0.61 mag. kpc^{-1} if one were looking in random directions within 1 kpc of the Sun (Spitzer, 1968). This selection effect is a consequence of our being restricted to stars whose brightness near 1216 Å is above a certain observational limit. Since $E(1216-V)$ is generally as much as 6 times $E(B-V)$ (Bless and Savage, 1972), stars with little reddening are given a distinct advantage in appearing on our observing list.

The strong attenuation of ultraviolet radiation by dust also precludes our verifying whether or not the ratio of H I to dust decreases in very dense clouds (with $A_V \gtrsim 3$), as some past studies have indicated (Van de Hulst *et al.*, 1954; Heeschen, 1955; Bok *et al.*, 1955; Varsavsky, 1968; Garzoli and Varsavsky, 1966; Gosachinskii, 1966; Kerr and Garzoli, 1968; Mesaros, 1968). This decrease has been interpreted as indirect evidence that the formation of H_2 may be depleting the H I in dark clouds. It is

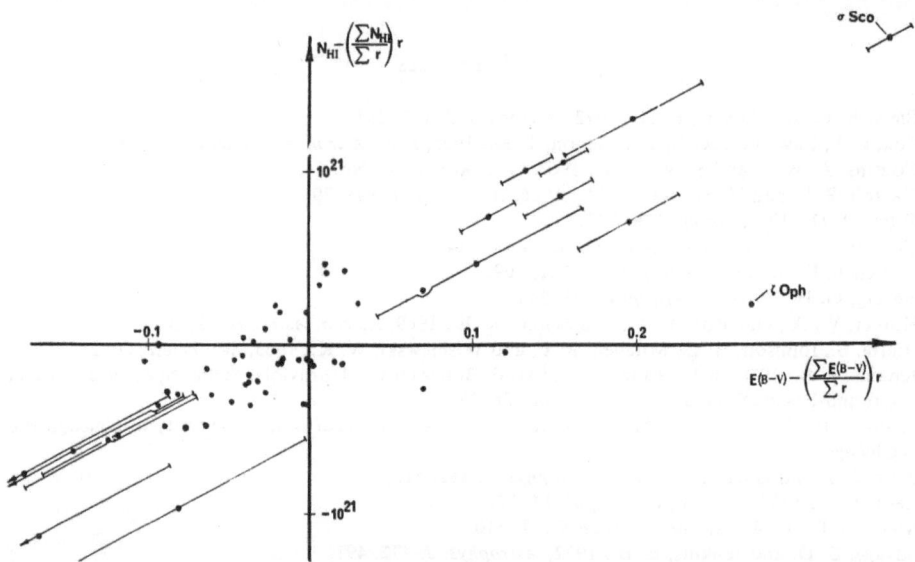

Fig. 2.   Excursions in the values of N_{HI} (in atoms cm^{-2}) and $E(B-V)$ from the mean amounts expected over the distance r to each star. Those points which are more crucial in showing the correspondence of gas and dust have error bars which indicate the displacement which could result from the expected error of 30% in r.

tempting to speculate, however, that the relative deficiency for ζ Oph (see Figures 1 and 2) could be attributed to molecule formation; this star exhibits a good number of interstellar molecular features in its visible spectrum (Herbig, 1968).

One might question whether the apparent correlation of gas and dust is simply a result of our observing stars at various distances. It is natural to expect more distant stars to have both more reddening and more H I, even if there were no relation between the collections of gas and dust in space. However if we examine the behavior of deviations of N_{HI} and $E(B-V)$ from the expected values over the distance to each star, we find a convincing demonstration for the physical association of gas with dust.

Figure 2 is a plot of $N_{H\,I} - (\sum N_{H\,I} / \sum r)r$ vs $E(B-V) - (\sum E(B-V) / \sum r)r$ for all 61 stars (the 8 emission-line stars were not included). The correlation coefficient for this scatter diagram is equal to 0.85. In principle, there could be no actual relationship in the fluctuations of gas and dust density, but one might arrive at the false impression that such a correlation existed if there were large *random* errors in the distances to the stars. However we estimate our error in distance to be around 30%, and this much uncertainty could not produce the large displacements which convince us that H I and reddening are correlated. The principal contribution to the distance error is from the error in estimating the star's absolute magnitude.

Note added in proof. The discussion in the last paragraph and Figure 2 were not presented at the Conference but were prepared afterwards in response to a question from the audience.

References

Bless, R. C. and Savage, B. D.: 1972, *Astrophys. J.* **171**, 293.
Bok, B. J., Lawrence, R. S., and Menon, T. K.: 1955, *Publ. Astron. Soc. Pacific* **67**, 108.
Cousins, A. W. J. and Stoy, R. H.: 1963, *Bull. Roy. Obs.*, No. 64.
Garzoli, S. L. and Varsavsky, C. M.: 1966, *Astrophys. J.* **145**, 79.
Gilra, D. D.: 1971, *Nature* **229**, 237.
Gosachinskii, I. V.: 1966, *Soviet Astron. AJ* **9**, 714.
Heeschen, D. S.: 1955, *Astrophys. J.* **121**, 569.
Herbig, G. H.: 1968, *Z. Astrophys.* **68**, 243.
Hiltner, W. A., Garrison, R. F., and Schild, R. E.: 1969, *Astron. Astrophys.* **2**, 202.
Iriarte, B., Johnson, H. L., Mitchell, R. I., and Wisniewski, W. K.: 1965, *Sky Telesc.* **20**, 21.
Jenkins, E. B.: 1970, in L. Houziaux and H. E. Butler (eds.), 'Ultraviolet Stellar Spectra and Related Ground-Based Observations', *IAU Symp.* **36**, 281.
Johnson, H. L.: 1963, in K. Aa. Strand (ed.), *Basic Astronomical Data*, University of Chicago Press, Chicago, p. 204.
Kerr, F. J. and Garzoli, S.: 1968, *Astrophys. J.* **152**, 51.
Lesh, J. R.: 1968, *Astrophys. J. Suppl.* **17**, 371.
Mesaros, P.: 1968, *Astrophys. Space Sci.* **2**, 510.
Savage, B. D. and Jenkins, E. B.: 1972, *Astrophys. J.* **172**, 491.
Spitzer, L., Jr.: 1968, *Diffuse Matter in Space*, Interscience, New York, p. 67.
Van de Hulst, H. C., Muller, C. A., and Oort, J. H.: 1954, *Bull. Astron. Inst. Neth.* **12**, 117.
Varsavsky, C. M.: 1968, *Astrophys. J.* **153**, 627.

GAS AND DUST IN INTERSTELLAR CLOUDS

R. D. DAVIES

University of Manchester, Nuffield Radio Astronomy Laboratories, Jodrell Bank, U.K.

Abstract. Detailed measurements have been made of the H, OH and H_2CO lines formed in the L1534 dense dust cloud; some H and H_2CO observations have been made of the dust cloud lying behind the Orion cluster. These indicate a similar distribution of the 3 species throughout the dust clouds. Studies of H, OH and H_2CO absorption features in clouds lying in front of bright radio sources provide estimates of the physical conditions in regions of molecule formation.

1. Introduction

There are a number of situations in the interstellar medium where an association between gas and dust can be observed or directly inferred. One such situation is found in the dense dust clouds as listed for example in the Lynds (1962) catalogue. Here the molecule OH is seen in emission and the formaldehyde (H_2CO) is an absorption against the 2.7 K background. Also the neutral hydrogen in many of these dust clouds is cool and can be observed in absorption against the galactic background of neutral hydrogen which is at a higher temperature. The other situation where gas and dust may be studied together is in the dense interstellar neutral hydrogen clouds which produce absorption when seen against strong continuum radio sources. These clouds also contain OH and H_2CO; their dust content can be inferred from the total optical obscuration in the line of sight to the optical counterpart of the radio source.

2. Dense Dust Clouds

Since the discovery of OH emission (Heiles, 1968) and H_2CO absorption (Palmer *et al.*, 1969) in dense dust clouds a number of investigations have been made of these objects. The present observations (Cohen *et al.*, 1972) were aimed at obtaining detailed information about conditions of the neutral hydrogen in two such dust clouds.

One cloud lies in the Orion nebula region. It is the object in which Kutner and Thaddeus (1971) found H_2CO in absoprtion in a region $\sim 0.5°$ in extent. Our more extensive observations show H_2CO absorption in an even larger area surrounding the Orion nebula. Examination of the Palomar Sky Survey prints shows a large obscuring cloud crossing the Orion nebula region. A preliminary study indicates 3–4 mag. of visual absorption in the cloud. When this is compared with the interstellar absorption in front of the Orion cluster which amounts to ~ 0.5 mag. it is clear that this dense cloud lies behind the Orion cluster and nebula. This explains the absence of H_2CO absorption in front of the Orion A continuum radio source. Neutral hydrogen measurements with the Mark IA radio telescope (beamwidth 11′) show much fine structure

Greenberg and Van de Hulst (eds.), Interstellar Dust and Related Topics, 251–255.

in this area, some of which may be attributable to cool absorbing hydrogen in this dust cloud.

The dense dust cloud numbered 1534 in Lynd's catalogue (Cloud 2 in Heiles' nomenclature) has been the subject of much study in the radio lines of H, OH and H_2CO At Jodrell Bank an attempt is being made to compare the line profile in each of these species with similar beam widths. Hydrogen profiles have been taken with the Mark IA radio telescope (beamwidth 11'), H_2CO profiles have been taken with the Mark II radio telescope (beamwidth 9') and soon OH profiles will be obtained with the Mark IA (9'). The present comparison of the H and H_2CO is made with OH spectra

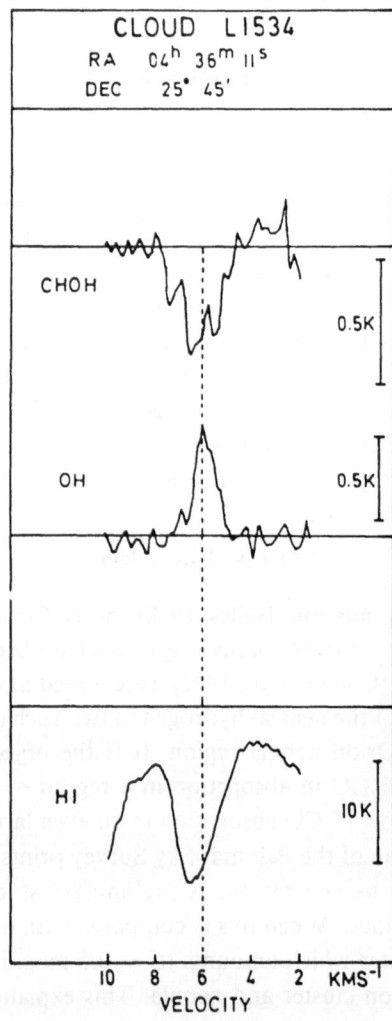

Fig. 1. H, OH and H_2CO spectra at RA = $04^h36^m11^s$, Dec = $25°45'$ in cloud 1534 of Lynds' (1962) catalogue. The intensities are given as antenna temperatures. The structure in the H_2CO spectrum is believed to be due to noise fluctuations only. Observing beamwidths for the H_2CO, OH and H lines are 9', 18' and 11' respectively.

taken by Heiles (1970) with the NRAO 140 ft telescope (beamwidth 18′). Spectra in the 3 lines are shown for the point RA = $04^h36^m11^s$, Dec = 24°45′ in Figure 1. The H_2CO absorption is produced against the 2.7 K Universal blackbody background radiation. The OH emission is the 'normal' 1667 MHz emission line. The neutral hydrogen profile shows narrowband absorption in a broader background emission profile; this absorption occurs only at the position of the dust cloud and is absent in the unobscured regions in the vicinity.

The central velocity of the features in each of the lines is closely similar which suggests that they arise in the same parts of the dust cloud. The width of the OH emission feature is less than that of the H and H_2CO absorption feature. This difference may arise because the OH beam area is about a factor of 3 greater and may well be sampling a different velocity. This is likely because our H observations show significant velocity changes over distances of 10′ or so.

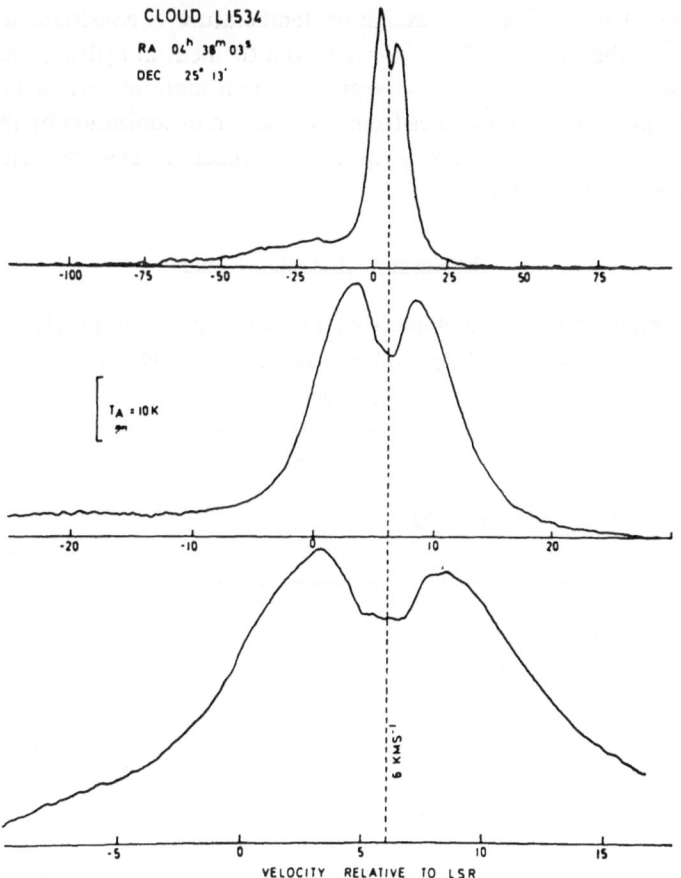

Fig. 2.   Neutral hydrogen spectra at RA = $04^h38^m03^s$, Dec = 25°13′ in cloud 1534 of Lynd's (1962) catalogue. The velocity resolutions are (from top to bottom) 1.86, 0.46 and 0.23 km s^{-1}. The integration times were increased in the narrower bandwidth observations to give similar rms noise on each spectrum.

An estimate can be made of the kinetic temperature of the neutral hydrogen in
L1534 by examining the thermal broadening in the absorption features. Many of the
profiles exhibit structure which indicates the presence of more than one velocity
feature. The H-line profile in Figure 1 is asymmetrical with its steepest edge on the
high velocity side. Another example of structure in the H-line absorption is shown in
Figure 2 which illustrates the spectrum at RA = $04^h38^m03^s$, Dec = $25°13'$ taken with
different overall bandwidths and resolutions. At least two absorption components are
present in this spectrum. An upper limit for the kinetic temperature derived from the
broadening of such profiles is 20 K, assuming the broadening is entirely thermal.
Clearly macroscopic motions exist as revealed by the OH and H_2CO spectra. Actual
kinetic temperatures as low as 10 K would be compatible with the observations. Heiles
estimated an excitation temperature of 5.5 K for the OH by a comparison of the 1665
and 1667 MHz transitions. Theoretically this temperature is expected to lie between
the kinetic temperature and the 2.7 K Universal background; its actual value depends
upon the gas density. This OH excitation temperature is consistent with a kinetic
temperature in the range 10–20 K inferred from the neutral hydrogen data.

The dense dust clouds are clearly regions of rich molecule production. The high
dust content protects the molecules from dissociation or ionization by the interstellar
ultraviolet flux; the high gas density ensures high reaction rates for molecule production (Hollenbach *et al.*, 1971).

3. Normal Absorbing Clouds

A more common type of cloud in interstellar space wich contains molecules and dust is
the cloud seen in absorption at neutral hydrogen wavelengths against continuum

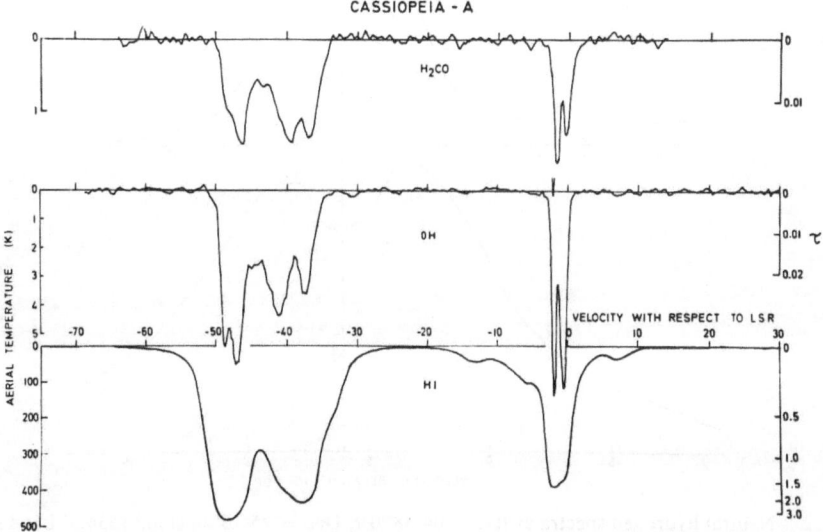

Fig. 3. The absorption spectra of Cassiopeia A in the lines of H, OH and H_2CO taken with the
Mark II radio telescope.

radio sources. An example is the group of clouds seen in absorption against Cassiopeia A shown in Figure 3. The H, OH and H_2CO spectra were all taken with the Mark II radio telescope. Although thermal broadening in the H-line spectrum makes the identification of some components less certain, it is possible to see the stronger H-line components in both the OH and H_2CO spectra. The same situation is found in the absorption spectra of Cyg A and Tau A (Davies and Matthews, 1972).

An important parameter of these clouds, as regards the processes of molecule formation, is the amount of obscuring matter in each cloud. This determines the destruction rate of molecules by UV radiation. The optical absorption in front of the optical remnants of the supernovae Cas A and Tau, and of the extragalactic Cyg A optical object are given in the literature. The total obscuration through each of the clouds, is 0.4 to 0.8^m assuming it is divided evenly amongst all the H-line components. The mean parameters of the molecule bearing absorption clouds are given in Table I.

TABLE I

Parameters of molecule bearing absorption clouds

$\int n(H)\,dl = N(H)$	1.1×10^{21} cm^{-2}
Temperature (H)	50–100 K
$N(OH)/N(H)$	1.1×10^{-7}
$N(H_2CO)/N(H)$	2×10^{-9}
Velocity dispersion	0.74 km s^{-1}
Cloud diameter	10 pc
Mean neutral hydrogen density	36 cm^{-3}
Mean cloud mass	640 $M_\odot$
Visual absorption at cloud centre	0.2^m–0.4^m
Magnetic field	2–20 μG

The low obscuration in these clouds means that the lifetimes against photoinduced destructive reactions are $\sim$ 1000 years for OH and 100 yr for H_2CO. The lifetime of an interstellar cloud on the other hand is $\sim 10^7$ yr, the time between cloud collisions. Thus the molecular destruction time is 10^{-5} to 10^{-4} of the lifetime of the interstellar clouds. This argues strongly that the molecules should be continually formed. In fact a gas atom such as oxygen will collide with a dust grain on timescale similar to the destruction timescales mentioned above. However no satisfactory process for molecule formation, during the atom-grain encounter has yet been worked out.

References

Cohen, R. J., Davies, R. D., and Wilson, A. J.: 1972, in preparation.
Davies, R. D. and Matthews, H. E.: 1972, *Monthly Notices Roy. Astron. Soc.* **156**, 253.
Heiles, C.: 1968, *Astrophys. J.* **151**, 919.
Heiles, C.: 1970, *Astrophys. J.* **160**, 51.
Hollenbach, D. J., Werner, M. W., and Salpeter, E. E.: 1971, *Astrophys. J.* **163**, 165.
Kutner, M. and Thaddeus, P.: 1971, *Astrophys. J.* **168**, L67.
Lynds, B. T.: 1962, *Astrophys. J. Suppl.* **7**, 1.
Palmer, P., Zuckerman, B., Buhl, D., and Snyder, L. E.: 1969, *Astrophys. J.* **156**, L147.

ON THE GENETIC RELATION BETWEEN
INTERSTELLAR CLOUDS AND DUST CLOUDS

FRANK H. SHU*

State University of New York at Stony Brook, N.Y., U.S.A.

Abstract. We propose that many of the denser dust clouds represent transient stages between the initiation of gravitational collapse of ordinary interstellar clouds by passage through a spiral galactic shock and the appearance of a stellar association. The interstellar gas is not processed at too rapid a rate (1) because the time spent between galactic shocks is appreciably longer than the free-fall collapse time of a typical dust cloud, and (2) because only a small fraction of the total mass of gas of a given collapsing cloud is converted into stars – the rest is presumably redispersed by various dynamical processes which accompany the formation of massive stars. We note that the residue of atomic hydrogen observed to exist in dark dust clouds is compatible with the interpretation that they are transient objects whose lifetimes are not longer than $\sim 10^7$ yr.

It can hardly be disputed that the self-gravity of a dust cloud is considerable. What has not been settled, either observationally or theroretically, is whether the darker dust clouds (with $A_v > 1$ or 2) represent stable dynamical objects, which are supported by rotation or by other means, or whether they represent transient objects, which exist only as temporary stages in the history of the formation of a cluster of stars from ordinary interstellar clouds. I shall take the latter view here.

In particular, I wish to pursue the possibility that isolated dust clouds are in state of dynamical collapse, whereas dust clouds in regions of active star formation may eventually be redispersed by various dynamical processes. In any case, the lifetimes of such clouds cannot greatly exceed 10^7 yr.

Such a conclusion would be important not only for its implication on the problem of star formation but also for the natural time scale it would impose on physical processes such as the formation of interstellar molecules. To date, theoretical calculations of the formation of molecules in dark clouds have generally assumed that the molecular concentrations of each species has sufficient time to reach an equilibrium in which the rate of formation is just balanced by the rate of destruction. In fact, the time scale associated with the rate-limiting reactions are often comparable or longer than the free-fall collapse time for a typical dust cloud. We shall see later that this circumstance allows, in principle, a 'dating technique' to investigate whether isolated dust clouds are, in fact, undergoing collapse.

1. Rate of Processing of Interstellar Gas

Let us first summarize the primary indirect arguments for and against the view that isolated dust clouds are collapsing gravitationally. The following argument *against* collapse, brought to my attention by G. B. Field, is perhaps the most serious.

* Alfred P. Sloan Foundation Fellow, 1972–74.

Greenberg and Van de Hulst (eds.), Interstellar Dust and Related Topics, 257-262.
All Rights Reserved. Copyright © 1973 by the IAU.

Hollenbach *et al.* (1971) have estimated from Beverley Lynds' (1962) catalogue that 20–40% of the gaseous matter in the local interstellar medium is in the form of dark clouds with visual extinction $A_v \geqslant 1$ mag. The free-fall collapse times for such clouds can be estimated to be about a few million years. A major dilemma would obviously result if 20–40% of the interstellar medium were processed into stars every few million years.

I believe the resolution of this dilemma can be found by examining a photograph of an external spiral galaxy such as M51. As is well-known (Lynds, 1970, 1972), the dark lanes in such galaxies are typically not found everywhere but are generally confined to the inside edges of the optical spiral arms. In the density-wave theory of spiral structure, the explanation is as follows.

The spiral structure owes its existence to a spiral density wave. In the frame which rotates with the speed of the wave pattern, the interstellar gas flows through the pattern and is strongly concentrated in the spiral arms – in many cases, as shown by Roberts (1969), by galactic shocks. The high space density reached in the post-shock region is associated with the dark lanes.

It is tempting to suppose that dense clouds are *formed* in the high-compression region behind the galactic shock; however, the existence of the dark lanes does not, *by itself*, provide a proof. There are, logically, two separate aspects of the large-scale compression: (1) the concentration of an aggregate of cloud centers into a smaller volume, and (2) the compression of each individual cloud to greater internal densities. If the gas to dust ratio remains constant, only the first process leads to an increase of the *average space density* and, thus, to an increase of the visual extinction. The compression of *individual clouds*, leading in some cases to gravitational collapse and to the formation of stars, is inferred from the concentration of giant H II regions a little downstream from the dark lanes.

A specific calculation of these effects in the context of the two-phase model of the interstellar medium was carried out recently by a number of us (Shu *et al.*, 1972). The results for the cloud phase along a streamline passing through the solar neighborhood is shown in Figure 1. Immediately after a galactic shock, the ordinary interstellar clouds have an average internal density of ~ 50 cm^{-3} if we ignore the effects of their self-gravity. They occupy $\sim 5\%$ of the volume of interstellar space. With the usual gas and dust parameters, this results in a visual extinction of ~ 4 mag. kpc^{-1}. (A minimum extinction of ~ 0.4 mag. kpc^{-1} is reached in between spiral arms.) At the post-shock density of 50 cm^{-3}, a 400 $M_\odot$ spherical cloud would provide ~ 0.5 mag. of extinction along a diameter. This would not qualify it as an especially dark cloud. However, if we consider the self-gravity of this same cloud, the calculated internal temperature of ~ 30 K would not be sufficient, in the absence of other stabilizing influences, to prevent the cloud from collapsing under the load of the ambient inter-cloud medium. Indeed, our calculations indicate the mass threshold for collapse in the post-shock region to be only 120 $M_\odot$ if only the gas kinetic pressure is available to support the cloud.

It is not unreasonable, then, to suppose that a significant fraction of the ordinary

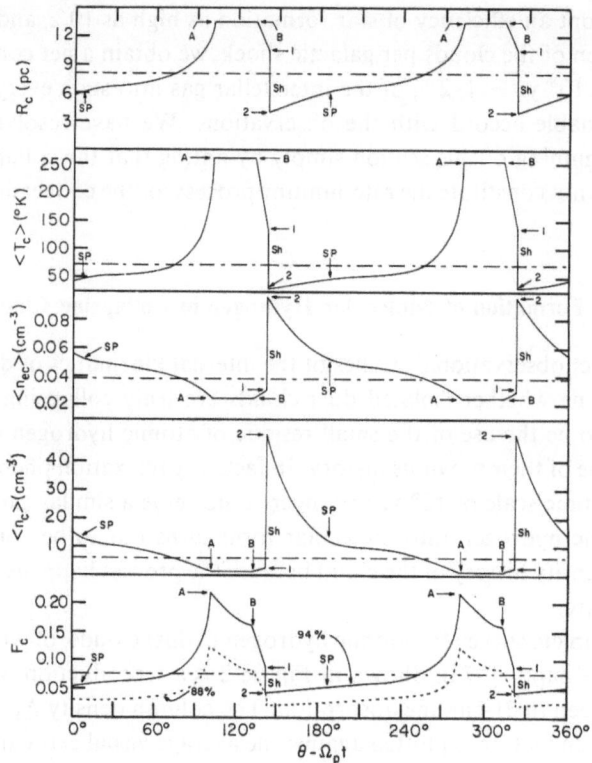

Fig. 1. The variation of the properties of the cloud phase along a streamline which passes through the solar neighborhood. From top to bottom, the curves give, respectively, the radius of a 400 $M_{\odot}$ cloud, the average internal temperature of the clouds, the average electron density inside the clouds, the average internal density, and the fraction of volume of interstellar space occupied by the aggregate of clouds. The solid curve for F_c corresponds to a space density of 0.5 cm^{-3} when averaged over the streamtube; the dashed curve, to half this value.

interstellar clouds are induced to collapse gravitationally – say, for sake of argument, a fraction even as large as to constitute 20–40% of the gaseous mass of the local inter-stellar medium. (The number, 20–40%, is illustrative. The Sun is probably not located immediately behind a primary galactic shock, and the dust clouds in the immediate neighborhood of the Sun may owe their origin to the presence of the local spiral arm.)

It would still be embarrassing if we were to conclude that 20–40% of the interstellar medium is converted into stars per galactic shock since the time spent between galactic shocks is only 2.7×10^8 yr at the solar circle. Fortunately, radio observations show that the mass of the gas associated with young stellar associations is often more than ten times larger than that contained in the form of stars (cf. Menon, 1962; Raimond, 1965; and others). This supports various theoretical considerations which suggest that the process of star formation is not highly efficient so that the collapse of any given cloud results in only a few percent of the total gaseous matter being converted into stars. Presumably the rest is redispersed by expanding H II regions, by supernovae explo-sions, and by other dynamical effects which accompany the formation of massive stars.

If we now adopt an efficiency of star formation as high as 10% and the collapse of a healthy fraction of the clouds per galactic shock, we obtain a net conversion of 10% $\times 20$–$40\%/2.7 \times 10^8$ yr $= 1$–2% of the interstellar gas into stars every 10^8 yr. Such a rate is in reasonable accord with the observations. We have resolved the dilemma posed at the beginning of this section simply by noting that the collapse of *individual* dust clouds does not constitute the rate-limiting process for the conversion of interstellar gas into stars.

2. Formation of Molecular Hydrogen in Collapsing Clouds

Apart from direct observational studies of the internal kinematics of dust clouds, how may we determine whether isolated dust clouds are truly collapsing ? A promising method seems to be the use of the small residue of atomic hydrogen detected in dust clouds as a probe of their previous history. In fact, any relaxation phenomenon having a characteristic time scale of 10^6 yr or longer could serve a similar capacity. The conversion of atomic hydrogen into molecular form turns out to be convenient for ascertaining the density history of the cloud because the process happens to be insensitive to the temperature.

The most complete survey for atomic hydrogen in dust clouds, of which I am aware, is contained in Knapp (1972). Shown in Figure 2 is a reproduction of her results for 28 dust clouds seen in 21-cm line absorption. The column density N_H deduced from a narrow absorption feature is plotted against the average visual extinction A_v produced by the cloud. Knapp's analysis ignores the possible presence of foreground hydrogen in emission over the frequency interval of the absorption feature; thus, her data points may underestimate the amount of atomic hydrogen actually present in the dust clouds.

The curve labelled $N_H = 2 \times 10^{21} A_v$ represents the value expected from simply extrapolating the gas to dust ratio found in the general interstellar medium. The difference between it and the plotted points presumably represents the amount of atomic hydrogen converted to molecular form.

The other curves labelled UV, CR, FFC, etc. have been superimposed by me to illustrate possible theoretical explanations. In the curves labelled UV, CR, and FFC, formation of H_2 is assumed to take place on grain surfaces with the rate given by Hollenbach *et al.* (1971). The curve labelled UV, which is taken from their paper, gives the column density of atomic hydrogen to be expected for a 2000 $M_\odot$ dust cloud at various densities if the cloud is assumed to be in equillibrium with the ultraviolet radiation fiield present in interstellar space providing a dissociation of H_2. Since 2000 $M_\odot$ corresponds to a fairly large dust cloud, it is probably safe to conclude that the residue of atomic hydrogen present in the darker clouds effectively rules out the possibility that these dust clouds could be in static equilibrium with UV providing the main mechanism of dissociation of H_2.

The dashed curve labelled CR gives the equilibrium column density of atomic hydrogen for a 120 $M_\odot$ cloud when low-energy cosmic rays are assumed to provide the mechanism of dissociation of H_2. The computation proceeds along the lines of

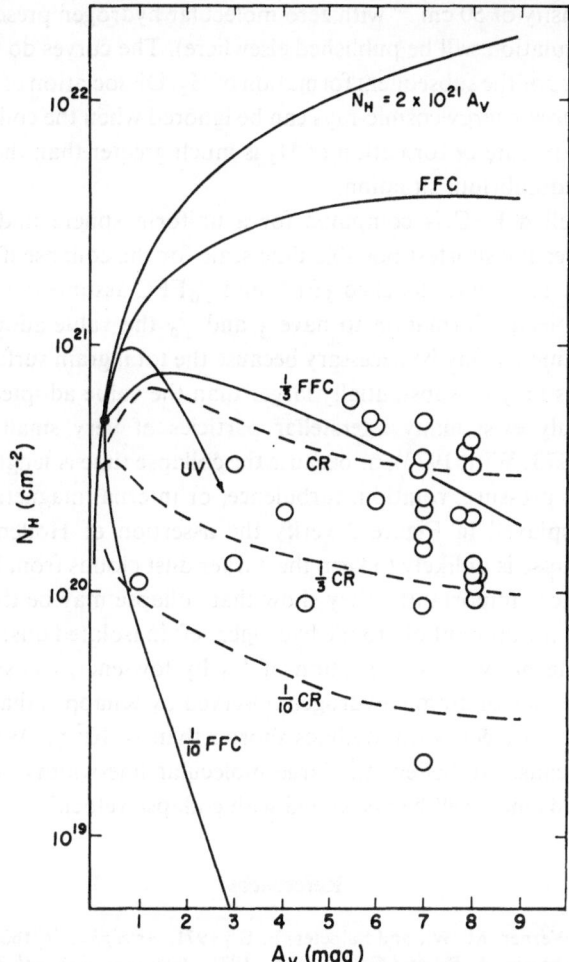

Fig. 2. The column density of atomic hydrogen along an average line of sight (spanning a length equal to $\frac{4}{3}$ times the radius if the cloud is spherical) as a function of the average visual extinction produced by the dust cloud. The curve labelled $N_H = 2 \times 10^{21} \, A_v$ is the relation found for the general interstellar medium; the data points are from Knapp (1972); the curved labelled UV is computed for a 2000 $M_\odot$ cloud; and the other curves are computed for a 120 $M_\odot$ cloud. See the text for other explanations.

Solomon and Werner (1970), but the flux of cosmic rays has been taken to have the value deduced by Hughes *et al.* (1971) from their study of the ionization level in ordinary interstellar clouds. The other dashed curves labelled $\frac{1}{3}$CR and $\frac{1}{10}$CF assume the ratio of the rates of cosmic-ray dissociation to molecule formation to have $\frac{1}{3}$ and $\frac{1}{10}$ the value adopted in the curve CR. Although self-shielding of the dust clouds from the external flux of low-energy cosmic rays may lower the dashed curves appreciably, we see that the curves labelled CR and $\frac{1}{3}$CR give a reasonable accounting of Knapp's observations.

The remaining curves give the locus of collapsing clouds of 120 $M_\odot$ which start from rest at a density of 50 cm^{-3} with zero molecular hydrogen present initially. (The details of the calculation will be published elsewhere). The curves do not follow $N_H = 2 \times 10^{21} A_v$ because of the subsequent formation of H_2. Dissociation of H_2 by ultraviolet radiation and by low-energy cosmic-rays can be ignored when the collapse is sufficiently rapid because the rate of formation of H_2 is much greater than the rate of destruction in this non-equilibrium situation.

The curve labelled FFC is computed for a uniform sphere undergoing free-fall collapse. This gives the shortest possible time scale for the collapse if external loading is not considered. The curves labelled $\frac{1}{3}$FFC and $\frac{1}{10}$FFC assume the ratio of the rates of collapse to molecule formation to have $\frac{1}{3}$ and $\frac{1}{10}$ the value adopted in the curve FFC. Such adjustments may be necessary because the total grain surface area available to form molecules may be substantially larger than the value adopted by Hollenbach *et al.* if there truly exist many interstellar particles of very small dimensions (cf. Lillie and Witt, 1973; Witt, 1973), or because the collapse time is lengthened by partial support from gas pressure, rotation, turbulence, or internal magnetic fields.

The results displayed in Figure 2 verify the assertion of Hollenbach *et al.* that gravitational collapse is unlikely to keep the darker dust clouds from becoming mostly molecular; but more importantly, they show that collapse may be the primary factor which determines the amount of atomic hydrogen left in isolated dust clouds. Thus, an interpretation, alternative to dissociation of H_2 by low-energy cosmic-rays, for the relatively large residue of atomic hydrogen observed by Knapp is that the darker dust clouds are transient objects with lifetimes shorter than $\sim 10^7$ yr. We favor the latter interpretation because we believe the large molecular line-widths observed in many isolated dust clouds may well be associated with collapse velocities.

References

Hollenbach, D. J., Werner, M. W., and Salpeter, E. E.: 1971, *Astrophys. J.* **163**, 165.
Hughes, M. P., Thompson, A. R., and Colvin, R. S.: 1971, *Astrophys. J. Suppl.* **23**, 323.
Knapp, G. R.: 1972, unpublished Ph.D. Thesis, University of Maryland.
Lillie, C. F. and Witt, A. N.: 1973, this volume, p. 115.
Lynds, B. T.: 1962, *Astrophys. J. Suppl.* **7**, 1.
Lynds, B. T.: 1970, in W. Becker and G. Contopoulos (eds.), 'The Spiral Structure of Our Galaxy', *IAU Symp.* **38**, 26.
Lynds, B. T.: 1972, in D. E. Evans (ed.), 'External Galaxies and Quasi-Stellar Objects', *IAU Symp.* **44**, 56.
Menon, T. K.: 1962, *Astrophys. J.* **135**, 394.
Raimond, E.: 1965, *Bull. Astron. Inst. Neth.* **18**, 191.
Roberts, W. W.: 1969, *Astrophys. J.* **158**, 123.
Shu, F. H., Milione, V., Gebel, W., Yuan, C., Goldsmith, D. W., and Roberts, W. W.: 1972, *Astrophys. J.* **173**, 557.
Solomon, P. M. and Werner, M. W.: 1971 *Astrophys. J.* **165**, 41.
Witt, A. N.: 1973, this volume, p. 53.

INTERSTELLAR REDDENING AT HIGH
GALACTIC LATITUDES

A. G. DAVIS PHILIP

Dudley Observatory

and

State University of New York at Albany, N.Y., U.S.A.

Abstract. Measures in the Strömgren four-color and $H\beta$ systems provide an accurate way to determine color excesses of early-type stars. Fourteen areas at high galactic latitude have now been searched for faint A stars which are then measured photoelectrically to obtain the color excesses. Non-main sequence A stars, which are easily detected by means of the four-color photometry, are not included in the analysis. Within 40° of each pole, the reddening is essentially zero, $E_{b-y} = 0.00$ north of the galactic plane and $E_{b-y} = 0.01$ south of the plane.

1. Introduction

Studies are being made currently to determine the stellar density distribution perpendicular to the galactic plane. An important by-product is the determination of the interstellar reddening in each of the regions under study. At the present time, data on color excesses is available in 26 regions (12 regions in the stellar density program, 5 globular clusters and 9 open clusters) at high galactic latitudes.

2. The Observations

Four-color and $H\beta$ photometry of early-type stars allow one to calculate their intrinsic $b-y$ colors according to a formula,

$$(b-y)_0 = 2.943 - \beta - 0.1(\delta m_1 + \delta c_1)$$

derived by Crawford and Barnes (1970). The quantities δm_1 and δc_1 are the differences between the observed c_1 or m_1 index and the c_1 or m_1 index that the star would have if it were on the zero age main sequence [$\delta m_1 = m_{1\,\mathrm{Hyades}} - m_{1\,\mathrm{obs}}$, $\delta c_1 = c_{1\,\mathrm{obs}} - c_{1\,\mathrm{ZAMS}}$]. The formula applies to A stars with $2.70 < \beta < 2.88$ for which $\delta c_1 < 0.28$. The color excess is determined by subtracting the calculated $(b-y)_0$ from the observed $b-y$. Color excesses have been determined in this way for the 12 regions in the stellar density program and the 5 regions surrounding globular clusters. The reddenings to the 9 open clusters were determined either by the four-color and $H\beta$ method or by means of UBV photometry.

The reddening data are presented in Table I. The galactic coordinates are listed in columns 1 and 2, the name of the region is listed in column 3, the method used is listed in column 4, the estimate of the color excess, E_{b-y} is listed in column 5, and the source of the reddening estimate is listed in column 6.

Greenberg and Van de Hulst (eds.), Interstellar Dust and Related Topics, 263–267.

TABLE I

Reddening in high galactic latitude regions

l	b	Region	Method	Color excess E_{b-y}	Source
—	+90	NGP	4-C, Hβ	0.000	Philip and Tifft (1971)
—	+90	NGP	UBV	0.000	McClure and Crawford (1971)
—	+90	NGP	4-C, Hβ, (GHK)	0.000	Feltz (1972)
221	+84	Coma	4-C, Hβ	−0.006	Crawford and Barnes (1969a)
221	+84	Coma	UBV	0.000	Hagen (1970)
42	+78	M3	4-C, Hβ	0.004	Crawford and Barnes (1969c)
4	+47	M 5	4-C, Hβ	0.030	Tifft (1972)
0	+45	3 HLF 3	4-C, Hβ	0.072	Philip (1973c)
110	+45	Ursa Major	4-C, Hβ	0.007	Crawford and Barnes (1969a)
59	+41	M13	4-C, Hβ	0.011	Crawford and Barnes (1969c)
206	+33	Praesepe	UBV	0.000	Hagen (1970)
206	+33	Praesepe	4-C, Hβ	−0.002	Crawford and Barnes (1969b)
216	+32	M67	UBV	+0.042	Hagen (1970)
180	+30	4 HLF 1	4-C, Hβ	0.009	Philip (1972)
123	+23	NGC 188	UBV	0.035	Hagen (1970)
351	+16	M4	4-C, Hβ	0.25	Philip (1973a)
180	−22	Hyades	UBV	0.000	Hagen (1970)
137	−23	NGC 752	UBV	0.021	Hagen (1970)
137	−23	NGC 752	4-C, Hβ	0.027	Crawford and Barnes (1970)
167	−24	Pleiades	UBV	0.028	Hagen (1970)
337	−26	NGC 6752	4-C, Hβ	0.056	Tifft (1972)
0	−30	3 HLF 2	4-C, Hβ	0.030	Philip (1973a)
76	−30	1 HLF 2	4-C, Hβ	0.047	Philip (1973a)
290	−30	2 HLF 2	4-C, Hβ	0.015	Philip (1973a)
279	−34	Bok Cluster	4-C, Hβ	0.014	Philip (1973c)
0	−45	3 HLF 4	UBV	0.000	Drilling (1971)
180	−45	4 HLF 4	UBV	0.056	Drilling (1971)
290	−45	2 HLF 4	4-C, Hβ	0.049	Philip (1973a)
76	−60	1 HLF 6	4-C, Hβ	0.002	Philip (1973a)
180	−60	4 HLF 6	4-C, Hβ	0.009	Philip (1973a)
—	−90	SGP	4-C. Hβ	0.013	Philip (1973c)
—	−90	SGP	Average of various methods	0.028	Rodgers (1971)

3. The Distribution of Color Excesses with Position in the Galaxy

The data displayed in Table I are plotted in Figures 1 and 2. In each figure galactic longitudes are indicated along the edges of the figure; galactic latitudes are indicated by concentric circles (in intervals of 20° in b). The color excess, E_{b-y}, determined in any region is indicated by a two digit number (the color excess in hundredths of a magnitude) which is plotted at the position of the center of that region. In this way the distribution of color excesses in regions at high galactic latitudes can be determined. Regions north of the galactic plane are shown in Figure 1. For $b > 50°$, no area was found with a color excess greater than a few thousandths of a magnitude. In Figure 2, there are only two regions for which color excesses have been determined for latitudes

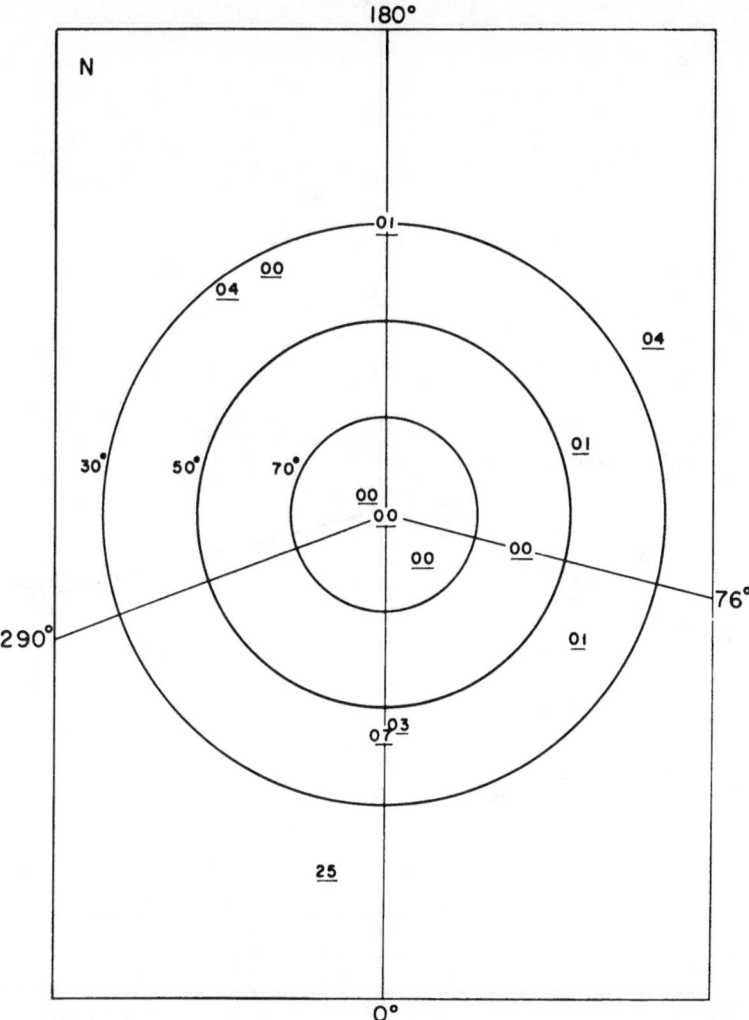

Fig. 1. The underlined two digit numbers represent the color excesses E_{b-y} in hundredths of a magnitude in areas at various galactic longitudes (indicated at the edges of the figure) and galactic latitudes (indicated by concentric circles). North of the galactic plane, for $b > 50°$ there are no color excesses greater than 0.00.

south of $-50°$; their average color excess is 0.01 mag. Thus it can be seen that regions within 40 deg of either galactic pole are essentially unreddened. Reddenings up to a few hundredths of a magnitude are found 60 deg away from each pole.

The stellar density distribution program will allow reddenings to be determined every 15 deg in galactic latitude along the four longitudes indicated in Figures 1 and 2. Reddenings are also being determined in front of globular and open clusters by a number of investigators. In a few more years, the number of high galactic regions in which reddenings have been determined will be greatly increased. Dr Stock and I have

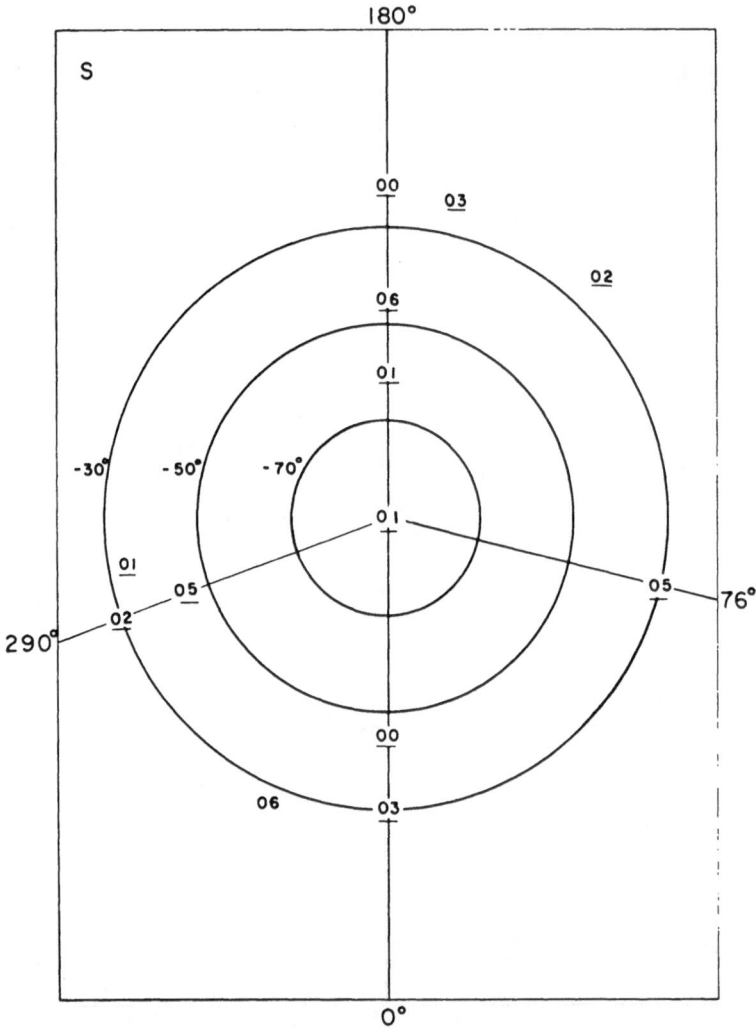

Fig. 2. Same as Figure 1, but for areas south of the galactic plane, the average reddening for $b > 50°$ south is 0.01.

in progress a program of identifying faint early-type stars in 5° strips. The first one (covering six hours of right ascension) covers a strip in front of the SGP (Philip and Stock, 1972); the second extends from the SGP to the NGP along $l = 290°$. Photometry of the early-type stars in these two lists will yield important data concerning the variation of interstellar reddening with galactic latitude.

Acknowledgements

The observations in this program were made at Kitt Peak National Observatory, Gerro Tololo Inter-American Observatory, Tonantzintla Observatory, and Warner and Swasey Observatory. The directors of these observatories are thanked for their assignment of telescope time. The project has been supported by the National Science Foundation and the SUNY Research Foundation.

References

Crawford, D. L. and Barnes, J. V.: 1969a, *Astron. J.* **74**, 407.
Crawford, D. L. and Barnes, J. V.: 1969b, *Astron. J.* **74**, 818.
Crawford, D. L. and Barnes, J. V.: 1969c, *Astron. J.* **74**, 1008.
Crawford, D. L. and Barnes, J. V.: 1970, *Astron. J.* **75**, 946.
Drilling, J. S.: 1971, *Astron. J.* **76**, 1072.
Feltz, K. A.: 1972, *Publ. Astron. Soc. Pacific* **84**, 497.
Hagen, G. L.: 1970, *Publ. David Dunlap Obs.* **4**, 1.
McClure, R. D. and Crawford, D. L.: 1971, *Astron. J.* **76**, 31.
Philip, A. G. D. (ed.): 1972, 'The Evolution of Population II Stars', *Dudley Obs. Report*, No. 4.
Philip, A. G. D.: 1973a, in Ch. Fehrenback and B. E. Westerlund (eds.), 'Spectral classification and Multicolour Photometry', *IAU Symp.* **50**, 295.
Philip, A. G. D.: 1973b, *Astrophys. J.* **180**, 421.
Philip, A. G. D.: 1973c, unpublished.
Philip, A. G. D. and Stock, J.: 1972, *Bol. Obs. Tonantzintla y Tacubaya*, in press.
Philip, A. G. D. and Tifft, L. E.: 1971, *Astron. J.* **76**, 567.
Rodgers, A. W.: 1971, *Astrophys. J.* **165**, 581.
Tifft, L. E.: 1972, private communication.

THE 4830 MHz FORMALDEHYDE LINE
OBSERVATIONS OF DARK CLOUDS IN IC 1795 (W3)

Y. K. MINN*

Max-Planck-Institut für Radioastronomie Bonn, Germany

and

J. MAYO GREENBERG

State University of New York at Albany,

and

Dudley Observatory U.S.A.

Abstract. A survey of H_2CO absorption in dark clouds in IC 1795 (W3) shows that the material in the direction of W3 can be distinctly separated into three groups: one, immediately surrounding W3, has a uniform velocity of about -41 km^{-1} s; the second, associated with the W3 (OH) source has a velocity of about -49 km^{-1} s; the third, as a velocity of about -20 km^{-1} s and may be associated with an interarm spur in the foreground.

1. Introduction

As a part of the survey of dark clouds in the galactic plane in the l_{11}–l_{10}, 4830 MHz transition of H_2CO (Minn and Greenberg, 1973), we observed dark clouds in the region of IC 1795 (W3). IC 1795 is the youngest object of the group of three H II regions in the Perseus arm, the others being IC 1805 (W4) and IC 1848 (W5). It has OH and H_2CO emission sources 17′ southeast of the main radio continuum source (Cudaback *et al.*, 1966; Rogers *et al.*, 1966; Knowles *et al.*, 1969) and large extinction in front of, as well as in, areas surrounding the main radio source. Lynds (1962) lists 13 dark clouds subtending from 0.016 to 0.2 sq deg in the vicinity of IC 1795. The area has been mapped in detail by Mezger and Henderson (1967), Mezger *et al.* (1967), and Wynn-Williams (1971) at various wavelengths using both the single dish and the interferometer. Mezger *et al.* (1967) found that a compact H II region is associated with the OH emission source (W3 (OH), G133.9 + 1.0) which is completely hidden from view in the Hα photograph. Wynn-Williams (1971) derived the magnitude of the extinction in front of W3. It is enormous, − 14 mag. in Hα, and it hides some of the exciting stars completely. An infrared star is also reported in the region within 30″ from the OH emission source (Raimond and Eliasson, 1969).

It is the purpose of this work to obtain the radial velocities of dark clouds in the area and to determine their distribution and relation with H II regions and the OH emission source.

Observed positions are indicated in Figure 1 in which is also shown a contour map of the continuum antenna temperature at 5 GHz taken from Mezger *et al.* (1967).

* Present address: Dept. of Physics and Astronomy, University of Alabama.

Greenberg and Van de Hulst (eds.), Interstellar Dust and Related Topics, 269–275.

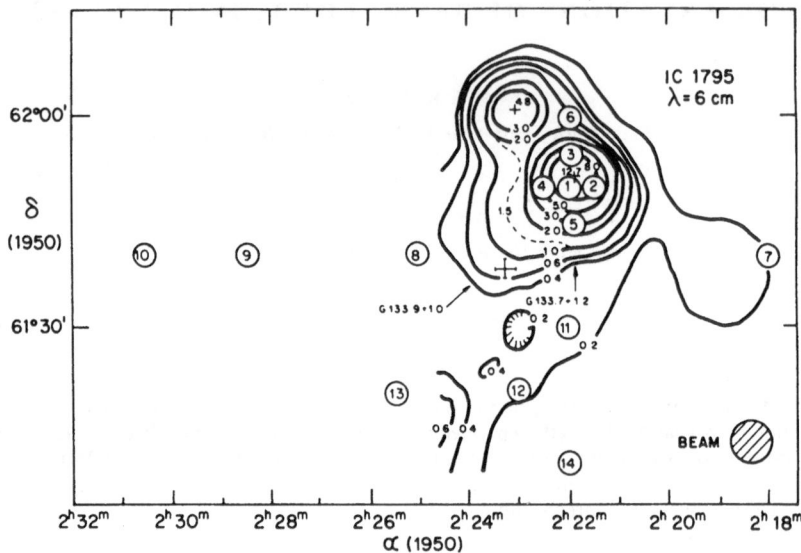

Fig. 1.   Positions at which 6-cm formaldehyde observations were made, shown against 6-cm continuum contour map of IC 1795 taken from Mezger *et al.* (1967). 1 unit = 1 K T_A.

Due to limited observing time we restricted our observations to the darkest areas in the representative regions. In the W3 continuum center area observations were made close to each other in order to determine the velocity variation and the fall-off of the H_2CO line against the radio source background. For most other positions, the continuum background contribution from the radio source is so low, generally less than 0.3 K, that the absorption is mostly that against the universal microwave background 2.8 K.

The observations were carried out with the 140-ft telescope at the National Radio Astronomy Observatory* equipped with a cooled 6-cm parametric amplifier and the 413-channel autocorrelation receiver. The total system noise temperature was typically about 100 K and a spectral resolution of either 6.5 kHz (0.4 km s^{-1}) or 3.2 kHz (0.2 km s^{-1}) was used. The data were taken in the 'total power mode'.

2. Results and Discussion

Observational results are presented in Table I and Figure 2. In Figure 2 representative line profiles are shown: one against the W3 radio source and two against the universal microwave background – one with strong lines and the other with weak lines. Column 1 of Table I gives the cloud number; Columns 2 and 3 are equatorial coordinates of the observed positions; Column 4 is the radial velocity of the line with respect to the local standard of rest (Allen, 1964); Column 5 is the measured antenna temperature of the line; Column 6 is the full line width at half-intensity; Column 7 is

---

* Operated by Associated Universities, Inc., under contract with the National Science Foundation.

TABLE I

Summary of the observations

Position number	$\alpha(1950)$	$\delta(1950)$	V_{lsr} (km s^{-1})	T_l (K)	W (km s^{-1})	T_c (K)
1.	$2^h\ 22^m\ 00^s$	$61^\circ\ 50'\ 00''$	-41.2	-1.17	3.5	12.7
			-22.2	-0.30	1.0	
2.	2 21 30	61 50 00	-40.5	-0.80	3.5	8.0
3.	2 22 00	61 55 00	-41.2	-1.10	3.3	8.0
4.	2 22 30	61 50 00	-41.2	-0.70	3.3	5.0
5.	2 22 00	61 45 00	-40.5	-0.20	0.3	3.0
6.	2 22 00	62 00 00	$-51.3?$	-0.25	0.3	1.0
7.	2 18 30	61 41 00	$-52.4?$	-0.14	0.4	< 0.3
8.	2 25 00	61 41 00	-48.9	-0.25	0.3	< 0.3
9.	2 28 30	61 41 00	-48.9	-0.27	0.3	< 0.3
10.	2 30 30	61 40 00	-49.0	-0.26	0.3	< 0.3
11.	2 22 00	61 30 00	-48.3	-0.29	0.3	< 0.3
12.	2 23 00	61 20 00	-49.3	-0.34	3.4	< 0.3
			-22.0	-0.21	2.3	
13.	2 25 30	61 20 00				< 0.3
14.	2 22 00	61 10 00				< 0.3

the 6-cm continuum antenna temperature, either taken from Mezger *et al.* (1967) or measured immediately before the line observation.

Lines were detected at most positions even though they are weak, indicating a wide distribution of the molecule in the region. At a few positions in less dark regions, the line strenghts and widths were reduced to a very weak level. In such cases, the line was confirmed by a careful tracing of the continuation of the line in adjacent positions as shown by the lower two traces in Figure 2. The narrow and weak features are typical characteristics of the H$_2$CO absorption against the universal microwave radiation in dark clouds as surveyed by the same authors (1973). Some of the lines observed here, however, are narrower than the lines typically found in dark nebulae. The average H$_2$CO line width in dark clouds is about 1.1 km s^{-1}. The lines observed here bear a resemblance to the weak high velocity lines ($V > 20$ km s^{-1}) observed in the dark cloud H$_2$CO survey. The kinematic distance of IC 1795 derived from the radio observation of the H 109α line is given as 3.1 kpc by Reifenstein *et al.* (1969). If the dark clouds are really associated with this bright nebula, they may be the most distant clouds identified so far.

At positions near the W3 radio continuum center, where the continuum background contribution from the radio source is high, line profiles are very deep and wide. They are very similar to those observed by Zuckerman *et al.* (1970) and Wilson (1972) against the radio continuum sources. The fall-off of the line against the *continuum source* background, away from the peak seems to be very rapid. At position 5 where the continuum antenna temperature is reduced to one-fourth of the peak antenna temperature of the source, the line almost disappears.

At position 1, which is in front of W3, and position 12, which is 30' south of W3, absorption lines other than the main absorption (around -40 and -49 km s^{-1}) were

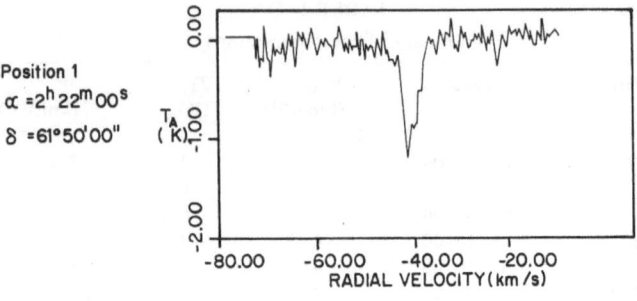

Position 1
α =2ʰ22ᵐ00ˢ
δ =61°50'00"

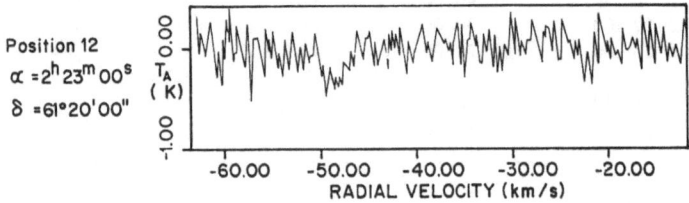

Position 12
α =2ʰ23ᵐ00ˢ
δ =61°20'00"

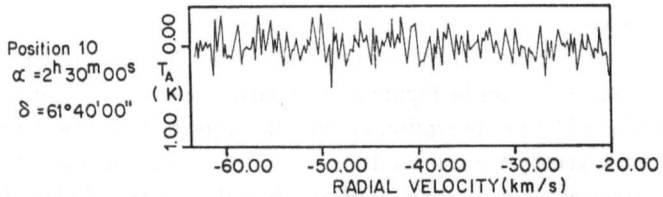

Position 10
α =2ʰ30ᵐ00ˢ
δ =61°40'00"

Fig. 2. Representative line profiles. Position 1 is against the W3 radio continuum background and positions 12 and 10 are against the universal microwave background.

detected at around -22 km s^{-1}. This feature was also detected at -21.6 km s^{-1} by Wilson (1972) in the direction of the W3 peak radio continuum. The large difference of velocities (about 20 km s^{-1}) of these lines from the main absorption suggests that these might be foreground clouds.

The velocity distribution of the H$_2$CO absorption lines is displayed in the plane of the sky in Figure 3. It is found that line velocities within the area of the W3 continuum center differ quite significantly from those in the surrounding regions. The velocity difference between the two areas is about 10 km s^{-1}. The velocity variation within each area seems to be negligible indicating there is little rotational or shearing motion. This result is consistent with the H109α-line survey in the region done by Rubin and Mezger (1970). Other molecular absorption lines in the direction of the OH emission source W3(OH) (which is between positions 8 and 11) show a similar velocity difference

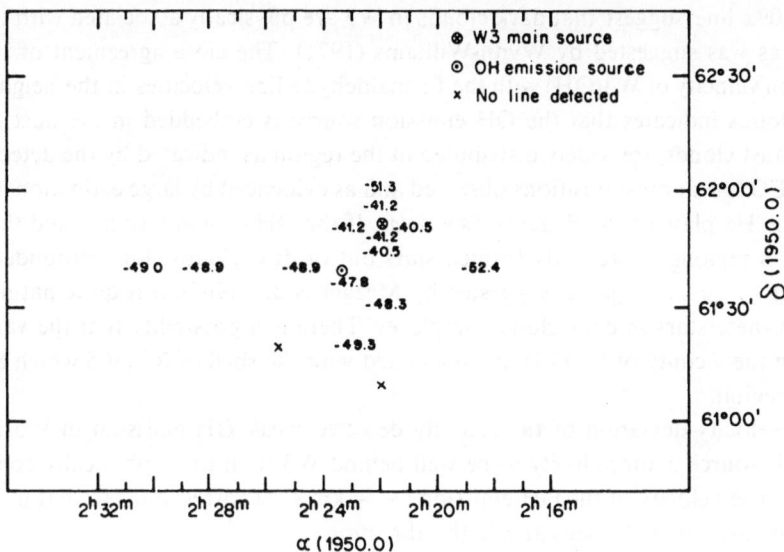

Fig. 3.   The velocity distribution of the main absorption lines. Positions of the W3 center and OH
emission source W3(OH) are also given.

from those of W3. All the known molecular and H 109α line velocities in both W3
and W3(OH) are given in Table II. Different line velocities for a source agree well with
each other with the exception of the weak OH emission in W3. Velocities of the
molecular and hydrogen lines in W3 listed in Table II, are also comparable with those
of the H_2CO lines we observed in the W3 radio source area, while the H_2CO line
velocities in the surrounding dust areas agree well with the molecular line velocities
in W3(OH). The close agreement between the velocities of molecular lines which are
supposed to be coming from the interstellar cold dust clouds (same authors, 1973) and

TABLE II

Radial velocities of radio lines

Line	W3 G133.7+1.2	W3 (OH) G133.9+1.0	Reference
H_2CO	−41.1[1]	−47.8[1]	(1) Zuckerman et al. (1970)
(absorption)	−40.1[2]	−45.9[2]	
OH	−54.2[4]	−46.5[3]	(2) Wilson (1972)
(emission)		−49.1[3]	(3) Barrett and Rogers (1966)
OH	−40.3[4]		(4) Turner (1970)
(absorption)	−40.0[5]		
			(5) Weaver, Dieter and Williams (1968)
H_2O (emission)	−40.7[6]	−46.6[6]	(6) Knowles et al. (1969)
H 109α	−41.8[7]		(7) Reifenstein et al. (1969)
H_2CO (140 GHz emission)	−	−48.0[8]	(8) Thaddeus et al. (1971)
HCN (88 GHz emission)	−	−49.0[9]	(9) Snyder and Buhl (1971)

− observed but not detected

the H109α line suggest that dark clouds in W3 are physically associated with the H II region as was suggested by Wynn-Williams (1971). The close agreement of the OH emission velocity of W3(OH) with the formaldehyde line velocities in the neighboring dark clouds indicates that the OH emission source is embedded in the dust clouds. These dust clouds are widely distributed in the region as indicated by the detection of the H_2CO line at most positions observed and as evidenced by large extinction indicated in the Hα plate of the *Palomar Sky Atlas*. If the OH emission source and the compact H II regions are recently formed stars out of dust clouds still surrounded by a 'cocoon' of dust and gas as suggested by Mezger *et al.* (1967), it is quite natural that we find these stars in dark cloud complexes. There is a possibility that the vast dark areas in the vicinity of IC 1795 are associated with the shell of IC 1805 which extends to this region.

The velocity deviation of the recently detected weak OH emission in W3 implies that this source is more likely to be well behind W3 than to be physically connected with it. The velocity of the OH emission (-54 km s^{-1}) corresponds to that of the H I peak intensity in the Perseus arm in this direction.

The velocity difference of about 10 km s^{-1} between W3 and its surroundings seems large enough to suggest that W3 (including the thick dark clouds in front of it) is physically separated from the surrounding features.

The additional absorption features with velocities of about -20 km s^{-1} are observed at only the W3 continuum center and at position 12 which is considerably separated

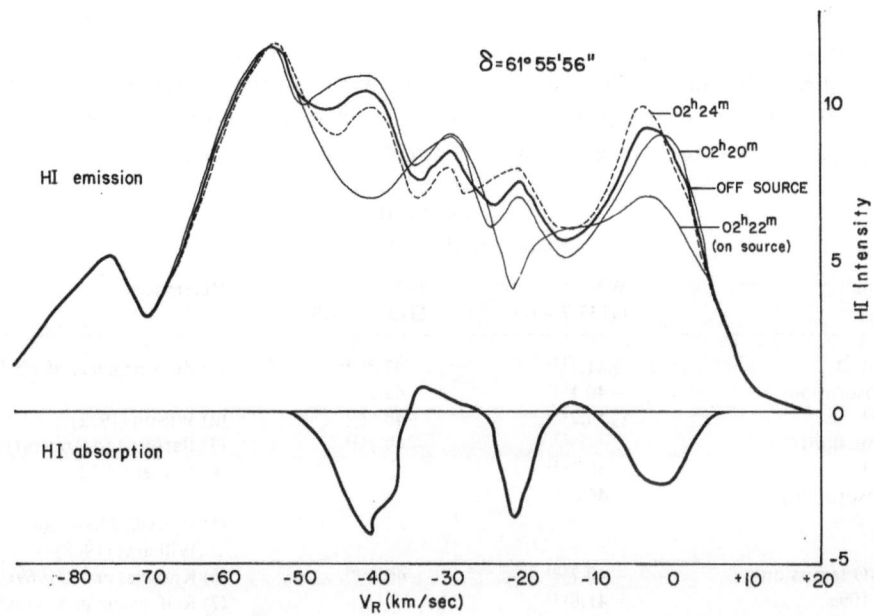

Fig. 4. H I emission and absorption line profiles constructed from the Maryland-Green Bank 21-cm Line Survey (Westerhout, 1969). The H I intensity unit in the ordinate corresponds to an antenna temperature of about 5 K.

from W3 and has no continuum contribution from it. Why this line is observed at only two contrasting and far separated positions is not known, but from their similarities in velocity it seems that they are located at foreground positions at about the same distance from us. A velocity of -20 km s^{-1} corresponds to an interarm spur of the H I distribution between the Orion and Perseus arms.

Strong H I absorptions were found in this direction in the Maryland-Green Bank 21-cm Line Survey (Westerhout, 1969). Absorption line profiles are constructed and presented in Figure 4. The H I survey is taken in the form of drift scans at constant declination. One such drift scan passes through $\delta = 61° 55'56''$ which is 4' north of the radio center position of W3 and well within the continuum peak. For the expected line profile, we took the average of the H I intensities at 4 min east and west of the source. Three large cold H I clouds are found at velocities of about -2, -21 and -40 km s^{-1} respectively. Recent interferometric observations confirm this result (Wilson *et al.*, 1972).

The H I absorption at -2 km s^{-1} is not observed in the H$_2$CO line, but the other two agree well enough with the H$_2$CO absorption features to show that H$_2$CO dust and cold H I occupy the same space. If this is the case, part of the extinction observed in the direction of W3 and its vicinity is the result of the foreground dust associated with the H I spur.

References

Allen, C. W.: 1964, *Astrophysical Quantities*, Athlone Press, London.
Barrett, A. H. and Rogers, A. E. E.: 1966, *Nature* 210, 188.
Cudaback, D. D., Read, R. B., and Rougour, G. W.: 1966, *Phys. Rev. Letters* 17, 452.
Knowles, S. H., Mayer, C. H., Cheung, A. C., Rank, D. M., and Townes, C. H.: 1969, *Science* 163, 1055.
Lynds, B. T.: 1962, *Astrophys. J. Suppl.* 7, 1.
Mezger, P. G. and Henderson, A. P.: 1967, *Astrophys. J.* 147, 471.
Mezger, P. G., Altenhoff, W., Schraml, J., Burke, B. F., Reifenstein III, E. C., and Wilson, T. L.: 1967, *Astrophys. J. Letters* 150, L157.
Minn, Y. K. and Greenberg, J. M.: 1973, *Astron. Astrophys.* 22, 13.
Raimond, E. and Eliasson, B.: 1969, *Astrophys. J.* 155, 817.
Reifenstein, E. C. III, Wilson, T. L., Burke, B. F., Mezger, P. G., and Altenhoff, W.: 1970, *Astron. Astrophys.* 4, 357.
Rogers, A. E. E., Moran, J. M., Crowther, P. P., Burke, B. F., Meeks, M. L., Ball, J. A., and Hyde, G. M.: 1966, *Phys. Rev. Letters* 17, 450.
Rubin, R. H. and Mezger, P. G.: 1970, *Astron. Astrophys.* 5, 407.
Snyder, L. E. and Buhl, D.: 1971, *Astrophys. J. Letters* 163, L47.
Thaddeus, P. Wilson, R. W., Kutner, M., Penzias, A. A., and Jefferts, K. B.: 1972, *Astrophys. J. Letters*, in press.
Turner, B. E.: 1970, *Astrophys. Letters* 6, 99.
Weaver, H. F., Dieter, N. H., and Williams, D. R. W.: 1968, *Astrophys. J. Suppl.* 16, 219.
Westerhout, G.: 1969, *Maryland-Green Bank Galactic 21-cm Line Survey*, Univ of Maryland, College Park.
Wilson, T. L.: 1972, *Astron. Astrophys.*, in press.
Wilson, T. L., Webster, W. J., Riegel, K. W., and Minn, Y. K.: 1972, in preparation.
Wynn-Williams, C. G.: 1971, *Monthly Notices Roy. Astron. Soc.* 151, 397.
Zuckerman, B., Buhl, D., Palmer, P., and Snyder, L. E.: 1970, *Astrophys. J.* 160, 485.

MOLECULAR DISTRIBUTIONS IN GALACTIC SOURCES

HÉLÈNE R. DICKEL

University of Illinois Observatory, Ill., U.S.A.

Abstract. The 4830 MHz absorption line of formaldehyde has been observed in the SNR complex W44 and the H II region complex W3. The results are presented as maps of apparent optical depth. The formaldehyde is concentrated toward the eastern edge of the shell and reaches a maximum apparent optical depth of 1.5 compared with 0.2 for OH; furthermore the centers of the H_2CO and OH concentrations are displaced by about 6'. The velocities of the H_2CO and OH agree and both molecular lines have the same velocity gradient across the source.

The H_2CO absorption profiles for W3 (IC 1795) exhibit at least two narrow components whose average velocity becomes more negative toward the NE. The greatest optical depth occurs where the optical obscurration is high but not at the position of maximum A_v. A general agreement between the velocities of other molecules and the H_2CO values is indicated.

In order to make models of molecular regions to explain their observed density and velocity structure and to determine their evolution as well as to gain insight into the molecular formation and destruction processes it is necessary to intercompare the distributions of a variety of different molecules in a given source and also compare them relative to the associated radio continuum, infrared sources and dust lanes. Thus far I have mapped the OH and H_2CO absorption in a number of extended continuum sources. The formaldehyde results are now presented for the SNR complex W 44 and the H II region complex W 3 (IC 1795), (Figures 1 and 2).

The formaldehyde observations were made with the NRAO* 140-ft radio telescope equipped with a cooled 6 cm parametric amplifier and the 413-channel autocorrelator. Measurements were made at 40 positions in W 44 and 24 positions in W 3 with integration times of $\frac{1}{2}$ to 1 h and with 3' separation between grid points. Concurrent profiles were obtained with 8 kHz and 16 kHz resolution. The corresponding separation between data points was 0.5 km s^{-1} and 1.0 km s^{-1}.

The W 44 profiles were adequately fit by one Gaussian curve and the resulting map of apparent optical depth

$$\left(\tau_{H_2CO} = -\ln\left[1 - \frac{|\Delta T_L|}{T_c} \right] \right)$$

is shown in Figure 1 (solid lines) superposed on an abbreviated version of a 6 cm continuum map** (dashed lines). For comparison two of the contours from the apparent optical depth map for the 1667 MHz OH absorption (Goss *et al.*, 1971) are

---

* Operated by Associated Universities, Inc. under contract with the National Science Foundation.
** The 6 cm contours are an adaptation of Milne's map (1969) which was convolved to an effective resolution of 6' and converted to T_A by applying the beam effciency factor of 0.81 for the NRAO 140-ft telescope. The formaldehyde line temperatures are on the same temperature scale.

Greenberg and Van de Hulst (eds.), Interstellar Dust and Related Topics, 277–280.

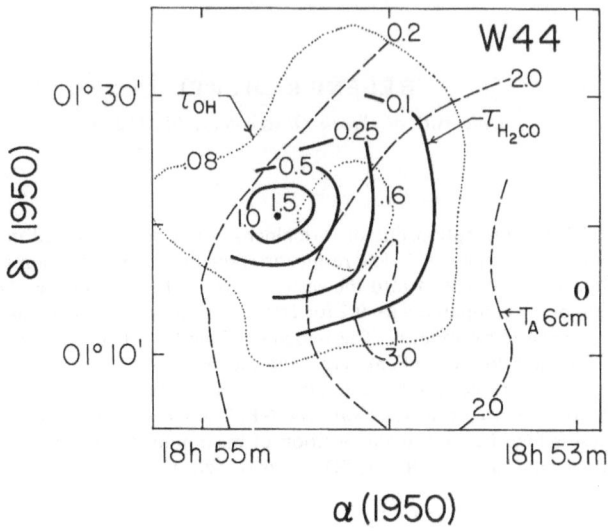

Fig. 1.   Map of apparent optical depth for formaldehyde in W 44 (see text).

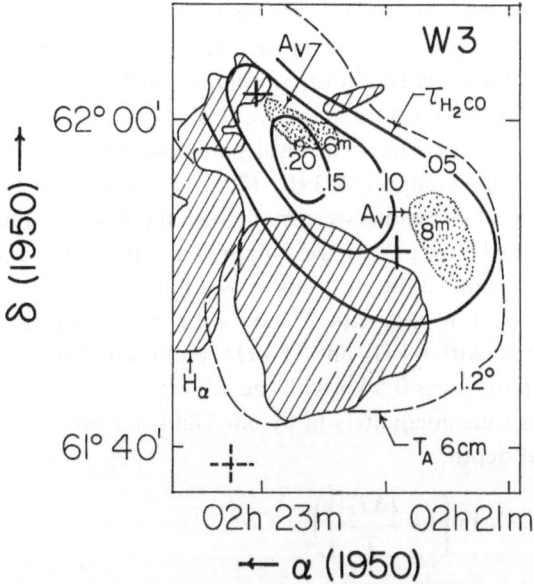

Fig. 2.   Map of apparent optical depth for formaldehyde in W 3 (see text).

shown by dotted lines. The formaldehyde is concentrated toward the eastern edge of
the shell where the gas compression is highest and no absorption was detected at the
secondary continuum peak on the western side (marked by O on Figure 1). The
maximum τ_{H_2CO} is 1.5 whereas τ_{OH} does not exceed 0.2. The centers of the H_2CO and
OH concentrations are displaced by about 6'. (These values have not been corrected
for resolution effects and may consequently be somewhat altered.) With more limit-

ed data Whiteoak and Gardner (1972) obtain similar results. The observed H_2CO and OH velocities agree surprisingly well, to within the errors of the measurements. There is a velocity gradient across the source with the values becoming more positive toward the edge of the shell as the optical depth increases. Goss *et al.* (1971) attribute this to a rotation of the cloud of molecules.

Most of the H_2CO absorption profiles for W 3 show two narrow features with half-intensity widths of about 2.2 km s^{-1} and a separation on the order of 2.0 km s^{-1}. There may be an additional weaker feature at the high velocity side. There is a velocity gradient across the source with the more negative velocities toward the NE. The optical depth maps of the individual velocity components are very similar. Thus the solid curves in Figure 2 are only the τ_{H_2CO} contours derived from a one-component Gaussian fit to the profiles and ignore the velocity gradient. The dashed line is the 1.2 K contour of the 6 cm continuum map of Rubin and Mezger (1970) with the main components *A*, *B* and *C* marked by crosses. The hatched area represents the H_α emission and the two speckled regions indicate where the visual absorption A_v exceeds 6 and 8 mag. respectively (Ishida and Kawajiri, 1968). The greatest optical depth for formaldehyde occurs in the optically obscured region but does not coincide with the highest values of A_v. The velocity gradient from *A* and *B* is similar to that found by Rubin and Mezger (1970) for the H 109α recombination line but is less complex; the total velocity spread is less and the H_2CO velocities are generally more positive by about 1.5 km s^{-1}. In spite of the different spatial resolutions it is noteworthy that the velocities of the other molecules (H – Chaisson, 1972; OH – Turner, 1970; H_2O – Knowles *et al.*, 1969; CO – Penzias *et al.*, 1970; X-ogen – Buhl and Snyder, 1972) near the main 6 cm continuum peak (*A*) either agree with the average H_2CO velocity or with one of the components of the line.

Further analysis and interpretation of these two sources is under way. I am assessing the effects of the finite resolution on the W 44 results. For W 3 I will investigate whether or not the distribution of formaldehyde can be related to the small scale structures of the molecular emission and IR sources (see Wynn-Williams' high resolution maps) (Wynn-Williams *et al.*, 1973). Thus far it appears that a rotating cloud model fits the W 44 data and may also apply to the W3 complex. This is in contrast to the model of an expanding shell of compressed material which we find for the Carina nebula (Dickel and Wall, 1971 and 1972). My planned studies of a number of molecules in many more sources should clarify the variety of conditions existing in these molecular clouds near the different kind of continuum sources and contrast these to the situation in dust clouds which are isolated from any continuum source.

Acknowledgements

I thank Tom Cram and David Buhl for their generous assistance in processing the data. I appreciated the dedication of George Behrens for keeping the equipment operating.

This research was supported in part by the National Science Foundation.

References

Buhl, D. and Snyder, L.: 1972, private communication.

Chaisson, E. J.: 1972, *Astron. Astrophys.* **18**, 149.

Dickel, H. R. and Wall, J. V.: 1971, *Bull. Am. Astron. Soc.* **3**, 459.

Dickel, H. R. and Wall, J. V.: 1972, 'Carina Nebula II: 18 cm OH Observations and Comparison of the Distribution of OH and H_2CO Molecules', in preparation.

Goss, W. M., Caswell, J. L., and Robinson, B. J.: 1971, *Astron. Astrophys.* **14**, 481.

Ishida, K. and Kawajiri, N.: 1968, *Publ. Astron. Soc. Japan* **20**, 95.

Knowles, S. H., Mayer, C. H., Cheung, A. C., Rank, D. M., and Townes, C. H.: 1969, *Science* **163**, 1055.

Milne, D. K.: 1969, *Australian J. Phys.* **22**, 613.

Penzias, A. A., Jefferts, K. B., and Wilson, R. W.: 1970, *Astrophys. J.* **165**, 229.

Rubin, R. H. and Mezger, P. G.: 1970, *Astron. Astrophys.* **5**, 407.

Turner, B. E.: 1970, *Astrophys. Letters* **6**, 99.

Whiteoak, J. B. and Gardner, F. F.: 1972, *Astron. Astrophys.* **21**, 159.

Wynn-Williams, C. G., Becklin, E. E., and Neugebauer, G.: 1973, this volume, p. 459.

PART VI

PHYSICAL PROCESSES, THEORY AND EXPERIMENT

COSMIC X-RAYS AND INTERSTELLAR DUST

SATIO HAYAKAWA

Dept. of Physics, Nagoya University, Nagoya, Japan

Abstract. Observational results of cosmic diffuse X-rays are reviewed with particular emphasis on soft X-rays. The intensity distribution of soft X-rays over the celestial sphere indicates that the diffuse component of soft X-rays consists of an extra-galactic and a galactic component. The absorption of the soft X-rays in the interstellar medium results in heating and ionization of interstellar matter. The ionization rate by X-rays is estimated as about 10^{-16} s^{-1} per H atom.

The scattering of X-rays by interstellar dust grains produces a halo of an X-ray source and smears out the pulsation of X-ray emission. The scattering coefficient and the halo size are given for some typical grain models.

The possibility that the dust grains gain relativistic energy is suggested. It is speculated that the relativistic dust grains in metagalactic space may be responsible for cosmic rays of ultrahigh energies and also for the diffuse X-rays by the interactions with cosmic black-body radiation.

1. Introduction

Since the discovery of cosmic X-rays ten years ago, little attention has been paid to the relevance of cosmic X-rays to interstellar dust. This is due to the facts that the energy density of X-rays in interstellar space is much smaller than that of radiation in a longer wavelength range, and that the interstellar medium is practically transparent for the X-rays of energies greater than 1 keV.

The extension of observation of cosmic X-rays to lower energies has changed this situation. It has been noted that the amount of energy dissipated by soft X-rays in interstellar matter may be comparable to or even greater than that by other agencies such as low energy cosmic rays. Further, the scattering of soft X-rays by interstellar dust grains may be of some astronomical relevance to studies of cosmic X-rays and interstellar dust.

The present article deals with three rather separate topics. The first part is devoted to a review of the diffuse component of cosmic X-rays and its contribution to the heating and ionization of interstellar matter. Recent observations have indicated that a substantial part of soft X-rays are generated in our Galaxy and then absorbed by the interstellar matter. The ionization rate is estimated to be comparable to that required for maintaining the temperature of the interstellar matter.

In the second part the scattering of X-rays by interstellar dust grains is discussed. The scattering results in a halo of a point X-ray source and in smearing out X-ray pulsations thereof. Hence the observations of soft X-rays from point sources will provide means of obtaining their distances as well as the size and the density of the dust grains.

The third part is so speculative that its validity remains to be examined by further

Greenberg and Van de Hulst (eds.), Interstellar Dust and Related Topics, 283–296.
All Rights Reserved. Copyright © 1973 by the IAU.

investigation. It is speculated that dust grains are accelerated in active galaxies with compact nuclei. Relativistic dust grains injected into intergalactic space may be responsible for cosmic ray air showers of very large size and may also produce X-rays by interactions with cosmic microwave radiation.

2. Diffuse Component of Cosmic X-Rays

Cosmic X-rays consist of the diffuse component and the component associated with individual sources. The diffuse component may result from a superposition of as yet unresolved sources, but its distribution over the celestial sphere is essentially isotropic at energies above 2 keV in contrast to the strong concentration of galactic X-ray sources towards the galactic plane. Hence it is inferred that galactic sources contribute little to the diffuse component, whereas extragalactic sources may contribute considerably to the diffuse component. This implies that the diffuse component belongs to the metagalaxy and its interactions with our Galaxy are very weak.

The spectrum of the diffuse component has been measured over a wide energy range. Since the observational results in the energy range 1–30 MeV remain to be confirmed, our discussions are restricted to the range below 1 MeV. Some controversy still exists in this energy range, but the spectrum shown in Figure 1 gives its general trend.

The spectrum is represented by a power law with a change in its slope near 40 KeV. The energy at which the slope changes and the rate of change are not well established, since most observations at energies above 25 KeV have been made at balloon altitudes and the correction for the scattering effect in the atmosphere is complicated, but the steepening of the spectrum is unmistakable from the low absolute intensity around 100 KeV.

Below 40 KeV the spectrum would appear to be represented by a single power law of $E^{-1.8}$, where E is the energy of an X-ray photon. However, more detailed examination shows that a flatter spectrum $E^{-1.4}$ is favoured in the energy range 1.5 to 10 KeV, and that the spectrum at energies below 1 KeV depends on the direction of sight.

In the soft X-ray region the diffuse component is not isotropic. A summary of several sky surveys (Kato, 1972) has shown that the intensity of soft X-rays generally increases with galactic latitude except in several selective regions. Recent sky surveys (Bleeker et al., 1973; Hayakawa, 1972b; Davidsen et al., 1972) have also confirmed this tendency. The latitude dependence is most clearly observed in the anticenter region, where few enhanced regions are found. In other directions the intensity distribution is complex and several enhanced regions are observed. The enhancement towards the galactic center may be contaminated by sources. This decreases as energy decreases and vanishes in the energy range 0.23–0.37 KeV. The enhancement around $l \sim 330°$ and $b \sim 20°$ is strong in the entire energy range below 1 KeV. The distributions in the energy ranges 0.37–0.65 keV and 0.65–0.90 KeV are given by Hayakawa (1972b) and the distribution in the range 0.23–0.37 KeV is shown in Figure 2, all based on the LEINAX sky survey by the Leiden-Nagoya collaboration experiment. A similar result is obtained by Davidsen et al. (1972) in the anticenter region.

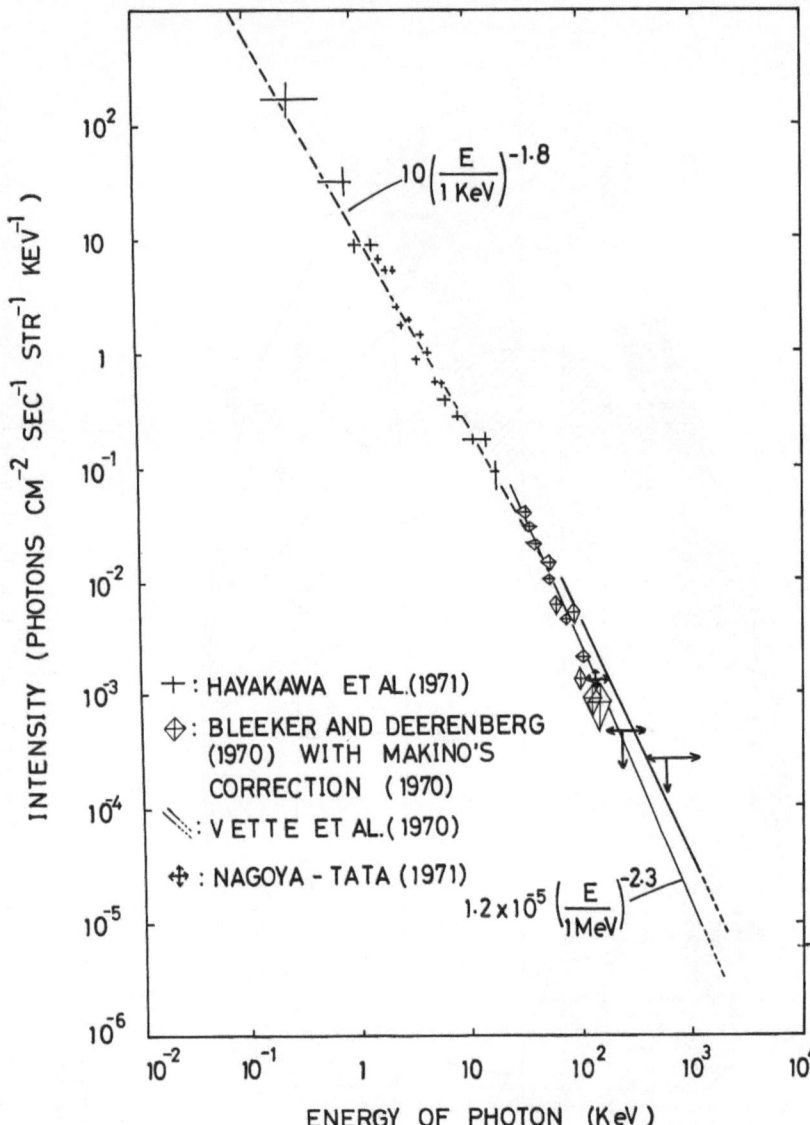

Fig. 1. The energy spectrum of cosmic diffuse X-rays. Two lowest energy points are taken from the intensities at high galactic latitudes. As described in the text, the intensity in such a low energy region depends on the direction of sight.

The galactic latitude dependence of the soft X-ray intensity may be interpreted in terms of the absorption of extragalactic X-rays by interstellar matter (Kato, 1972; Hayakawa, 1972b; Davidsen *et al.*, 1972). Further analysis of the LEINAX result supports the above conclusion for the hydrogen column densities greater than 10^{21} atoms cm^{-2}, as shown in Figure 3. At greater hydrogen column densities the intensity observed is higher than that expected from the interstellar absorption of extragalactic

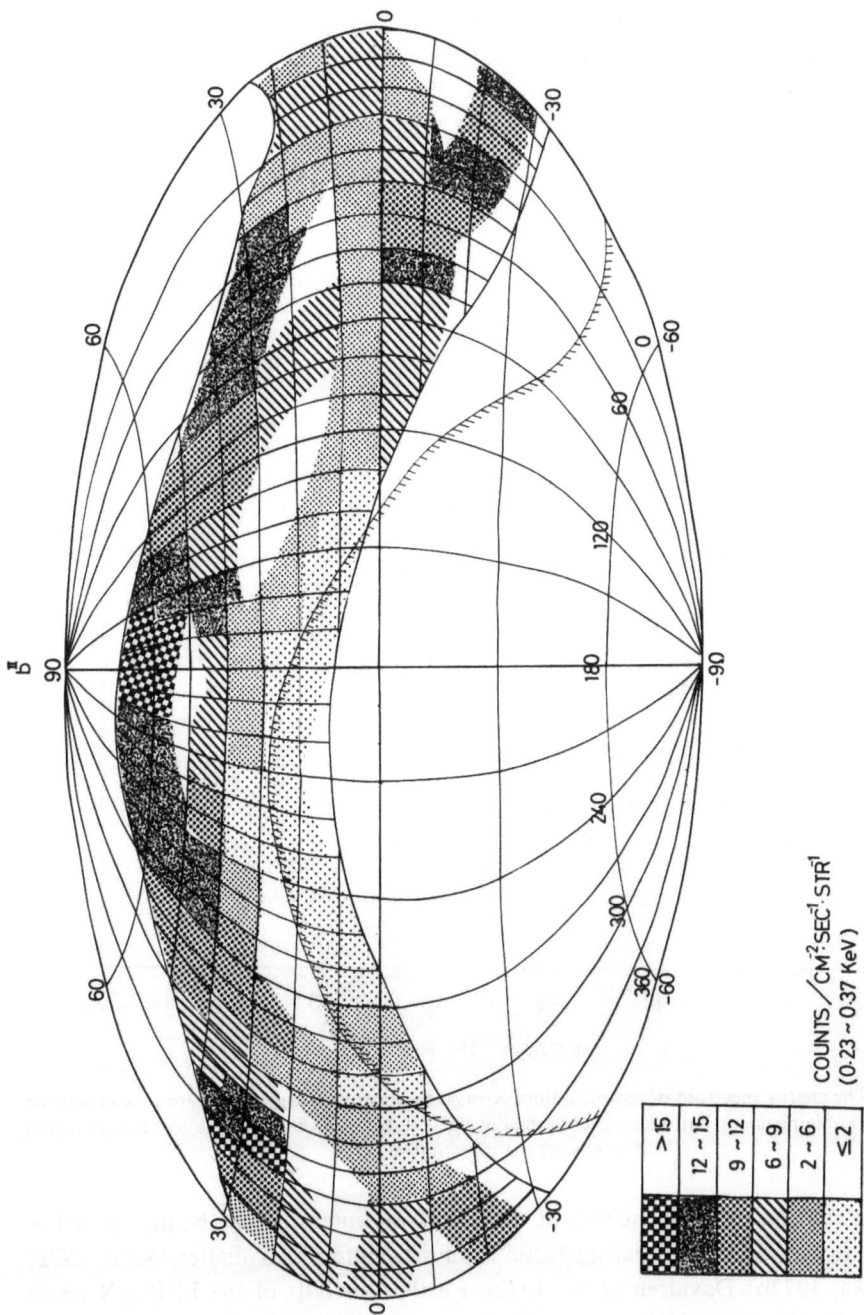

Fig. 2. The distribution of soft X-rays over the celestial sphere, prepared by T. Kato, based on the Leiden-Nagoya collaboration experiment. The energy range 0.23–0.37 keV represents the pulse height range of the counters. The energy range of X-rays is greater, since the energy resolution of the counters at 0.28 keV is 64%.

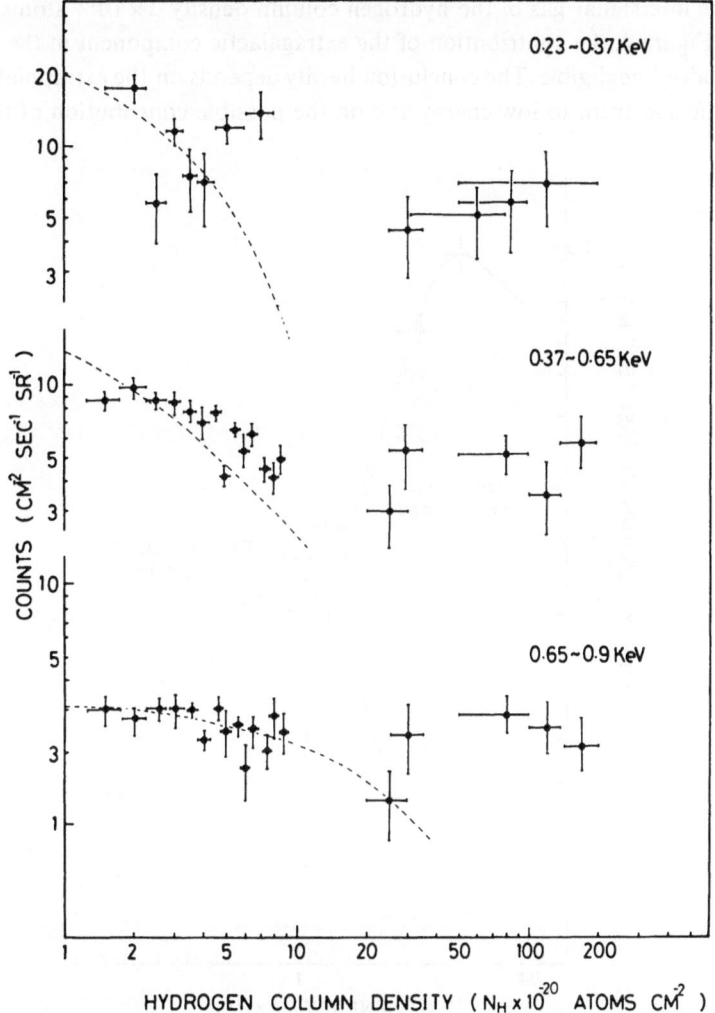

Fig. 3. The X-ray intensity versus the hydrogen column density. The energy ranges indicated represent the pulse height ranges. The dashed curves represent the theoretical absorption curves for extragalactic X-rays.

X-rays. The excess intensity at low galactic latitudes can be attributed to the emission of soft X-rays in our Galaxy.

The distribution of the soft X-ray emissivity in our galaxy remains to be further investigated; it seems to form a disk a little thicker than the gas disk. At low galactic latitudes the extragalactic component and the galactic component generated far from the galactic plane are negligible because of the interstellar absorption and, consequently, the emissivity of the galactic component can be obtained nearly independently of the thickness of the X-ray disk. The intensity of the extragalactic component is obtained by modifying the $E^{-1.4}$ spectrum extended to low energy by the absorption through the interstellar gas of the hydrogen column density 3×10^{21} atoms cm^{-2}. As seen from Figure 4, the contribution of the extragalactic component in the soft X-ray region is indeed negligible. The conclusion hardly depends on the extrapolation of the extragalactic spectrum to low energy and on the possible contribution of the galactic

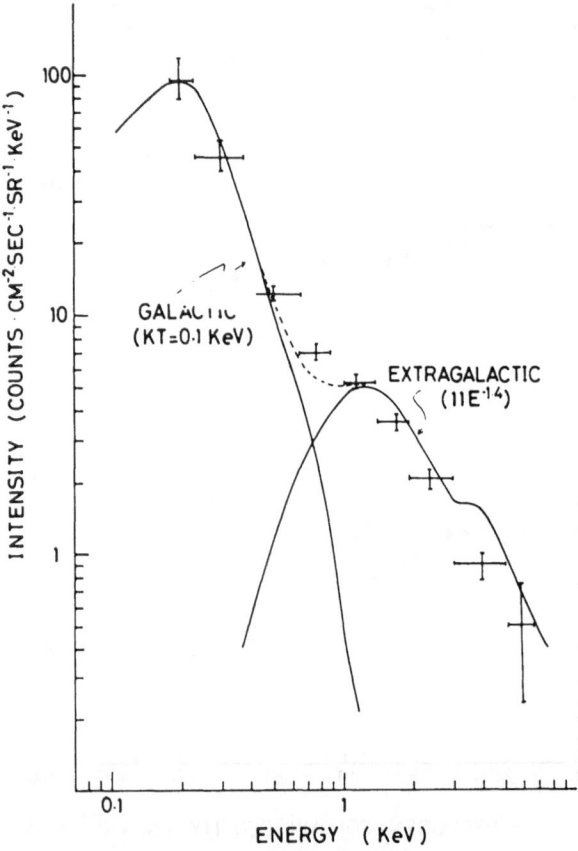

Fig. 4.   The energy spectrum of diffuse X-rays at a low galactic latitude. The solid curves represent the pulse height spectra for the $E^{-1.4}$ extragalactic spectrum and the $E^{-1} \exp(-E/kT)$ galactic spectrum at the hydrogen column density of 3×10^{21} cm^{-2}. The dashed curve represents their sum.

X-rays generated far above the galactic plane. The difference between the intensities observed and of the extragalactic component gives the intensity of the galactic component, which is also given in Figure 4.

The X-ray emissivity is derived from the intensity of the galactic component, taking the interstellar absorption into account. The solid line drawn for the spectrum of the galactic component is obtained from the emissivity

$$g(E) = g_0 n_H E^{-1} \exp(-E/kT) \, \text{photons cm}^{-3} \, \text{s}^{-1} \, \text{keV}^{-1} \tag{1}$$

with n_H = number of H atoms per cm^3.

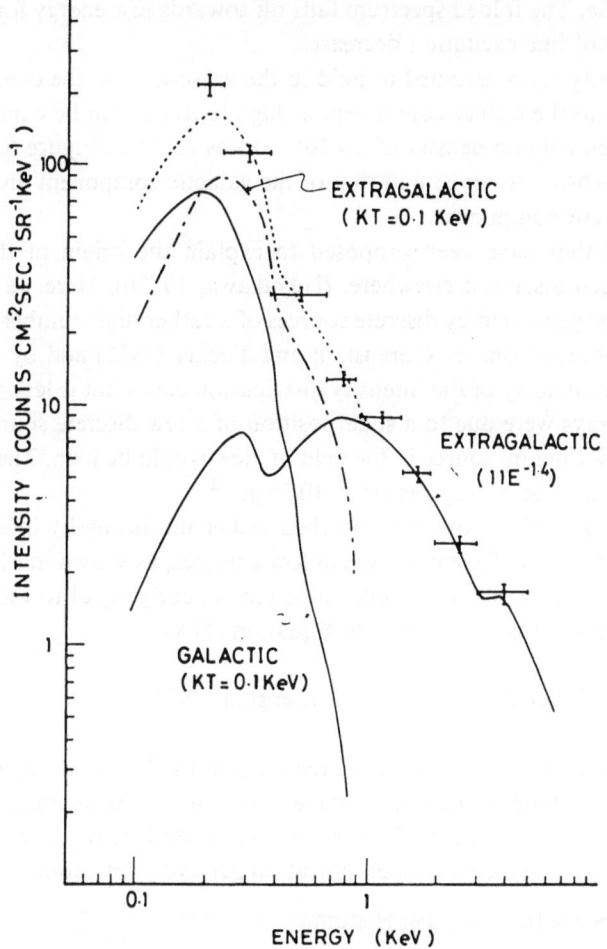

Fig. 5. The energy spectrum of diffuse X-rays at high galactic latitude. The solid curves are the same as in Figure 4, but different in the hydrogen column density, 3×10^{20} cm^{-2}. The dot-dashed curve represents the extragalactic E^{-1} exp $(-E/kT)$ spectrum subject to the interstellar absorption. The dashed curve represents the sum of these three components.

$$g_0 = 3 \times 10^{-17} \text{ photons s}^{-1}, \qquad kT = 0.1 \text{ keV},$$

and taking into account the efficiency and the energy resolution of the counters.

The expression (1) simulates the free-free emission. Actually, however, the emission lines are stronger by an order of magnitude than the continuum at such a low temperature. Hence the expression (1) should be regarded as an empirical relation holding for $E \gtrsim 0.2$ KeV, since the generation spectrum for $E < 0.2$ keV is obscured by the interstellar absorption. Indeed, the spectrum obtained by folding the emission lines by the energy resolution of the counters is approximately expressed by the expression (1) with a slightly higher temperature, except in a few narrow ranges in which strong emission lines lie. The folded spectrum falls off towards low energy for $E \lesssim kT$, since the probability of line excitation decreases.

If the emissivity (1) is assumed to hold in the galactic disk, the contribution of the galactic X-rays to the diffuse component at high latitudes can be computed, as given for the hydrogen column density of 3×10^{20} atoms cm^{-2} in Figure 5. The difference between the intensity observed and that of the galactic component gives the intensity of the extragalactic component.

The theories that have been proposed to explain the origin of the soft galactic X-rays have been discussed elsewhere. (Hayakawa, 1972b). Here we only point out that they may be generated by discrete sources of a rather high number density. This is based on the observations by Gorenstain and Tucker (1972) and by Davidsen et al. (1972) that a granularity of the intensity distribution exists but is less than expected if the galactic X-rays were due to a superposition of a few discrete sources, so that the probability of finding no source in the field of view would be high. The source density is thus estimated to be as large as 10^{-2}–10^{-3} pc^{-3}.

Accordingly the soft X-rays are absorbed rather uniformly by interstellar matter. Since the absorption coefficient is large at low energies, as shown in Figure 6, the rate of energy dissipation in the interstellar medium is nearly equal to the rate of energy generation. The latter is obtained from Equation (1) as

$$G = \int Eg(E)\,dE \simeq 5 \times 10^{-27}\, n_{\mathrm{H}}\, \mathrm{erg\, cm}^{-3}\, \mathrm{s}^{-1}. \tag{2}$$

An overestimate due to the extension of the integral to $E < kT$ is nearly compensated by the omission of bumps in excess of the expression (1) due to emission lines.

The energy dissipation rate, which is equal to G, is divided by the energy per ionization, 6×10^{-11} erg, to give the rate of ionization per hydrogen atom

$$\zeta_{\mathrm{X}} \simeq 1 \times 10^{-16}\, \mathrm{s}^{-1}\, \text{per H atom}. \tag{3}$$

This is greater than the rate of ionization by cosmic rays, which, on the assumption that the spallation of cosmic ray nuclei colliding with interstellar matter gives the galactic abundances of Be and B, is estimated at

$$\zeta_{\mathrm{CR}} \simeq 10^{-17}\, \mathrm{s}^{-1}\, \text{per H atom}.$$

The value of ζ_{X} must be compared to the ionization rate required for the ionization and thermal equilibria for H I regions (Spitzer and Scott, 1969)

$$\zeta_{\mathrm{H}} \simeq 5 \times 10^{-16}\, \mathrm{s}^{-1}\, \text{per H atom}. \tag{4}$$

The difference between ζ_{X} and ζ_{H} may not be serious in view of uncertainties in the values of parameters necessary for deriving the value of ζ_{X}. Since the value of ζ_{X} in Equation (3) is based on the X-ray intensity in the anticenter region, a larger value is obtained in the regions where the X-ray intensity is stronger than in the anticenter region. Although the conclusion is not yet quantitative, the soft X-rays are responsible for a substantial part of the heating and ionization of interstellar matter.

3. Scattering of X-Rays by Dust Grains

The absorption coefficient in Figure 6 is obtained under the assumption that all interstellar atoms are in the gas phase. If some of them form solid grains and their size is
comparable to or greater than the absorption mean free path, the extinction coefficient
for X-rays is reduced since the absorption in the grain is saturated. However, this is
not the case since the interstellar dust grains are believed to be of submicron size,
which is smaller than the absorption mean free path of X-rays with energies down to
0.1 keV.

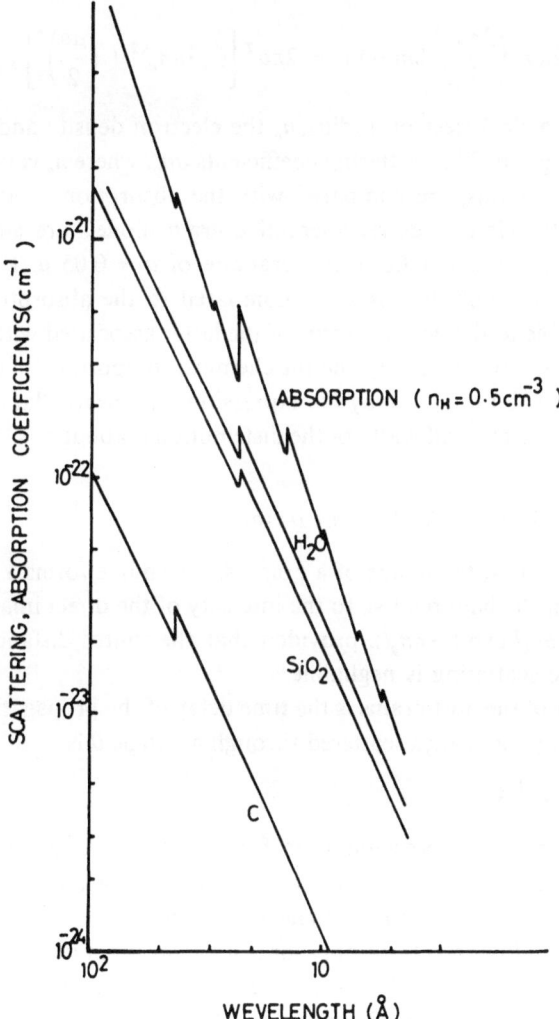

Fig. 6.   The scattering and the absorption coefficients of X-rays in the interstellar medium. The
absorption coefficient is based on the calculation for the hydrogen density $n_H = 0.5$ cm^{-3} by Brown
and Gould (1970). The scattering coefficients by dust grains are based on the calculation for C
$(a = 0.05 \mu)$, SiO$_2$ $(a = 0.08 \mu)$ and H$_2$O $(a = 0.14 \mu)$ by Hayakawa (1970).

The scattering of X-rays by dust grains has been discussed from various aspects by Overbeck (1965), by Slysh (1969), by Naranan and Shah (1970), by Ryter (1970), and by Hayakawa (1970). The differential scattering cross section for an X-ray of wavelength λ is obtained by the Rayleigh-Gans approximation as

$$\frac{d\sigma}{d\Omega} = 2a^2 \left(\frac{2\pi a}{\lambda}\right)^4 |m-1|^2 \left|\frac{j_1(x)}{x}\right|^2 (1+\cos^2\theta), \tag{5}$$

where $j_1(x)$ is the spherical Bessel function of the first order, $x = (4\pi a/\lambda)\sin(\theta/2)$ and θ is the cattering angle. Here the dust grain is assumed as a sphere of radius a and with the complex diffraction index m. The total scattering cross section is given by

$$\sigma = 2\pi a^2 \left(\frac{2\pi a}{\lambda}\right)^4 |m-1|^2 = 2\pi a^2 \left[(r_e \lambda a n_e)^2 + \left(\frac{\mu a}{2}\right)^2\right], \tag{6}$$

where r_e is the classical electron radius, n_e the electron density and μ the absorption coefficient in the grain. The scattering coefficients σn_g, where n_g is the number density of the interstellar grains, are compared with the absorption coefficient in Figure 6 (Hayakawa, 1970). Here three representative grain models are adopted, i.e., ice of $a = 0.14\ \mu$, silicate of $a = 0.08\ \mu$ and graphite of $a = 0.05\ \mu$. For $n_g = 1 \times 10^{-12}$ cm^{-3}, the scattering coefficient is small compared to the absorption coefficient and may be comparable to the possible error of the latter associated with the uncertainties in the interstellar hydrogen density and the chemical composition of interstellar matter.

The angular distribution given by the expression (5) is nearly flat for $x < 1$ and falls off as θ^{-4} for $x \gg 1$. The half width of the distribution is about $x = 1.8$, corresponding to

$$\theta_h = 7.4 \times 10^2 (1\ \text{keV}/E)(0.1\mu/a)\,\text{arc s}. \tag{7}$$

This demonstrates that the image of a point X-ray source forms a halo of radius θ_h. The brightness of the halo relative to the intensity of the direct image of the source is proportional to $\sigma n_g l/\exp(-\sigma n_g l)$, provided that the source distance l is not so large that the multiple scattering is negligible.

Another effect of the dust grains is the time delay of the propagation of X-rays. The average time delay for X-rays scattered through an angle θ is

$$\Delta t = l\theta^2/4. \tag{8}$$

Since the time delay is appreciable even for a scattering angle as small as one arc second, this effect is observable even when the angular resolution is good enough to resolve the halo from the core in the image of a source.

These effects are important not only for X-ray astronomy but also provide a novel means of studying the interstellar dust. Since the scattering cross section increases as the X-ray energy decreases, the effects are dominant in the soft X-ray region, in particular in the energy range just below the absorption edge of oxygen. The halo size gives direct information on the grain size. Once the size of known, the brightness of the halo and the decrease of the relative intensity of the pulsating component give the value of

$\varrho^2 n_g l$, where ϱ is the density of the grain. These effects can be observed by the technique available at present. Further development in technology will make it possible to observe further detail of interstellar dust grains.

4. Dust Grains of Relativistic Energies

Although the dust grains will mainly move with the interstellar gas, because of the short mean free path of collisions between dust grains and atoms in a gas, a small fraction of the dust grains may acquire high energies, as about 10^{-9} of the interstellar diffuse atoms form cosmic rays. Nearly a quarter century ago, Spitzer (1949) suggested that heavy nuclei discovered in primary cosmic rays could be explained by relativistic dust grains of meteoritic composition that impinged into the atmosphere and then evaporated into atoms. He suspected that such grains could be accelerated by a strong radiation pressure caused by supernova explosions. Although the acceleration to relativistic energy was found to be difficult because of the retardation effect, acceleration by moving magnetic fields may operate after the injection to semi-relativistic energy by radiation pressure, since the dust grains are supposed to be charged. The behaviour of a dust grain after it gains energy exceeding the critical injection energy may be essentially the same as that of nuclear particles and electrons. Below this energy the dust grain may be decelerated or may even be destroyed by collisions with ambient matter.

The injection by radiation pressure may take place in the compact cores of active galaxies, such as quasistellar objects and Seyfert galaxies. If the radiation is as strong as 10^{45} erg s^{-1} and is emitted from a compact source as small as several light months, the Lorentz factor of a grain, γ, becomes appreciably greater than unity, though it hardly exceeds $\gamma = 10$. Since these objects are known to be strong radio sources, the magnetic acceleration responsible for generating relativistic electrons which emit radio waves by the synchrotron radiation may also be responsible for the further acceleration of the dust grains. The magnetic regidities of the dust grains accelerated may exceed 10^{17} V, if their charge is 10^2 to 10^3 times the unit electric charge. The acceleration to such high rigidity is considered to be possible, since the magnetic fields in the active galaxies may be stronger than in our Galaxy and the sizes of the radio emitting regions are larger than the dimension of our Galaxy. If these dust grains escape into intergalactic space, they form a part of metagalactic cosmic rays.

Thus Hayakawa (1972a) has speculated that the relativistic dust grains are responsible for extensive air showers of sizes greater than 10^{10}. It has been generally believed that such large extensive air showers are produced by metagalactic protons of energies greater than a few times 10^{19} eV, because of their essential isotropy and of the difficulty of being trapped in our Galaxy. A serious objection has, however, been raised against this interpretation. In the rest system of the protons, the energies of photons of the cosmic black-body radiation are high enough to produce mesons and, as a result, the protons would lose their energy so rapidly that the spectrum of the protons should fall off steeply beyond a few times 10^{19} eV. On the other hand, the size spectrum of extensive air showers does not fall off but maintains it slope beyond the size of 10^{11}.

If the extensive air showers are originated by the dust grains, the energy loss by meson production does not take place, since the Lorentz factors are of the order of 10^3. The total energy of a dust grain of $\gamma \simeq 10^3$ is large enough for the production of huge extensive air showers, since a grain of radius 0.1 μ contains about 10^{10} nucleons. The flux of the relativistic dust grains is estimated under the assumptions that the intensity of metagalactic cosmic rays is 10^{-3} times that of the galactic ones, and that the grain to proton ratio at a given value of γ is the same as that in the interstellar medium, that is, 10^{-12}. Since the proton intensity observed for $\gamma \geqslant 10^3$ is about 10^{-5} cm^{-2} s^{-1} sr^{-1}, the flux of metagalactic dust grains of $\gamma \geqslant 10^3$ is expected to be about 10^{-20} cm^{-2} s^{-1} sr^{-1}. This is nearly the value required for the frequency of extensive air showers of the sizes greater than 5×10^{10}.

It should, however, be remarked that the relativistic dust grains lose energies by interaction with the cosmic black-body radiation, and that the scattering of the black-body photons and the thermal emission of the grains contribute to metagalactic X-rays.

The average energy of the black-body photons is $\varepsilon \sim 10^{-3}$ eV. In the rest system of the dust grain the photon energy is of the order of $\gamma\varepsilon$. Since this is in the visible and near infrared ranges, these photons are scattered and absorbed with a cross section close to the geometrical one. The photons scattered have energies as large as $\gamma^2\varepsilon$, which is about 1 keV for $\gamma = 10^3$. The temperature of the grain is of the order of 10^2 K and the thermal photons for $\gamma = 10^3$ is as high as some tens of eV. Since the density of the black-body photons is of the order of 10^3 cm^{-3}, the scattering cross section of 10^{-10} cm^2 gives the X-ray flux of about 10 cm^{-2} s^{-1} sr^{-1}, if the contribution over the cosmic radius of 10^{28} cm^2 is taken into account. The thermal emission gives a higher flux in the XUV region.

A quantitative analysis, which will be published elsewhere, gives the following results. The energy loss of a relativistic dust grain is dictated by absorption for $\gamma < 10^3$, whereas the scattering is a dominant cause for $\gamma > 10^3$. This is because the scattering cross section increases steeply with the energy of the incident photon and levels off in the visible region, whereas the absorption cross section is supposed to be proportional to the photon energy over a wide range of the energy. The fractional energy loss over the cosmic age approaches unity at $\gamma = \gamma_c$, so that the spectrum of the dust grains becomes steeper for $\gamma > \gamma_c$. In this range the scattering is dominant over the absorption, and the grains of inital γ greater than γ_c are degraded to γ_c.

The flux of the dust grains is estimated under the assumptions that the intensity of primary particles producing extensive air showers of sizes greater than 5×10^{10} is 3×10^{-20} cm^{-2} s^{-1} sr^{-1}, and that the energy required for a shower particle is 2×10^{10} eV, taking into account the fact that the energy of a primary nucleon is as low as 10^{12} eV. Thus the differential spectrum of the relativistic dust grains is approximately expressed as

$$j(\gamma) = j_0 \gamma^{-\alpha} + j_c \delta(\gamma - \gamma_c) \quad \text{for} \quad \gamma \leq \gamma_c, \tag{9}$$

with

$$j_0 = 1 \times 10^{-16}(0.1\mu/a)^{3(\alpha-1)}\,\text{cm}^{-2}\,\text{s}^{-1}\,\text{sr}^{-1},$$
$$j_c = j_0/(\alpha-1)\gamma_c^{\alpha-1}, \quad \gamma_c = 2 \times 10^3(0.1\mu/a), \quad \alpha \simeq 1.6,$$

where a is the radius of the grain. The term with the δ-function may oversimplify the actual spectrum, the latter being peaked at about γ_c and having some width.

The contribution to the scattered photons comes mainly from the δ-function term, and the spectrum of photons is a black-body spectrum with a temperature of about $\gamma_c^2 T_b$, where T_b is the temperature of the cosmic black-body radiation. The maximum of the spectrum appears at

$$E_m \simeq 0.8(0.1\mu/a)^2\,\text{keV}, \tag{10}$$

and the flux at E_m is

$$f_s(E_m) \simeq 0.2(0.1\mu/a)^{1.2}\,\text{cm}^{-2}\,\text{s}^{-1}\,\text{sr}^{-1}\,\text{keV}^{-1}. \tag{11}$$

If a is as small as $0.03\,\mu$, this agrees with the observed X-ray flux, but the spectrum falls off toward high energy more rapidly than the observed one.

The thermal emission gives a flux

$$f_{th}(E) \simeq 1 \times 10^2(0.1\mu/a)^{3.3}\,\text{cm}^{-2}\,\text{s}^{-1}\,\text{sr}^{-1}\,\text{keV}^{-1}, \tag{12}$$

and its spectrum is nearly flat for $E \lesssim 0.1\,(0.1\mu/a)^{1.2}$ keV. This is in approximate agreement with the flux of the soft extragalactic component described in Section 2.

These results rather critically depend on the optical properties of the dust grains in the near infrared region. The artificial separation of the spectrum in two parts, as shown in Equation (9), is due partly to the lack of knowledge about the optical properties. The spectra obtained in Equations (11) and (12) are greatly affected by this artificial assumption as well as by the simplification in the kinematics of scattering and thermal emission. Hence they should be regarded as giving rough values of the X-ray fluxes in the two energy regions.

The existence of the relativistic dust grains discussed in this section is merely speculative and its reality will have to be explored by future studies. It may, however, be instructive to bring up this speculative theory, since it emphasizes the important bearing of the dust grains on various astrophysical phenomena.

Acknowledgements

The author would like to express his thanks to the members of the Leiden-Nagoya X-ray group who permitted him to use unpublished results described in Section 2 and also to Dr T. Kato for her help in the preparation of Section 2.

References

Bleeker, J. A. M. and Deerenberg, A. J. M.: 1970, *Astrophys. J.* **159**, 215.
Bleeker, J. A. M., Deerenberg, A. J. M., Yamashita, K., Hayakawa, S., Kato, T., and Tanaka, Y.: 1973, in preparation.

Brown, J. and Gould, R. J.: 1970, *Phys. Rev.* **D1**, 2252.

Danjo, A., Hayakawa, S., Ideka, M., Makino, F., Tanaka, Y., Agrawal, P. C., Gokhale, G. S., and Sreekantan, B. V.: 1970, *Space Research* **XI**, 1373.

Davidsen, A., Shulman, S., Fritz, G., Meekins, J. F., Henry, R. C., and Friedman, H.: 1972, *Astrophys. J.* **177**, 629.

Gorenstein, P. and Tucker, W. H.: 1972, *Astrophys. J.* **176**, 333.

Hayakawa, S.: 1970, *Prog. Theor. Phys.* **43**, 1224.

Hayakawa, S., Kato, T., Makino, F., Ogawa, H., Tanaka, Y., Yamashita, K., Matsuoka, M., Miyamoto, S., Oda, M., and Ogawara, Y.: 1971, *Astrophys. Space Sci.* **12**, 789.

Hayakawa, S.: 1972a, *Astrophys. Space Sci.* **16**, 238.

Hayakawa, S.: 1972b, in H. Bradt and R. Giacconi (eds.), 'X- and Gamma-Ray Astronomy', *IAU Symp.* **55**, 235.

Kato, T.: 1972, *Astrophys. Space Sci.* **16**, 478.

Makino, F.: 1970, *Astrophys. Space Sci.* **8**, 251.

Naranan, S. and Shah, G. A.: 1970, *Nature* **225**, 836.

Overbeck, J. W.: 1965, *Astrophys. J.* **141**, 864.

Ryter, Ch.: 1970, *Nature* **226**, 1040.

Slysh, V. I.: 1969, *Nature* **224**, 159.

Spitzer, L.: 1949, *Phys. Rev.* **76**, 583.

Spitzer, L. and Scott, E. H.: 1969, *Astrophys. J.* **158**, 161.

Vette, J., Gruber, D., Matteson, J. L., and Peterson, L. E.: 1970, *Astrophys. J.* **160**, L161.

OPTICAL MEASUREMENTS ON SOLIDS OF POSSIBLE
INTERSTELLAR IMPORTANCE

DONALD R. HUFFMAN and JAMES L. STAPP*

Physics Dept., University of Arizona, Tucson, Ariz. 85721, U.S.A.

Abstract. Optical constants of olivine – $(Mg, Fe)_2SiO_4$ and magnetite – Fe_3O_4 are presented for the wavelength range from near infrared to far ultraviolet. A feature occurs in the optical constants of olivine at about 1460 Å, but no structure that could give rise to a 2200 Å interstellar feature is found. The most peculiar characteristic of the magnetite results is the large change of optical properties with temperature in the infrared.

1. Introduction

Optical constants of solids over wide wavelength ranges needed for Mie scattering calculations relating to interstellar and circumstellar solids have not been readily available in the scientific literature. We have begun a program of measuring optical constants of likely interstellar and circumstellar solids from far infrared to far ultraviolet. Results for two solids, olivine and magnetite, are presented in this paper. These are perhaps the best possibilities from the two groups of solids of interest to us, i.e., silicates and oxides of iron.

The reason for interest in silicates, apparent from some of the papers being presented in this symposium, is the likely identification of silicates in various circumstellar regions by their characteristic absorption bands in the 10 μ region. In addition it has been suggested (Huffman and Stapp, 1971) that enstatite silicate in certain size ranges could produce a 2200 Å band corresponding to the prominent interstellar feature. The extinction 'bump' expected in small particles of enstatite is produced by the onset of strong absorption near 2100 Å. Whether this is a general characteristic of silicates has not been known because of the lack of ultraviolet measurements on other solids.

Our interest in iron oxides is for the following reasons: (1) the possibility that some of the diffuse interstellar bands may be due to Fe^{+++} in such solids, and (2) the alignment of grains necessary to produce interstellar polarization may require ferromagnetic solids. The Fe^{+++}-diffuse-band-theory was suggested several years ago by different authors independently (Manning, 1970; Dorschner, 1970; Huffman, 1970). Manning and Dorschner discussed the possibility of iron ions in various silicates. Our calculations and measurements show, however, that the oscillator strength of Fe^{+++} is generally too low to produce a band of the observed strenght of λ 4430, unless the bands are due to iron ions in a crystal with cooperative magnetic ordering (ferromagnetism, antiferromagnetism, or ferrimagnetism). In magnetic iron oxides such as α-Fe_2O_3, γ-Fe_2O_3, Fe_3O_4, and $MgFe_2O_4$, the oscillator strengths of the absorption

---

* Present address: Kirtland A.F.B., Albuquerque, N.M.

Greenberg and Van de Hulst (eds.), Interstellar Dust and Related Topics, 297–301.

bands can be as much as three orders of magnitude greater than in non magnetic solids because of the internal magnetic interaction coupling adjacent iron atoms (Huffman, 1969, and unpublished results). Furthermore, this cooperative effect of neighboring iron atoms in the solid is just the feature that Purcell and Spitzer (1971) have concluded is necessary for alignment by a Davis-Greenstein-type mechanism in 10^{-6} G interstellar magnetic fields.

2. Determination of Optical Constants

For these reasons, we have made measurements to determine optical constants from near infrared to far ultraviolet for magnetite – Fe_3O_4, and olivine – $(Mg, Fe)_2SiO_4$. The samples were carefully selected natural specimens that were cut, polished, and etched to provide specularly reflecting surfaces with very small-scale surface roughness. Reflectance measurements were made at near-normal incidence from about 3 μ to about 400 Å in wavelength, and the optical constants determined by a Kramers-Kronig analysis of the reflectance data (Stern, 1963). Figures 1 and 2 show the energy dependence of the real and imaginary parts of the complex optical constant N, defined as

$$N = n + ik.$$

n is the index of refraction and k is commonly called the extinction coefficient.

Although the combination of specular reflectance measurements and Kramers-Kronig analysis is the most widely used and probably the best method for obtaining

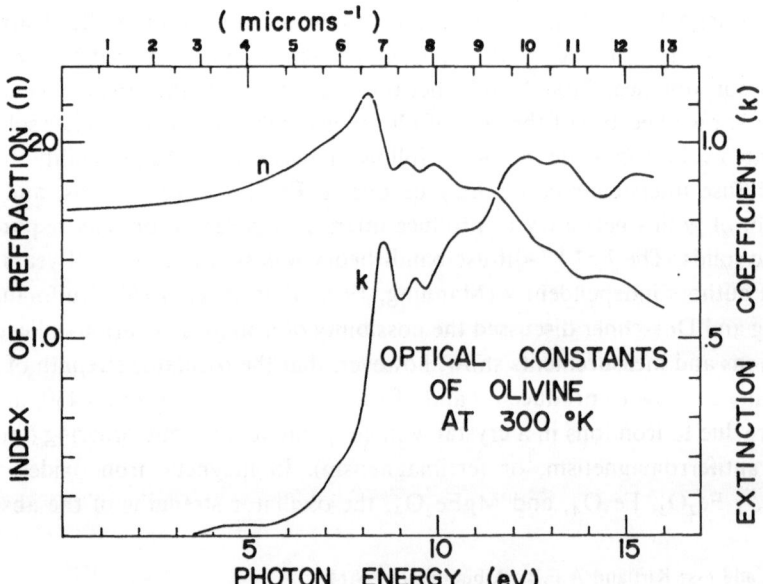

Fig. 1.   Optical constants of Olivine $(Mg, Fe)_2SiO_4$ at about 300 K.

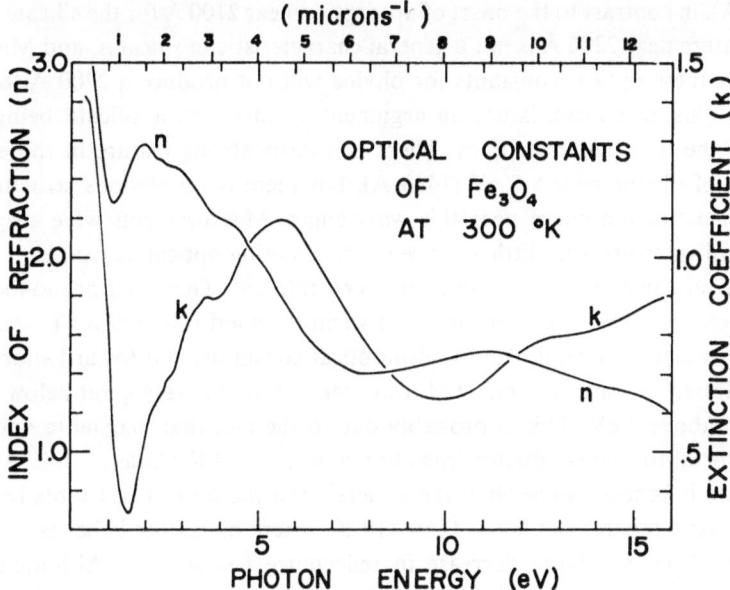

Fig. 2.   Optical constants of magnetite Fe_3O_4 at about 300 K. Note that the scales for both optical constants do not start at 0.

optical constants in the highly absorbing ultraviolet regions, various possibilities for error exist. The most difficult error to circumvent may be caused by the extrapolation of the reflectance data to infinite energy required by the Kramers-Kronig integral (Stern, 1963). Although the wide range of data used here (extending to 30 eV) minimizes this problem, the unknown extrapolation to infinite energy can still cause appreciable error. In addition, surface films, residual surface roughness, and amorphous layers introduced by polishing can cause severe problems in the far ultraviolet. For these reasons we feel that it would be a mistake to trust any determinations of this kind to better than 10% in highly absorbing regions ($k > 0.1$). A revealing example of the kind of errors involved can be obtained by comparing results on MgO obtained independently by two well-respected groups of workers in this field (Williams and Arakawa, 1967; Roessler and Walker, 1967). Although MgO can be artificially grown as a very pure single crystal which cleaves easily, leading to atomically smooth and clean surfaces, discrepancies between the two independent measurements of optical constants of 10–20% are present. Despite possible uncertainties in magnitude, the method is usually quite reproducible in revealing structure in the optical constants curves.

3. Discussion of Results

The main value of these results is that they make possible realistic scattering calculations on two new materials. A few comments on the results are perhaps in order. Regarding the olivine, strongly increasing absorption does not set in until about 7 eV

($\sim$ 1800 Å), in contrast to the onset of absorption near 2100 Å for the silicate enstatite. Thus a feature near 2200 Å is not a general characteristic of silicates, and Mie calculations using these optical constants for olivine will not produce a 2200 Å 'bump' for any sizes. This, of course, is not an argument against such a silicate being a component of the interstellar medium. There is a fairly strong feature in the extinction coefficient of olivine near 8.5 eV (1460 Å), but there is no obvious structure in the interstellar extinction curve near this wavelength. Measurements were also made at 100 K on olivine with very little change in the resulting optical constants.

The optical constants of magnetite are very different. There is a pronounced minimum in both n and k in the near infrared when the solid is at 300 K. Toward longer wavelengths there is a rapid rise in both optical constants, and toward shorter wavelengths a lesser increase. The effect of lower temperature is very great below 1 eV and very small above 2 eV. This is probably due to the fact that magnetite undergoes a so-called metal-to-semiconductor transition at about 119 K (Adler, 1968). Above this temperature it behaves somewhat like a metal, and the optical constants below 1 eV show this. At temperatures lower than 119 K, where magnetite behaves like a semiconductor, there is a large decrease in reflectance below 1 eV. Although the low temperature studies of magnetite in the infrared have not progressed to where we can determine optical constants below 1 eV, preliminary indications are that k will decrease toward 0 and n will level off. These variations of optical properties with temperature are among the largest occurring in common solids, and could have important consequences if magnetite is present in space. There is some fine structure in the visible region in magnetite which is not apparent on this energy scale and at 300 K. This is the structure (due to iron ions in the solid), that might provide explanations for some diffuse interstellar features. Further work is being done at low temperatures to explore these small (on this scale) but potentially significant variations of optical constants.

Further work now in progress at this laboratory will provide optical constants measurements comparable to Figures 1 and 2 for other possible interstellar solids. We also have begun detailed determinations of optical constants for silicates and other solids in the 10 μ and 20 μ regions.

Acknowledgements

We would like to gratefully acknowledge the collaboration in making the far ultra-violet measurements of H. Fritzsche, J. Freeouf, J. W. Osmun, and K. Murase of the University of Chicago. This research was supported in part by USAF Office of Scientific Research, contract number F44620-70-C-0029.

References

Adler, D.: 1968, in *Solid State Physics*, Vol. 21, Academic Press, New York, p. 109.
Dorschner, J.: 1970, *Astron. Nachr.* **292**, 107.

Huffman, D. R.: 1970, *Astrophys. J.* **161**, 1157.
Huffman, D. R.: 1969, *J. Appl. Phys.* **40**, 1334.
Huffman, D. R. and Stapp, J. L.: 1971, *Nature Phys. Sci.* **229**, 45.
Manning, P. G.: 1970, *Nature* **226**, 829.
Purcell, E. M. and Spitzer, L.: 1971, *Astrophys. J.* **167**, 31.
Roessler, D. M. and Walker, W. C.: 1967, *J. Opt. Soc. Am.* **57**, 835.
Stern, F.: 1963, in *Solid State Physics*, Vol. 15, Academic Press, New York, p. 300.
Williams, M. W. and Arakawa, E. T.: 1967, *J. Appl. Phys.* **38**, 5272.

OPTICAL AND PHOTOEMISSION PROPERTIES
OF GRAPHITE GRAINS

R. F. WILLIS, B. FEUERBACHER, and B. FITTON

Surface Physics Division, European Space Research Organisation, Noordwijk, Holland

Abstract. The optical properties and collective excitations are discussed for graphite and vitreous carbon. The bulk resonance energy for the π-plasmon is seen to be shifted by 1.4 eV from the crystalline to the amorphous case, indicating that the degree of structural disorder is an important parameter in the optical properties of carbon grains. Data are presented on work function, photoelectric yield, and photoelectron energy distribution for both graphite and vitreous carbon. Under solar irradiation, photoelectrons from graphite are emitted with an average energy of about 1 eV (12 000 K).

Multi-component models of interstellar grains (Bless and Savage, 1972; Hoyle and Wickramasinghe, 1962; Wickramasinghe and Nandy, 1970; Gilra, 1971) have been widely used to match the observed absorption and polarisation features of interstellar extinction. Observations in the far UV (Stecher, 1965; Bless and Savage, 1972) have revealed a pronounced structure in the extinction curve, located at 2175 Å wavelength. This structure has been attributed to absorption by small graphite particles (Stecher and Donn, 1965; Gilra, 1971), which are assumed to be one of the major constituents of the interstellar matter. Until recently the basic optical properties of bulk graphite were not completely understood, which led to difficulties in predicting the optical response of small graphite particles. In this paper, results of experimental studies on the optical and photoemission properties of graphite and of vitreous carbon, which is a highly disordered or glassy form of carbon, are presented. These experiments were performed in order to provide data to support calculations of the extinction due to graphite grains and the grain charging due to photoelectric emission.

1. The Optical Properties of Graphite Grains

The optical properties of graphite for light polarized with the electric field vector normal to the c-axis ($E \perp c$) have been measured over a wide energy range by Taft and Philipp (1965). Optical measurements for $E \,//\, c$ are available only up to 10 eV (1240 Å). In order to determine the optical constants for higher energies, Tosatti and Bassani (1970) applied a Kramers-Kronig analysis to the electron energy-loss data of Zeppenfeld (1968) and derived a strong $E \,//\, c$ absorption peak near 11 eV (1130 Å). However recent experimental studies of the energy band structure of graphite (Willis *et al.*, 1971) do not endorse this assignment of an $E \,//\, c$ electronic transition at 11 eV. The lowest transition for $E \,//\, c$ that could give rise to optical absorption in this direction of polarization was found to lie at 14.5 eV (855 Å). This result is in agreement with a recent band-structure calculation by Painter and Ellis (1970), and with optical measurements performed with a synchrotron light source (Klucker and Skibowski,

Greenberg and Van de Hulst (eds.), Interstellar Dust and Related Topics, 303–309.

unpublished). From these results it is concluded that crystalline graphite is essentially transparent for light polarized with $E // c$ at least up to the interstellar cutoff at 13.6 eV.

Stecher and Donn (1965) have calculated the extinction due to small graphite particles using Mie's theory and the optical constants of Taft and Philipp (1965). They found an absorption peak near 2200 Å in close agreement with the observed bump in the extinction curve. This structure has been considered as due to an electronic interband transition of the π-electrons in graphite (Stecher, 1969). However accurate thermoreflectance measurements (Anderegg *et al.*, 1971) have located the $\pi - \pi$ transition at 2580 Å rather than at 2200 Å. so that additional effects must be contributing to the observed absorption features. As pointed out by van de Hulst (1957), the role of collective electron oscillations (plasmons) has to be considered. In fact, collective processes may become the dominant absorption feature in cases when the size of the particles becomes small compared to the wavelength of the light, as shown for example by Genzel and Martin (1973). While of course the plasmon treatment does not give results different from the Mie theory, Gilra (1973) has shown that a study of the peak position of the plasma resonance allows an easier means of drawing conclusions on size, shape, and coatings of the grains, at least in the small particle limit. Using dielectric constants of graphite as given by Taft and Philipp (1965) he concluded that the particles should be very small (< 150 Å), spherical, and free of any coating.

There is some uncertainty concerning the degree of crystallographic order in interstellar grains. While it is not expected that the grain will be free of defects, mainly because of the bombardement by energetic particles and cosmic rays (Watson, 1972), it is also unlikely that the grains are composed of highly disordered vitreous carbon. Optical data on both these limiting cases of structure are available. Measurements on vitreous carbon have been published by Taft and Philipp (1965) and extended recently by Williams and Arakawa (1973). Electronic interband transition features are found at the same energies as with graphite for $E \perp c$ though somewhat broadened. Marked differences however are observed for the plasma resonance frequencies. The bulk plasmon energies for collective excitation of the π-electrons and of the π- and σ-electrons, observed at 7 eV (1770 Å) and 25 eV for graphite, are found to be shifted to 5.6 eV (2210 Å) and 20.4 eV for vitreous carbon. This shift in plasma resonance frequency depends on the electron density in the amorphous material. From this point of view, some of the stringent limitations imposed on size, shape, and coatings deduced from the plasmon model by Gilra (1973) have to be treated with care. Given various degrees of micro-structural disorder one can arrive at a variety of possible sizes and shapes for interstellar carbon grains which could fit the observed extinction curve in the region around 2200 Å.

2. Photoemission and Charging of Graphite Grains

The problem of photoemission from interstellar grains has received little attention to date, primarily due to a lack of information on the galactic UV radiation field and the photoemission properties of grain materials. Previous calculations of the charge on grain surfaces indicated negatively charged particles as a result of the equilibrium between the sticking of electrons and positive ions in the ambient plasma (Spitzer, 1948, Wickramasinghe, 1967). However recent investigations (Watson, 1974), in which an estimate of the photoelectron yield has been made, indicate positively charged grains.

Attempts to calculate the electrostatic surface charge require information concerning the photoelectric threshold and quantum yield of the grain material, the energy distribution of the photoemitted electrons, and the ambient charged particle flux density, sticking probability, and kinetic temperature.

Figure 1 shows the measurement of the work function (photoelectric threshold) of graphite, using the Fowler method. A straight line extrapolation of the square root of the yield plotted against photon energy gives a work function value of 4.7 eV (2640 Å) for graphite. A work function of 4.75 eV was obtained for the vitreous carbon sample by the same method. The value of the work function is a particularly important parameter for the determination of the grain potential in those regions where the photon flux is large compared to the incident particle flux onto the grain.

The photoelectric yield in terms of emitted electrons per incoming photon is given in Figure 2 as a function of photon energy for both graphite and vitreous carbon. The yield of graphite is found to be exceptionally small compared to other materials (Feuerbacher and Fitton, 1972), reaching a maximum value of only 0.03 at 15 eV. In contrast, metals may have a value ten times this in the same energy range, whilst the yield for insulators can approach unity. In the range up to 13.6 eV the photoelectric yield of graphite does not exceed 2%, and in the 1000–2000 Å region the average value is about 10^{-4}. Slightly higher values are found for vitreous carbon. The inset shows the observed reflectance in the same energy range. It can be seen that the

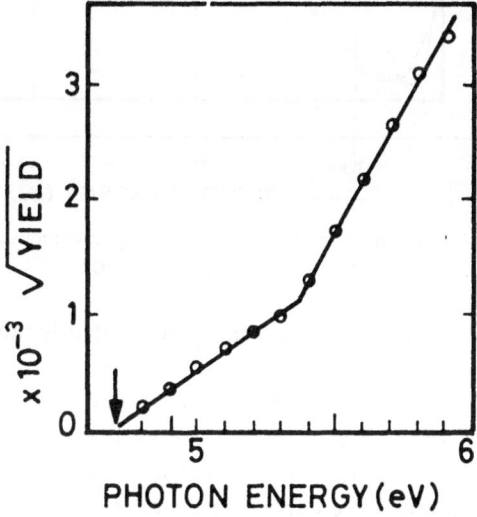

Fig. 1. Determination of the work function of graphite using Fowlers method.

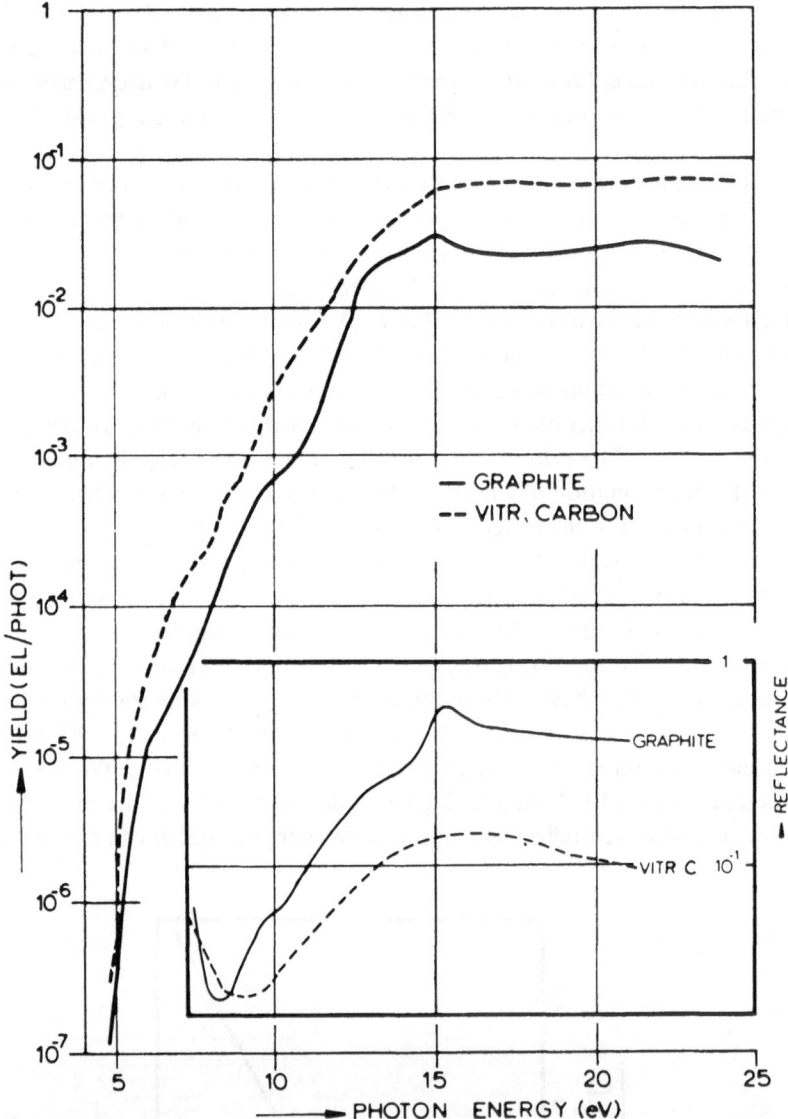

Fig. 2. Photoelectric yield per incident photon for graphite and vitreous carbon. The inset gives the reflectance in the same photon energy scale.

higher yield of vitreous carbon may be partly accounted for by the lower reflectivity of this material.

A set of photoelectron energy distribution curves for graphite is shown in Figure 3. Each curve has been scanned at a photon energy that may be read from the right hand scale. The number of electrons emitted in an energy interval is plotted as a function of electron energy. The curves are normalized such that the integral under each spectrum equals unity. All spectra are seen to peak at relatively low electron

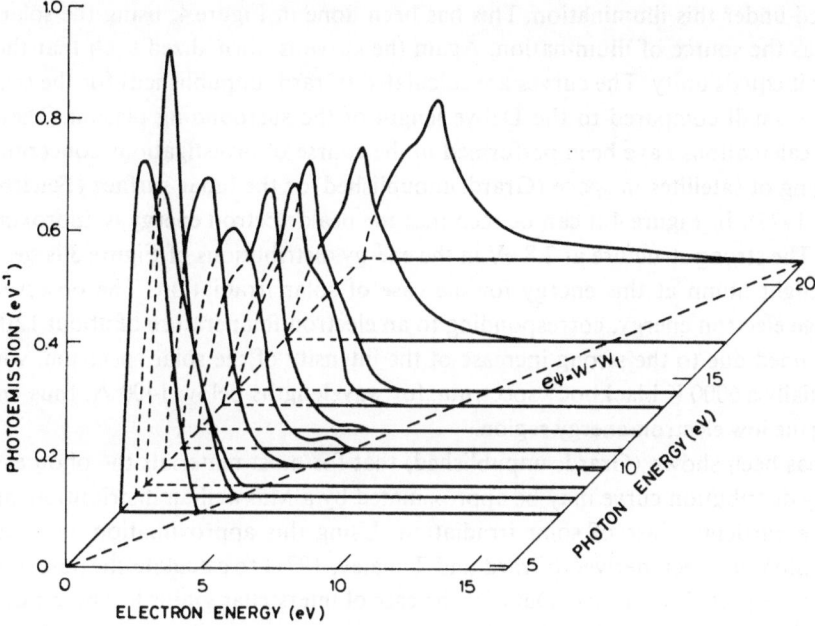

Fig. 3.   Electron energy distribution of photoelectrons emitted from graphite for various photon energies as given on the right hand scale. All spectra are normalized to unity area under the curves.

energies. Spectra taken with photon energies above 11 eV show a pronounced peak at 2.8 eV. It is important to note that very few electrons are emitted with energies close to the highest possible energy, which is the photon energy minus the work function, given as the dashed straight line in Figure 3.

Providing the spectrum of the light incident on the particle is known, the curves of Figure 3 may be integrated to obtain the total energy distribution of the electrons

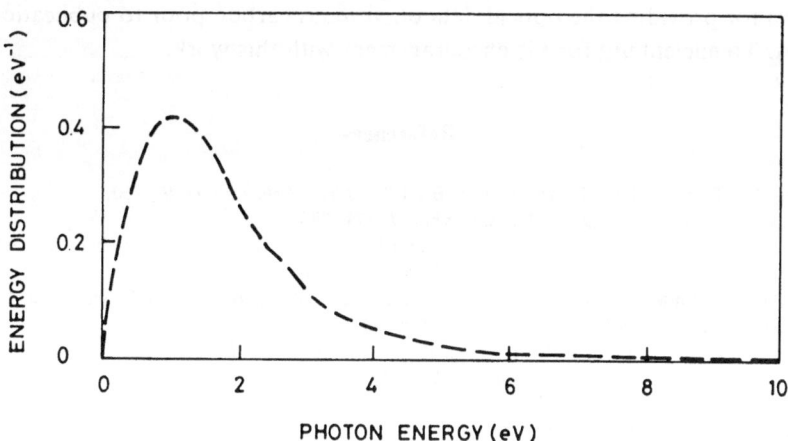

Fig. 4.   Total energy distribution of photoelectrons emitted from graphite under solar irradiation at 1 AU, normalized to unity area.

emitted under this illumination. This has been done in Figure 4, using the solar spectrum as the source of illumination. Again the curve is normalized such that the area under it equals unity. The curves are calculated (Grard, unpublished) for the case of a particle small compared to the Debye length of the surrounding plasma. These and other calculations have been performed in the course of investigations concerning the charging of satellites in space (Grard, unpublished) or the lunar surface (Feuerbacher et al., 1973). In Figure 4 it can be seen that the peak electron energy is approximately 1 eV. The strong structure at 2.8 eV in the energy distributions of Figure 3 is seen only as a slight hump at this energy for the case of solar irradiation. The observed low average electron energy, corresponding to an electron temperature of about 12 000 K, is obtained due to the strong increase of the intensity of the solar spectrum, which is essentially a 6000 K blackbody spectrum, for wavelengths below 1400 Å, thus emphasizing the low electron-energy region.

It has been shown (Grard, unpublished) that for most materials the photoelectron energy distribution curve may be approximated by a Maxwellian distribution, at least for the particular case of solar irradiation. Using this approximation an analytical expression has been derived (Grard and Tunaley, 1971) to calculate the potential of a body or of particles in space. Data for the case of interstellar grains has been published (Feuerbacher et al., 1973).

In summary, the optical properties of small grains are likely to be dominated by collective electron effects and modified somewhat by structural disorder. The photoemission from graphite grains will be abnormally low compared to that of most materials with a maximum in the emission energy of about 1 eV for the particular case of solar irradiation.

Acknowledgements

We thank Dr R. J. L. Grard and Dr D. P. Gilra for helpful discussions, Dr E. T. Arakawa for providing the optical data on vitreous carbon prior to publication, and Dr E. A. Trendelenburg for his encouragement with this work.

References

Anderegg, M., Feuerbacher, B., and Fitton, B.: 1971, *Phys. Rev. Letters* **26**, 760.
Bless, R. C. and Savage, B. D.: 1972, *Astrophys. J.* **171**, 293.
Carter, J. G., Huebner, R. H., Hamm, R. N., and Birkhoff, R. D.: 1965, *Phys. Rev.* **137**, A639.
Feuerbacher, B. and Fitton, B.: 1972, *J. Appl. Phys.* **43**, 1563.
Feuerbacher, B., Anderegg, M., Fitton, B., Laude, L. D., Willis, R. F., and Grard, R. J. L.: 1972, *Proc. 3rd Lunar Sci. Conf. Geochim. Cosmochim. Acta*, Suppl. 3, **3**, 2655.
Feuerbacher, B., Willis, R. F., and Fitton, B., 1973, *Astrophys. J.* **181**, 101.
Genzel, L. and Martin, T. P.: 1973, *Surface Sci.* **34**, 33.
Gilra, D. P.: 1971, *Nature* **229**, 237.
Gilra, D. P.: 1971, this volume, p. 517.
Grard, R. J. L. and Tunaley, J. K. E.: 1971, *J. Geophys. Res.* **76**, 2498.
Grard, R. J. L.: ESTEC IWP 663, European Space Research Organization, Noordwijk, Holland, unpublished.

Greenaway, D. L., Harbeke, G., Bassani, F., and Tosatti, E.: 1969, *Phys. Rev.* **178**, 1340.
Hoyle, F. and Wickramasinghe, N. C.: 1962, *Monthly Notices Roy. Astron. Soc.* **124**, 417.
Klucker, R. and Skibowski, M.: DESY, Hamburg, Germany, unpublished.
Painter, G. S. and Ellis, D. E.: 1970, *Phys. Rev.* **B1**, 4747.
Spitzer, L.: 1948, *Astrophys. J.* **107**, 6.
Stecher, T. P.: 1965, *Astrophys. J.* **142**, 1683.
Stecher, T. P.: 1969, *Astrophys. J.* **157**, L125.
Stecher, T. P. and Donn, B.: 1965, *Astrophys. J.* **142**, 1681.
Taft, E. A. and Philipp, H. R.: 1965, *Phys. Rev.* **138**, A197.
Tosatti, E. and Bassani, F.: 1970, *Nuovo Cimento* **65**, 161.
Van de Hulst, H. C.: 1957, *Light Scattering by Small Particles*, John Wiley and Sons, New York.
Watson, W. D.: 1972, *Astrophys. J.* **176**, 103 and 271.
Watson, W. D.: 1973, this volume, p. 335.
Williams, M. W. and Arakawa, E. T.: 1973, to be published.
Willis, R. F., Feuerbacher, B., and Fitton, B.: 1971, *Phys. Rev.* **B4**, 2941.
Wickramasinghe, N. C.: 1967, *Interstellar Grains*, Chapman and Hall, London.
Wickramasinghe, N. C. and Nandy, K.: 1970, *Nature* **227**, 51.
Zeppenfeld, K.: 1968, Z. *Physik* **211**, 391.

THE THERMAL ACCOMMODATION COEFFICIENT
OF GRAPHITE

KENRICK L. DAY

*Astronomy Dept., The Ohio State University, Columbus, Ohio 43210, U.S.A.**

Abstract. An experiment has been performed to determine the thermal accommodation coefficients for the gases hydrogen, oxygen, methane, and carbon dioxide on graphite, in the temperature range 273–77 K. The experimental results are compared to those indirectly predicted in a theoretical paper by Hollenbach and Salpeter, and agreement is found to be satisfactory. A sticking coefficient for atomic hydrogen on a graphite grain is derived which would support the conclusion that recombination of hydrogen on grain surfaces could be a significant process. The very high measured thermal accommodation coefficients for the heavier gases supports the popular assumption that retention of such gases on grain surfaces should be considered extremely likely.

1. Introduction

The thermal accommodation coefficient (hereafter abbreviated TAC) is a measure of the efficiency of energy transfer between a gas and a solid. It is of interest in discussing the condensation of various gases on interstellar grains, which in this paper are assumed to be graphite. The TAC has often been confused with the sticking coefficient, which measures the actual fraction of atoms condensing, while the TAC gives the fractional energy transfer. Both these quantities tend to rise with decreasing temperature, with the TAC generally higher than the sticking coefficient. For a cold surface, or one on which the surface adsorptive potential is greater than the thermal energy, the main variable influencing the TACs and sticking coefficients is the energy of the incoming gas particle.

2. Experimental Details

The experiment was performed via the hot-wire, low pressure method. For a comprehensive discussion, the reader is referred to Wachmann (1962) and Thomas (1967). Basically, a tube of cylindrical symmetry is constructed with the substrate material in the form of a fine filament running axially. A graphite filament was secured from the Poco Graphite Company of Decatur, Texas. This was a polycrystalline material graphitized from coke at 2500 °C. The manufacturers estimate that at this temperature of graphitization, approximately half the coke may be considered to have been converted to graphite. Impurities present amount to about 200 ppm, the primary ones being Fe, Si, and V. The material may be regarded as nearly perfectly isotropic, with grain sizes under one mil. L_c is given to be 340 Å, and the BET area is approximately $0.336 \text{ m}^2 \text{ g}^{-1}$.

The basic procedure is to heat the filament electrically in a vacuum by passing a

---

* Current address: Physics Dept., University of Arizona, Tucson, Ariz. 85721, U.S.A.

Greenberg and Van de Hulst (eds.), Interstellar Dust and Related Topics, 311–315,
All Rights Reserved. Copyright © 1973 by the IAU.

small current through it, and measure the power loss in the filament. A small amount of sample gas at a measured pressure (0.01 to 0.05 torr) is then introduced and the filament is returned to its previous temperature. This will require more power than before due to heat loss to the gas. By subtracting the vacuum power loss, one finds the net power loss to the gas; this measured quantity, when compared to a theoretical maximum calculated from kinetic theory, gives the TAC.

A few experimental considerations must be mentioned. It is necessary that the filament be of fine diameter so that it does not appreciably disturb the temperature distribution of the gas particles impinging on it; this allows one to assume that all particles striking the filament are in thermal equilibrium with the walls of the experimental tube, which is placed in a constant temperature bath. The filament diameter in this case was just under 9 mils. A filament would tend to lose heat into its end supports. The amount of this loss is unfortunately not the same in vacuo and in the sample gas, because the gas changes the heat distribution along it. The reader is referred to Thomas and Olmer (1943). This was avoided by placing small auxiliary heaters at either end of the tube, which held the end supports at a constant temperature equal to the filament's. Thus there was no heat lost to the ends either in vacuo or in the sample gas. This temperature was monitored by means of iron-constantan thermocouples.

The temperature of the filament itself was found by measuring its resistance at any given moment and then comparing this to a large scale graph of temperature vs resistance which was previously prepared for this sample. The resistance was found by measuring the voltage drop across the filament and across a standard resistor in series with the filament and then applying Ohm's law. For most experimental runs the graphite was maintained at ten degrees Centigrade above it surroundings.

Considerable inconsistancy in TAC measurements has been noted in the past, and much of it has apparently been due to variations in surface cleanliness. Current cleaning techniques include flashing the filament or the use of various getters. Because of several problems, this was not possible with the sample used in this experiment, so that the surface must be regarded as having an adsorbed layer of sample gas on it during measurement. A molecular sieve sorption pump was used to minimize external contamination. It was also noted that no progressive alteration on the surface was taking place as the measured TAC values were repeatable over a period of several months.

3. Results of the Experiment

For H_2 on graphite, the following average TACs were found. At a gas temperature of 273 K, TAC = 0.36. At 195 K, TAC = 0.38. At 77 K, TAC = 0.82.

Results for oxygen, methane, and carbon dioxide were obtained only at the temperature of water ice, or about 273 K. The following results were noted. Oxygen, TAC = 0.96. Methane, TAC = 0.78. Carbon dioxide, TAC = 0.90. For all the above, the p.e. is ± 0.009.

4. Interpretation of Experimental Data and Conclusions

It is informative to campare the measured TACs for H_2 on graphite to theoretical predictions by Hollenbach and Salpeter (1970). These authors did not actually make predictions for the hydrogen/graphite combination, but developed an analytic formula which should fit many situations in which the energy of the substrate material is small compared to the surface adsorption potential. This should be the case in an interstellar cloud and for the temperatures used for this paper in the laboratory. For this circumstance, the variable of note is the energy of the incoming gas particle.

In order to make a prediction it is necessary to know the adsorption potential for H_2 on graphite. This is of critical importance to the problem, and existing data is badly in need of confirmation. The value adopted is that calculated by Ross and Oliver (1964) for carbon black graphitized at 2700 K, which is 960 cal mole^{-1}. The results of the experiment and the curve calculated from the work of Hollenbach and Salpeter (1970) are shown in Figure 1. It is noted that the fit is fairly reasonable, being just a little too shallow.

Hollenbach and Salpeter (1970) produce formulae for both sticking and thermal accomodation coefficients in terms of a single parameter which they call gamma. If one accepts the fit of the theoretical curve to the TAC data, it is an indication that gamma is not far from being correct.

To plot a curve for the sticking coefficient of hydrogen vs temperature, gamma must be modified somewhat. One would generally be interested in the case of impinging atomic, rather than molecular, hydrogen. However, the adsorptive potential for atomic hydrogen on graphite has never been measured. According to Augason (1970), we may assume that the ratio of the adsorption potentials should be the same

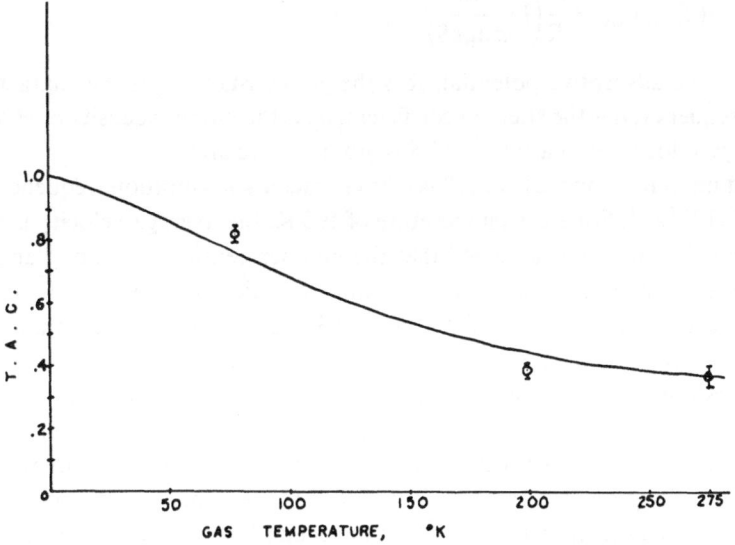

Fig. 1.   TAC vs temperature for H_2. – Theory. ◌ Experimental values.

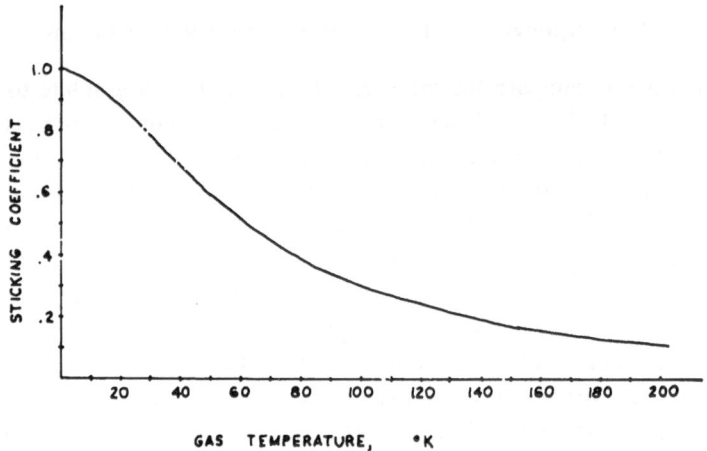

Fig. 2.   Computed sticking coefficient for atomic hydrogen on a graphite grain.

as the ratio of the polarizabilities, 0.84, so the adsorption potential of H on graphite should be 810 cal mole^{-1}.

Using this figure for the adsorption potential, the curve appearing in Figure 2 has been generated, which should represent the sticking probability of atomic hydrogen on graphite. The sticking coefficient is of use in determining under which conditions one might expect to find production of H_2 by recombination on grain surfaces. A basic requirement is that new hydrogen atoms stick to the surface at a rate equal to or greater than the departure rate. This may be used to derive a maximum grain temperature for which the condition holds. According to Augason (1970), this is:

$$(T_{grain})_{max} = \frac{U}{R}\left(\ln\frac{v_0}{a\,n_H v S}\right)^{-1},$$

where U is the adsorptive potential, R is the gas constant, v_0 is the characteristic vibration frequency, a is the sticking coefficient, n_H is the number density of H atoms, v is the average velocity of H atoms, and S is grain surface area.

According to Ross and Oliver (1964) the characteristic vibration frequency for H on graphite is 10^{13} s^{-1}. For a gas temperature of 100 K, the average velocity is 1.45×10^5 cm s^{-1}. Further, it will be assumed that the number density is 10 cm^{-3} and that the grain surface area is 10^{-8} cm^2. The sticking coefficient derived in this work for an incident particle temperature of 100 K is 0.30. This yields a maximum allowable grain temperature of 11.53 K.

The question of grain temperatures has been debated for some time, but in the opinion of the author the work of Field (1969) is the most comprehensive to date. Taking into account all known cooling mechanisms, he derives minimum possible temperatures for graphite grains of from 8.9 to 9.3 K. This would appear to leave a significant range of temperatures in which molecular hydrogen could form on graphite grains.

The experimental data for the gases oxygen, methane, and carbon dioxide tend to confirm what people have been assuming at least since Van de Hulst (1949). This is that the sticking coefficients for these gases should be unity. The measured TACs for these gases at 273 K range from 0.78 to 0.96. While the sticking coefficients would probably be lower by a factor of about a third at this temperature, both coefficients rise with decreasing temperature so that by 100 K a coefficient of nearly unity should prevail. It would appear that the conclusions of Williams (1968) are correct; molecule accretion by graphite grains may actually begin before the grain is completely free of the atmosphere of the parent star, and nearly all impinging *molecules* should stick.

Acknowledgements

The author wishes to acknowledge helpful letters from Drs B. Donn and G. C. Augason, the support of Dr T. P. Roark and Mr Roy S. Tucker of the Ohio State University, and partial funding by NSF grant GP-7616.

References

Augason, G. C.: 1970, *Astrophys. J.* **162**, 463.
Field, G. B.: 1969, *Monthly Notices Roy. Astron. Soc.* **144**, 411.
Hollenbach, D. and Salpater, E. E.: 1970, *J. Chem. Phys.* **53**, 79.
Ross, S. and Oliver, J. P.: 1964, *On Physical Adsorption*, Interscience Publishers, New York, p. 236.
Thomas, L. B.: 1967, in Saltsburg *et al.* (eds.), *Fundamentals of Gas-Surface Reactions*, Academic Press, New York, p. 346.
Thomas, L. B. and Olmer, F. G.: 1943, *J. Am. Chem. Soc.* **65**, 1036.
Van de Hulst, H. C.: 1949, *Recherches Astronomiques de l'Observatoire d'Urecht* **XI**, Part 2, 1.
Wachman, H. Y.: 1962, *J. Am. Rocket Soc.* **32**, 2.
Williams, D. A.: 1968, *Astrophys. J.* **151**, 935.

SPECTROSCOPIC SIMULATION OF DIFFUSE INTERSTELLAR LINES AND A LABORATORY MODEL FOR INTERSTELLAR CHEMICAL SYNTHESIS

F. M. JOHNSON

Dept. of Physics, California State University, Fullerton, Calif. 92634, U.S.A.

and

D. T. BAILEY and P. A. WEGNER

Dept. of Chemistry, California State University, Fullerton, Calif. 92634, U.S.A.

Abstract. The interstellar existence of dipyridyl magnesium tetrabenzporphin (I) is supported by the spectroscopic coincidence of laboratory generated and astronomically observed diffuse interstellar lines. A potential route for the synthesis of I in space from molecules which exist or are expected to exist there will be discussed.

Initially, we wish to review some of the spectroscopic evidence for the identification of the diffuse interstellar lines as being due to dipyridyl magnesium tetrabenzporphin (I). Laboratory spectra (Johnson, 1972, a, b) of this molecule taken at 77 °K (matrix isolation, Skpol'skii technique) resulted in the spectra shown in Table I. Also shown in this table is data taken by Sevchenko *et al.* (1967) corroborating some of our spectra.

TABLE I

Laboratory spectra of dipyridyl magnesium tetrabenzporphin (I) compared with corresponding diffuse interstellar astronomical data

Laboratory data			Astronomical data[a]	
MgTBP Johnson (1972a,b)	width (Å)	MgTBP Sevchenko *et al.* (1967)	wavelength (Å)	width (Å)
f *6663* Å	1–2		*6661*	1
f 6633vw				
f 6628w		6623 Å		
f *6614*	1–2		*6614*	1
f 6610vw		6570		
f 6334vw				
		6404		
		6377	*6376*	2
f 6289s				
f *6284s*	1–2	6282	*6284*	4
abs *6174*	14 (total)		*6175*	30
abs *4428vs*	40		*4428vs*	20
6274[b]				
6278[b]				

f = fluorescence data.
[a] Herbig (private communication).
[b] Probable assignment: dihydrotetrabenzporphin.

Greenberg and Van de Hulst (eds.), Interstellar Dust and Related Topics, 317–321.
All Rights Reserved. Copyright © 1973 by the IAU.

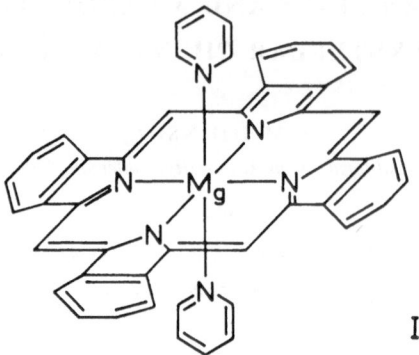

Fig. 1.   The interstellar molecule $\chi(MgC_{46}H_{30}N_6)$.

Note the coincidence of the laboratory spectra with the astronomical data, in particular the strongest interstellar band at 4428 Å coincides with the strongest laboratory generated band of I. There are now six coincidences to ± 2 Å and correlation of line width to within a factor of two for all the identified laboratory and astronomical data. Additional corraboration for this identification is based on the infrared measurement of I, whereby IR absorption data of this molecule, characterizing some of the dominant vibrational modes, can be used to construct an energy level diagram which shows excellent coincidences with the observed astronomical spectroscopic data. Further more, all of the 25 diffuse interstellar lines, kindly obtained from Herbig (private-communication), fit to within ± 2 cm^{-1} of the laboratory observed vibrations of I. The identification if I is in fact more reliable than that of a microwave molecule based on a single line coincidence. Two questions posed by the existence of I are: (1) how is I produced in the interstellar medium and (2) why do most of the astronomically observed transitions arise from excited states? This paper gives a plausible stellar synthesis of I and gives a possible explanation for excited state transitions.

Microwave techniques (Rank *et al.*, 1971; Snyder, 1972) have now identified a series of organic molecules containing acetylenic and cyano functional groups (Table II). Additionally, Table II contains some symmetric molecules, not observable by microwave spectroscopy, which in view of their close similarity to already identified

TABLE II

Structure of accepted interstellar acetylene and cyano derivatives

$H-C \equiv C-CH_3$		$H-C \equiv C-C \equiv N$
methylacetylene		cyanoacetylene
$H-C \equiv N$	$CH_3-C \equiv N$	$\cdot C \equiv N$
hydrogen cyanide	acetonitrile	cyanide radical

Likely symmetrical interstellar acetylene and cyano derivatives

$H-C \equiv C-H$	$N \equiv C-C \equiv N$	$H-C \equiv C-C \equiv C-H$
Acetylene	cyanogen	diacetylene

molecules almost certainly should also exist in space. These postulated molecules are acetylene, cyanogen and diacetylene. Although not observable by microwave techniques these molecules should be seen in the infrared and UV region of the spectrum (possibly in an excited state, in analogy with I and some of the 'microwave molecules').

A synthetic route to I requires a brief review of some known chemistry of acetylenic and cyano group functions with emphasis on the well known cyclization reactions of these groups. Acetylene and some substituted acetylenes undergo a facile condensation reaction to form benzene derivatives (Equation (1)). This is a very general type of reaction (Bird, 1967) with little restriction on the nature of the R group.

$$\text{(reaction scheme)} \qquad \text{where } R = \text{H, alkyl, aryl} \quad (1)$$

The reaction occurs in the temperature range of 500 to 700 °K or catalytically at lower temperatures. Using the eight molecules in Table II (now identified in space or likely to be found there) and the cyclization chemistry just emphasized we would predict the following large molecules susceptable to interstellar synthesis: benzene (II), toluene (III), benzonitrile (IV) and phenylacetylene (V). This list is of course not exhaustive but is exemplary of the type of benzene derivatives which should be searched for in astronomical studies. Another class of molecules susceptable to 'interstellar' synthesis are pyridine derivatives. Again using molecules from Table II and the cyclization reaction of Equation (1), we suggest the possible 'interstellar' existence of pyridine (VI), 2-methylpyridine (VII), 2-cyanopyridine (VIII), and 2-ethynylpyridine (IX).

$$3 \text{ H–C}\equiv\text{C–H} \longrightarrow \text{II} \qquad (2)$$

$$2 \text{ H–C}\equiv\text{C–H} + \text{H–C}\equiv\text{C–CH}_3 \longrightarrow \text{III (CH}_3\text{)} \qquad (3)$$

$$2 \text{ H–C}\equiv\text{C–H} + \text{H–C}\equiv\text{C–C}\equiv\text{N} \longrightarrow \text{IV (C}\equiv\text{N)} \qquad (4)$$

$$2 \text{ H–C}\equiv\text{C–H} + \text{H–C}\equiv\text{C–C}\equiv\text{C–H} \longrightarrow \text{V (C}\equiv\text{C–H)} \qquad (5)$$

$$2 \ H{-}C{\equiv}C{-}H + H{-}C{\equiv}N \ \longrightarrow \ \text{[pyridine]} \qquad (6)$$

VI

$$H{-}C{\equiv}C{-}H + H{-}C{\equiv}N + H{-}C{\equiv}C{-}CH_3 \ \longrightarrow \ \text{[2-methylpyridine]} \qquad (7)$$

VII

$$H{-}C{\equiv}C{-}H + H{-}C{\equiv}N + H{-}C{\equiv}C{-}C{\equiv}N \ \longrightarrow \ \text{[2-cyanopyridine]} \qquad (8)$$

VIII

$$H{-}C{\equiv}C{-}H + H{-}C{\equiv}N + H{-}C{\equiv}C{-}C{\equiv}C{-}H \ \longrightarrow \ \text{[2-ethynylpyridine]} \qquad (9)$$

IX

Using the condensation reaction shown in Equation (1) and three molecules from Table II, *o*-cyanophenylacetylene (**X**) can be prepared (Equation (10)). This molecule is isoelectronic to phthalonitrile (**XI**)

$$(10)$$

X

which undergoes a well known condensation reaction with a variety of metals to yield metallo-phthalocyanine (**XII**) (Cotton and Wilkinson, 1972) (Equation (11)). Assuming a similar condensation of **X**, magnesium tetrabenzporphin (**XIII**) could be synthesized

$$(11)$$

XI   XII

(Equation (12)). Addition of pyridine above and below the ring

$$4 \quad \overset{C\equiv C-H}{\underset{C\equiv N}{\bigcirc}} \quad \xrightarrow{Mg} \qquad\qquad\qquad (12)$$

X   XIII

would generate **I**. The only step of the entire sequence for the synthesis of **I** which is not based on completely known chemical reactions is the final condensation of **X** to form magnesium tetrabenzporphin. We are presently examining this reaction.

The stability of **I** rests on its conjugated bonds and the stability of the benzenes attached to the pyrrole structures. The resonance energy (Longo *et al.*, 1970) of such a molecule is in excess of 400 °K calories.

Finally, the reason why a large number of observed diffuse interstellar lines arise from excited states may be due to the astronomical observation of these molecules under conditions of formation (Johnson, 1972b) whereby the molecule is in an excited state upon completion of the exothermic cyclization process. A similar situation is observed in other chemical reactions, i.e. thermo-luminescence arising from the gas phase synthesis of a molecule (Klemperer, 1971). We therefore believe (1) a synthetic pathway for the formation of **I** in space exists and (2) that this pathway embodys an exothermic condensation reaction explaining the observation that **I** exists in excited vibrational states.

Acknowledgements

We wish to acknowledge the support of this work by NASA Grant NGR 05-071-002.

References

Bird, W. C.: 1967, *Transition Metal Intermediates in Organic Synthesis*, Logos Press, London.
Cotton, F. A. and Wilkinson, G.: 1972, *Advanced Inorganic Chemistry*, 3rd ed., John Wiley & Sons, New York, p. 626.
Herbig, G. H.: private communication.
Johnson, F. M.: 1972a, *Ann. N.Y. Acad. Sci.* **187**, 186.
Johnson, F. M.: 1972b, *Mém. Soc. Roy. Sci. Liège, 6th Ser.* **III**, 391.
Klemperer, W.: 1971, in C. de Jager (ed.), *Highlights of Astronomy*, p. 421.
Longo, F. R., Finarelli, J. D., Schmalzbach, E., and Adler, A. D.: 1970, *J. Phys. Chem.* **74**, 3296.
Sevchenko, A. N., Solov'ev, K., Gradyushko, A., and Shkirman, S.: 1967, *Soviet Phys.-Dokl.* **11** (7), 587–590.
Rank, D. M., Townes, C. H., and Welch, W. J.: 1971, *Science* **174**, 1083.

A LABORATORY EXPERIMENT ON THE NUCLEATION
OF INTERSTELLAR DUSTS

F. KAMIJO, Y. NAKADA, T. IGUCHI, M. FUJIMOTO,
M. TAKADA, and S. TAKAJO

Dept. of Astronomy, University of Tokyo, Japan

Abstract. Small particles of silica, carbon, and iron are produced in the laboratory for the eventual
purpose of studying their optical properties.

Very fine powders of some silicates, carbon and iron are produced by the gas evaporation technique in argon gas. After evacuating the working chamber of an evaporation
unit, argon is introduced to a given pressure. A tungsten heater in the chamber is then
gradually heated to evaporate the material. The evaporated atoms are cooled by
collision with those of the argon gas and when the temperature of the vapour becomes
lower than its dew point, fine particles are formed in the atmosphere of the argon gas.

To investigate the size and shape of the particles, several standard grid meshes coated
with carbon film for electron microscopy are put in the chamber. Generally the size
becomes smaller for lower temperature and also for lower pressure. The largest size

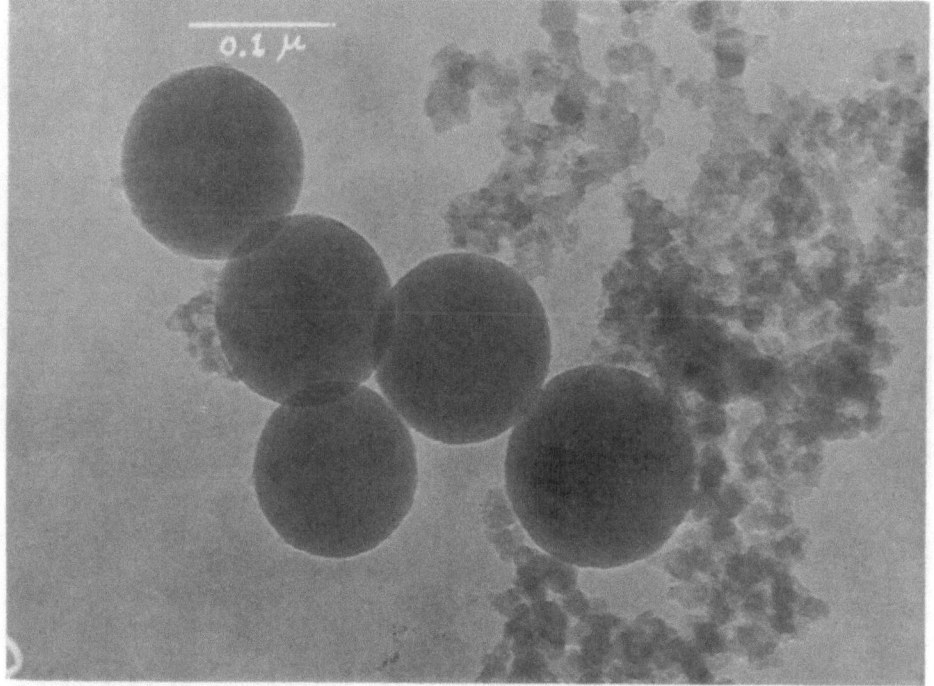

Fig. 1. SiO_2.

Greenberg and Van de Hulst (eds.), Interstellar Dust and Related Topics, 323–325.
All Rights Reserved. Copyright © 1973 by the IAU.

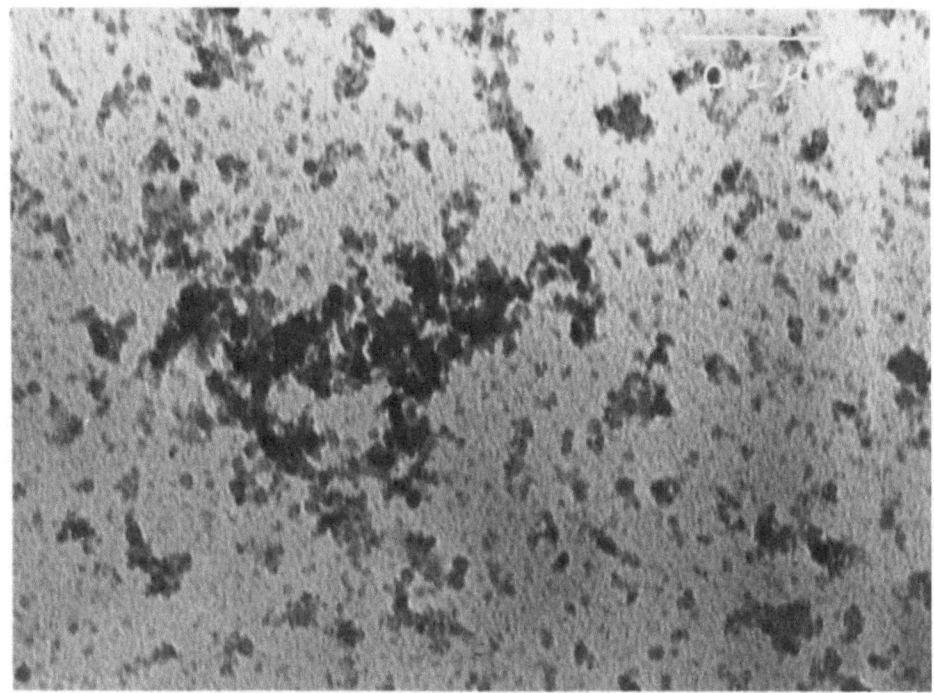

Fig. 2. C.

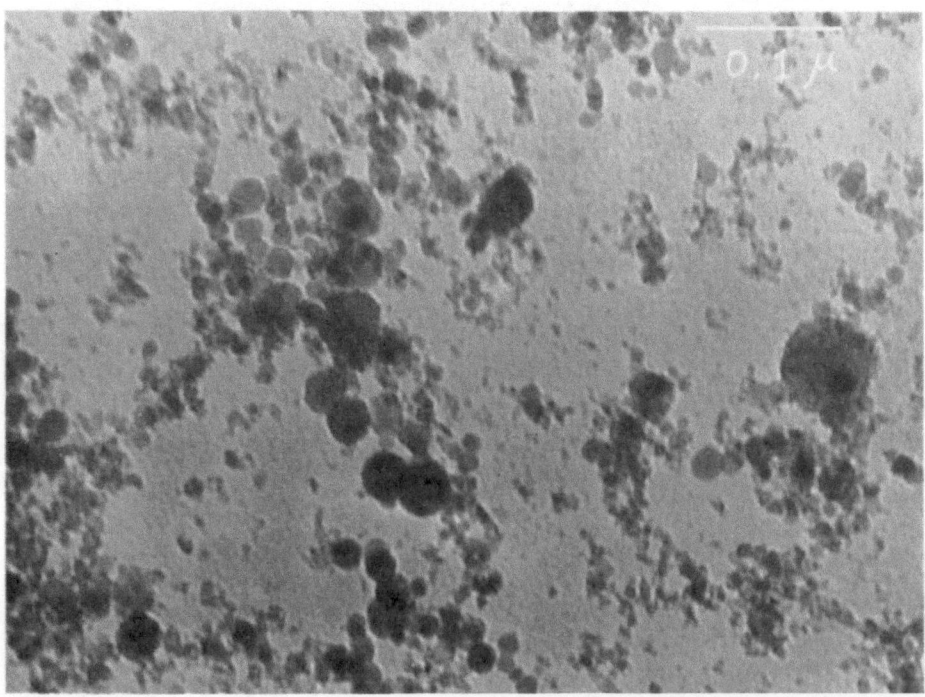

we have obtained is about 0.1 μ for silica and the smallest one is about 40 Å for carbon. (see Figures 1 and 2).

1. Silica (Figure 1)

Instead of the evaporation unit, an ordinary plasma jet is used to produce silica particles. A broken piece of silica glass is placed on the water cooled copper hearth. Then the working chamber of the system is filled with argon gas of one atmosphere. While the plasma jet gun is fired, the hearth is operated carefully so that the flame shoots the specimen. Silica glass melts immediately. Fine particles are produced in the argon atmosphere and deposit on the inside walls of the chamber. As we expect from their spherical shape, they show no sign of crystal structure by electron diffraction.

2. Carbon (Figure 2)

Fine particles of carbon are obtained by striking the arc in argon gas of 9 torr. They also show no crystal structure.

3. Iron (Figure 3)

Iron particles prepared by the arc in argon gas of 6 torr show characteristic hexagonal shape. It is said that they are rhombic dodecahedron.

A detailed report on the experiment including the UV and IR absorption of the particles will be published in *Publ. Astron. Soc. Japan.*

OPTICAL AND CHEMICAL STUDIES ON SIMULATED
INTERSTELLAR GRAIN MATERIALS

J. D. McCULLOUGH, G. R. FLOYD, R. H. PRINCE, and W. W. DULEY

Centre for Research in Space Science, and Physics Dept., York University, Toronto, Canada

Abstract. The possibility that many different diffuse interstellar absorption features may be produced by the same type of absorbing atom in different hydrocarbon matrices on interstellar grains has been examined experimentally. The present study shows that absorption bands due to Na atoms in various hydrocarbon matrices can occur within the wavelength range 5400–5800 Å. A study of the molecules generated from the radiation-induced polymerization of C_2H_2 at 55 K is also reported. It is shown that C_6H_6 is an abundant product of this polymerization and may therefore be an important constituent of interstellar grains.

The 26 or so diffuse interstellar absorption bands and lines likely provide a definitive clue to the chemical nature of interstellar dust. Laboratory studies which provide simulations of these bands will therefore eventually result in the construction of more plausible models for interstellar grains.

For the past several years we have been examining the spectra of astrophysically abundant metal atoms such as Ca, Ti, etc., trapped in solid hydrocarbons at temperatures approaching 55 K (Duley, 1968; Duley and Graham, 1969; Graham and Duley, 1971). This model presupposes that some component of interstellar dust consists of pure hydrocarbon or core-hydrocarbon mantle grains. It is this component which is responsible for the occurrence of the diffuse interstellar bands. In this paper we examine the possibility that many different diffuse bands may be produced by the presence of one type of atom in hydrocarbon matrices on interstellar grains. The observation of many diffuse bands then reflects the fact that such grains are composed of a complex mixture of hydrocarbon molecules with relatively few optically active atomic species.

It has been noted previously (Duley, 1969) that Na atoms trapped in hydrocarbon grains are expected to produce spectral features in the 5000–6000 Å region due to the shifted Na Doublet at 5890–96 Å. Here we examine this proposal in more detail.

Figure 1 shows some spectra obtained recently of Na-hydrocarbon solids at 55 K. These solids are all of the substituted benzene type. We have previously found that such molecules may, because of their chemical stability, be present on interstellar grains (Duley and Graham, 1969; Graham and Duley, 1971). Here the Na atom interacts with one nearby hydrocarbon molecule in the solid to give a single absorption band with peak wavelength between 5600–5500 Å and full width at half maximum intensity of between $\leqslant$ 20 Å and 100 Å. Chemically, the complexes formed in these solids are similar to the well known benzene-I_2 complexes. (Andrews and Keefer, 1964; Walkley *et al.*, 1960). Since a weak chemical bond is formed between Na and a neighbouring hydrocarbon molecule, these complexes are stable thermodynamically

Greenberg and Van de Hulst (eds.), Interstellar Dust and Related Topics, 327–333.

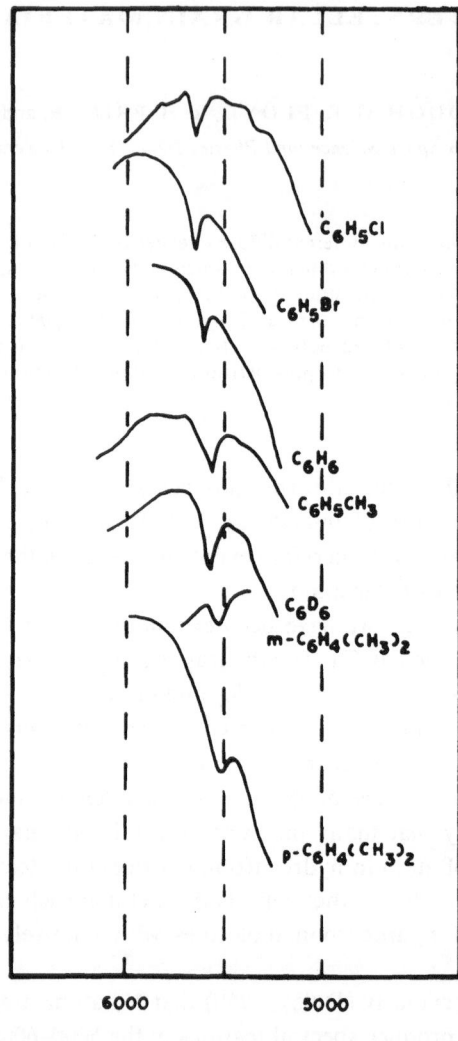

WAVELENGTH (Å)

Fig. 1. Absorption spectra of Na-hydrocarbon solids at 55 K.

to approximately 200 K. In addition, the spectra show no evidence of photobleaching when irradiated with unfiltered light from a tungsten filament.

In Figure 2 we have plotted the peak wavelength of the absorption band in each of these systems against the ionization potential of the hydrocarbon molecule with which the Na atom is interacting. We see empirically that a simple relationship exists with the characteristic absorption band shifting to longer wavelengths with increasing ionization potential. The molecules complexing with Na studied to date have a rather narrow range of ionization potentials (8–9 eV). Since other molecules have ionization

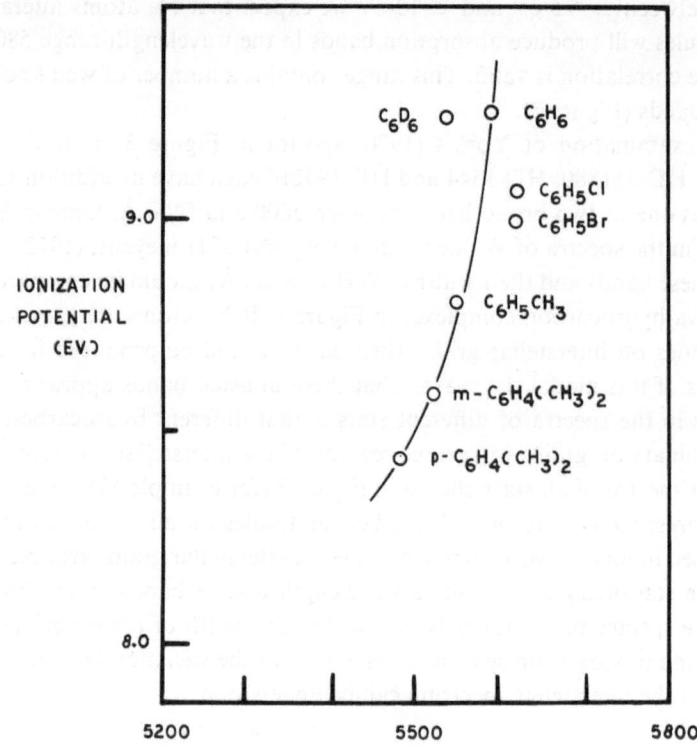

Fig. 2. Wavelength of Na-hydrocarbon band in Figure 1, plotted against the ionization potential of the hydrocarbon molecule.

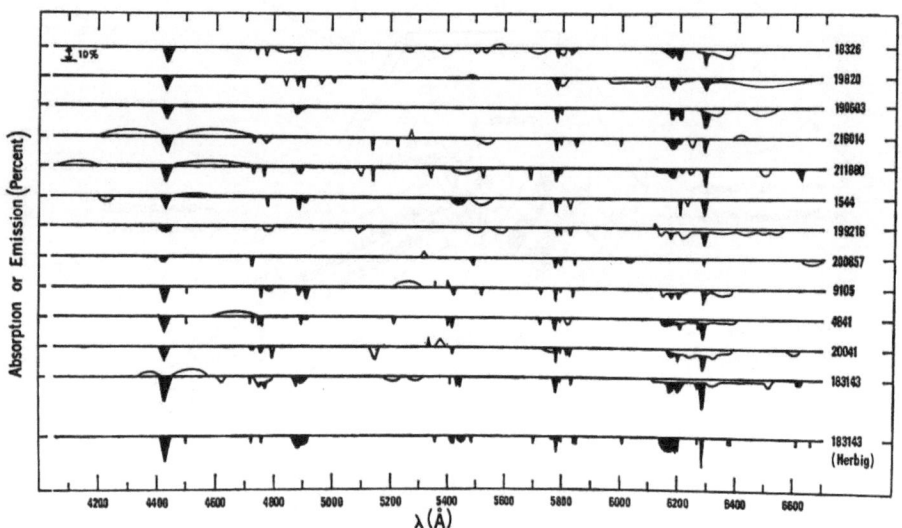

Fig. 3. Diffuse absorption and spectra after York (1970).

potentials between $\simeq$ 6.5 eV and $\simeq$ 13 eV we expect that Na atoms interacting with other molecules will produce absorption bands in the wavelength range 5800–5400 Å if this simple correlation is valid. This range contains a number of well known diffuse interstellar bands (Figure 3).

A closer examination of York's (1970) spectra in Figure 3 show that the stars HD 216014, HD 211880, HD 1544 and HD 199216 each have in addition to the usual diffuse bands one or two broad bands between 5600 and 5400 Å. Similar features are to be found in the spectra of Walker *et al.* (1969) and of Honeycutt (1972). The wavelengths of these bands and their width (FWHM $\simeq$ 50 Å) should be compared to those shown for Na-hydrocarbon complexes in Figure 1. If Na atoms are present in hydrocarbon mantles on interstellar grains then bands would be produced in this region. On the basis of this model the reason that these unusual bands appear at distinctive wavelengths in the spectra of different stars is that different hydrocarbon molecules may predominate on grains in various regions of the interstellar medium. Stationary bands in the spectra of all stars shown in Figure 3 (for example 5780 and 5796 Å) are due to the presence of certain hydrocarbon molecules on all grains. In this way Na atoms trapped in mixed hydrocarbon mantles on interstellar grains are able to account for both the stationary and variable wavelength diffuse bands in the 5800–5400 Å region of the spectrum. It should be noted that the width of the experimental bands has been found to vary from system to system. Thus the variation in width of bands in this region of the interstellar spectrum can be understood.

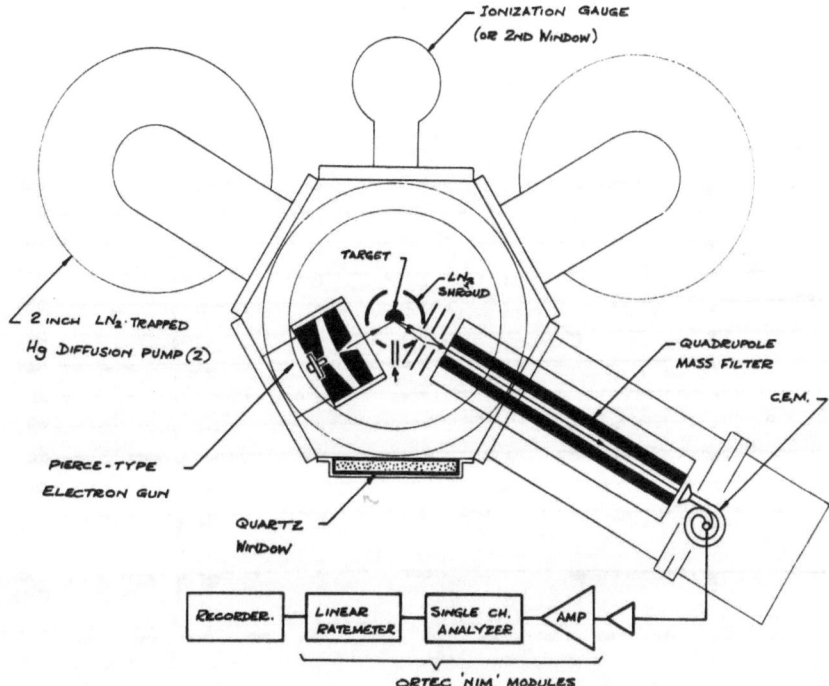

Fig. 4.   Schematic diagram of the system used to study products of C_2H_2 irradiation at 55 K.

It is interesting to speculate how molecules as complex as C_6H_6 could be produced in the interstellar medium. In this regard we suggest that the simple reaction

$$3(C_2H_2) + \text{energy} \rightarrow C_6H_6$$

may play an important role. This reaction has been shown to occur efficiently in the gas phase (Lind, 1961) with γ, β, α or ultraviolet radiation sources. Realizing that gas phase C_2H_2 densities in the interstellar medium are unlikely to be high enough to promote polymerization, we have designed a system which allows us to examine this reaction in solid C_2H_2 at 55 K. This system is shown schematically in Figure 4.

C_2H_2 is condensed on a finger cooled to 55 K in high vacuum. The solid is then irradiated with low (10–100 eV) electrons or short wavelength radiation from a Lα source and the ejected molecular ions analysed with a quadrupole mass filter. Prior to an experiment the background gas in the cryostat is mass analyzed by directing electrons from the gun onto the target. Typically background counts in the region

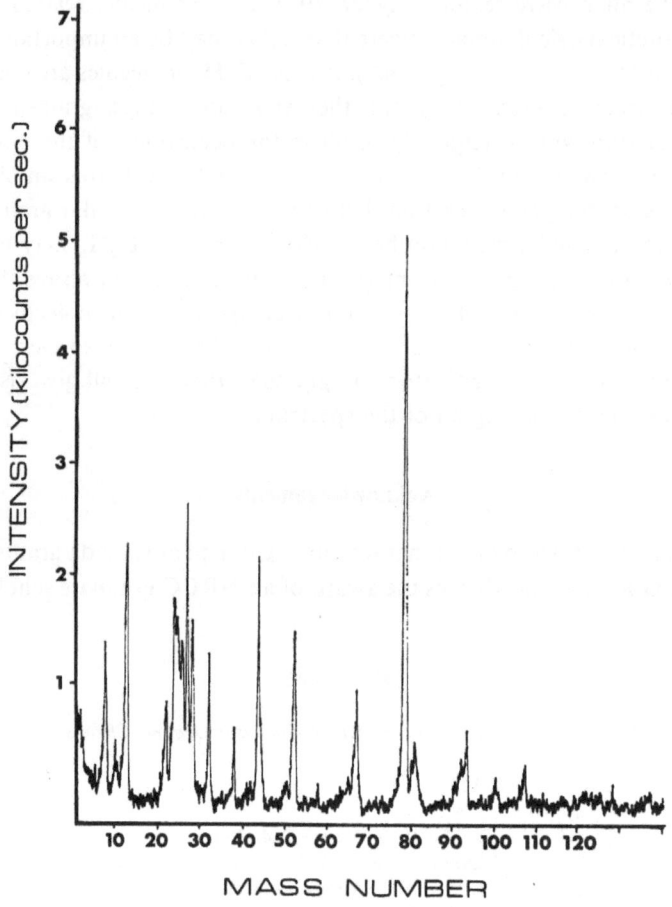

Fig. 5.   Mass spectrum of products ejected from solid C_2H_2 during irradiation with low energy electrons.

$m > 50$ amu are $\lesssim 40$ counts/sec. This reflects the fact that the system was constructed and assembled without the artificial introduction of any organic solvents, other than methyl alcohol. We feel that a 'clean' system such as this is necessary if experiments on the generation of complex molecules from simple reactants are to provide unambiguous results.

When C_2H_2 has been condensed on the target at 55 K, a complex spectrum is obtained as shown in Figure 5. The mass peak at $m/e = 78$ amu is attributed to $C_6H_6^+$. Other peaks at lower masses are characteristic of the cracking pattern of C_6H_6. In other experiments to be described more fully elsewhere (Floyd et al., 1972), we have obtained further evidence that the peak at mass 78 does in fact correspond to C_6H_6. It should be noted that C_6H_6 is far from the only product produced in the polymerization of C_2H_2. Figure 5 indicates that the resulting mixture is complex and probably includes members of the alkane, alkene and alkyne series.

Although the existence of C_2H_2 in the interstellar medium has yet to be confirmed, much circumstantial evidence for its presence exists. Many of the molecules detected recently in the microwave region (Snyder, 1973) are chemically related to C_2H_2. In addition theoretical calculations indicate that C_2H_2 may be an important component of some stellar atmospheres (Tsuji, 1964). If these C_2H_2 molecules are coated at some stage on the surface if interstellar grains then exposure of these grains to almost any source of radiation will subsequently result in the occurrence of the polymerization reaction given above and in the formation of C_6H_6. Obviously this simple reaction is only indicative of the type of reaction that can occur on interstellar grains when they are exposed to interstellar radiation fields. We suggest that C_6H_6 may be one of the *simpler* molecules to be found on interstellar grains. As noted above, hydrocarbon grains containing a mixture of various types of hydrocarbon molecules will, when exposed to a background gas of Na, Na^+, Ca and other atomic species, accrete these atoms. The presence of impurity atoms in grains of this sort will give rise to absorption features in the visible region of the spectrum.

Acknowledgements

This work was supported by National Research Council of Canada grants to W.W.D. and R.H.P.. G.R.F. acknowledges the award of an NRCC graduate scholarship.

References

Andrews, L. J. and Keefer, R. M.: 1964, *Molecular Complexes in Organic Chemistry*, Holden-Day Inc., San Francisco.
Duley, W. W.: 1968, *Nature* 218, 153.
Duley, W. W.: 1969, *Physica* 41, 134.
Duley, W. W. and Graham, W. R. M.: 1969, *Nature* 224, 785.
Floyd, G. R., Prince, R. H., and Duley, W. W.: 1972, unpublished work.
Graham, W. R. M. and Duley, W. W.: 1971, *Nature Phys. Sci.* 232, 43.
Honeycutt, R. K.: 1972, *Astron. J.* 77, 24.
Lind, S. C.: 1961, *Radiation Chem. of Gases*, Reinhold Publ. Corp., New York.

Snyder, L. E.: 1973, this volume, p. 351.
Tsuji, T.: 1964, *Ann. Tokyo Obs.* **IX**, No. 1.
Walker, G. A. H., Hutchings, J. B., and Younger, P. F.: 1969, *Astron. J.* **144**, 921.
Walkley, J., Glew, D. N., and Hildebrand, J. H.: 1960, *J. Chem. Phys.* **33**, 621.
York, D. G.: 1970, *Astrophys. J.* **166**, 65.

PHOTOELECTRON EMISSION FROM INTERSTELLAR GRAINS*

WILLIAM D. WATSON**

Center for Radiophysics and Space Research

and

Laboratory for Atomic and Solid State Physics, Cornell University, Ithaca, N.Y., U.S.A.

Abstract. Photoelectric emission from interstellar dust grains due to ultraviolet starlight is reinvestigated. Ejection of photoelectrons can make a substantial contribution to the heating of standard interstellar clouds. Both positive and negative grain charges are shown to be possible.

1. Introduction

The emission of photoelectrons from interstellar dust grains in H I regions by absorption of ultraviolet starlight (energy $\approx$ 10–13.6 eV) is of chief interest at present because it influences the electric charge on the grain and provides a mechanism for converting starlight energy into gas kinetic energy. Also, the force on a grain resulting from photoemission can be comparable or greater than that due to the usual radiation pressure in both H I and H II regions.

The charge on grains is determined by an equilibrium resulting from the sticking of positive charges (mainly protons or carbon ions) and electrons, and the ejection of electrons by the photoelectric effect. Since the early investigation of Spitzer (1948), photoemission has been considered to be negligible so that the charge on interstellar grains has been thought to be slightly negative $e\phi/kT_{gas} \approx -2.5$, where ϕ is the potential of the grain, T_{gas} is the gas temperature and k is Boltzmann's constant. This analysis was however based on the photoemission data at relatively low photon energies ($\gtrsim$ 5 eV) available at that time. Since then, laboratory investigations have established that photoemission is much more efficient at ultraviolet energies ($\approx$ 10–13.6 eV). If interstellar grains have appreciable positive charges, molecule formation on grains (see Watson and Salpeter, 1972a, b) involving atoms that are ionized in the interstellar gas is inhibited since the sticking of these ions to grains is reduced by a factor exp $(-e\phi/kT_{gas})$. Carbon is an important example. Photoelectrons typically have kinetic energies of a few eV and this can represent a source of heating to the gas since $kT_{gas} \approx$ 0.01 eV in interstellar H I clouds. This heating may be significant in 'standard' clouds. The properties of the grain material that determine the charge and heating are then the photoemission yield (photoelectrons/absorbed photon), the photoelectron energy distribution, and the sticking coefficient S for low energy (0.01 eV) electrons to attach themselves to a grain. In addition a knowledge of the average interstellar starlight

* Supported in part by U.S. National Science Foundation Grants GP-26068 and GP-27355.
** Present address: Departments of Physics and Astronomy University of Illinois, Urbana, Ill. 61801, U.S.A.

Greenberg and Van de Hulst (eds.), Interstellar Dust and Related Topics, 335-339.

flux in the (10–13.6) eV range, the electron density N_e and the total absorption cross section of grains $Q_{abs}N_g\sigma_g\,N_H$ cm^{-2} per H atom is necessary. We discuss these factors in the remainder of the paper, as well as in a previous article (Watson, 1972, hereafter referred to as Paper I) in somewhat more detail.

2. Relevant Properties of Interstellar Grains

Although neither the material composition of the grains is known, nor is photoemission data available for all potential grain materials, some generalizations fortunately are possible. The most frequently mentioned grain materials seem to be silicates, oxides, silicon carbide, graphite and ices, of which the last is most likely to occur as a

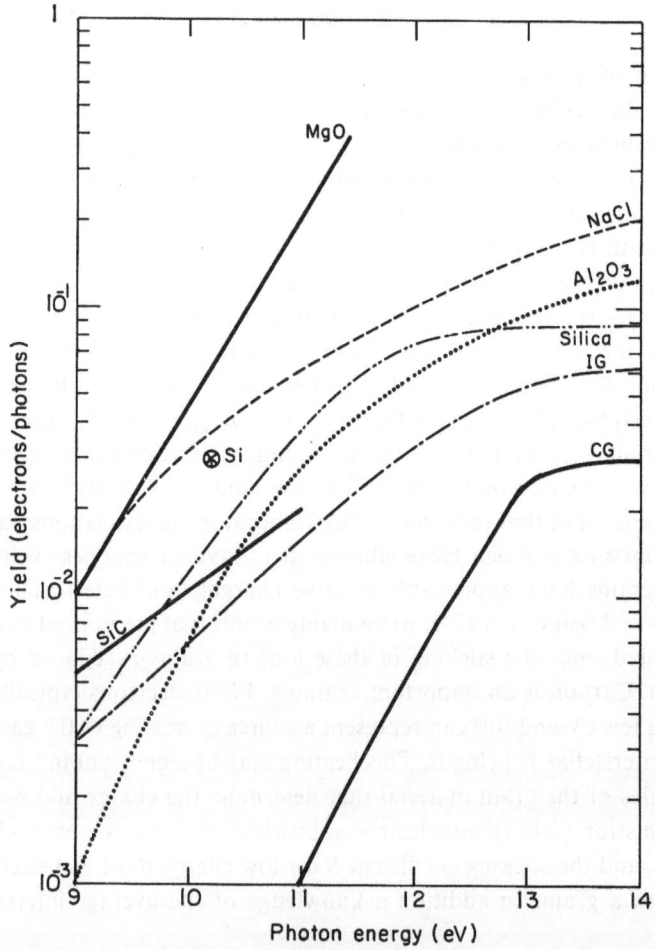

Fig. 1. Photoemission yields for representative grain materials, sources: MgO, Stevenson and Hensley (1961); NaCl, Taylor and Hartmann (1959); Al₂O₃, Pong (1969); silica, 'irregular' graphite (IG), and clean graphite (CG), Feuerbach and Fitton (1972); Si, Pierce and Spicer (1971); SiC, see Sommer (1968).

mantle on a core of one of the other materials. Frozen CH_4, NH_3, and especially H_2O are good candidates for the ices. These have thresholds for ionization near 13.6 eV were the interstellar radiation is cut off, so that photoemission from these is expected to be unimportant. There is some observational evidence against frozen H_2O on grains (Knacke *et al.*, 1969). For the other materials, except perhaps graphite, photoemission yields of $\approx$ (0.03–0.2) electrons/photon are expected (see Paper I and Figure 1). Data have recently become available for graphite and the yield in the pure form is much lower than anticipated by analogy with other material having small band-gaps. The actual yield of graphite is increased by irregularities in the crystal structure and probably by impurities, both of which most likely occur in interstellar grains (see Paper I). In Figure 1 we have summarized the data on yields from various materials representative of interstellar grains. These laboratory data are for radiation normally incident onto large plane surfaces. Because the size of a grain is comparable to ultraviolet wavelengths, the yield from a grain is not necessarily the same as from a plane. In Paper I and elsewhere (Watson, 1973) we have examined this effect. Under some conditions (especially for small grains) the yield can be enhanced by a factor of two or more over laboratory data. Representative distributions for the kinetic energy of the ejected electrons are shown in Figure 2.

Little data are available for the sticking coefficient S for very low energy ($\gtrsim$ 1 eV) electrons from the gas onto surfaces. We are mainly interested in the case where photoemission is efficient and here the grain will assume an appreciable positive

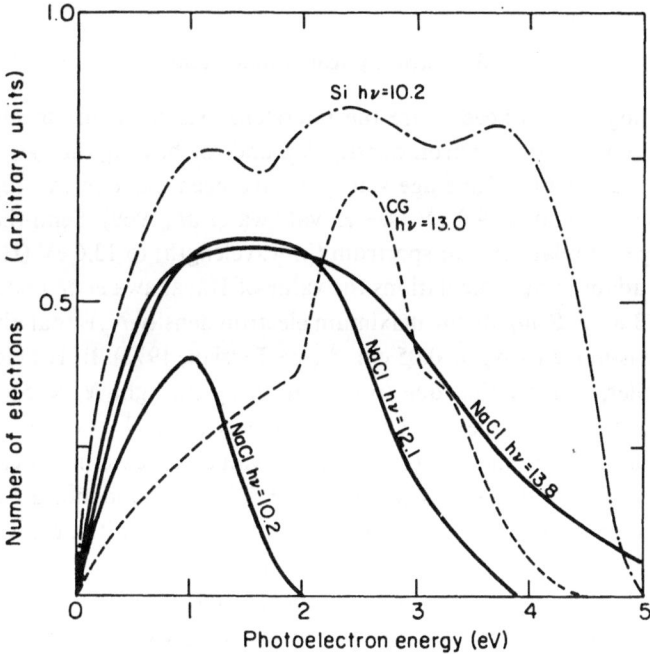

Fig. 2. Typical electron energy distributions for photoemission at representative photon energies *hν* (see sources and notation in Figure 1).

potential (a few tenths eV) so that $e\phi \gg kT_{gas}$. On neutral surfaces, data suggest that for incident electron energies of about an eV, $S \gtrsim \frac{1}{3}$ (see Paper I). Hence for $e\phi \gg kT_{gas}$, probably $S \approx 1$.

In principle, the electron energy distributions should be integrated over all photon energies and electron energies E for $E > e\phi$ to obtain the effective yield and energy input into the gas. However such a detailed treatment is not warranted in view of the uncertainties in the various factors. Hence we introduce in Paper I an approximate model for the heat input H and the photoemission yield y that incorporates the above information.

$$H = 0.1(N_g\sigma_g Q_{abs}/N_H)F_{eV}\frac{(3.5 - e\phi)^2}{2}$$

(eV per H atom s^{-1}).

and

$$y = 0.1, \qquad h\nu > 10\,\text{eV} + e\phi$$
$$= 0, \text{ otherwise}$$

($h\nu$ = photon energy) for materials other than ice. The photon flux per electron volt at (10–13.6) eV is given by F_{eV}, and N_g, N_H, σ_g and Q_{abs} are number densities of grains and hydrogen, the grain cross section, and the absorption efficiency for these ultraviolet photons.

3. Astrophysical Applications

The remaining factors needed are the interstellar starlight flux in H I regions at energies (10–13.6) eV, the electron density N_e, and for heating the gas $N_g\sigma_g Q_{abs}/N_H$. No direct measurements of average starlight have been made in this region, though there are measurements to 9.2 eV (see Hayakawa et al., 1969). Semi-theoretical calculations suggest a relatively flat spectrum (in wavelength) to 13.6 eV (Habing, 1968). We therefore adapt in our calculations the value of Hayakawa et al. and assume a flat spectrum to 13.6 eV. Roughly the maximum electron density N_e is that given by pulsar dispersion measurements $N_e \approx 0.05$ cm^{-3} (see Terzian, 1972). In H I clouds N_e may be much smaller, roughly that due to ionization by starlight $N_e \approx 5 \times 10^{-4}$ N_H for cosmic abundances (see Paper I). The grain density ($N_g\sigma_g Q_{abs}/N_H$) is also uncertain by factors of about three and two representative values are used in our calculations.

The results of these calculations are presented in Paper I, and indicate that standard H I clouds can be maintained at temperatures in agreement with observations by this heating mechanism. The grain voltages are always positive and equal to a fraction of an electron volt. Because of the possible importance of graphite as a grain material, we have performed detailed calculations using the yield data for clean, ordered graphite (Feuerbach and Fitton, 1972). Any enhancement of y due to the small size of the particle is ignored and the above photon flux is used. For this very low photoemission yield (which is not likely to be the actual case), the grain is essentially neutral for $N_e \approx 0.05 \text{ cm}^{-3}$ and $S \approx 1$.

Ejection of photoelectrons causes a force on a grain as a result of the momentum given to the photoelectron. For a symmetric particle in an isotropic radiation field the net force is zero. However, the radiation may not be isotropic due to shielding by a nearby cloud or another grain, or because the grain is near a hot star (an H II region). The grain is then accelerated (see Spitzer, 1968). In the average radiation field, the ratio of the net force on a grain due to photoemission to that due to the usual radiation pressure is $\approx \frac{1}{2} (Q_{uv}/Q_{opt}) (f_{uv}/f_{opt})$. The ultraviolet and optical absorption efficiencies Q_{uv} and Q_{opt}, and the factors by which the radiation is anisotropic in the ultraviolet and in the optical are such that the ratio is almost certainly greater than one (the ultraviolet is more readily absorbed and shielded). In H II regions the relative flux of ultraviolet radiation ($hv > 10 \text{ eV}$) is much greater than in the average interstellar radiation field, so that photoemission is always the dominant force on grains resulting from absorption of photons in H II regions.

References

Feuerbach, B. and Fitton, B.: 1972, *J. Appl. Phys.* **43**, 1563.

Habing, H. J.: 1968, *Bull. Astron. Inst. Neth.* **19**, 421.

Hayakawa, S., Yamashita, K., and Yoshioka, S.: 1969, *Astrophys. Space Sci.* **5**, 593.

Knacke, R. F., Cudaback, D. D., and Gaustad, J. E.: 1969, *Astrophys. J.* **158**, 151.

Pierce, D. T. and Spicer, W. E.: 1971, *Phys. Rev. Letters* **27**, 1217.

Pong, W.: 1969, *J. Appl. Phys.* **40**, 1733.

Sommer, A. H.: 1968, *Photoemissive Materials*, Wiley, New York.

Spitzer, L., Jr.: 1948, *Astrophys.* **107**, 6.

Spitzer, L., Jr.: 1968, *Diffuse Matter in Space* Wiley, Interscience, New York.

Stevenson, J. R. and Hensley, E. B.: 1961, *J. Appl. Phys.* **32**, 166.

Taylor, J. W. and Hartmann, P. L.: 1959, *Phys. Rev.* **113**, 1421.

Terzian, Y.: 1972, in A. M. Lenchek (ed.), *Physics of Pulsars*, (Gordon and Breach, New York).

Watson, W. D.: 1972, *Astrophys. J.* **176**, 103 (Paper I).

Watson, W. D.: 1973, *J. Opt. Soc. Am.* **63**, 164.

Watson, W. D. and Salpeter, E. E.: 1972a, *Astrophys. J.* **174**, 321.

Watson, W. D. and Salpeter, E. E.: 1972b, *Astrophys. J.* **175**, 659.

DESTRUCTION OF DIRTY ICE MANTLES BY SPUTTERING

P. A. AANNESTAD

Institute for Space Studies, Goddard Space Flight Center, NASA, N.Y., U.S.A.

Abstract. Dirty ice mantles are destroyed efficiently by sputtering of He atoms when clouds encounter shock velocities greater than 13–15 km s^{-1}. Destruction due to grain-grain collisions is found to be about 10^{-3} times less efficient. Sputtering in the hot intercloud medium should make intercloud grains smaller than cloud grains.

We have studied the processes of grain destruction and molecule formation due to the sputtering of dirty ice grain mantles as the grains move relative to the gas in hot shock waves. Such waves are set up when interstellar clouds encounter expanding H II regions or when they collide with each other. The model for the grains is a core of either graphite or silicate with a radius of 0.05 μ surrounded by a mantle composed primarily of H_2O and CH_4. The sputtering rates of these mantle materials have been calculated as a function of gas temperature and dust to gas velocity, employing a theory of oblique angle incidence sputtering. Helium is found to be the main sputtering agent, and the decoupling of gas and dust through the shock front is found to increase the sputtering yields substantially.

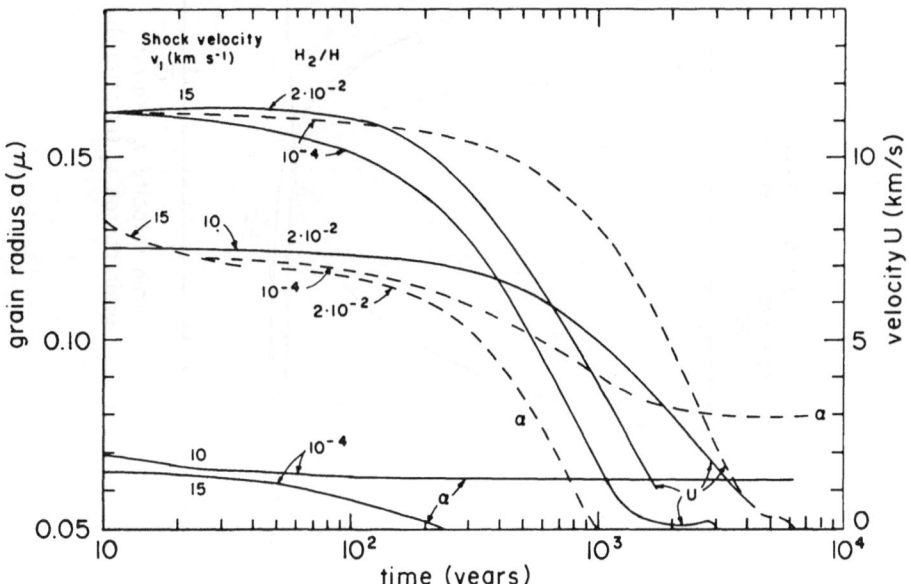

Fig. 1. The relative velocity between grain and gas and the grain radius as functions of time. The solid curves are for a cloud where the initial grain radius is 0.07 μ, and the dashed curves for a grain radius of 0.14 μ.

Greenberg and Van de Hulst (eds.), Interstellar Dust and Related Topics, 341–344.

In Figure 1 we show how the grain radius and the grain to gas velocity change as functions of time behind the shock front. The solid curves are for a cloud where the initial grain radius is about 0.07 μ, and the dashed curves for a grain radius of about 0.14 μ. We find that dirty ice mantles are destroyed efficiently by shock velocities greater than about 13–15 km s^{-1} if the abundance of molecular hydrogen is about one per cent or more, the corresponding life time of a mantle is about 5×10^7 yr. This leads to an equilibrium size distribution with an average grain size of 0.09 μ–0.07 μ, the smallest value roughly satisfying the limit on the amount of ice in interstellar grains as found by Knacke *et al.* (1969).

In Figure 2 we show how the grain velocity, the relative grain density, the grain charge and the hydrogen density vary with the distance behind the shock front. The shock velocity is 10 km s^{-1} and H$_2$/H is 2×10^{-2}. We see that the grains, starting out with a velocity of 10 km s^{-1} are decelerated to about 1 km s^{-1} within about 10^{-3} pc from the shock front.

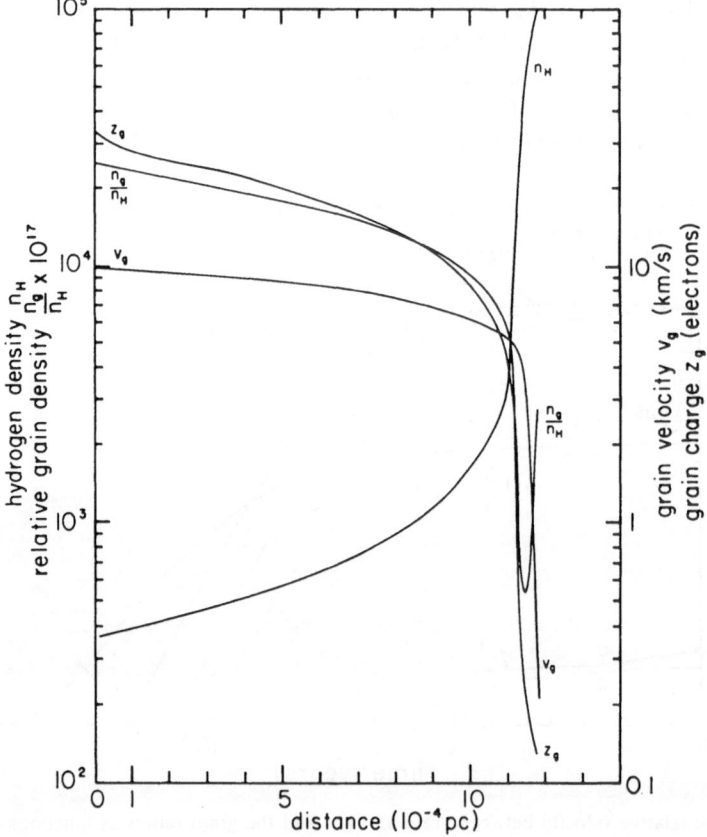

Fig. 2. The velocity of the grain with respect to the shock front, the relative grain density, the grain charge and the gas density as functions of the distance behind the shock front. The shock velocity is 10 km s^{-1} and H$_2$/H $= 2 \times 10^{-2}$.

Since the mean free path for grain-grain collisions in this region is about 1 pc and only about 10% of the mass of a typical cloud is in the interpenetration region of two clouds, we conclude that the probability of mantle destruction due to grain-grain collision is about 10^{-4}. This is a factor of more than 10^3 smaller than the destruction probability due to the sputtering process.

Since the growth of dirty ice mantles will cause the interstellar gas to lose cooling elements, clouds may tend to heat up and make a sort of phase transition into the hot and rarified state of the intercloud gas. If we use the thermal sputtering yields for He and H sputtering of H_2O shown by the solid lines in the Figure 3 (the dashed lines are for CH_4), we find a yield of about 0.1 for an intercloud temperature of 7500 K. With a gas density of 0.2 cm^{-3} a grain may be sputtered from 0.15 μ to 0.05 μ in about 3×10^7 yr.

It thus appears that sputtering processes are fast enough to effectively clean the interstellar ice grains if they are exposed to the conditions in the intercloud medium. Intercloud grains should thus be somewhat smaller than cloud grains. From the ob-

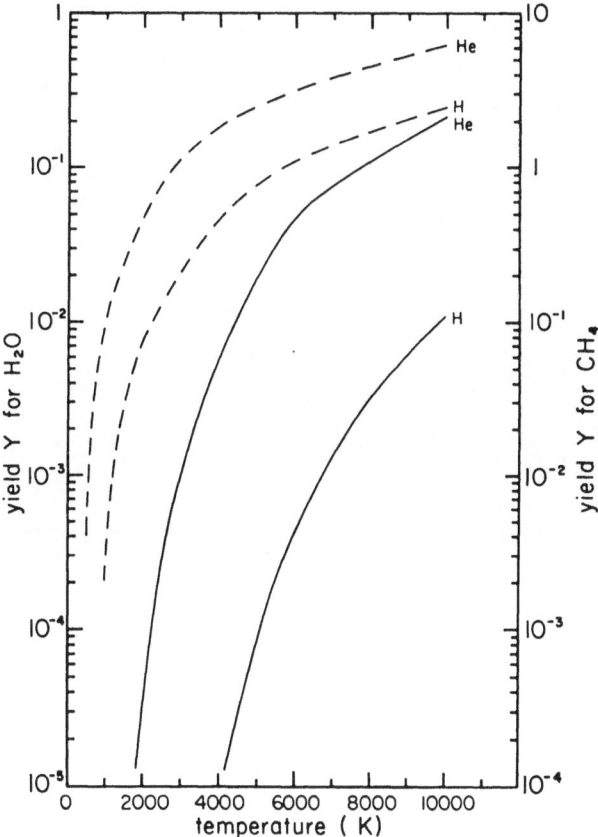

Fig. 3.   Thermal sputtering yield as a function of temperature. The solid curves are for sputtering of H_2O, and the dashed curves for sputtering of CH_4. The sputtering agents are H and He as indicated.

servations by FitzGerald (1968), assuming a ratio of total to selective absorption of 3, we find that the ratio of visual extinction to column density of hydrogen is at least a factor of two smaller for regions outside the clouds, indicating an effect of the kind we have proposed here.

References

FitzGerald, M. P.: 1968, *Astron. J.* **73**, 893.
Knacke, R. F., Cudaback, D. D., and Gaustad, J. E.: 1969, *Astrophys. J.* **158**, 151.

EJECTION OF GRAINS FROM COOL STARS

N. C. WICKRAMASINGHE

Dept. of Applied Mathematics, University College, Cardiff, Wales, U.K.

Abstract. Dust grains expelled from cool stars are injected into interstellar clouds with typical velocities of $\sim 10^3$ km s^{-1}. The interaction of such high-speed grains with gas is discussed. Grains are preferentially stopped within clouds of number density $n_H \approx 10-30$ cm^{-3}. Collisions of grains with gas atoms could lead to heating and ionization of clouds and to the production of displacement defects in grains.

The process of grain formation in cool stars and the ejection of such grains into inter-stellar space has been discussed for many years. (Hoyle and Wickramasinghe, 1962; Gilman, 1969). I wish to consider briefly a few consequences of this stellar grain ejec-tion process which have not hitherto been considered. A more detailed discussion is published elsewhere (Wickramasinghe, 1972). To begin with we note that the ratio of radiation pressure to gravity for most grains near a giant star exceeds unity by a large factor. Values of this ratio for various types of spherical particle are set out in Table I for a giant star of $T_{\text{eff}} = 3000$ K, $g = 1$ cm s^{-2} and $R_* = 3 \times 10^{13}$ cm

TABLE I
P/G
($T_* = 3000$ K; $g = 1$ cm s^{-2}; $R_* = 3 \times 10^{13}$ cm)

a/μ	Iron	Silicate	Graphite
0.01	5.59×10^2	1.57×10	1.62×10^3
0.03	9.18×10^2	2.32×10	1.83×10^3
0.05	1.70×10^3	5.04×10	2.46×10^3
0.08	2.84×10^3	1.44×10^2	3.90×10^3
0.10	3.25×10^3	2.32×10^2	4.72×10^3
0.30	1.96×10^3	9.55×10^2	3.49×10^3
0.50	1.02×10^3	9.44×10^2	1.84×10^3
1.0	3.34×10^2	4.20×10^2	6.24×10^2

Due to these high P/G ratios grains are readily expelled from the regions where they form. Although there would be a strong frictional coupling to the gas near in to the star, grains and gas which expand outwards will quickly decouple. Once grains are driven clear of the photospheric layers they accelerate to very high speeds. The ter-minal velocity in the absence of an appreciable density of ambient gas is given by

$$v_t = \left[2\left(\frac{P}{G}\right) g R_0 \right]^{\frac{1}{2}}, \tag{1}$$

where g is the surface gravity and $R_0 \simeq R_*$ is the radial distance at which grain-gas

Greenberg and Van de Hulst (eds.), Interstellar Dust and Related Topics, 345–347.

decoupling occurs. Ejection velocities of iron, graphite and silicate grains of various radii from a giant star with $T_* = 3000$ K, $g = 1$ cm s^{-2} and $R_* = 3 \times 10^{13}$ cm have been calculated and are set out in Table II below.

TABLE II

Ejection velocities in km s^{-1}; and $\frac{1}{2}m_H V^2$ keV^{-1} within parentheses

a/μ	Graphite	Silicate	Iron
0.01	3.11×10^3 (50.66)	3.07×10^2 (0.49)	1.17×10^3 (7.21)
0.03	3.31×10^3 (57.24)	3.73×10^2 (0.73)	1.45×10^3 (11.02)
0.05	3.84×10^3 (77.16)	5.50×10^2 (1.58)	1.96×10^3 (20.11)
0.08	4.83×10^3 (122.04)	9.30×10^2 (4.51)	2.60×10^3 (35.24)
0.10	5.32×10^3 (147.82)	1.18×10^3 (7.28)	2.86×10^3 (42.77)
0.30	4.58×10^3 (109.46)	2.39×10^3 (29.93)	2.51×10^3 (32.94)
0.50	3.32×10^3 (57.60)	2.38×10^3 (29.57)	1.84×10^3 (17.59)
1.00	1.94×10^3 (19.56)	1.59×10^3 (13.16)	1.08×10^3 (6.04)

Grains thus leave their parent stars with typical speeds 10^3 km s^{-1}. The stopping distance of such grains (neglecting disruptive effects) in a cloud of density n_H is

$$\frac{L}{pc} \cong 3 s \left(\frac{a}{10^{-5}\,cm}\right)\left(\frac{1\,cm^{-3}}{n_H}\right) \ln\left(\frac{v_t}{km\,s^{-1}}\right), \tag{2}$$

where s is the density of grain material. A grain of radius $a \cong 3 \times 10^{-6}$ cm, density $s \cong 3$ g cm^{-3} and initial speed $v_t \cong 1000$ km s^{-1} is stopped within ~ 3 pc inside a typical interstellar cloud of number density $n_H \sim 10$ cm^{-3}. The stopping distance of the same grain in a tenuous intercloud medium with $n_H \sim$ cm^{-3} is ~ 300 pc. Since the mean distance between clouds (near the galactic plane) is considerably smaller, ~ 40 pc, this feature would ensure the trapping of grains preferentially within clouds. A fairly natural explanation is thus provided for the observed correlation between H I and dust in normal clouds (Lilley, 1955; Heiles, 1967).

The slowing of grains within clouds is attended by several effects. Gas atoms impinging on grains have initial energies $\sim 10^2$ keV. Such impacts lead to sputtering of the grains, but it may be shown that their radii are reduced, only by a small factor. A more important effect is the accumulation of solid-state defects. It can be shown that for most types of solid material grains are saturated with a defect concentration of $\sim 1\%$. Such a defect concentration could produce important optical effects in the grains – e.g. introduce colour centres and increase their infrared absorbtivity.

Finally, I would like to discuss a few implications of the grain slowing process for the interstellar medium itself. A hydrogen atom striking a grain with relative energy $\sim 10^2$ keV is stopped inside it and is eventually re-evaporated at the grain temperature. The re-evaporated atoms/ions have energies $\sim 10^2$ keV and this energy dissipated in the cooler cloud could lead to significant heating and ionization.

The average mass density of grains in a cloud of density n_H is

$$\varrho_g \simeq 10^{-2} n_H m_H\,g\,cm^{-3} \tag{3}$$

(see, for example, Wickramasinghe, 1967). According to the ideas discussed here a large fraction of this mass density may be assumed to be supplied from cool stars with initial grain speeds corresponding to

$$\tfrac{1}{2}m_{\mathrm{H}}v^2 \simeq 10^2 \,\mathrm{keV}. \tag{4}$$

The total kinetic energy initially associated with the mass density (3) is

$$U \simeq \tfrac{1}{2}\varrho_g v^2 \simeq 1.6 \times 10^{-9} n_{\mathrm{H}} \,\mathrm{erg\,cm}^{-3}. \tag{5}$$

An appreciable fraction of this energy is converted firstly to kinetic energy of ejected protons, and thence, through their interaction with the gas, into thermal energy of the gas cloud. Assuming the grains are re-supplied with a typical turnover time of $\sim 10^9$ yr we have a heat input from this process given by

$$\Gamma_{\mathrm{gr}} \simeq 5 \times 10^{-26} n_{\mathrm{H}} \,\mathrm{erg\,cm}^{-3}\,\mathrm{s}^{-1}. \tag{6}$$

The associated ionization rate is obtained by dividing (6) by ~ 30 eV corresponding to each ionization (the kinetic energy liberated by primary collisions of protons of energy ~ 100 keV with H atoms (Bethe, 1933)). Thus we have an ionization rate

$$\zeta_{\mathrm{gr}} \simeq 10^{-15} n_{\mathrm{H}} \,\mathrm{cm}^{-3}\,\mathrm{s}^{-1}. \tag{7}$$

The above values of Γ and ζ exceed those derived from earlier arguments of sub-cosmic ray heating (Hayakawa et al., 1961; Spitzer and Tomasko, 1968; Spitzer and Scott, 1969) and of X-ray heating (Silk and Werner, 1969) by a factor ~ 10. Allowing even for an efficiency factor as low as 0.1 it is clear that the present process could well provide the dominant source of heating and ionization of interstellar gas.

References

Bethe, H.: 1933, *Handbuch der Physik* **24**, 491.
Gilman, R. C.: 1969, *Astrophys. J.* **155**, L185.
Heiles, C.: 1967, *Astrophys. J.* **148**, 299.
Hayakawa, S., Nishimura, S., and Takayanagi, K.: 1961, *Publ. Astron. Soc. Japan* **13**, 184.
Hoyle, F. and Wickramasinghe, N. C.: 1962, *Monthly Notices Roy. Astron. Soc.* **124**, 417.
Lilley, A. E.: 1955, *Astrophys. J.* **121**, 559.
Silk, J. and Werner, M. W.: 1969, *Astrophys. J.* **158**, 185.
Spitzer, L. and Scott, E. H.: 1969, *Astrophys. J.* **158**, 161.
Spitzer, L. and Tomasko, M. G.: 1968, *Astrophys. J.* **152**, 971.
Wickramasinghe, N. C.: 1967, *Interstellar Grains*, Chapman & Hall Co., London.
Wickramasinghe, N. C.: 1972, *Monthly Notices Roy. Astron. Soc.* **159**, 269.

PART VII

MOLECULES, THEORY AND OBSERVATIONS

SOME IMPLICATIONS OF THE DISTRIBUTION OF INTERSTELLAR MOLECULES

LEWIS E. SNYDER, JR.

Astronomy Dept., University of Virginia, Charlottesville, Va. 22903, U.S.A.

Abstract. It is suggested that molecular observations of interstellar clouds can be used to determine the stage of cloud evolution. However, before observational data can reliably be used to investigate the evolutionary path leading from a tenuous cloud of gas and interstellar dust to a star cluster, mechanisms of molecular formation, excitation, and destruction must be taken into account. The evidence is growing that interstellar dust grains have an important role in molecular formation and destruction.

1. Introduction

In the past few years a number of interesting and informative review articles on interstellar molecules have been published (Buhl, 1971; Buhl and Snyder, 1971; Heiles, 1971; McNally, 1968; Rank *et al.*, 1971; Snyder, 1972; Snyder and Buhl, 1970; Turner, 1970) and are recommended reading for background information. In addition, many of the papers presented at this meeting represent a significant contribution to the study of the interrelationship between dust and molecules. In this review, I have been asked to discuss the distribution of interstellar molecules. Since the true galactic distribution of most of the twenty-six known molecules is not yet known, I will approach this problem by suggesting tentative classifications for molecular clouds which I hope will influence those of you who are planning future molecular survey work. In particular, it is important for astronomers to begin to think of interpreting molecular survey data in terms of the evolution of dust clouds. Before this can be done meaningfully, however, it becomes necessary to correctly interpret abundances and excitation and to understand the dominant molecular formation mechanisms.

2. Molecular Abundances

Tables I and II list the known inorganic and organic interstellar molecules. In each table, the first column gives the chemical formula and name, the second the approximate number of known sources in our Galaxy, the third column the range of projected densities in cm^{-2} (number density integrated along the line-of-sight); the fourth column lists the spectral region where the molecule has been observed; and the last column indicates whether the molecular line is observed in absorption, emission, or both. A molecular source is defined as a given direction in the Galaxy which typically contains multiple molecular clouds along the line of sight. Hence the source counts listed in column two of both Tables I and II may be regarded as lower bounds on the actual number of molecular clouds observed.

Greenberg and Van de Hulst (eds.), Interstellar Dust and Related Topics, 351–361.

TABLE I
Inorganic interstellar molecules

	No. reported sources	Projected density (No cm^{-2})	Spectral region	Absorption (A) or emission (E)
Diatomic molecules				
H$_2$ – molecular hydrogen	1 (ξ Per)	$\sim 10^{20}$	Far UV	A
OH – hydroxyl	> 200	10^{12}–10^{16}	$^+$R	A, E
SiO – silicon monoxide	1 (Sgr B2)	$\sim 10^{13}$	R	E
Triatomic				
H$_2$O – water	> 50	–	R	E
H$_2$S – hydrogen sulfide	> 9	10^{13}–10^{14}	R	E
Four-atomic				
NH$_3$ – ammonia	< 10	$\sim 10^{16}$	R	E

$^+$R = Radio Detection

TABLE II
Organic interstellar molecules

	No. reported sources	Projected density (No cm^{-2})	Spectral region	Absorption (A) or emission (E)
Diatomic molecules				
CH$^+$	~ 60	$\sim 10^{13}$	$^+$O	A
CH	40	$\sim 10^{13}$	O	A
CN – cyanogen radical	$\sim 14/3$	$\sim 10^{12}/10^{15}$	O/R	A, E
CO – carbon monoxide	Many/1 (ζ Oph)	10^{17}–$10^{19}/10^{15}$	R/O	E/A
CS – carbon monosulfide	4	10^{13}–10^{14}	R	E
Triatomic				
HCN – hydrogen cyanide	~ 10	10^{14}–10^{15}	R	E
OCS – carbonyl sulfide	1 (Sgr B2)	$\sim 10^{15}$	R	E
Four-atomic				
H$_2$CO – formaldehyde	Many (> 100)	10^{12}–10^{16}	R	A (6 cm line), E (all other lines)
HNCO – isocyanic acid	1 (Sgr B2)	$\sim 10^{14}$	R	E
H$_2$CS – thioformaldehyde	1 (Sgr B2)	–	R	A
Five-atomic				
H$_2$CNH – methylenimine	1 (Sgr B2)	–	R	E
HCOOH – formic acid	1 (Sgr B2)	< 10^{13}	R	E
HC$_3$N – cyanoacetylene	2 or more	$\sim 10^{16}$	R	E
Six-atomic				
CH$_3$OH – methyl alcohol	3 or more	–	R	E
CH$_3$CN – methyl cyanide	2	$\sim 10^{14}$	R	E
HCONH$_2$ – formamide	2 or more	> 10^{11}	R	E
Seven-atomic				
CH$_3$C$_2$H – methylacetylene	1 (Sgr B2)	$\sim 10^{14}$	R	E
HCOCH$_3$ – acetaldehyde	2	–	R	E
Other				
X-ogen (CCH?)	8	–	R	E
'HNC' (hydrogen isocyanide)	> 5	–	R	E

$^+$O = Optical detection

The majority of interstellar molecules have been observed in the radio region of the spectrum. Two inherent advantages are gained by using radio techniques which usually more than offset the disadvantages of small line strengths and the molecular polarity required for pure rotational transitions. The first advantage is that radio waves are capable of penetrating dense dust clouds where complex molecules presumably are afforded the maximum degree of protection against photoionization and photodissociation. Optical observers, on the other hand, must rely on a bright background star as an absorption source when searching the intervening clouds for interstellar molecules; thus such observations are necessarily limited to low opacity clouds which don't offer much protection for molecules with low photodestruction thresholds. The second advantage is the low excitation required by pure rotational transitions. The molecular excitation, E, may be written

$$E = E_{rot} + E_{vib} + E_{el},$$

where for molecules of interstellar interest the rotational energy, E_{rot}, may be only 10^2 cm^{-1} or less; compare this with typical vibrational energy (E_{vib}) values of 10^3 cm^{-1} or more and electronic excitation energies (E_{el}) of $\sim 10^4$ cm^{-1}. Hence radio transitions (E_{rot}) may be excited by low energy sources such as collisions or radio radiation from H II regions but infrared, optical and ultraviolet transitions require a background star with intervening cloud. Due to these physical differences in excitation mechanisms, the few molecules listed in Tables I and II which have been detected by optical or ultraviolet spectroscopy have been observed in low opacity clouds against hot O and B stars and tend to have simpler chemical composition than those molecules detected by radio techniques in denser dusty regions such as the Sagittarius complex or the circumstellar region associated with IRC + 10216.

The projected densities listed in Tables I and II are representative samples taken from various interstellar regions. Comparison of molecular densities observed in the same region is particularly difficult for radio observations due to beam dilution and nonhomogeneous excitation across the dimension of a typical beamwidth. For example, Figure 1 illustrates typical telescope beamwidths projected on the Orion Nebula. The ideal halfpower beamwidth (HPBW) is given for a parabolic radio telescope (in minutes of arc) by

$$\text{HPBW} = 4125.4 \cdot v(\text{GHz})^{-1} \cdot D(\text{ft})^{-1} = 138.6 \cdot \lambda(\text{cm}) \cdot D(\text{ft})^{-1},$$

where the observational frequency v is in GHz, the observational wavelength λ is in cm, and the telescope diameter D is in ft. Thus, as shown in Figure 1, an OH observation with a single element telescope at 18 cm wavelength covers a much broader region of the molecular source than an HCN observation at 3 nm wavelength. Hence the correct interpretation of molecular excitation can be difficult for radio molecules.

Probably it is safe to say that none of the radio molecules detected to date is in thermal equilibrium. We might think of the following three general categories of molecules, characterized by their energy level structures, which are useful for assigning relative reliabilities to reported abundances of radio molecules.

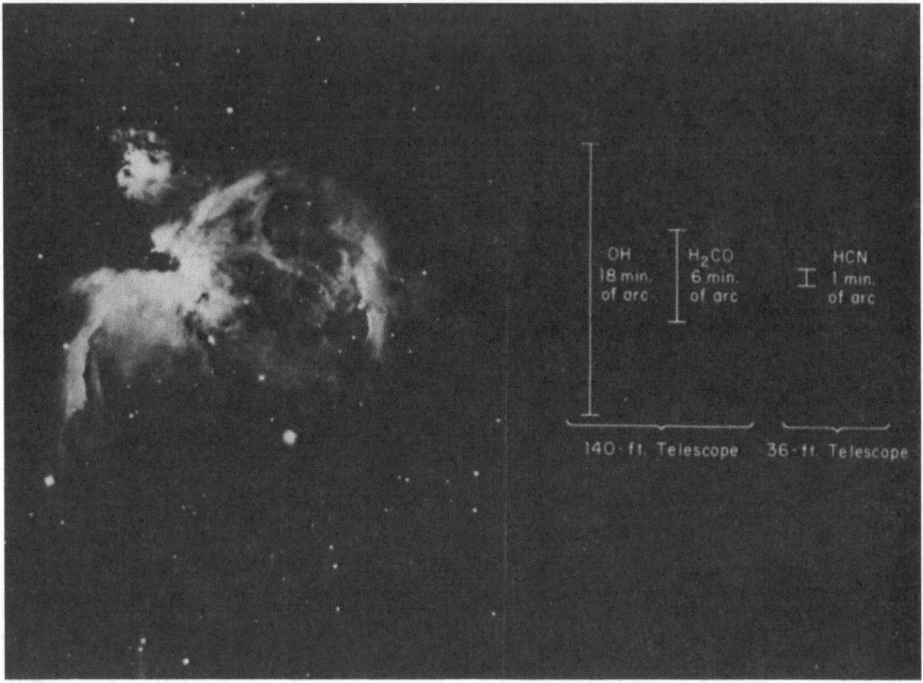

Fig. 1. The spatial resolution (HPBW) of the NRAO 140-ft telescope is illustrated against the Orion Nebula for the 18 cm transitions of OH and the 6.2 cm transitions of formaldehyde (H₂CO). Compare the relatively large HPBW obtained for OH and H₂CO with the small HPBW of the NRAO 36-ft telescope used to observe the 3.4 mm transition of hydrogen cyanide (HCN). Hence it is difficult to compare radio observations of the same region which are made at different wavelengths (Buhl and Snyder, 1971).

(A) Extreme Maser Action. These molecules have energy level structures with accompanying selection rules which permit both radiative and collisional pumping to occur. Typical examples are H_2O, an asymmetric rotator, and OH, a diatomic molecule with lambda-type doubling. Figure 2 illustrates the fairly complex ground state energy level schemes of OH, H_2O and the recently discovered (Thaddeus *et al.*, 1972) interstellar H_2S.

(B) Mild Maser Action. These molecules have selection rules which allow strong radiative (electric dipole) transitions only between immediately adjacent levels in the ground electronic state. Hence radiative pumping is much more difficult but collisional pumping – which usually is much weaker – may still be allowed. These molecules are either diatomics with energy levels uncomplicated by Coriolis or quadrupole interactions (e.g. CO and CS) or symmetric rotators such as CH_3C_2H (Figure 3) where all energy levels with a given angular momentum projection quantum number K may behave exactly as a simple diatomic molecule. Due to its extremely small dipole moment (~ 0.1 D) and simple level structure, CO is probably closer to equilibrium with the surrounding gas than any other molecule detected so far. Hence if saturation

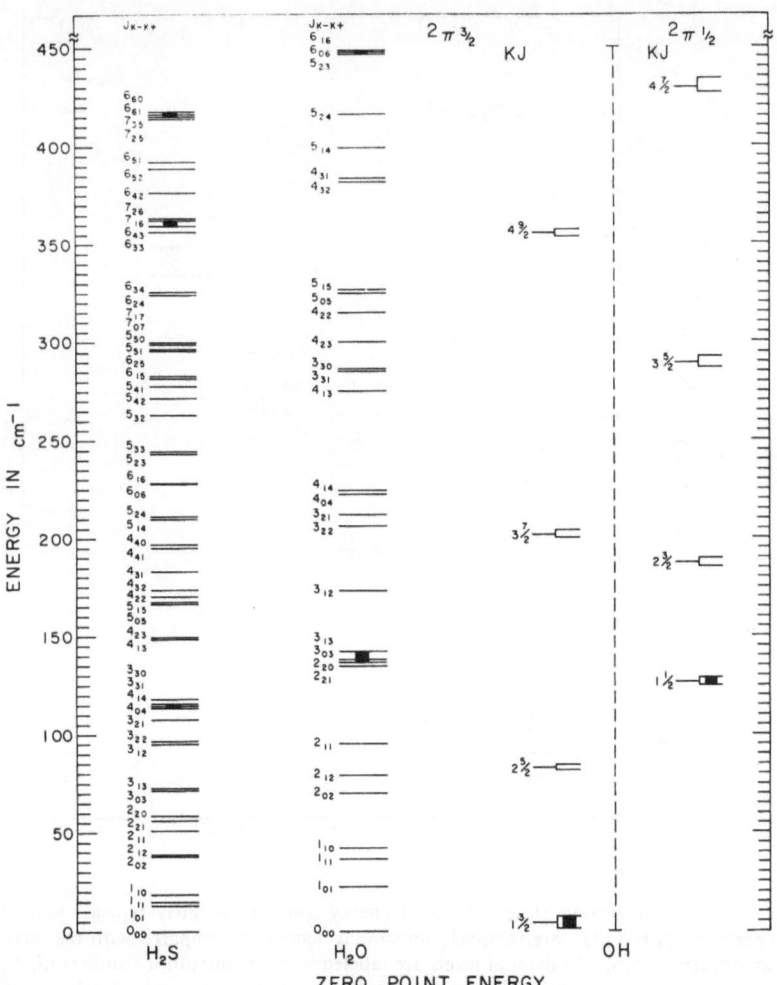

Fig. 2.   The ground state rotational energy levels of several molecules which might be expected to exhibit masering are compared. Here it can be seen that the energy levels of hydrogen sulfide (H_2S) bear very little resemblance to those of water (H_2O) and neither molecule has the relatively simple level structure of OH. The bars indicate the permitted radio frequency transitions $6_{60} \rightarrow 7_{35}$, $7_{16} \rightarrow 6_{43}$ and $3_{31} \rightarrow 4_{04}$ in H_2S (all currently undetected in the interstellar medium); $6_{16} \rightarrow 5_{23}$ (detected) and $3_{13} \rightarrow 2_{20}$ (undetected) in H_2O; and two of the typical lambda-doublet type transitions in OH. The OH lambda-doublet hyperfine splitting is not illustrated.

effects are properly taken into account, CO abundances should be among the most re-liable of those listed in Tables I and II.

(C) *Intermediate Maser Action.* This class of molecules has not yet been thoroughly studied but is possibly the most interesting of the three. Here, due to the molecular geometry, we find symmetry groups of rotational energy levels with selection rules and excitation similar to both class A and class B molecules. An example might be HNCO which has pseudo 'parastates' (Figure 4) which may have excitation similar to a simple

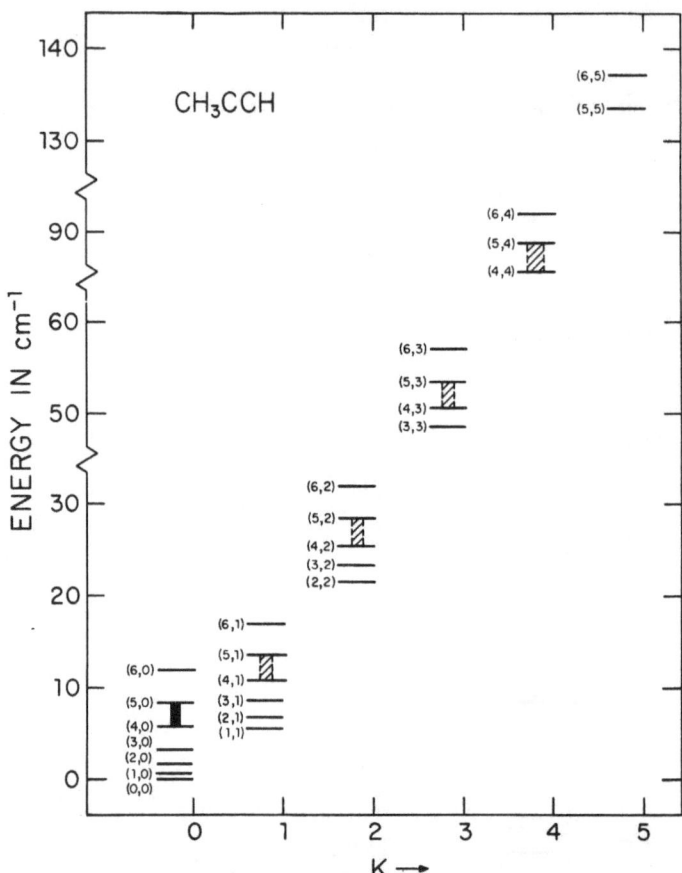

Fig. 3. The lowest ground state rotational energy levels of a fairly typical symmetric rotator, methylacetylene (CH$_3$C$_2$H), are relatively uncomplicated when compared with the analogous levels in an asymmetric rotator. Individual levels are labeled with the quantum numbers (J, K) and, in the absence of strong collisional interactions, each set of levels with the same K value is effectively isolated from every other set and hence resembles the level scheme for a simple diatomic molecule. The solid bar represents the most intense component of the detected interstellar line at 3.5 mm; the dashed bars represent weaker transitions.

class B molecule while the 'orthostates' are structured similarly to H$_2$CO, a class A molecule, and hence may be strongly pumped. Other examples may be HCN and HC$_3$N which have linear structures and hence would be expected to fit into class B. However, the nitrogen atom introduces quadrupole splitting of the rotational levels which may contribute to non-LTE population distributions when the line widths are less than the quadrupole separation.

3. Molecular Cloud Classifications

The molecular cloud observations of the past few years have shown that several types of molecular clouds exist which vary greatly in chemical composition, density and

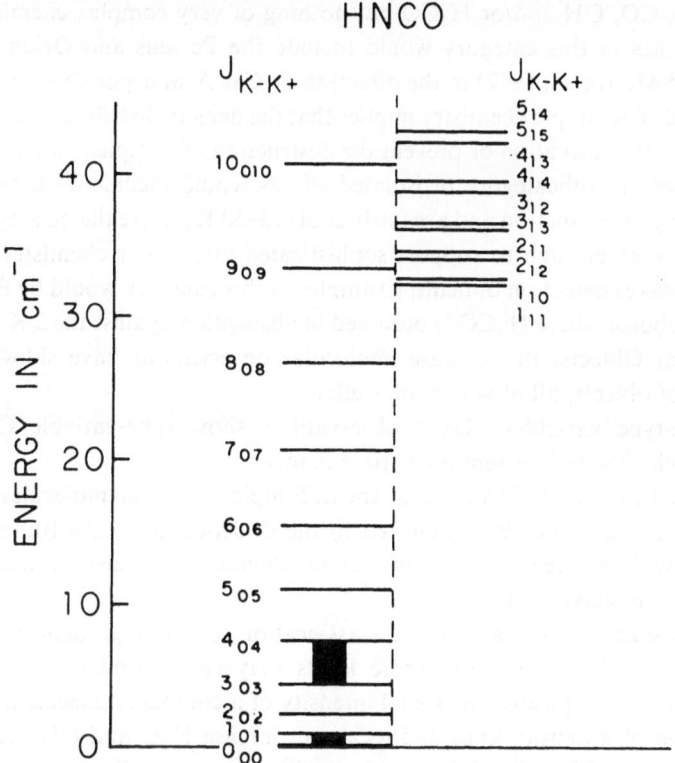

Fig. 4.   The lowest ground state rotational energy levels of isocyanic acid (HNCO) are illustrated with the pseudo 'parastates' on the left-hand side and 'orthostates' on the right. Although the level structure is similar to formaldehyde (H$_2$CO) and thioformaldehyde (H$_2$CS), the ortho and para separation does not really exist for HNCO because relatively weak radiative transitions are allowed across the dotted line.

excitation. In some cases, it is possible to distinguish between prestellar and stellar regions of excitation; furthermore, we might invoke the following molecular cloud classifications:

(A) Clouds associated with H II regions. These molecular clouds have nearby H II regions and may contain infrared objects which often are believed to be recently formed. The neutral particle densities found from collisional excitation calculations indicate that these clouds are very massive (10^3–10^6 $M_\odot$). In addition, high temperature knots of 100 K or greater are occasionally found. Clouds of this type are of interest because they contain detectable abundances of the most complex molecular species in our Galaxy. Good examples are the galactic center clouds (in the direction of Sgr A and Sgr B2), the clouds associated with the Orion Nebula and the W51 region. Clouds in this category may contain both prestellar and stellar components.

(B) Isolated clouds not physically associated with excitation sources. There may be two subcategories of isolated molecular clouds; both are prestellar.

(1) The first category would include those of moderate mass ($< 10^3$ $M_\odot$) and kinetic temperatures of 50–100 K. The detectable molecules may include species such

as CN, CH, CO, OH and/or H_2CO but nothing of very complex chemical composition. Examples in this category would include the Perseus and Orion arm regions (Davies and Matthews, 1972) in the direction of Cas A and possibly the well-known ζ Oph cloud. The simple chemistry implies that the density distribution does not effectively support the formation or prevent the destruction of complex molecules.

(2) The second subcategory of isolated clouds would include dark nebulae which have 6–8 mag. of extinction and are fairly cool (15–50 K). Here the density distribution is probably high enough to support sophisticated interstellar chemistry but the low excitation makes detection difficult. Examples in this category would be Bok Globules and dark nebulae where H_2CO is observed in absorption against the 3 K background.

(C) Stellar Objects. In this case, molecular observations have shown three subcategories of objects, all of which are stellar.

(1) Mira-type variables. Radio observations show time-variable OH and H_2O signals which change in a semipredictible manner.

(2) Carbon stars with circumstellar shells. Simple but stable molecules such as CO, CN, HCN and CS have been detected in the direction of IRC + 10216 and several other stars with infrared excess. However, the chemical complexity is not yet as interesting as that of class A objects.

(3) 'Protostars'. This is a tentative classification for the high density knots within class A clouds. In some cases, these knots may correspond to recently detected infrared sources. Typically, the signal intensity of a complex molecule may double in the direction of a density knot and localized, intense H_2O and OH maser emission may be observed. Thus the high densities and unusual excitation suggest that regions of this type may be in the process of undergoing stellar formation.

The foregoing discussion illustrates that there is no such thing as a 'typical' molecular cloud but instead several cloud categories, each containing specific information. Possibly a solution to this problem would be a broad cloud classification scheme from which we would attempt to deduce the stage of evolution of interstellar clouds. In general, we would only be completing our picture of the stellar life cycle by trying to categorize the physical properties of dust clouds beginning sometime after stellar death and continuing to the initial stages of stellar formation. Thus we could think of a general cloud classification scheme as the other half of a very complete stellar evolutionary track in an HR diagram. An invaluable tool for this classification will be the kinds and abundances of molecules found in the clouds. In order to use interstellar molecules as a measure of cloud evolution, however, it is also critical to understand molecular formation processes.

4. Comments on Some Suggested Formation Processes

From the wide variety of molecules found (Tables I and II) it is probable that more than one formation mechanism is responsible for the observed species. Over the years numerous formation processes have been suggested which deserve some comment in light of molecular observations.

(1) Vapor-Phase Reactions. Two-body radiative association and inverse predissociation reactions have been used to predict (Solomon and Klemperer, 1972, Julienne et al., 1971; Stecher and Williams, 1973) the abundances of diatomics such as CH, CH^+, CN, OH and CO – usually for the interstellar cloud associated with ζ Oph. This type of calculation is important but it should be emphasized that the polyatomic molecules are not being found in the ζ Oph cloud and the difficult chemistry lies in addressing questions such as why isocyanic acid (HNCO) densities reach 10^{-5} cm^{-3} in the direction of Sgr B2 and why molecules such as acetaldehyde ($HCOCH_3$) and methylacetylene (CH_3C_2H) can be formed at all in the interstellar clouds. Current observational evidence suggests that formation mechanisms more efficient than vapor-phase reactions are needed to explain the abundances of the heavy polyatomics. However, in view of the extremely high H_2 densities inferred from studies of collisional excitation of microwave lines, it appears to be important to study reactions which utilize H_2 to form simple polyatomics such as H_2CO, H_2O and NH_3. In addition, the potential role of CH_4 in the formation of CH_3CN and CH_3C_2H as well as the importance of C_2H_2 to the formation of HC_3N and CH_3C_2H should be studied through vapor-phase formation models.

(2) Stellar Ejection of Molecules. Although this mechanism for molecular formation has been discussed in the literature many times, at present there is no radio evidence that this process contributes significantly to the observed abundances. There are, however, radio observations of several circumstellar shells (discussed by Buhl (1971), Buhl and Snyder (1971), Heiles (1971), NcNally (1968), Rank et al. (1971), Snyder (1972), Snyder and Buhl (1970), and Turner (1970); hence this process may be important for a very restricted class of objects (e.g. molecular cloud classifications Cl and C2, discussed previously).

(3) Dissociation of More Complex Molecules. It is possible that some of the molecular species are direct byproducts of more complex molecules. Carbon monoxide (CO), for example, can be formed by both vapor-phase reactions or by the photodistruction of formaldehyde (Glicker and Stief, 1971). So it is not unlikely that large accumulations of some of the structurally simpler interstellar molecules mark the 'graveyard' of more complex species. But the dissociation approach to molecular formation tends to beg the question of how the more complex parent molecules are formed.

(4) Reactions Involving Grains. Until recently, there was a tendency to downgrade the role of grains in the molecular formation process. Perhaps we are at a turning point in astronomical thinking regarding the interrelationship between dust grains and molecular formation when we see the papers presented at this meeting which quantitatively employ the grains as a source of interstellar molecules (Greenberg, 1972, 1973; Salpeter and Watson, 1973; Greenberg and Yencha, 1973). Introducing grains to explain the astrochemistry appears to be the most promising approach for satisfying the conditions beginning to emergy from the observations. Observationally we see strong (but not yet conclusive) suggestions of preferred reaction routes for molecular formation. Some of these are:

(a) Structural correlations – Some or all of these may be due to observational selection effects but it is noteworthy that numerous species of the form HCOX have been observed where X = H (formaldehyde), OH (formic acid), NH$_2$ (formamide), or CH$_3$ (acetaldehyde). Another structural correlation based on formaldehyde is of the form H$_2$CY where Y = O (formaldehyde), S (thioformaldehyde) or NH (methylenimine). At first glance, these structural correlations appear to support vapor-phase formation reactions of the form HCO + X → HCOX. However, such simple two-body reactions are not probable because radio observers have failed to detect the HCO radical.

(b) Molecular intermediates and unidentified molecules – molecular intermediates such as methylenimine (CH$_2$NH) and thioformaldehyde (H$_2$CS) are quickly destroyed by wall collisions in terrestrial laboratories. Unidentified compounds such as X-ogen and 'HNC' may be molecules with such short terrestrial lifespans that we may eventually classify them as 'nonterrestrial' molecules. Both classes of short-lived molecules suggest that low temperature matrix-phase chemistry (where a grain serves as the matrix) is important to interstellar formation.

(c) Negative results – not many negative observations are yet regarded as truly definitive, but it is worth mentioning that the failure to find simple compounds containing a nitrogen-oxygen bond (e.g. NO) lends substance to the idea that there are strongly preferred formation reactions for interstellar molecules. This perhaps would be surprising if vapor-phase reactions were the only allowed formation process.

5. Conclusion

In this review, I've attempted to relate observed molecular species and their abundances to the idea of cloud evolution. Before we can classify clouds and build meaningful models, we have to know more about the formation and excitation of interstellar molecules. The evidence is growing that interstellar dust grains are central to the understanding of molecular formation and hence play an important role in understanding cloud evolution. In turn, the molecules probably are our best tool for the quantitative analysis of the grain composition.

Acknowledgement

I wish to acknowledge partial support from National Science Foundation Grant GP-34200.

References

Buhl, D.: 1971, *Nature* **234**, 332.
Buhl, D. and Snyder, L. E.: 1971, *Tech. Rev.* **73**, 54.
Davies, R. D. and Matthews, H. E.: 1972, *Monthly Notices Roy Astron. Soc.* **156**, 253.
Glicker, S. and Stief, L. J.: 1971, *J. Chem. Phys.* **54**, 2852.
Greenberg, L. T.: 1973, this volume, p. 413.
Greenberg, J. M.: 1972, in M. A. Gordon and L. E. Snyder (eds.), *Molecules in the Galactic Environment*, Wiley Interscience, New York.

Greenberg, J. M. and Yencha, A. J.: 1973, this volume, p. 369.

Heiles, C.: 1971, *Ann. Rev. Astron. Astrophys.* **9**, 293.

Julienne, P. S., Krauss, M., and Donn, B.: 1971, *Astrophys. J.* **170**, 65.

McNally, D.: 1968, *Adv. Astron. Astrophys.* **6**, 173.

Rank, D. M., Townes, C. H., and Welch, W. J.: 1971, *Science* **174**, 1083.

Salpeter, E. E. and Watson, W. D.: 1973, this volume, p. 363.

Snyder, L. E.: 1972, in D. A. Ramsay (ed.) *MTP International Review of Science, Physical Chemistry Series One, Volume 3: Spectroscopy*, Butterworth & Co. Ltd., London, p. 193.

Snyder, L. E. and Buhl, D.: 1970, *Sky Telesc.* **40**, 267.

Solomon, P. M. and Klemperer, W.: 1972, *Astrophys. J.* **178**, 389.

Stecher, T. P. and Williams, D. A.: 1973, to be published.

Thaddeus, P., Kutner, M. L., Penzias, A. A., Wilson, R. W., and Jefferts, K. B.: 1972, *Astrophys. J. Letters* **176**, L73.

Turner, B. E.: 1970, *Publ. Astron. Soc. Pacific* **82**, 996.

ABUNDANCES OF INTERSTELLAR MOLECULES
AND THEIR FORMATION ON GRAIN SURFACES[*]

E. E. SALPETER and W. D. WATSON

Cornell University, Ithaca, N.Y., U.S.A.

Abstract. Surface phenomena on interstellar dust grains, which are relevant for molecule formation, are summarized. For various molecular species in the interstellar gas, the dependence of abundance on gas density and the degree of shielding of starlight is predicted. These predictions seem to fit with recent observations on carbon-monoxide, but there seems to be a discrepancy for formaldehyde.

1. Introduction

We report here briefly on two recent papers. In one (Watson and Salpeter, 1972a, hereafter referred to as WS1) surface phenomena on interstellar dust grains which are relevant for molecule formation were investigated; in the second paper (Watson and Salpeter, 1972b, hereafter referred to as WS2) we discussed destruction rates for molecules and predicted molecular abundances for the interstellar gas. This work is most relevant to 'normal' H I regions, i.e., to interstellar clouds of modest density and light to moderate shielding from starlight. We first summarize the results of these two papers and then discuss comparisons with more recent observational results. The relevance of our work to the very densest 'molecular clouds' is not yet clear; but we end with some pleas to the observationalists for further measurements.

2. Regions with Light or Moderate Shielding

Let n be the total (atomic plus molecular) number density of hydrogen in an interstellar cloud and τ_v the optical depth (in extinction) in the visible from the outside to the center of the cloud. The ratio of gas density (or of number-density of dust grains) to the density of attenuated UV photons from the general interstellar radiation field is approximately proportional to the parameter

$$\xi \equiv (n/100\,cm^{-3})e^{2.5\tau_v}. \tag{1}$$

We consider in this section clouds where neither the shielding τ_v nor the density are particularly great, so that $\xi < 10^4$ (as will be discussed subsequently, we do not understand how the molecules are prevented from condensing out of the gas into grains if $\xi > 10^4$).

Most atoms and molecules (such as N, O, CO, etc.) in the interstellar gas are expected to be neutral under normal 'H I-conditions'. For any such neutral radical or molecule hitting an interstellar dust grain, WS1 predict a high probability for it to

[*] Supported in part by the U.S. National Science Foundation Grant GP-26068.

stick to the grain surface long enough to form some kind of a molecule by recombining with another adsorbed radical. Saturated molecules involving hydrogen (CH_4, NH_3, H_2O, H_2S, etc.) are most simply made in this manner. However, reactions between a molecule in an excited state and another molecule on the surface of the grain can also produce more complex molecules.

Some of the molecules formed on a grain surface are ejected into the gas during the process of formation, but some fraction of them remain adsorbed to the surface. An important problem is then how the adsorbed molecules are eventually ejected from the surface. The growth of a *pure* dust-grain is still problematical, but the case of a molecule adsorbed to a grain made of material different from the molecule is actually less uncertain. For lightly shielded clouds the UV photons with energies between about 7 eV and 13.6 eV (915 to roughly 2000 Å) are probably the most efficient ejection mechanism. WS1 estimate that this mechanism should prevent the interstellar gas from condensing out onto grains as long as ξ in Equation (1) is less than about 10^4. The critical value of ξ for condensation is numerically uncertain by an order of magnitude or so, but experimental results from the Berkeley group (Greenberg, 1973) to be presented at this Symposium tend to corroborate at least qualitatively the efficiency of photo-ejection.

To predict molecular abundances for the interstellar gas from the formation rates on grain surfaces, estimates for the destruction rates for molecules in the gas phase are needed. For some molecular species (Solomon and Klemperer, 1972; WS2) destruction by exchange reactions (e.g., $OH + C^+ \rightarrow CO + H^+$) competes with photodissociation of the molecule. In such a case (e.g., for OH) the dependence of the predicted abundance on the value of ξ is then of the form

$$\frac{[OH]}{[H]} \approx \frac{a\xi}{1 + b\xi}. \tag{2}$$

For the particular case of OH (and assuming the cosmic abundance ratio of $[O]/[H] \approx 10^{-3}$) WS2 estimate $a/b \approx 2 \times 10^{-8}$ and $b \approx 5$. Observational data by Heiles (1971) and by Davies and Matthews (1972) indicate that $[OH]/[H] \approx 10^{-7}$ for $\xi \gg 0.1$ and that $[OH]/[H] \lesssim 2 \times 10^{-8}$ for $\xi < 0.04$. The data are consistent with the functional form of Equation (2) with $a/b \approx 10^{-7}$ and $b \approx 5$. This can be considered reasonable agreement in view of the uncertainties involved.

For clouds with moderately heavy shielding, $100 < \xi < 10^4$, the atomic carbon in the gas is neutral and photons beyond 11.3 eV energy are eliminated by its ionization edge, as well as by absorption from molecular hydrogen. This sharply decreases the photodissociation of CO, so that much of the carbon in the cloud is in the form of gaseous CO. Other molecules have lower photodissociation thresholds. The theory thus predicts a rapid increase of the CO-abundacne with increasing ξ and (through further grain reactions) increased abundances of complex molecules containing CO, such as formaldehyde and alcohols. Observations indeed indicate a high abundance of CO in dense clouds, but not a formaldehyde abundance (Heiles, 1971) as large as we predict theoretically for large ξ.

The theoretical situation is complicated for carbon compounds in lightly shielded clouds, $\xi \ll 100$, where the carbon atoms in the gas are mainly ionized. In these low density clouds the dust grains are likely to be positively charged (Watson, 1972), so that the rate at which C^+-ions stick to grain surfaces would be greatly reduced. The rate for surface-production of CH, CH_4, etc. should then also be greatly reduced. Lightly shielded clouds are the most favorable setting for the formation of CH^+ by radiative recombination in the gas phase from C^+ and H (Solomon and Klemperer, 1972). This gas phase reaction thus complements the surface reactions, but theoretically it is difficult to produce more complex carbon compounds in lightly shielded clouds. In particular, we seem to predict smaller formaldehyde abundances for small ξ than the observations (Davies and Matthews, 1972) indicate – opposite to the situation for large ξ. Further detailed observations of the molecular abundances in directions of little optical extinction would be most helpful.

3. General Discussion

We have mainly discussed so far the formation of molecules on the surfaces of interstellar dust grains (besides the radiative recombination of CH^+). One sometimes hears discussions on alternative, or 'rival', production mechanisms. We suggest here that one should also make a clear distinction between 'rival conditions' for molecule formation in interstellar space, since these conditions largely determine which of various rival production mechanisms should dominate.

We have already introduced a one-dimensional classification scheme for interstellar clouds, namely the parameter ξ in Equation (1). If one attempts to equate the optical depth τ_v with the observed mean visual extinction for passage through a cloud or a cloud complex, additional factors that affect the physical conditions are ignored. The molecular abundances in the interior of a cloud also depend on (i) the shape of the cloud (if it has a very irregular surface, more photons can penetrate into the interior), (ii) whether young stars are imbedded in the cloud complex, (iii) whether turbulence can produce shock-wave heating (Bar-Nun et al., 1970) or can 'turn the cloud inside out' within a relaxation time for condensation of gas molecules onto grains, etc. Our first plea to observers is to measure systematically abundances for various molecules in clouds with different density and shielding. Our second plea is to measure these abundances for spatial fine-structure, i.e., determine the variation of molecular abundances within a single cloudcomplex.

For a cloud-complex with moderate shielding, $\xi \sim \xi_0 \ll 10^4$, the theory would predict a greater abundance for many molecules in the deep interior where the shielding is greatest. Observational abundance ratios from the center to the edge of such a cloud will be welcome. Theory predicts the opposite difficulty for a very dense cloud complex with $\xi \gg 10^4$. There is no difficulty in making molecules in the most shielded regions but it seems that they should be condensed out as grain mantles. In this case molecules in the gas phase will be more abundant near the boundaries of the cloud complex. Thus it will be particularly interesting to study observationally whether there really is a correlation between abundances of some molecules and local H II-regions inside a large cloud complex. If the complex is highly shielded from the average interstellar radiation field but has some hot stars embedded in it, there should be a 'spherical shell' just *outside* the 'Strömgren sphere' surrounding each star where the hydrogen is neutral but some photons from the star between $\approx$ 7 eV and 13.6 eV are present. These photons could prevent the molecules from condensing out onto grains without destroying all the molecules. Hence molecular abundances could peak in such a 'spherical shell' around each star.

Molecule formation in the very dense, contracting nebula surrounding a proto-star (a generalized 'solar nebula') has been discussed by various authors (Herbig, 1971; Anders, 1972; Sagan, 1972). Molecule formation in the atmosphere of an existing, cool star will be discussed at this Symposium (Shimizu, 1973). There is certainly no difficulty in making molecules under these high-density conditions. It has been pointed out (Salpeter, 1971) that an 'average atom' passes through such conditions only about once in the life-time of the Galaxy, whereas it collides with dust grains a few hundred times during this period. For many molecules under 'average conditions' one would expect formation on grain surfaces to dominate, but this certainly will not be universally true and it will be useful to watch for characteristic differences in the molecular abundances predicted for formation under the various conditions.

For clouds with light and moderate shielding we have stressed the formation of molecules on grain surfaces (plus radiative recombination of CH^+), but we are really dealing with a cycle for which the UV photons (energies of about 10 eV) provide the 'driving force', i.e., the source of energy and entropy. These photons not only eject molecules from grains and destroy the molecules directly, but by maintaining an appreciable abundance of atoms or radicals in the gas they also insure indirectly that many electron volts of excitation energy are available for the surface reactions required to form some molecules. As a consequence, the ratio of the rates of formation of two different molecules with comparable numbers of constituent C, N, and O-atoms depends on the properties of the surface and does not involve a Boltzmann factor. Although it is difficult to predict this ratio, it might be near unity for some molecules even though their binding energies differ by a few eV. In contrast these abundance ratios exhibit a Boltzmann factor involving the difference in binding energies for molecule production in stellar atmospheres (Shimizu, 1973). For example, production in a stellar atmosphere with a temperature of $\approx$ 2500 K leads to a ratio $\approx 10^4$ when the binding energies differ by 2 eV – a representative energy difference (the actual observed ratio may be further increased by rearrangement collisions as the gas flows away from the star and cools).

In the 'solar-type nebula' surrounding a proto-star, reactions on grain-surfaces are also considered, but the details are drastically different because UV photons are absent and the kinetic temperature is a few hundred degrees. Instead of a helter-skelter production of many different molecules, one relies on a chain of catalytic reactions. This will single out a subset of all possible types of molecules, somewhat as in meteorites and in terrestrial Fischer-Tropsch reactions (Anders, 1972).

Finally we emphasize that the major uncertainty in understanding the large molecule abundances in dense clouds does not seem to be how and where the molecules are formed, but why they are not condensed onto the dust grains that cause the observed extinction.

References

Anders, E.: 1972, in *Proceedings of the Symposium on Interstellar Molecules* (Charlottesville, Virginia, October 1971), in press.

Bar-Nun, A., Bar-Nun, N., Bauer, S., and Sagan, C.: 1970, *Science* **168**, 470.

Davies, R. D. and Matthews, H. E.: 1972, *Monthly Notices Roy. Astron. Soc.* **145**, 249.

Greenberg, L. T.: 1973, this volume, p. 413.

Heiles, C.: 1971, *Ann. Rev. Astron. Astrophys.* **9**, 293.

Herbig, G. H.: 1971, in C. de Jager (ed.), *Highlights of Astronomy*, D. Reidel, Dordrecht.

Sagan, C.: 1972, in *Proceedings of the Symposium on Interstellar Molecules* (Charlottesville, Virginia, October 1971), in press.

Salpeter, E. E.: 1971, in C. de Jager (ed.), *Highlights of Astronomy*, D. Reidel, Dordrecht.

Shimizu, M.: 1973, this volume, p. 405.

Solomon, P. M. and Klemperer, W.: 1972, *Astrophys. J.* **178**, 389.

Watson, W. D.: 1972, *Astrophys. J.* **176**, 103 and 271.

Watson, W. D. and Salpeter, E. E.: 1972a, *Astrophys. J.* **174**, 321 (WS1).

Watson, W. D. and Salpeter, E. E.: 1972b, *Astrophys. J.* **175**, 659 (WS2).

Finally we emphasize that the major uncertainty in understanding the large-scale ... phenomena... clouds... not... to be ... and where the molecules ... are not condensed onto the dust grains that cause the observed opacities.

References

...

EXPLODING INTERSTELLAR GRAINS AND COMPLEX MOLECULES

J. MAYO GREENBERG

State University of New York at Albany

and

Dudley Observatory

and

ANDREW J. YENCHA

State University of New York at Albany

Abstract. The triggering of an explosive chemical reaction in a grain which consists of free radicals frozen in a matrix of photolyzed dirty ice material is seen as a possible source of complex interstellar organic molecules.

1. Introduction

The interstellar dust appears to play a critical role in the formation of the interstellar molecules. As is shown elsewhere in this volume the rate of formation of complex molecules directly out of the gas fails by orders of magnitude to compete with the rate of destruction in maintaining the observed abundances. Exactly what the mechanism of molecule production via dust may be is not clearly defined. But it seems clear that the chemical and physical character of the dust are critical in distinguishing between the relative importance of the various conceivable processes.

The distribution of the molecules under consideration is sufficiently far removed from stars that they can only have reached the regions in times much larger than their typical decay times as determined by the ultraviolet (or other disruptive) radiation.

A molecule with a radiative life limited to 100 yr in a radiation field like that suggested by Habing would travel, at a speed of 10 km s^{-1}, a distance of merely 10^{-3} pc before being destroyed. The existence of reasonably complex molecules well beyond this distance from the center of dark clouds seems to preclude the likelihood of their formation in the atmosphere of protostars.

Molecules may be formed on or in grain surfaces. The surface phenomenon has been a subject of both theoretical and experimental studies (Breuer, 1973). The possibility of molecules being an intrinsic part of the grain and appearing in space as the result of some grain disruption process or event is the focal point of this investigation. Both laboratory as well as theoretical investigation bearing on this process are considered.

2. Radiation Effects on Dust

For our basic dust model we consider those interstellar grains which have formed by

Greenberg and Van de Hulst (eds.), Interstellar Dust and Related Topics, 369-373.

accretion of O, C. and N atoms in combination with hydrogen to form a dirty ice mantle on some core – generally a silicate.

The significance of radiation effects on the dust is determined by the time it takes for all the molecules in a dust grain to have absorbed a photon compared with characteristic dust lifetimes which are of the order of 10^7 yr. The number of photons impinging per second on a dust grain of radius a is

$$\frac{dN_{h\nu}}{dt} = \frac{\pi a^2}{h} \int_{912}^{\lambda_t} u_\lambda d_\lambda \qquad (1)$$

which, for the distribution we are using $u\lambda = 40 \times 10^{-18}\, \text{erg cm}^{-3}$, $912 > \lambda > 2000$ Å, (approximated from Habing, 1968), gives

$$\frac{dN_{h\nu}}{dt} = 30\,\pi a^2(\lambda_t^2 - 912^2), \qquad (2)$$

where λ_t is in ångströms.

Thus for $\lambda_t \leq 2000$ Å the number of photons impinging on a 0.1 μ size grain is

$$\frac{dN_{h\nu}}{dt} \leq 33 \times 10^3\,\text{s}^{-1}. \qquad (3)$$

If we assume that the absorption efficiency of the grains is of the order of one, the total time for every molecule to have been subjected to a bondbreaking photon is $(dN_{h\nu}/dt/n_{mol})^{-1}$ where n_{mol} is the number of molecules in the grain. For an average molecular size $\simeq 2$ Å, we find the saturation time to be of the order of 130 y. This is indeed very short compared with 10^7 yr.

As a consequence of the ultraviolet irradiation of the interstellar grains a number of phenomena take place (Bass and Broida, 1960), among which are the formation and trapping of free radicals, and the formation of complex organic molecules. This latter problem has been investigated by many workers largely inspired by the initial work on the prebiogenic synthesis of organic molecules in the Earth's atmosphere (see R. M. Lemon, 1970 for a bibliography). The very low temperature solid state conditions prevailing in interstellar grains have not been considered in this connection for obvious reasons. However, low temperature solids have sufficiently low diffusion rates that they are most ideally suited for the trapping of free radicals.

The variety of possible processes to be considered are: (1) Creation of free radicals by either bond-breaking or ionization (we will limit ourselves here to bond-breaking), (2) immediate recombination of the original molecule, (3) migration of the free radical, (4) recombination of different radicals, (5) combination of radicals with molecules. The latter process is like polymerization.

Schematically the system is represented by (as modified from Morawetz, 1960):

radical production $m \rightarrow r$ rate: p,

growth of large unsaturated molecules: $M + r \rightarrow r$ rate: $k_g\,[M]\,[r]$,

termination to saturated molecules: $r + r \rightarrow S$ rate: $k_t\,[r]^2$, where $[M]$ and $[r]$ are the number of reacting molecules and free radicals.

In steady state

$$\frac{d[r]}{dt} = p - k_t[r]^2 = 0 \tag{4}$$

$$[r] = (p/k_t)^{\frac{1}{2}}.$$

The rate of consumption of 'reactive' molecules is

$$\frac{d[M]}{dt} = (k_g/k_t^{\frac{1}{2}})[M]p^{\frac{1}{2}} \tag{5}$$

and the average number $\bar{s}$, of 'reactive' molecules in the large molecules, S, is given by

$$\bar{s} = -\frac{d[M]/dt}{d[S]/dt} = fk_g(k_t p)^{-\frac{1}{2}}[M], \tag{6}$$

where f is a fraction varying between $\frac{1}{2}$ and 1.

We have already shown that $p \approx 2.5 \times 10^{-10}$. In order to achieve a reasonable concentration of free radicals we see that the recombination rate must also be of this order. This rate will depend on the molecular composition already achieved because it will be significantly less as the molecule and radical sizes increase to produce reduced diffusion rates. In some cases ESR measurements have indicated that, subsequent to irradiation, free radical concentration remained unchanged for a month (Bijl and Rose-Innes, 1955) implying quite small values of k_t.

The two extreme possibilities for a grain are that either it maintains a high concentration of free radicals or it forms predominantly saturated complex molecules. It should be noted that from a purely theoretical point of view the absolute maximum of free radical concentration cannot be greater than about 10% (Jackson and Montroll, 1958). Experimentally the maximum values achieved are less than 1%. However, there does not appear to be sufficient data or theory at the present to exclude the possibility of substantial free radical concentration in the grains sufficient to give rise to exotic optical properties (Greenberg et al., 1971). In any case the formation of large organic molecules is certain to take place during irradiation. It is interesting to note that if a grain containing 1% free radical concentration is triggered to ignition of a chain chemical reaction, the total energy released is sufficient to evaporate the grain.

With this process of grain disruption by evaporation we may anticipate the presence of a wide range of complex organic molecules, many appearing originally with extremely high molecular weights. Subsequent degradation of the initially very large molecules would occur as a result of photon disruption. On this basis the molecules observed would be the resulting products of large molecule breakup rather than the growth from smaller molecules either in space or on grain surfaces. The equilibrium density of molecules is then given by equating the creation rate for grain disruption with the destruction rate by radiation. We have then equality between

$$\left(\frac{dN_{mol}}{dt}\right)_{creation} = n_{mol}N_d/\tau_d$$

and   (7)

$$\left(\frac{dN_{mol}}{dt}\right)_{destruction} = N_{mol}/\tau_{mol},$$

where n_{mol} = number of molecules per dust grain, N_d = number of dust grains per cm^3, τ_d = disruption lifetime for dust and τ_{mol} = radiative lifetime of a typical molecule, and N_{mol} = number of molecules per cm^3 in space.

Using a grain size of 0.1 μ and a ratio of dust to hydrogen density of 10^{-2} we get

$$\tau_d = 4 \times 10^{-3} \tau_{mol} N_H/N_{mol}.$$ (8)

In dense clouds, where molecules are found, τ_{mol} may be of the order of 10^4 yr or greater and $N_H/N_{mol} \approx 10^{-8}$. For these values τ_d is of the order of 4×10^9 yr which is extremely large compared with the formation and growth lifetimes of dust, and would not deplete the dust density significantly.

Photochemical experiments have been made on simulated interstellar materials. In the initial stages of our studies we have irradiated solid mixtures of water, methane (or ethane), and ammonia with non-ionizing radiation between 1400 and 2000 Å wavelength and have looked at the product distribution using a mass spectrometer. The experiments are carried out by preparing various mixtures of gaseous water, methane or ethane and ammonia in 'cosmic abundance' proportions of oxygen, carbon, and nitrogen, and then depositing the mixture on a cold finger in a cryostat. After depositing approximately 20 mg of the mixture onto the cold finger, which results in a 'dirty ice' pattern of about 25 mm in diameter, the radiation source (Cary hydrogen arc lamp fitted with a sapphire window) is turned on. The pressure in the system is maintained below 10^{-6} torr throughout the irradiation process. We have performed experiments at 28, 42, and 77 K with irradiation times varying between 4 and 6 h. The experiments at 77 K required the use of ethane instead of methane since methane has an appreciable vapor pressure at this temperature. After irradiation is complete the cold finger is allowed to warm up to room temperature and the sample cryogenically pumped into an evacuated tube and subsequently injected into a mass spectrometer (AEI ms 902 Cl) for analysis. The mass spectra of the samples clearly indicate the presence of high molecular weight (mass > 106 and possibly some with mass > 200) organic material. In order to insure against organic contamination we have employed a mercury diffusion pump with liquid nitrogen cold trapping between it and the mechanical pump and the system itself. Even so, we routinely run blanks before each experiment to check for possible contamination. In general, two blanks are taken. One consists of depositing a sample onto the cold finger and then removing it without any radiation. The second blank obtained by irradiation of the cold finger without any sample and subsequent collection of any material. Both of these blank samples are run on the mass spectrometer immediately prior to the analysis of our irradiated sample and the results compared with the mass spectrum of the irradiated sample.

The results obtained thus far indicate the presence of numerous organic materials, the total amount of products being of the order of 10 μg (implying a photon efficiency of the order of 0.1 %.) We have not been able to make positive identification of these species. The degree of complexity of the final composition of the material does not appear to depend sensitively on the temperature and the implication is that complex products will be produced at temperatures of 10 K or less with efficiencies which make the process astronomically important.

At present we are conducting our experiments using nitrogen-15 ammonia, deuterated methane, and deuterated water in further attempts to clarify our results. In addition, we have interfaced a gas chromatograph (Hewlett Packard 5750) with the mass spectrometer to enable us to separate our products prior to being mass analyzed. This should prove to be very helpful in our attempts to identify the irradiation products.

References

Bass, A. M. and Broida, H. P., (eds.): 1960, *Formation and Trapping of Free Radicals*, Academic Press, New York.

Bijl, D. and Rose-Innes, A. C.: 1955, *Nature* **175**, 82.

Breuer, H. D.: 1973, this volume, p. 399.

Greenberg, J. M., Yencha, A. J., Corbett, J. W., and Frisch, H. L.: 1972, *Mem. Soc. Roy. Sci. Liège*, 6ème Sér. 3.

Jackson, J. L. and Montroll, E. W.: 1958, *J. Chem. Phys.* **28**, 1101.

THE SURPRISING CONSTANCY OF
RELATIVE CHEMICAL ABUNDANCES IN DIFFERENT REGIONS
OF INTERSTELLAR SPACE – PARTICULARLY
FOR FORMALDEHYDE

CARL HEILES

Astronomy Dept., University of California, Berkeley, Calif., U.S.A.

and

M. A. GORDON

NRAO, Charlottesville, Va., U.S.A.

Abstract. Many of the places in which radio astronomers search for molecules are located in exotic regions lying near the galactic center, near H II regions or other continuum sources, or in high density regions. Such regions have wide variations in their physical environment. Here we wish to compare two types of region whose environments are thought to be well-understood. One of these is dust clouds; the other is unshielded (from UV) cold clouds in front of bright radio sources. For the sake of interest, we include without comment results for molecular regions near the Galactic center and near Orion A.

Comparison of dust clouds and unshielded clouds is of interest because the volume density differs by a factor of 10^4 or more and the extinction differs by 8 mag. or more. Neither type of cloud lies near H II regions or other sources of intense radiation. Even though their properties differ widely, we will see that the abundances of OH and H_2CO relative to hydrogen are approximately the same in the two. Although this is understandable for OH, no current theory would predict this situation for H_2CO.

1. Dust Clouds

For dark dust clouds we have long known $\tau > 8$ (e.g. Bok, 1956; Heiles, 1968), and $n_{H_2} > 10^3$ cm^{-3} (Heiles, 1968). However, we now know that the density of molecular hydrogen must be at least the 10^4 cm^{-3} required to collisionally pump the CO molecule which has been observed in emission in dust clouds by Penzias *et al* (1972). The following paragraph argues that the density of H_2 is actually typically near 10^4 cm^{-3} rather than being much larger.

CO is not observed in every dark dust cloud. However, in those clouds showing an observable line its optical depth is always very large, of order 30 (with a large fraction of the cloud's carbon in CO). Thus dust clouds fall into two classes, one which shows a well-saturated CO line and one which shows no lines at all. There are no other apparent differences between the two classes. If a cloud shows no line, it could be either because it has no CO or because the H_2 density is too small to collisionally excite the CO. If the reason is related to the abundance of CO we would be forced to conclude that there are two classes of cloud in which the relative abundance of CO varies by a factor of over 50, with no intermediate cases. It seems much less unreasonable to assume that the H_2 volume density varies by an order of magnitude or so from one

Greenberg and Van de Hulst (eds.), Interstellar Dust and Related Topics, 375–379.

class to the other, the exact value being dependent on the state of dynamical evolution of the particular cloud. Since some clouds show CO and some don't, we would then conclude that the average H_2 density is about 10^4 cm^{-3}, with substantial fluctuations on either side of this value from one cloud to another.

The situation regarding relative abundances in dust clouds has been discussed by Heiles (1971), who assumed $n_{H_2} = 10^3$ rather than 10^4 cm^{-3}. Furthermore, the situation regarding OH and H_2CO was not well known at that time. Since then Heiles (1972, 1973) has derived H_2CO excitation temperatures and Heiles and Gordon (1973) have carefully observed OH, H I, and H_2CO in dust clouds, properly accounting for the differences in telescope beam size. We find that the relative abundance of OH and H_2CO varies substantially, by a factor of at least five, from one position to another within the same dust cloud (Cloud 2 of Heiles (1968) in the Taurus region). There is no correlation of the 21 cm line of H I (seen in self-absorption) with the OH or H_2CO lines. Therefore in speaking of relative abundances of OH and H_2CO we can only take some average value which seems characteristic of the cloud as a whole, realizing that the numbers vary. Furthermore, we should keep in mind that there is no way of measuring the H_2 density directly; thus, we are somewhat unsure of its constancy, and less its reliability. The adopted values of relative abundance, along with physical conditions, are given in Table I.

TABLE I

	Dust clouds	Unshielded cold clouds	Galactic center[b]	Orion A
Hydrogen volume density	$\simeq 10^4$(H_2)	$\simeq 20$ (H I)	$> 10^3$(H_2)	10^5(H_2)[c]
Optical depth to center	> 4	$\simeq 0.5$	?	?
ζ^a	10^6	1	?	?
[OH]/[H or H_2]	$\sim 10^{-7}$	$\sim 10^{-7}$	$< 5 \times 10^{-6}$	3×10^{-10}[d]
[H_2CO]/[H or H_2]	$\sim 3 \times 10^{-10}$	$\sim 2 \times 10^{-9}$	$< 2 \times 10^{-7}$	2×10^{-9}[c]
[H_2CO]/[OH]	$\sim 2 \times 10^{-3}$	$\sim 2 \times 10^{-2}$	$\simeq 4 \times 10^{-2}$	6

[a] See text and Watson and Salpeter (1972).
[b] See Rank, D. M., et al. (1971)
[c] See Thaddeus et al. (1971).
[d] See Goss (1968).

2. Unshielded cold Clouds in Front of Bright Radio Sources

The observational situation has been recently reviewed by Davies and Matthews (1972). Several distinct clouds at different velocities typically lie in front of these sources. Total optical extinctions are reasonably well-known and the edge-to-edge optical extinction per cloud is usually less than 1 mag. Hydrogen volume densities are not too uncertain because the sources have been observed interferometrically in the 21 cm line (see Clark, 1965), and of course column densities of all species are determined easily given a reasonable estimate of the excitation temperature. The possibility that H_2 exists in these regions is poor from a theoretical standpoint (Hollenback et al., 1971), although no observations exist; we neglect this possibility.

Davies and Matthews (1972) find a possibly significant dependence of relative molecular abundance on the hydrogen column density. However, since their observed range of molecular abundance amounts to less than the variation found in dust clouds mentioned above, we question the statistical significance of their derived variation. More sensitivity in measuring weak molecular adsorption features in the sources should decide the issue, and one of us (CH) is presently conducting such measurements. We have taken representative values from Davies and Matthews (1972) and list them in Table I along with the values for dust clouds.

3. Discussion: Comparison with Watson and Salpeter (1972)

The differences in density and UV extinction between the two types of region compared herein are explicitly and succinctly summarized in terms of the parameter introduced by Professor Salpeter earlier today:

$$\xi = \frac{n_H}{100} e^{2.5\tau_{vis}},$$

where n_H is the atomic or molecular hydrogen volume density cm^{-3} and τ_{vis} is the visible extinction to the cloud center in magnitudes. The value of this parameter is about 1 for the unshielded cold clouds in front of radio sources, and about 10^6 for dark dust clouds (although as pointed out by Watson and Salpeter (1972), this value may be an overestimate if the clouds are not spherical in shape).

The parameter ξ is a measure of the ratio of the binary collision rate to the intensity of UV starlight. Relative abundances will correlate well with this parameter if indeed these processes are the relevant ones for molecular formation and destruction. The data of Table I indicate that the relative abundances do not correlate with this parameter, but instead are approximately the same for unshielded and shielded clouds; furthermore, the higher relative abundance of H_2CO in unshielded clouds, if real, is quite surprising. The only existing theory which might possibly account for these abundances is that of Watson and Salpeter (1972).

3.1. OH

Watson and Salpeter (1972) conclude that formation on grains followed nearly instantaneously by UV photoejection is the most efficient production mechanism in unshielded ($\xi \cong 1$) and moderately shielded ($\xi < 10^4$) clouds. Destruction occurs mainly by exchange reactions with C^+ if carbon is mainly ionized ($\xi < \xi_0 \cong 100$) and otherwise by exchange reactions with other species. Thus the equilibrium relative abundance of OH is independent of density within the two regimes of ξ. For $\xi < \xi_0$, $[OH]/[H] \cong 3 \times 10^{-8}$; for $\xi > \xi_0$, $[OH]/[H] \cong 2 \times 10^{-7}$. For $\xi > 10^4$, UV radiation becomes too weak for photoejection so that grains are expected to accrete molecules with a resultant decrease in molecular abundances relative to hydrogen.

Observational data in Table I are not drastically different from these theoretical predictions. The apparent discrepancy for unshielded cold clouds should not be taken seriously because the derived OH abundance varies linearly with the assumed excitation temperature. The data in Table I from Davies and Matthews (1972), was derived assuming an excitation temperature of 10 K. This value is equal to the upper limit derived by Rogers and Barrett (1968) for the OH in front of Cas A. and is therefore certainly too large (but not by more than a factor of three!).

3.2. H_2CO

Watson and Salpeter (1972) find that H_2CO formation occurs efficiently only on grains. The rate of this process is limited by the rate at which a saturated molecule and a radical can find each other on a grain surface, since they are usually much more likely to instead encounter an H atom on the surface. However, if H_2 preponderates in the gas phase it will also preponderate on the grain surface and since radicals are much less likely to react with H_2 the formation rate of H_2CO is enhanced under such conditions. For large values of ξ CO radicals hitting dust grains will likely form H_2CO by two H-attachment reactions. Therefore, under all conditions the formation rate is proportional to the collision rate. Since destruction occurs by interaction with UV, the equilibrium abundance is given by

$$[H_2CO]/[H \text{ or } H_2] \cong 10^{-9} \xi\eta,$$

where η is an efficiency factor accounting for the above-mentioned difficulty with hydrogen atoms on the grain surface. For unshielded regions ($\xi < 30$) $\eta \cong 0.03$, while for moderately shielded regions η can be taken as unity. This leads to relative abundances $[H_2CO]/[H]$ of 10^{-11} for unshielded regions and 10^{-5} for the dust cloud in Table I (taking $\xi = 10^4$ instead of 10^6 to generously allow for a possible nonspherical dust cloud (see Watson and Salpeter, 1972)).

Here the discrepancy between theory and observation is serious. It cannot be ascribed to uncertainties in the excitation temperatures. For the unshielded clouds, Davies and Matthews (1972) assumed an excitation temperature of 3 K for the H_2CO, which is certainly correct to within a factor of two (Roberts and Gordon, 1971). The excitation temperature for H_2CO in dust clouds is known, in some cases, to within 20% (Heiles, 1972). Furthermore, the smaller relative abundance of H_2CO in dust clouds as compared with that in unshielded clouds is completely contradictory to theoretical expectations; although one might wish to explain it on the basis of grain accretion in clouds with $\xi > 10^4$, one would then expect that OH would be similarly depleted, which is just not the case.

In order that the H_2CO abundance be so large in the unshielded region, where the volume density is small, its formation process must be concerned with some agent other than binary collisions. The photo-destruction rate in unshielded regions is so large (Steif *et al.*, 1972) that the lifetime of a given molecule is less than 100 yr. It is therefore tempting to conclude that UV is also responsible for forming these molecules, since its intensity varies by a factor of about 10^4 between dust clouds and the unshielded clouds. Perhaps interaction of UV with grains can form H_2CO at a sufficient rate. If so, grain lifetimes would be limited to $\sim 10^8$ yr in typical regions of interstellar space; alternatively, perhaps the grains could be only a temporary source of supply, their ability as molecule sources being restricted to a special surface condition.

The problem of the large H_2CO abundance in unshielded regions is not new; a comparison similar to the present one was made earlier by Heiles (1971). It is a vexing one, however, because it illustrates our lack of ability to make even qualitative predictions about some molecular abundances. It is interesting to speculate on the possibility that, given the data in Table I, the process which is most important in H_2CO formation occurs only outside of dust clouds and that the H_2CO observed at present in dust clouds was formed long ago and is 'frozen in'. Present abundances would then be more a result of historical accident than statistical equilibrium at the present. The large variations in $[OH]/[H_2CO]$ found by Heiles and Gordon (1973) would not be inconsistent with this picture. If this is the case, our inability to detect H_2CO and OH in unshielded clouds which do not happen to lie in front of radio sources would be due to its excitation temperature rather than its low relative abundance.

Finally, we note that OH is more abundant relative to hydrogen in our two classes of cloud than in Orion A, and not necessarily less abundant than in the galactic center. The relative abundance of H_2CO in Orion A is the same as in our clouds, again, the situation with the galactic center is unclear. It seems that the old concept that dust clouds, H II regions, and the galactic center sources are regions in which prodigous molecule formation occurs – i.e., 'molecule factories' – must be abandoned, at least for OH and H_2CO.

References

Bok, B. J.: 1956, *Astron. J.* **61**, 309.
Clark, B. G.: 1965, *Astrophys. J.* **142**, 1398.
Davies, R. D. and Matthews, H. E.: 1972, *Monthly Notices Roy. Astron. Soc.* **156**, 253.
Gordon, M. A. and Roberts, M. S.: 1971, *Astrophys. J.* **170**, 277.
Goss, W. M.: 1968, *Astrophys. J. Suppl.* **151**, 131.
Heiles, C.: 1968, *Astrophys. J.* **151**, 919.
Heiles, C.: 1971, *Ann. Rev. Astron. Astrophys.* **9**, 293.
Heiles, C.: 1972, in M. A. Gordon (ed.), *NRAO Molecule Symposium*, October, 1971.
Heiles, C.: 1973, *Astrophys. J.*, to be submitted.
Heiles, C. and Gordon, M. A.: 1973, *Astrophys J.*, to be submitted.
Hollenbach, D. J., Werner, M. W., and Salpeter, E. E.: 1971, *Astrophys. J.* **163**, 165.
Penzias, A. A., Solomon, P. M., Jefferts, K. B., and Wilson, R. W.: 1972, *Astrophys. J.* **174**, 143.
Rank, D. M., Townes, C. H., and Welch, W. J.: 1971, *Science* **174**, 1083.
Rogers, A. E. E. and Barrett, A. H.: 1968, *Astrophys. J.* **151**, 163.
Stief, L. J., Donn, B., Glicker, S., Gentieu, E. P., and Mentall, J. E.: 1972, *Astrophys. J.* **171**, 21.
Thaddeus, P., Wilson, R. W., Kutner, M., Penzias, A. A., and Jefferts, K. B.: 1971, *Astrophys. J.* **168**, L59.
Watson, W. D. and Salpeter, E. E.: 1972, *Astrophys. J.* **175**, 659.

CYANOACETYLENE: NEW LINES AND SOURCES
IN THE GALAXY

M. MORRIS and PATRICK PALMER*

University of Chicago, Chicago, Ill. 60637, U.S.A.

B. E. TURNER

*National Radio Astronomy Observatory** Green Bank, W.V. 24944, U.S.A.*

and

B. ZUCKERMAN

University of Maryland, College Park, Md. 20740, U.S.A.

Abstract. Microwave emission from the $J = 8 \to 7$ through the $J = 11 \to 10$ rotational transitions of HC_3N has been detected in the directions of $Sgr(NH_3A)$, Sgr B2, and Orion A. A map of the $J = 8 \to 7$ emission around Sgr B2 shows that the emitting region extends at least 2′ in galactic longitude and 3′ in galactic latitude. Column densities may be comparable to those of HCN in some sources. This molecule is thus more abundant in the interstellar medium than would be expected from its complexity, and may provide clues to formation processes where it is found.

With a reasonable model for excitation of the rotational levels, we have used the ratio of antenna temperatures of these 2 lines to derive a value for the molecular density in the region where the lines originate. For Sgr B2, this yields $N_{H_2} \approx 10^6$, and for $Sgr(NH_3A)$, $N_{H_2} \approx 6 \times 10^4$.

The pursuit of information about physical conditions in opaque dust clouds has proceeded primarily through surveys of radio frequency molecular lines. In this presentation, we suggest that cyanoacetylene is especially useful as a probe of these clouds because of its large abundance and large number of detectable rotational lines. Four previously unobserved rotational lines of $HC_3N(H-C \equiv C-C \equiv N)$ were detected, with at least one line appearing in four of the eight sources surveyed. The implied abundances of cyanoacetylene are large enough that a study of its chemistry and formation processes may supply needed clues about why large interestellar molecules exist in detectable quantities.

Astrophysically speaking, cyanoacetylene is a relatively large molecule. With four heavy atoms, it has the greatest number of large atoms seen together in a single molecule. However, it is a linear molecule with a simple ladder of rotational energy levels corresponding to end over end rotation just like a diatomic molecule. The resulting rotational spectrum of HC_3N is uncomplicated and happens to be quite accessible to present radio receivers.

The molecule was discovered as an interstellar constituent in 1970, when Turner (1971) observed the $J = 1 \to 0$ line in the direction of Sgr B2. His search for the same transition in many other sources yielded negative results, probably because of the

---

* Alfred P. Sloan Foundation Fellow.

** The National Radio Astronomy Observatory is operated by Associated University, Inc., under contract with the National Science Foundation.

relatively low optical depth of this line for most excitation conditions. The three new sources reported here from observations of higher lying transitions are on his list of negative results. With various assumptions Turner derived a column density of 2×10^{16} cm^{-2} for Sgr B2. For comparison, this is larger than ammonia and an order of magnitude larger than formaldehyde in the same source.

We report observations of four successive rotational lines of HC$_3$N:

$$J = \begin{array}{rll} 8 \to 7 & \text{at} & 72.8 \text{ GHz} \\ 9 \to 8 & \text{at} & 81.3 \text{ GHz} \\ 10 \to 9 & \text{at} & 91.0 \text{ GHz} \\ 11 \to 10 & \text{at} & 100.1 \text{ GHz.} \end{array}$$

The observations were made in February, May and June, 1972 with the Kitt Peak 36′ radiotelescope of the National Radio Astronomy Observatory. At least one of these lines was detected in four of the eight sources studied. Full details of the observations will be published later.

In Sgr B2, all four lines were seen and although a complete analysis remains to be done, several things can be readily deduced. First, a density of at least 10^6 particles cm^{-3} is needed to produce the observed $J = 9 \to 8$ and $8 \to 7$ line intensities if

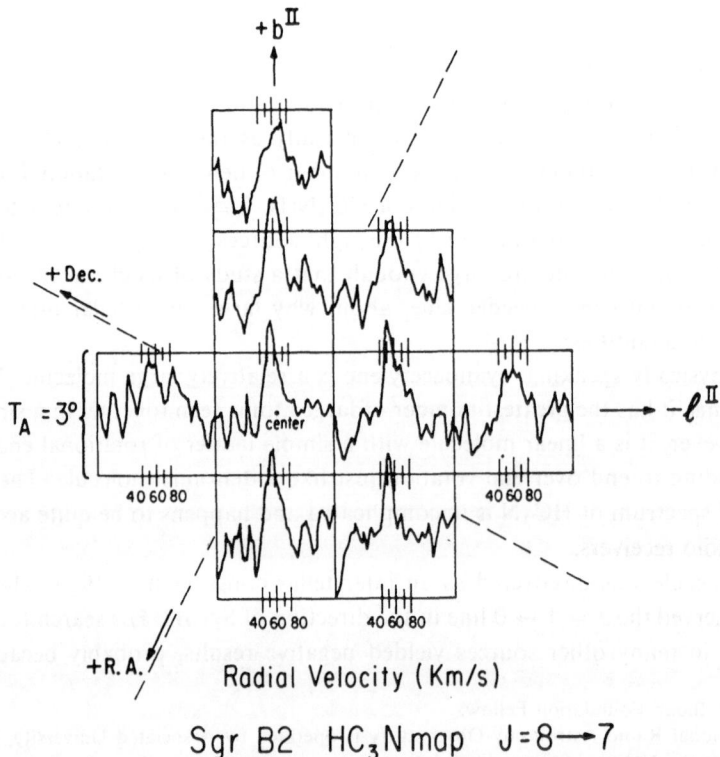

Sgr B2   HC$_3$N map $J = 8 \longrightarrow 7$

Fig. 1.   The $J = 11 \to 10$ transition of HC$_3$N in Sgr B2. *Ordinate*, antenna temperature; *abscissa*, radial velocity with respect to the local standard of rest. The velocity resolution is 3.6 km s^{-1}.

collisional excitation is the principal factor in populating the levels. Second, either the $J = 11 \to 10$ transition is 'anomalously' strong or the $J = 10 \to 9$ is 'anomalously' weak. Figure 1 shows the $J = 11 \to 10$ line in Sgr B2. It is displaced in the spectrum in order to facilitate a search for another molecular candidate. A density much higher than 10^6 cm^{-3} is required to significantly populate the $J = 11$ level by collisions. Perhaps another mechanism is at work here, as this line was also relatively strong in Orion A and possibly the only line seen in W51. The excitation is apparently not characterized by thermal populations. However, few interstellar molecules have shown thermal excitation. Statistical equilibrium analyses are almost always necessary, and are definitely needed in this case.

A limited map with points spaced by one arcminute intervals was made of the 72.8 GHz emission in Sgr B2. (Figure 2). The full beam width at half power for the 36 ft telescope at 73 GHz is about 95″, so the emission is clearly extended over a region of about $2' \times 3'$ and perhaps even larger. Given the high molecular hydrogen densities implied by the line ratios at the center point of the map, simple assumptions lead to a total cloud mass of $\geq 10^7$ solar masses. This lower limit can perhaps be increased further when similar and more complete surveys are made in all of these rotational lines. Observations of ammonia (2,1) inversion radiation (Zuckerman *et al.*, 1971) lend support to this surprisingly large total mass for this source. Simplifying assumptions allow calculations to be made which show that for this lower limit to the mass, the cloud is unstable to gravitational contraction. A second feature of our map is that

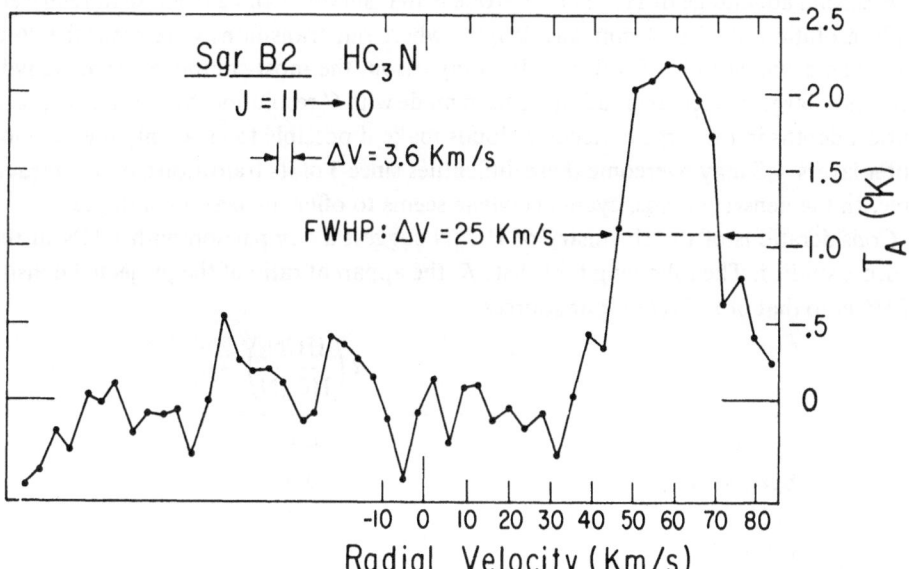

Fig. 2. Map in galactic coordinates of the $J = 8 \to 7$ emission of HC$_3$N in Sgr B2. The spectra displayed in each square are spaced by 1′ in each direction: galactic longitude increases from right to left, and galactic latitude increases fro bottom to top. Dashed lines show the equatorial axes for reference. The velocity resolution is 6.2 km s^{-1}.

the radial velocity changes from 54 km s^{-1} to 71 km s^{-1} in the direction of increasing galactic latitude. The same general velocity distribution has been noticed for other large molecules such as methyl alcohol in this source (Turner *et al.*, 1972).

The other three sources containing detectable amounts of cyanoacetylene are Sgr (NH$_3$A), Orion A, and W51. Projected densities were calculated assuming that the rotational levels are thermalized at 50 K. They range from 3×10^{13} cm^{-2} in Orion A to 3×10^{14} cm^{-2} in Sgr B2. These values are lower by a large factor than a previous estimate based on the $J = 1 \to 0$ transition in Sgr B2 (Turner, 1970) suggesting that the $J = 1 \to 0$ transition may be inverted. Deviations from thermal excitation in the states which we have observed will probably not affect the projected density estimates significantly.

Our observations make it apparent that HC$_3$N will be quite useful for obtaining a detailed picture of the physical conditions in large interstellar molecular clouds. First, with the plethora of available lines, the excitation of this molecule can be almost completely determined. This would present an opportunity for ascertaining the collisional and radiative contributions to the excitation, and thus very useful estimates of density, kinetic temperature, and radiation temperature throughout the source. The $J = 5 \to 4$ line will soon be searched for at Kitt Peak, and if it is seen, then a total of 8 of the 11 lowest lying rotational transitions will have been detected. This situation can be compared to that of ammonia, where the inversion radiation from within several rotational levels has been seen, although transitions between the rotational levels are inaccessible and at present undetected.

A second advantage of HC$_3$N as a probe is that surveys can be made with relatively high resolution. At 3 or 4 mm wavelengths where our transitions were seen, the 36-ft telescope beamwidth is of order 1'. In comparison, the most complete formaldehyde surveys of large molecular clouds have been made with 6' resolution, but the tremendous optical depths in the large molecular clouds make it possible to view only the extreme outer layers. CS may overcome these difficulties since 3 of its transitions are now tractable. In the densest clouds, cyanoacetylene seems to offer the best advantages.

Considerations of the chemistry of HC$_3$N suggest a comparison with HCN in the sources studied. The following table lists R, the apparent ratio of the projected density of HCN to that of HC$_3$N in four sources:

$$R\left(\frac{\text{HCN}}{\text{HC}_3\text{N}}\right)$$

Sgr B2	0.1
Sgr (NH$_3$A)	0.6
W51	0.8
Orion A	6.8

It is assumed that the $J = 1 \to 0$ HCN transition is thermalized and that the level populations fall off rapidly above the $J = 1$ state. Because the excitation of these molecules is uncertain, these ratios should be regarded as rough estimates. The significance of these values lies in showing that the abundance of HCN relative to cyano-acetylene may vary by almost 2 orders of magnitude. This is possibly a density effect related to either the chemistry or the excitation of these molecules. Under non-inter-stellar conditions at least, an appreciable fraction of HCN can be converted to HC_3N in the presence of acetylene via a surface reaction, though more realistic experiments remain to be done before this mechanism for formation can be presumed.

We also consider more saturated hydrocarbons related chemically to cyanoacetylene. Vinyl cyanide results with the addition of 2 hydrogen atoms. We searched for the $3_{13} \rightarrow 2_{02}$ transition of this molecule without success, though little can be said about the existence of a molecule from a single negative result.

Finally, we note the possibility of studying isotopic variations. The preferential placement of C^{13} in any of the three slots in HC_3N would be an important clue to the formation process, but this is expected to be small and probably awaits the advent of highly sensitive receivers in the millimeter-wavelength range. The rotational tran-sitions of $H^{12}C^{13}C^{12}CN$ and $H^{12}C^{12}C^{13}CN$ are nearly coincident in frequency. The velocity width of these lines in several sources causes a frequency overlap of the lines of these two species. This increases the optical depth by a determined amount. Ideally, observations of the three C^{13}-substituted variations will thus yield the optical depth. The C^{13}/C^{12} isotope ratio can then be obtained in sources where the line is strong such as the galactic center where this ratio is presently in question.

In summary, these observations suggest the potential use of cyanoacetylene for the study of both the physics and chemistry of large molecular clouds. The molecules which have accessible radio frequency transitions will collectively be excellent probes for interstellar conditions in regions that are optically obscured.

References

Turner, B. E.: 1971, *Astrophys. J. Letters* 163, L35.
Turner, B. E., Gordon, M. A., and Wrixon, G. T.: 1972, *Astrophys. J.* 177, 609.
Zuckerman, B., Morris, M., Turner, B. E., and Palmer, P.: 1971 *Astrophys. J. Letters* 169, L105.

CHEMICAL EVOLUTION OF DENSE CLOUDS

E. W. CHAPELLE, B. D. DONN, W. A. PAYNE, JR., and L. J. STIEF

Astrochemistry Branch, Laboratory for Extraterrestrial Physics,
NASA/Goddard Space Flight Center, Greenbelt, Md. 20771, U.S.A.

Abstract. Thermal reactions largely determine the chemical evolution of interstellar molecules during various phases associated with the evolution of interstellar clouds. A similar mechanism may be the source of the array of observed polyatomic molecules. If astronomical time scales are comparable to or less than chemical time scales, thermodynamic equilibrium will not be attained. Other factors also affect the validity of thermodynamic calculations of molecular composition. An experimental investigation of thermal reactions in a simplified system is under way. Preliminary results of the HCN–H_2 reaction at 1000 K indicate reaction times of several hours. These have been extrapolated to lower temperatures and fitted to models of the primordial solar nebula, proto stars and red giants. In these situations equilibrium would generally not be attained and reaction products would freeze out at temperatures near 700 K or greater.

1. Astronomical Problem

The origin and subsequent evolution of the complex array of molecules found in dense, heavily obscured clouds present a variety of significant astronomical and chemical problems. Several investigations, e.g., the Liege Symposium, *Stellar Evolution Before the Main Sequence*, (1970); Kaplan and Pikelner, (1970); and recent papers by Cameron (1969, 1972) have indicated that in many cases such clouds are gravitationally unstable and undergoing collapse. This process is the accepted view of the formation of young star clusters and planetary systems. During the collapse phase high temperatures and pressures occur, densities from 10^{10}–10^{16} cm^{-3} and temperatures to 2500 K appear likely. As the infall of matter ceases, the cloud temperature falls and the cloud becomes much colder again. It is during this latter stage that planet accumulation would primarily take place (Larimer, 1970; Anders, 1971).

This paper considers chemical processes that could determine the molecular composition of the cloud during the several stages of its evolution. In particular we examine thermal reactions at the relatively high interstellar densities expected during the process.

Three distinct circumstances occur to which our analysis applies.

1.1. THE ORIGIN OF THE OBSERVED MOLECULES

(A) If the molecules preceded collapse and star formation, the molecule formation mechanism is very uncertain but thermal processes are unlikely. However, there is a tendency for many of the observed polyatomic molecules in Orion to be concentrated in small regions associated with the infra-red source (Snyder and Buhl, 1971; Penzias *et al.*, 1971; Barrett *et al.*, 1971). In a paper at this symposium, Martin and Downes (1972) presented evidence for several regions of star formation in Sag B2. These are

Greenberg and Van de Hulst (eds.), Interstellar Dust and Related Topics, 387–393.

associated with dense, high temperature ionized regions which act as continuum sources.

(B) A possible source of the molecules in these complex cloud arrays is in expanding atmospheres of newly formed stars with subsequent ejection into the cloud. A combination of expansion and turbulence set up by the onset of radiation of hot OB stars may well be a mechanism to spread the molecule through a large volume of the surrounding cloud. Shimizu (1973) and in this symposium has considered the atmospheres of newly formed stars as molecular sources and attempted to calculate the composition. In this hypothesis thermal reactions would be a major source of molecules. The question arises as to the molecular composition when the temperature and density become so low that reactions effectively cease.

1.2. MOLECULAR EVOLUTION DURING COLLAPSE

According to hypothesis 1.1B, few (if any) polyatomic molecules would be present in the cloud during collapse. However, during the subsequent cooling stage, as temperatures drop from about 2000 K and densities are high, thermal reactions must occur. The cooled nebula would have a complex composition. Several detailed equilibrium calculations have been made for the primordial solar nebula (e.g. Urey, 1952; Lord, 1965; Larimer, 1967; Lewis, 1972). Closely related to this process is the evolution of precollapse molecules in hypothesis 1.1A. In this situation a complex array of molecules reacts thermally during the heating phase. At temperatures of 2000 K or above and densities of 10^{-3} to 1 atm equilibrium would be quickly established. As the cloud cooled, the composition would continuously change. We inquire as to the composition to be expected when the temperature has fallen to the value occurring when, in the primordial solar nebula, the various condensation and accumulation processes occurred (Anders, 1971). In particular, we consider the validity of thermodynamic equilibrium calculations of molecular composition.

1.3. THE ROLE OF ORGANIC MOLECULES IN PLANETARY CHEMICAL EVOLUTION

Although this aspect of molecular phenomena is somewhat removed from the main argument of the conference, non-thermodynamic equilibrium effects may be significant and some consideration will be given to it. Pre-collapse molecules may not survive the high temperature stage in their initial array but a different array should appear during the cooling phase as outlined in 1.2. Pre-collapse molecules may not exist but, as in hypothesis 1.1B, they arise in the primordial stellar nebula following star formation. Under any circumstance it thus appears that an array of organic compounds is present when planets accumulate around newly formed stars. What happens when massive, planetary objects form? At the final stage of accumulation the solid material will have accreted into a hierarchy of planetesimals (Savronov, 1964, 1966). These would impact on cool planetary surfaces causing local high temperatures but with rapid cooling. Organic compounds may tend to be destroyed as Sagan (1965) suggested but as in hypothesis 1.2, the composition of the vaporized constituents after cooling may well contain a disequilibrium excess of organic molecules. Oro's (1965) proposal for the initiation of biochemical evolution may have considerable merit according to this analysis.

As a model for the astronomical aspects of the collapse we use Cameron's (1972) recent analysis of the formation of the solar system. A massive cloud collapses and fragments into units containing up to ten solar masses. The Sun and planets form from a two solar mass fragment. Within 1 AU of the Sun the temperature reaches about 2000 K and cools with a time scale of ten years. At 10 AU the temperature is much less than a thousand degrees with a cooling time of 100–1000 yr.

2. Chemical Problems

2.1. THEORY

Molecular equilibrium calculations for cool stellar atmospheres, interstellar clouds, and primordial stellar and planetary nebula have assumed thermodynamic equilibrium. For several reasons the applicability of thermodynamic equilibrium to these objects is open to question. (1) Variations in temperature and pressure will invalidate such analyses if the time scale for attaining equilibrium is of the order of, or less than the time scale for astronomical changes. A determination of each time scale must be made. (2) Equilibrium calculations only yield results for those species that are initially included in the program. It does not predict species. Unexpected or non-equilibrium species regardless of concentration would not appear in the results nor would their absence cause any obvious effects.

In order to investigate the question of the rate of approach to equilibrium and the ultimate molecular composition we have begun a series of experiments at high temperatures. Because of the continuously changing temperature and pressure, equilibrium will not be attained in most cases except perhaps at the highest temperatures. An extremely complex set of reactions would determine the composition at each stage. As the temperature decreases during the cooling phase, each reaction and molecular species will behave differently. Chemical processes will be frozen out and metastable products will tend to be preserved. Because of these problems with a theoretical calculation, this experimental study was begun. This procedure develops its own set of difficulties which must be overcome. Experimental techniques are being developed to accomplish this.

The first series of experiments are designed to determine the reaction rate as time scale for approach to equilibrium and its temperature dependence. The measurement of an overall activation energy will permit extrapolation of the reaction rates to lower temperatures where rates are too slow for experimental determination.

2.2. EXPERIMENTS

Mixtures of HCN and H_2 in ratios of 1:100, 1:000, 1:3000 were prepared and heated in quartz reaction vessels at 1093 K for various times. The starting pressure at room temperature was approximately 300 torr. At the operating temperature, the pressure in the vessel was about 1.3 atm. As the final composition did not noticeably depend upon the $HCN:H_2$ ratio, the experiments were run at the 1:100 ratio to enhance product yield and improve the accuracy of the results. After the nominal duration of heating the furnace was turned off and allowed to cool. The gradual cooling process had a significant effect on the shortest (2 h) heating time only.

The fraction of HCN remaining is a measure of the completion of the reaction. Equilibrium calculations (Tsuji, 1964) show that at equilibrium the residual HCN concentration would be 0.1 % at 1090 K and 0.06 % at 1170 K. Figure 1 shows the residual HCN percentage as a function of heating time. From Figure 1, we obtain a half life (e^{-1}) of 18 h. $HCN:H_2 = 1:100$ mixtures were also reacted at 1170 K and 980 K. The decay times are listed in Table I for the three temperatures. The effective activation energy for the reaction, obtained from the dependence of reaction rate on temperature, is 20–25 kcal mole^{-1}. A value this low is unexpected as activation energies for reactions of stable molecules of this type generally fall between 40–100 kcal mole^{-1} (Bahn, 1968; Kondratiev, 1972).

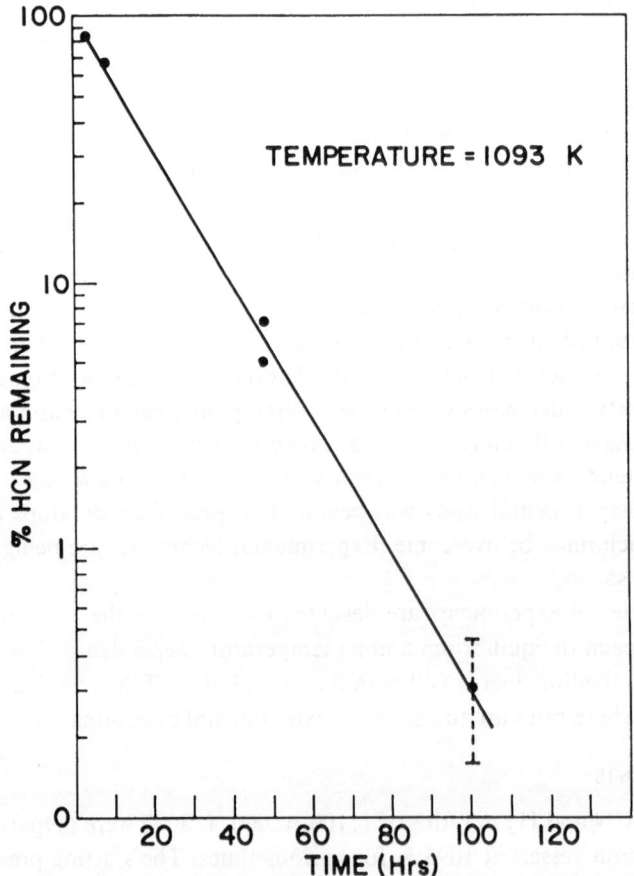

Fig. 1.   Dependence of $HCN-H_2$ reaction on time. The dashed vertical bar at 100 h is the uncertainty of measuring that quantity of HCN on the mass spectrometer.

A few runs were made with an oxygen or water concentration equal to that of hydrogen cyanide in the mixture. The essential result, the HCN decay time, was similar to the previous case.

There is a possibility that surface reactions made a significant contribution to the observed rate. A quartz vessel was filled with 20 quartz tubes, thereby increasing the surface to volume ratio by a factor of 9. In 10.5 h, the HCN:H_2 mixture at 1090 K showed a 2% HCN residue whereas from Figure 1, the anticipated residue was 55%. This indicates a major surface reaction contribution which would explain the low activation energy derived from the experiment. It also means that the 18 h half-life is a lower limit.

An additional factor affecting the experiment is contamination by gases driven from the walls of the reaction vessel at the temperatures of 1000 K over several hours. Significant hydrogen loss via diffusion through the walls occurs as well. For example, in forty eight hours 20% of the hydrogen was lost.

Work on these high temperature reactions related to molecular equilibrium in astrophysical systems is continuing in order to refine the experiments and extend the results to more systems and conditions.

2.3. APPLICATION TO CLOUD EQUILIBRIUM

We return to the question of the approach to thermodynamic equilibrium in a collapsing or cooling cloud or in an expanding stellar atmosphere or shell. Hydrogen cyanide is one of the more abundant interstellar molecules and also is an important reactive precursor in biochemical evolution (Calvin, 1969). Consequently the behavior of this compound not only gives some indication of the approach to equilibrium in a cloud but also yields information on the ultimate concentration of an important prebiotic specie.

The dependence of the rate of the HCN–H_2 reaction on surface to volume ratio indicates that the experimental half-life of 18 h at 1090 K is a lower limit. In Table I we use this value and extrapolate to lower teperatures using activation energies of 20 and 50 kcal mole^{-1}. The higher value yields results that may be more applicable in the absence of extensive surface reaction. Further experiments will seek to establish the gas phase activation energy.

Table II shows that if the effective activation energy were as low as 20 kcal mole^{-1}, equilibrium would take approximately 300 yr at 500 K. In Cameron's latest model of the solar nebula this temperature is reached between 5 and 10 AU. His calculated life-

TABLE I

Half life of HCN–H_2 reaction (measured)

T (K)	$\tau(e^{-1})$ (h)
1170	9
1093	18
980	53

times are comparable to the chemical lifetime raising a question of the approach to thermodynamic equilibrium. Half lives calculated using an activation energy of 50 kcal mole^{-1} suggest that there would only be sufficient time to approach equilibrium within about 1 AU.

TABLE II

Half life of HCN–H_2 reaction (theory)

| T (K) | Activation Energy | |
	$E-20$ (kcal mole^{-1})	$E-50$ (kcal mole^{-1})
2000	0.27 h	2 s
1500	1.48 h	140 s
1000	40 h	147 h
900	120 h	101 days
800	20 day	94 yr
700	123 day	7.2×10^3 yr
500	270 yr	1×10^9 yr

In expanding atmospheres of red giant stars or in proto-stars the astronomical time scale is days or months. The present analysis indicates temperatures close to 1000 K would be required to attain a close approach to equilibrium.

Reactions caused by impact of planetesimals on the earth's surface near the end of its accumulation are very complex to analyze. The rapid change in temperature and pressure in the vapor suggest an extremely non-equilibrium final composition. The present experiments give some support to the idea that organic compounds may have been present on the surface of the newly-formed Earth.

There is some observational evidence that suggest organic compounds were present when objects accumulated in the solar nebula. A number of carbon radicals, CH, CN, C_2 and C_3, occur in cometary atmospheres. More complex parent compounds were almost certainly present in the primordial nebula from which comets accreted. If not originally present in the inner region, molecules like HCN could have formed in the non-equilibrium impact reactions.

The chemical theory and experiments are preliminary and the astronomical models are also very uncertain. Hence, no definite conclusions can be reached. The present analysis does point out that non-equilibrium processes may become very important in determining the molecular composition of cosmic clouds. This possibility must be taken into account in drawing conclusions from calculated compositions.

A continuing study of this problem is under way. We will investigate more thoroughly the limits of validity of thermodynamic equilibrium in the astronomical systems considered here. Procedures for estimating molecular compositions when thermodynamic equilibrium do not apply are also being examined.

References

Anders, E.: 1971, *Ann. Rev. Astron. Astrophys.* **9**, 1.
Bahn, G. S.: 1968, *Reaction Rate Compilation for the H–O–N System*, Gordon and Breach, N.Y.
Barrett, A. H., Schwartz, R. P., and Waters, J. W.: 1971, *Astrophys. J.* **168**, L101.
Calvin, M.: 1968, *Chemical Evolution*, Oxford University Press, N.Y.
Cameron, A. G. W.: 1969, in S. S. Kuman (ed.), *Low Luminosity Stars*, Gordon and Breach, N.Y., p. 423.
Cameron, A. G. W.: 1972, preprint submitted to Icarus.
Kaplan, S. A. and Pikel'ner, S. B.: 1970, *The Interstellar Medium*, Harvard Univ. Press. Cambridge, Mass.
Kondratiev, V. N.: 1972, in R. M. Fiestrom (ed.), *Rate Constants of Gas Phase Reactions-Reference Book* (transl. by L. J. Holtschlag), Office Standard Reference Data, N.B.S., National Technical Information Service, Washington, D.C., U.S.A.
Larimer, J. W.: 1967, *Geochim. Cosmochim. Acta* **31**, 1215.
Lewis, J. S.: 1972, *Icarus* **16**, 241.
Liège Astrophysics Symposium: 1970, 'Stellar Evolution before the Main Sequence', see particularly introductory reports and references, Univ. of Liege, 1970.
Lord, H. C.: 1965, *Icarus* **4**, 279.
Oro, J.: 1965, in S. W. Fox (ed.), *The Origin of Prebiological Systems*, Academic Press, N.Y., p. 137.
Penzias, A. A., Solomon, P. M., Wilson, R. W., and Jefferts, K. B.: 1971, *Astrophys. J.* **168**, L53.
Sagan, C.: 1965, in S. W. Fox (ed.), *The Origin of Prebiological Systems*, see discussion following paper by Oro, p. 102, Academic Press, N.Y.
Savronov, V. S.: 1964, *Techtonophysics* **1**, 217.
Savronov, V. S. (1966) Soviet Astron. AJ **9**, 987.
Shimizu, M.: 1973, *Prog. Theor. Phys.* **49**, 153.
Snyder, L. E. and Buhl, D.: 1971, *Astrophys. J.* **163**, L47.
Tsuji, T.: 1964, *Ann. Tokyo Astron. Obs.* **9**, 1.
Urey, H. C.: 1952, *The Planets*, Yale Univ. Press, New York.

FORMATION MECHANISM FOR INTERSTELLAR MOLECULES

PAUL HARTECK, ARTHUR BEAUDOIN, and ROBERT REEVES

Chemistry Dept., Rensselaer Polytechnic Institute, Troy, N.Y. 12181, U.S.A.

Abstract. The gas phase production of molecules in dust clouds in interstellar space may occur largely as H_2 formation on the interstellar grains, with relatively minor formation in the gas phase. The formation of more complex molecules, however, may be augmented by gas phase processes including (1) negative ions formed by radiative attachment of electrons, (2) direct radiative association of molecules and atoms, and (3) reaction of atoms and molecules with H_3^+ molecules.

1. Introduction

The formation of molecules in interstellar space may be over a variety of different types of processes (McNally, 1968) which may be considered in two different categories, one for gas phase reactions and the other on interstellar grains. The latter type has been discussed in some detail (Watson and Salpeter, 1972). The gas phase reactions are probably playing a minor, but possibly crucial role in molecule formation by augmenting the production of the larger, more complex molecules. The simple molecular hydrogen may be largely produced on interstellar grains and evaporated into space. The more complex molecules if formed may not be so readily evaporated from the cold grains.

2. Discussion

In the region of the interstellar clouds, the density of hydrogen may be $10^{4 \pm 1}$ molecules per cubic centimeter and an order of magnitude less, about 10^3, for atomic hydrogen. Both H^+ and H_2^+ will be the major ions primarily formed with a typical steady state concentration of positive ions of about 0.05 per cubic centimeter ($= 0.05$ electrons cc^{-1}). These may be formed by photo ionization, cosmic rays, or other processes, but are probably produced at about equal rates, possibly more H_2^+, because of the greater amount of H_2 compared to H.

Under the rarified conditions the chemistry of the H_2^+ is interesting. The main reaction anticipated would be with H_2.

$$H_2^+ + H_2 \rightarrow H_3^+ + H \qquad k_1 = 6 \times 10^{-10} \tag{1}$$

Followed by:

$$H_3^+ + e^- \rightarrow 3H \qquad k_2 = 10^{-7}. \tag{2}$$

The rate of formation of H_3^+ can be balanced with the rate of loss:

Greenberg and Van de Hulst (eds.), Interstellar Dust and Related Topics, 395–398.

$$\frac{+dH_3^+}{dt} = k_1(H_2^+)(H_2) = \frac{-dH_3^+}{dt} = k_2(H_3^+)(e^-)$$

or

$$\frac{(H_3^+)}{(H_2^+)} = \frac{k_1(H_2)}{k_2(e^-)}.$$

Assuming $(H_2) = 10^4$ and $(e^-) = 0.05$, then

$$(H_3^+)/(H_2^+) = \frac{6 \times 10^{-10}}{10^{-7}} \times \frac{10^4}{(0.05)} = \sim 10^3.$$

The loss of H_3^+ would be in the order of some years, while the H_2^+ would be a few days.

(The rate of loss of molecular ions A_2^+ or $A_3^+ + e^-$ vs the rate of loss of atomic ions $A^+ + e^-$ is high. So the H^+ ions may be of an order of magnitude greater density than H_3^+. However, we don't know all processes by which the H^+ will be consumed. The reaction $H + H_2^+ \rightarrow H_2 + H^+$ would not normally compete with reaction 1, since $H_2 > H$). We assume here conservatively that the $(H_3^+) \sim 10^{-3}$.

A series of reactions which produce molecules or molecular ions may be illustrated with the carbon atoms.

$$C + H \rightarrow CH + \hbar v \qquad k_3 \sim 10^{-21} \tag{3}$$

$$C + e^- \rightarrow C^- + \hbar v \qquad k_4 \sim 10^{-16} \tag{4}$$

$$C + H_3^+ \rightarrow CH^+ + H_2 \qquad k_5 \sim 10^{-9} \tag{5}$$

and the relative rates*:

$$\frac{-dC}{dt} \cdot \frac{1}{C} = k_3(H) = 10^{-21} \times 10^3 = 10^{-18}$$

$$= k_4(e^-) = 10^{-16} \times (0.05) = 5 \times 10^{-18}$$

$$= k_5(H_3^+) = 10^{-9} \times (10^{-3}) = 10^{-12}.$$

Here it can be seen that reaction (5) is several orders of magnitude faster than the other two processes. Once formed the CH^+ may react with the molecular hydrogen:

$$CH^+ + H_2 \rightarrow CH_2^+ + H \tag{6}$$

followed by

$$CH_2^+ + H_2 \rightarrow CH_3^+ + H. \tag{7}$$

(The subsequent analogous reaction to form CH_4^+ is endothermic and cannot occur under these conditions). The ion-electron reactions can then result in neutral CH radicals, e.g..

---

* As fraction of carbon atoms lost per second.

$$CH_3^+ + e^- \rightarrow CH + H + H. \quad (or\ CH_2 + H).$$

The CH^+ ion can combine with an electron under these conditions in about 10 yr ($CH^+ + e^- \rightarrow C + H$), while it will probably react with H_2 in a much shorter time – if $k_6 \sim 10^{-11}$, then with $H_2 \sim 10^4$, it would be a few months.

In themselves the positive ion reactions do not add new bonds, but rather act to rearrange these bonds. A reaction of the type (Rank et al., 1971) forms a new bond, but for atom-atom interaction, it is known to be generally very slow. The more complex associations may be much faster. The reaction:

$$SO + O \rightarrow SO_2 + \hbar\nu \tag{8}$$

has been measured and is between 10^{-15} and 10^{-16} (Rolfes et al., 1965). Rates of this order of magnitude or even faster may be expected for reactions of this complexity, i.e. forming triatomic or larger polyatomic atoms in association. If the association process were then about five orders of magnitude faster than that indicated for reaction (3), it could well compete with the H_3^+ reactions of type (5).

Similarly the electron attachment process should be considered for the more complex molecule where attachment occurs with a higher probability. Such processes in the gas phase can yield molecules which are of relatively low boiling point, but could not reasonably evaporate if formed on interstellar grains which have relatively low temperatures.

The formation of nitrogen containing molecules may again be mainly on interstellar grains. Negatively ion reactions, however, may also contribute to the production of certain species. The reaction of C^- with N_2 can produce the CN^- ion fragment:

$$C^- + N_2 \rightarrow CN^- + N.$$

This reaction may be exothermic if the higher value (73.8 kcal) (Neale, 1964) of the electron affinity of CN is found correct, rather than the 64 kcal value sometimes cited (Napper and Page, 1963). The CN^- might react to produce HCN via

$$CN^- + H \rightarrow HCN + e^-.$$

The source of additional CN and HCN may be from the N-atom attack on carbon compounds, well-known in the extensive literature on active nitrogen.

The reaction of H_3^+ with molecular N_2 is known to yield HN_2^+ (Burt et al., 1970), but this probably does not lead to any observed nitrogen compounds. Further studies by these workers may shed light on other important reactions and their mechanisms.

The positive ion reaction:

$$H_3^+ + N \rightarrow NH^+ + H_2.$$

Seems to be endothermic and may be neglected. The alternate products might be considered since the reaction is then exothermic:

$$H_3^+ + N \rightarrow NH_2^+ + H.$$

Although to our knowledge, this has not been observed.

Lind (1961) was an advocate of ion clusters as the explanation of many reactions in radiation chemistry. In his monograph, he quotes the paper by Magee and Funabaski (1959) which came to the conclusion that clusters in general don't account for the chemical reactions induced by ionizing radiation under laboratory conditions. However, after an exhaustive treatment of all types of binding energies, Magee and Funabaski conclude that "It is possible that specific ion-catalyzed reactions could be obtained between clustered molecules in the sense originally suggested by Lind."

At interstellar conditions, there will exist regions where the temperatures and particle concentrations are favorable for the formation of ion clusters. The clustering is obviously enhanced at low temperatures. If the cluster contains in the order of a few dozen particles then in the case of neutralization, the electrostatic forces will disappear and the energy of neutralization may 'evaporate' the clustered molecules or contribute to their activation as a complex. So larger molecules may be formed in the gas phase which could not originate by evaporation from a grain at low temperatures. In this respect, clusters would act as a hybrid of pure gas and grain surface reactions.

Magee and Funabaski emphasize that the experimental base is very minor in their theoretical considerations and the matter becomes even more complicated when we try to apply this theory to interstellar problems. However, it seems necessary and valuable to reconsider the cluster theory in the light of interstellar molecular formation.

3. Conclusion

The reaction types discussed seem to be adaptable to a series of chemical species found in interstellar space, and they are also in line with the observed amounts of these species, total pressures and particle concentrations in the clouds. A more detailed discussion will follow in another paper.

Acknowledgement

This work was supported by a grant from the National Aeronautics and Space Administration, NGL 33-108-007.

References

Burt, J. A., Dunn, J. L., McEwan, M. J., Sutton, M. M., Roche, A. E., and Schiff, H. I.: 1970, *Chem. Phys.* **52**, 6062.
Lind, S. C.: 1961, *Radiation Chemistry of Gases* (American Chemical Society Monograph No. 151), Reinhold Publishing Corporation, New York.
Magee, J. L. and Funabaski, K.: 1959, *Radiation Res.* **10**, 622.
McNally, D.: 1968, *Adv. Astron. Astrophys.* **6**, p. 173.
Napper, R. and Page, F. M.: 1963, *Trans. Faraday Soc.* **59**, 1086.
Neale, R. S.: 1964, *Phys. Chem.* **68**, 143.
Rank, D. M., Townes, C. H., and Welch, W. J.: 1971, *Science* **174**, 1083.
Rolfes, T. R., Reeves R. R., and Harteck, P.: 1965, *J. Phys. Chem.* **69**, 849.
Watson W. D. and Salpeter, E. E.: 1972, *Astrophys. J.* **174**, 321.

THE DISTRIBUTION OF INTERSTELLAR MOLECULES IN DUST CLOUDS AS DERIVED FROM LABORATORY EXPERIMENTS

H. D. BREUER

Institut für Physikalische Chemie II, Universität des Saarlandes, Saarbrücken, W. Germany

Abstract. Irradiation of adsorbed simple gases leads to the formation of rather complex molecules. On the basis of the wavelength dependence and the cross sections for formation and desorption an average molecule density can be estimated. The wavelength dependence further shows that molecules can be formed in denser regions where they are protected from dissociating radiation. A simple model is proposed which shows the molecule distribution as a function of dust density.

1. Introduction

In the last few years several mechanisms have been proposed for the formation of interstellar molecules. For some of the diatomic radicals and molecules radiative association and charge exchange reactions in the gas phase seem to be mechanisms which at least in special regions of the interstellar space give satisfactory results (Solomon and Klemperer, 1973). The formation of more complex molecules consisting of four or more atoms can for several reasons not occur in the gas phase. Dust grains are involved and the reactions seem to be catalytical and possibly photo-catalytical. The purpose of this paper is to apply some laboratory results for molecule formation in photoinduced reactions at solid surfaces on the possibilities of molecule formation in interstellar dust clouds. From the wave length dependence of the formation process and of the destruction of molecules by photo dissociation a qualitative model of molecule distribution in dust clouds can be deduced.

2. Experimental

The experimental set up we used for studying molecule formation in photoreactions at solid surfaces has been described in detail elsewhere (Breuer, 1969; Moesta *et al.*, 1969). Two different experiments were performed. In one experiment we investigated the wave length dependence of these photoreactions; in the other experiment the photo products were identified by a mass spectrometer.

The surface used in theses experiments were either metal films or metal foils, the metals being nickel, tungsten or rhenium. The gases introduced in the reaction system were H_2, (D_2), N_2O_2, CO, NH_3, CH_4, (CD_4).

These gases were adsorbed in various combinations and ratios. The adsorption systems were irradiated either by an exploding wire light source (Moesta and Breuer, 1965) or by resonance lamps. The experiments were performed at room temperature.

Measuring the wave length dependence we found the most effective spectral region

Greenberg and Van de Hulst (eds.), Interstellar Dust and Related Topics, 399–404.

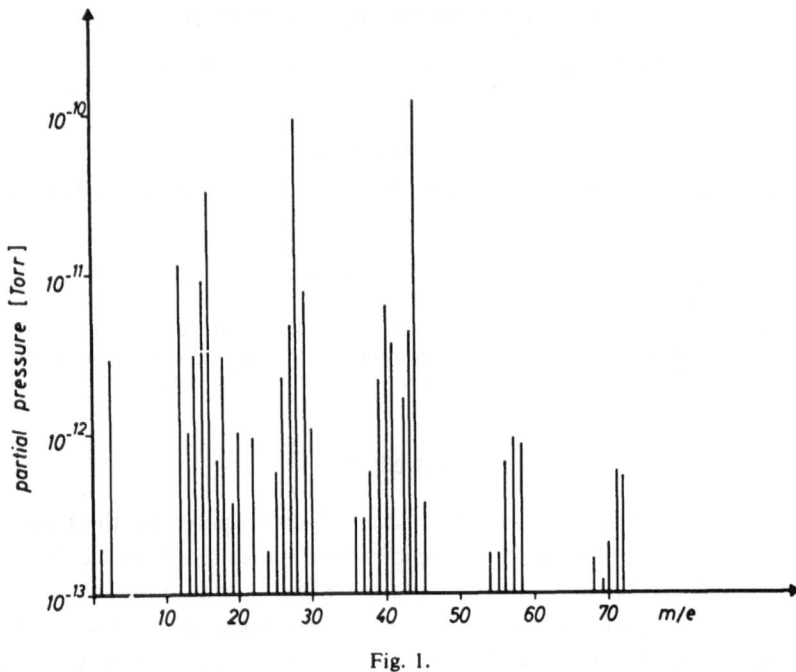

Fig. 1.

for changes in the chemical composition of the adsorption layer for all adsorption systems being between 2000 Å and 3000 Å.

As an example for the photo production of molecules Figure 1 shows the mass spectrum which is obtained by irradiating a mixture of CO and CH_4 adsorbed on tungsten. In this spectrum the peaks which are observed without irradiation are subtracted and only the photo peaks are shown. An analysis of this spectrum shows fragmentation peaks which are typical for the presence of aldehydes. Using CD_4 instead of CH_4 formaldehyde could be identified. Other photo products are hydrocarbon radicals and CO_2, C_2O, C_2O_2 and C_3O_2. Irradiating adsorption layers containing nitrogen we could identify the products listed in Table I.

For some molecules we were able to measure the cross sections for formation and desorption. These cross sections are shown in Table II.

3. Discussion

The experimental results show that by irradiating simple gases adsorbed at the surface of a solid rather complex organic molecules can be formed. The cross sections for formation and desorption indicate that photocatalytic reactions are a very operative mechanism for molecule formation. The formation of the observed molecules can only be understood if preadsorbed diatomic molecules, e.g. CO, are dissociated by the radiation. By the interaction of the adsorbed molecule with the surface CO can be electronically excited with $\lambda \sim 2000$ Å. From this excited state it dissociates to C and

TABLE I

Molecular weights

m/e	Molecule
26	CN
27	HCN
29	H_2H
30	NO, N_2H_2
38	C_2N
39	C_2HN
40	C_2H_2N
41	CH_3CN
43	HNCO
44	N_2O
46	NO_2
47	$HCONH_2$
51	HC_3N

TABLE II

Cross sections for formation
and desorption of molecules

$$\sigma_{CN} = 1.5 \times 10^{-18}\ [cm^2]$$
$$\sigma_{HCN} = 5 \times 10^{-19}\ [cm^2]$$
$$\sigma_{H_2CO} = 1 \times 10^{-16}\ [cm^2]$$
$$\sigma_{CH_3CN} = 1.5 \times 10^{-18}\ [cm^2]$$
$$\sigma_{HNCO} = 1 \times 10^{-18}\ [cm^2]$$
$$\sigma_{HCONH_2} = 1 \times 10^{-18}\ [cm^2]$$
$$\sigma_{HC_3N} = 5 \times 10^{-19}\ [cm^2]$$

O (Breuer, 1969). These atoms can react with neighboring H-atoms or already formed CH, CH_2 etc. to form the observed molecules.

Applying these laboratory results to the conditions in interstellar space seems to be risky since interstellar dust grains are certainly not made of pure nickel or tungsten. However, experiments using soil or pulverized vycor as substrates show that very similar processes occur on non-metal surfaces as well. The major result of our experiments is that all the observed photoreactions occur in the spectral region between 2000 Å and 3000 Å a region which is almost inoperative in normal photochemistry in the gas phase since all of the gases relevant for astrochemical purposes are transparent in this spectral region. The interstellar extinction curve, however, shows that this radiation can penetrate into the clouds at a higher rate than the radiation below 2000 Å. Since for most of the molecules dissociation starts at about 2000 Å appreciable life times can only be achieved if the molecules are shielded from the dissociating radiation by interstellar dust (Stief, 1971).

Calculations on the distances that molecules can cover during their life time show that they can only be formed in the regions where they are observed, i.e. in the dust clouds. If photocatalytical reactions are one of the processes responsable for molecule formation in interstellar space the conclusion which can be drawn from the experimental results should be at least in qualitative agreement with this fact.

The calculation of the molecular densities should in principle be possible by combining the experimental results with the calculated life times. The rate, R_i, at which molecules are formed and desorbed from the surface of a dust grain, is given by

$$R_i = N_{ads}\theta A \Sigma_{ph}\sigma_i \, [s^{-1}], \tag{1}$$

where N_{ads} is the number of available adsorption sites per cm^2, θ is the degree of coverage and is assumed to be unity in the following calculations. A is the area of the dust grain, Σ_{ph} is the photon flux $cm^{-2} s^{-1}$, and σ_i is the cross section for formation and desorption. Since the chemical composition of the adsorption layer is one of the factors which determine the way in which the reactions proceed, N_{ads} has to be divided up into the various gaseous constituents which can be involved in the reactions at the surface. The rate r_k at which k-atoms of k-molecules hit the surface and stick to it is given by

$$r_k = S_k u_k v_n A \, [s^{-1}]. \tag{2}$$

S_k is the sticking coefficient which for all atoms and small molecules in H I-regions is equal to or very close to unity, n_k is the number density of the species k and v_k the corresponding velocity. Using the abundances relative to hydrogen and putting v_k/v_H where v_H is the velocity of a hydrogen atom the composition of the adsorption layer can be estimated. However, if reactions proceed in the surface layer this composition can be changed in an unforseeable way since newly formed radicals and molecules can occupy a part of the surface sites. The number density of molecules produced by this reaction may be obtained by multiplying Equation (1) by τ_i, the life time of the molecule i and the number density of the dust grains.

Applying this equation to the conditions in interstellar space, one can assume an adsorption layer consisting of $\sim 10^{14}$ 'reactive particles' and a mean cross section of 10^{-18} cm^2. The other assumptions are: hydrogen density $n_H = 100$ cm^{-3}, dust density $n_d = 10^{-10}$ cm^{-3}, grain diameter $d_g = 0.1$ μ, grain surface $A = 3 \times 10^{-10} cm^2$, radiation density $U_\lambda = 4 \times 10^{-17}$ erg cm^{-3} $Å.^{-1}$ The photon flux in the spectral region between 2000 Å and 3000 Å is then 1.5×10^8 $cm^{-2} s^{-1}$. With an absorption in the visual of $A_v = 3$ the mean attenuation of the radiation is 10^{-3} and the corresponding mean life time is 10^{13} s (Stief, 1971). Inserting these numbers in the equation we obtain an integral molecular density of 4.5×10^{-6} cm^{-3}. This density has to be divided up into the different species according to their relative cross sections and life time.

As mentioned above the formation and the destruction of the molecules occur in different wave length regions. Photodissociation occurs in less protected regions mainly by photons having an energy of 6 eV or more, whereas the formation reactions need only photons in the 4 to 6 eV-region. Since both the formation and the destruction of the molecules are proportional respectively inversely proportional to the number density of the dust grains the following model is postulated for the distribution of molecules in dust clouds. Figure 2 shows the spatial distribution of a dust cloud. This distribution has not necessarily to be Gaussian as drawn in the figure. Due to the

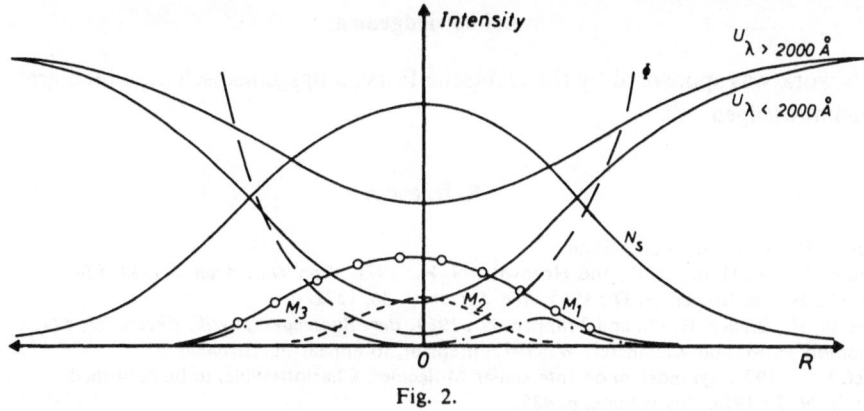

Fig. 2.

above mentioned proportionalities the model is valid for any shape of the cloud. The separation of the radiation in $\lambda > 2000$ Å and $\lambda < 2000$ Å is arbitrary. It shall only reflect the influence of the extinction curve on the radiation field. Since molecule formation is directly proportional to the dust concentration the highest molecule densitites can be expected in the center of the cloud. This is the case for the molecules M_1 and M_2. Having the same wave length dependence a higher concentration is obtained if $\sigma_1 > \sigma_2$. If photons of higher energies are necessary to form a molecule this species will be formed in a region of the cloud where the grain density permits the radiation to penetrate with sufficient intensity. Here the formation rate will be lower since in the outer shell of the cloud the grain density is lower. The formation in an outer shell of the cloud is shown for the molecule M_3. The spatial distribution of each molecular cloud is limited by the dissociation in the interstellar radiation field. The curve indicated by Φ is the integrated quantum yield for the photo dissociation of the molecules. Φ is unity in dust free regions and decreases as the molecules are protected from the dissociating radiation by the interstellar dust.

In this simple model the molecules M_1 and M_2 form concentric spheres and are mixed in the core of the cloud while in an outer shell M_1 and M_3 are mixed. Any irregular shape of the dust cloud will be reflected in the distribution of the molecules and M_1 for instance will resemble the dust distribution the same way as it resembles the gaussian shape of the dust cloud in Figure 2. Quantitative calculations of formation rates and molecular densities seem to be impossible at the moment for several reasons. One of them is that the chemical composition of the dust grains is still unknown although there is some evidence from recent results that silicates are the major constituents of interstellar dust clouds (Woolf, 1973). On these materials the reaction cross sections have to be measured at temperatures which are typical for interstellar grains. Calculating the number densities of interstellar molecules one also has to know the rates at which molecules are formed which do not have a permanent dipole momentum and thus can not be detected by radio spectroscopy. These molecules may consume a significant amount of the atoms and radicals which are available for reactions.

Acknowledgement

This work was sponsored by the Deutsche Forschungsgemeinschaft, which gratefully is acknowledged.

References

Breuer, H. D.: 1969, Thesis, Bonn.
Hubbard, J. S., Hardy, J. P., and Horowitz, N. H.: 1971, *Proc. Nat. Acad. Sci.* **68**, 574.
Moesta, H. and Breuer, H. D.: 1965, *Rev. Sci. Instr.* **36**, 1372.
Moesta, H., Breuer, H. D., and Trappen, N.: 1969, *Ber. Bunsenges. physik. Chemie* **73**, 879.
Solomon, P. M. and Klemperer, W.: 1973, preprint, to appear in *Astrophys. J.*
Stief, L. J.: 1971, Symposium on Interstellar Molecules, Charlottesville, to be published.
Woolf, N. J.: 1973, this volume, p. 485.

STELLAR ATMOSPHERES:
THE SOURCES OF INTERSTELLAR MOLECULES

MIKIO SHIMIZU

Institute of Space and Aeronautical Science, University of Tokyo, Komaba, Meguro-ku, Tokyo

Abstract. The analysis of the abundance of interstellar molecules in compact H II regions suggests that the molecules are formed in stellar atmospheres (possibly of protostars or of late type stars), transported to the location of the neutral clouds in the cpmpact H II regions, and shielded from decomposition due to stellar radiation by the dust in the clouds. Cometary nuclei and interstellar dust are argued from the astrochemical point of view to be dirty ice of the second kind (or a sort of frozen interstellar molecules). The chemical structure of the primordial solar nebula is discussed under the assumption that long-period comets consist of the most primordial substances of the solar system.

1. Introduction

Recently more than twenty kinds of interstellar molecules have been found by radio-astronomers in the compact H II regions of our galaxy. Emission lines from more than thirty kinds of molecules have been searched for but have not been detected. A careful survey on the physico-chemical character of each molecule might suggest the formation mechanisms of these molecules.

One of the most conspicuous feature of the interstellar molecules in the compact H II regions is that CO is so plenty as to contain almost all the C atoms in the neutral clouds in these regions. Furthermore, N atoms in the same clouds appear to be in the form of N_2 (which is difficult to be found in the radio wave region), because the amount of all the other detected nitrogen compounds such as CN, HCN, NH_3 etc. are very much smaller than the total amount of N atoms expected in these clouds from the 'cosmic abundance'. CO and N_2 are stable molecules at high temperature, say at 1500–4500 K (see Figure 1). This evidence suggests that interstellar molecules were formed in the stellar atmospheres. Excess of O atoms might form OH and H_2O at these temperatures. If the interstellar moleucles are formed on the dust grains in the neutral clouds, the most natural products under the conditions of low temperature (70–100 K) and of hydrogen rich environment might be CH_4 and NH_3, contrary to the above observational evidence.

It is easily found that the composition of cometary atmosphere resembles the abundance distribution of interstellar molecules in the compact H II regions: Even from the results disclosed *until present*, we know that OAO 2 has detected two kinds of new chemical species, H and OH, which might be expected in the compact H II region, at least in the form of H_2O. Emission bands of N_2^+ and CO^+ were already found in cometary tails. Some molecules found at optical wavelengths, such as C_2, CN, CH, NH, etc., may merely be the minor constituents of the cometary atmospheres. This evidence suggests that cometary nuclei were essentially frozen interstellar mole-

Greenberg and Van de Hulst (eds.), Interstellar Dust and Related Topics. 405–412.

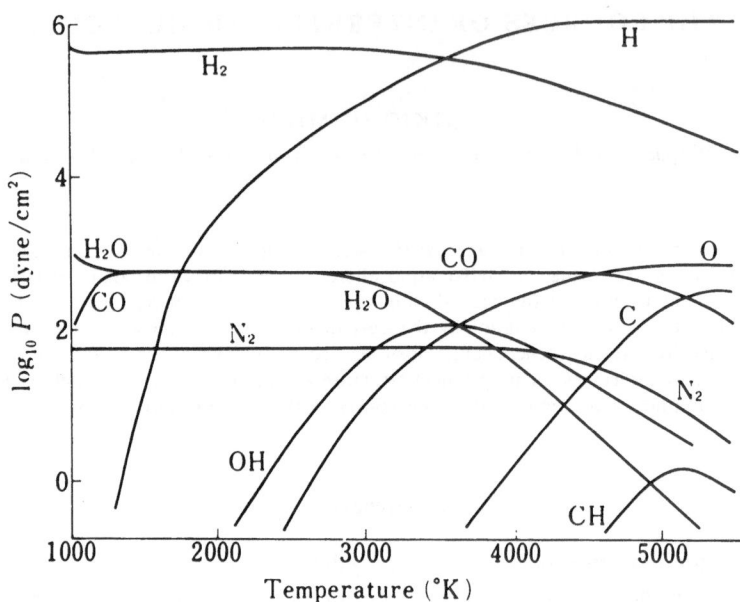

Fig. 1. Some examples of the abundances of various molecules at the temperatures of 1000–5500 K and at the pressure of 10^6 dyn cm^{-2}.

cules or 'dirty ice of the second kind' (the composition is H_2, CO, N_2, OH, H_2O, and some atoms, with some silicates and metallic compounds), in contrast to 'dirty ice of the first kind' (the composition is H_2O, CH_4, and NH_3 with some condensates) so far assumed.

Comets may be formed in the primordial solar nebula in the vicinity of Jovian orbit and be scattered to the present cometary reservoir at the fringe of the solar system by the perturbation of stars and giant planets (Oort, 1950). Or, they may be trapped from the interstellar space by the gravity of the Sun (Lyttleton, 1948). In these cases, the possible sources of cometary molecules may be the atmospheres of the protosun (more generally the prestellar atmospheres) and the atmospheres of late type stars, respectively, since these stars have some favorable properties for the ejection of molecules from their atmospheres to the dusty circumstellar envelopes, such as low temperature, high convection, and small gravities.

2. Abundances of Interstellar Molecules

At the high temperature such as that of stellar atmospheres, thermochemical equilibrium may easily be established. The advantage of thermochemical calculation is that it uses a few parameters, essentially temperature and pressure, to discuss the abundances of many kinds of molecules. We have calculated the abundances of molecules, including H, C, N, O, S, P, and Si atoms in them, at the temperature of 1000–6000 K and at the pressure of about 1 atm. After an averaging procedure over the mass distribution of stars in a galactic cluster, the calculated distributions were found to be in good agreement with the observed abundances (Shimizu, 1973). Here we shall refer to only a few examples which are necessary for the discussions in the following sections or the recent results that we have discussed in this symposium.

(1) Minor constituents of cometary atmospheres such as C_2, CN, NH, CH etc. are most stable at around 4000–5000 K, the effective temperature of the protosun (Narita *et al.*, 1970). At this temperature, the amounts of atoms are comparable with those of molecules. The presence of O atoms in the cometary coma were confirmed by the detection of its emission line at 6300 Å. A part of the O atoms may be in the photo-dissociation product of H_2O but some of them may be contained in the cometary nuclei from the beginning.

(2) The observed amount of recently detected H_2S in the compact H II regions is comparable with that of CS (P. Thaddeus, private communication), and is in a good agreement with the result of our calculation, again at the temperature of 4000 K. PN could not be observed with the sensitivity such that, if all P atoms are contained in the molecule, it should be detected (B. E. Turner, private communication). This may be due to the instability of this molecule at high temperature, contrary to N_2, because of its relatively small dissociation energy.

(3) In his review on interstellar molecules at this symposium Snyder suggested that Xogen at 89 GHz might be C_2H. This is very likely from our standpoint, because the dissociation energy in the $C_2 - H$ bond is 5.7 eV, much larger as compared with that of the usual $C - H$ bond, 3–4 eV. The detailed calculation of its abundance is not possible

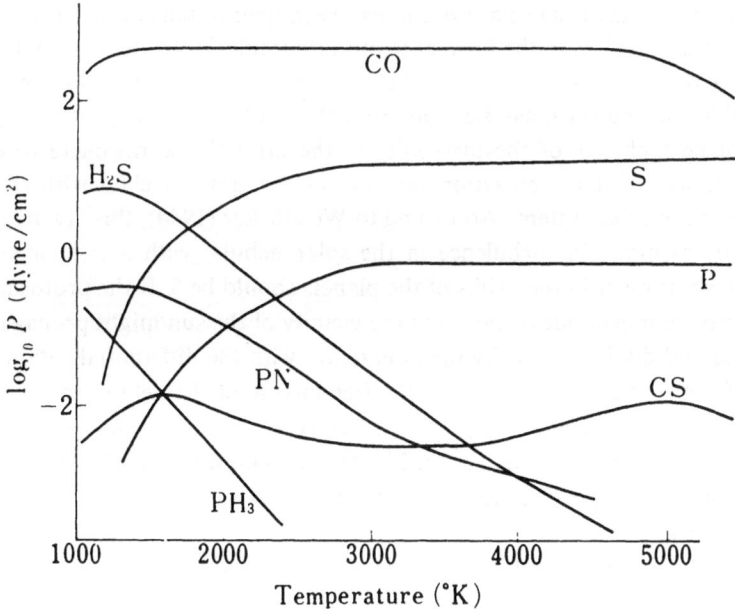

Fig. 2. Some examples of the abundances of the molecules including S and P at the temperatures of 1000–5500 K and at the pressure of 10^6 dyn cm^{-2}.

at present because of the lack of some molecular data of this molecule. A rough estimation by taking into account the dissociation energy only gives a rather large abundance of C_2H, say comparable with that of C_2, at the temperature of 4000 K. HC_3N is composed of C_2H and CN. Therefore, if the amount of C_2H is plenty, that of HC_3N might also be abundant. Furthermore, the bond between these radicals are stabilized by the resonance phenomenon called hyperconjugation between the two triple bonds in C_2H and CN (Mulliken, 1939). A rough thermochemical equilibrium estimate again gives a rather large amount of this molecule. The actual existence of HC_3N, a rather complex molecule, in many compact H II regions was reported by Morris *et al.* (1973).

3. Protostar Theory

One of the candidates for the location of molecular formation whose physical condition may be adequate for the ejection of interstellar molecules is the prestellar atmosphere, because recent theoretical studies on protostars suggest that they have very large radii and wholly convective atmospheres. An explanation of the similarity between interstellar molecules and cometary atmosphere might be that interstellar molecules ejected from the protosun were accreted to be comets in a very cool region of the solar nebula where the sunlight was completely shielded by the dense dust around the protosun. If the accreted mass could grow up to that of planets, the heat of accretion would have changed the composition of the molecules. Such a possibility seems to explain the fact that the present atmospheres of Jovian planets contains H_2, CH_4, and NH_3 in accord with their effective temperatures. On the other hand, comets scattered to the fringe of the solar system may keep their initial composition.

The reported mass loss in the form of neutral molecules from IRC + 10216 and red giants whose physical conditions are rather similar to that of protostars might suggest the possibility of neutral mass ejection from the protostar. If a protostar pulsates, there might be a chance of the instability in the prestellar atmosphere to eject the neutral molecules. Such a pulsation phenomenon might correlate with the Titius-Bodes' law in the solar system: According to Weizsäcker (1944), this law may well be explained by assuming a turbulence in the solar nebula, with a regular system of vortices whose number in the orbits of the planets should be 5. If the protosun pulsated, the hydrodynamical tide occurred in the vicinity of the sun might propagate in the solar nebula and dissipate in it by the interaction with the differential rotation of the nebula to form turbulence. This is not the free mixing of the nebula so far discussed (ter Haar, 1950), but forced mixing. Consequently the turbulence may continue much longer than the free mixing time, 10^3 yr. The above magic number, 5, might be explained in discussing the dispersion relation of the nebular waves. This point will be argued in another paper.

In relation to this argument, it is noteworthy that Herbig (1970) already discussed a possiblity for the formation mechanism of interstellar dust that they were blown off from the 'solar' nebula by the pressure of radiation and particles of the very young 'Sun'.

4. Late Type Star Theory

As suggested in the introduction, it may also be possible to explain the similarity of interstellar molecules with cometary atmospheres by the formation of molecules in the atmospheres of the late type stars. There is no spatial correlation of late type stars with compact H II regions. Consequently the molecules should be sufficiently shielded by some means from decomposition by the stellar ultraviolet radiation in the interstellar space during the transport from the stars to the dustry neutral clouds in the compact H II regions. This might be possible if the molecules are frozen in a sufficiently large lump which absorbs and scatters the stellar radiation at the surface.

There is some evidence that the neutral molecules are ejected from some kinds of late type stars, as we have seen in the previous section. Comets might be formed in the envelopes of these stars by a similar mechanism to that in the solar nebula. (The number of comets in our solar system is said to be decreasing during 4.5 b.y. though their loss to the interstellar space by planetary perturbation or through their decomposition in the vicinity of the Sun. If the supply of comets from the interstellar space is necessary to explain their present number, there should be places of cometary formation somewhere in our Galaxy.) According to this theory, interstellar dust is formed in the envelopes of late type stars and its composition is again dirty ice of second kind.

Witkowski (1970) suggested the possibility that the number of comets in our Galaxy was 0.5×10^{23} in his discussion of the origin of the comets. Then the total mass of comets may be comparable with that of fine interstellar dust grains in the Galaxy. Dust of intermediate mass between the micron-size dust and comets might also exist in interstellar space. A lot of such large-size dust grains may be contained in the dark neutral clouds and be melted to be observable in the form of molecules by the heat of protostars or of the shock waves accompanied with the gravitational collapse of the clouds.

5. Dirty Ice of the Second Kind and the Primordial Solar Nebula

In this section we shall attempt to explain some chemical features of cosmic dust, comets, and planets in terms of dirty ice of the second kind defined in the first section.

It is known that the life times of possible parent molecules of the radicals in the cometary coma estimated from the laboratory measurements of absorption cross-section and the observed solar ultraviolet flux are too large to explain the observed life times (Potter and Del Duca, 1964). One of the possibility to avoid this difficulty is to assume that radicals are contained in the cometary nucleus from the beginning and are evaporated by the heat of the sunlight at the fringe of an icy halo surrounding the nucleus (Delsemme and Wenger, 1970). The evaporation of dirty ice of the second kind easily produces an atmosphere similar to comets. The radical reactions in the nuclei might provide free energy for cometary bursts (Donn and Urey, 1956). If cometary atmospheres contained a great amount of H_2 and H, the main ions around the comet would be H^+ instead of CO^+ and N_2^+. Then the ion density in the cometary ionosphere

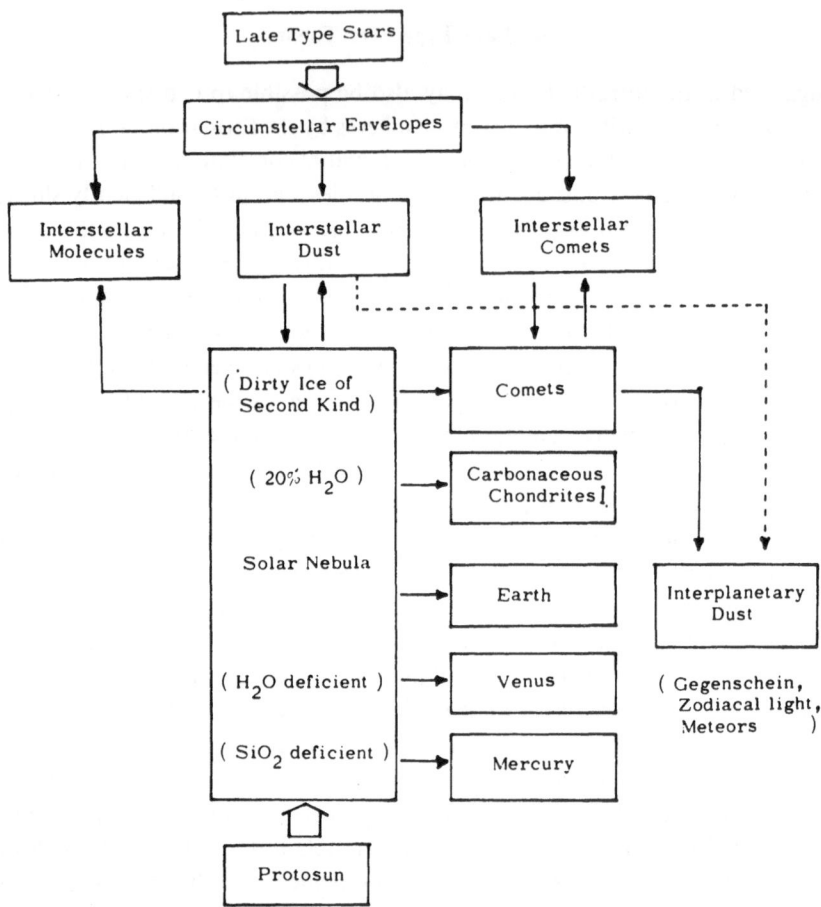

Fig. 3. A possible flow diagram among the highly convective stars, planets, dust, and molecules

should be large, due to the slow recombination rate of this atomic ions with electrons. Such a situation is similar to the case of the Jovian ionosphere (Shimizu, 1971).

The origin of interplanetary dust in the vicinity of the Earth is frequently ascribed to the debris of comets. The volatiles in it may evaporate due to the heat of the Sun and only metals and silicates may remain in the dust after a long time. Dust in the cometary tails might also be the same kind of substances.

Recently it was suggested that the chondrites were formed by the direct accretion of the substances in the primordial solar nebula (Anders, 1971). Under the assumption that the dirty ice of the second kind chemically represents the dust at the outer part of the solar nebula, we may obtain the following working hypothesis for the chemical structure of the solar nebula:. The primordial solar nebula was so opaque due to the dust in it that the intensity of solar radiation decreased according to an exponential law, instead of a r^{-2} law, before the accretion time. Mercury may have been formed by the accretion of dust which was deficient of volatile substances and even deficient of a part of the silicates due to the strong irradiation from the primordial Sun at its very active phase with the luminosity of 10 $L_\odot$ (Narita *et al.*, 1970). At the orbit of Venus, the solar radiation was still strong enough to evaporate most of the H_2O in the dust. (Sometimes the scantiness of H_2O on Venus is acribed to the escape of H from its exosphere after the photodissociation of H_2O. However, the accumulation of O_3 in the lower atmosphere should form a cold trap for H_2O which prevented the transport of water from the lower atmosphere to the photochemical level, as discussed by Berkner and Marshall (1965) in the case of the Earth.) There is a possibility that a fair part of the solar radiation was absorbed between Venus and the Earth. If Venus were ground into fine dust and put close together at its orbit, the thickness of the dust may roughly be estimated to be of the order of cm, which may be enough to absorb the solar radiation to keep some amount of water to form ocean on the Earth later, in the dust at 1 AU from the sun. Type I carbonaceous chondrites which contain 20% of water might be accreted at more distant location and comets might be formed in the completely dark nebula (Figure 3).

It should be remarked that the above suggestions have been obtained from the astrochemical consideration only. Details on the physical properties of the primordial solar nebula will be studied in the future.

6. Conclusions

It is found from a thermochemical calculation that interstellar molecules may be formed at places of high temperature and of high pressure. Such a physical condition is quite different from those in cool and tenuous interstellar space. Molecules may be formed in the atmospheres of some stars, possibly protostars and/or late type stars, transported to the dusty clouds in the compact H II regions by some means, and may survive there for a time long enough to be observable. If the interstellar molecules are formed around late type stars, the existence of a lot of cometary substances in the interstellar space and an exchange of matter between the solar system and interstellar space follows.

In the cool and dark regions of the solar nebula, the molecular condensates may be put close together to form comets consisted of dirty ice of second kind by some mechanism similar to the formation mechanism of planetesimals of the Jovian planets. The solar nebula in the vicinity of the protosun may be deficient of volatiles due to the high luminosity of the Sun (Narita *et al.*, 1970). This nebular model might explain the sucessive chemical properties in the Mercury-Venus-Earth-carbonaceous chondrites-comets sequence. Interstellar dust could also be the dirty ice of the second kind.

The above discussion might suggest that long-period comets were the most primordial substances in the solar system and that good clues for studying the origin of the solar system would be obtained from the new observation of comets at ultraviolet, far-infrared, and radio wavelengths by the space probes in the near future.

References

Anders, E.: 1971, *Ann. Rev. Astrophys. Astron.* **9**, 2.
Berkener, L. V. and Marshall, L. C.: 1965, *J. Atmospheric Sci.* **22**, 225.
Delsemme, A. H. and Wenger, A.: 1970, *Planetary Space Sci.* **18**, 709.
Donn, B. and Urey, H. C.: 1956, *Astrophys. J.* **123**, 339.
Herbig, G. H.: 1970, *Mém. Soc. Roy. Sci. Liège* **19**, 13.
Lyttleton, R. A.: 1948, *Monthly Notices Roy. Astron Soc.* **180**, 465.
Morris, M., Palmer, P., Turner, B. E., and Zuckerman, B.: 1973, this volume, p. 381.
Mulliken, R. S.: 1939, *J. Chem. Phys.* **7**, 339.
Narita, S., Nakano, T., and Hayashi, C.: 1970, *Prog. Theor. Phys.* **43**, 942.
Oort, J. H.: 1950, *Bull. Astron. Inst. Neth.* **11**, No. 408.
Potter, A. E. and Del Duca, B.: 1964, *Icarus* **3**, 103.
Shimizu, M.: 1971, *Icarus* **14**, 273.
Shimizu, M.: 1973, *Prog. Theor. Phys.*, **49**, 153.
Ter Haar, D.: 1950, *Astrophys. J.* **111**, 179.
Von Weizsäcker, C. M.: 1944, X. *Astrophys.* **22**, 319.
Witkowski, J. M.: 1970, in G. A. Chebotarev, E. I. Kazimirchak-Polonskaya, and B. G. Marsolen (eds.) 'The Motion, Evolution of Orbits, and Origin of Comets', *IAU Symp.* **45**, 419.

PHOTODESORPTION FROM INTERSTELLAR DUST GRAINS*

LAWRENCE T. GREENBERG

Dept. of Physics, University of California, Berkeley 94720, Calif., U.S.A.

Abstract. An experiment to measure the photodesorption of physisorbed gases from a cold fused quartz substrate by near UV photons is described. The observed yield (mol photon^{-1}) is as large as 10^{-5}, a higher yield is likely in the vacuum UV. Included is a discussion of the photodesorption process and some applications to the interstellar medium. The observed yield is enough to maintain molecular abundances in moderate density interstellar clouds in equilibrium.

1. Introduction

The abundance of molecules in the interstellar gas depends not only on the rates of the chemical reactions responsible for their formation and destruction but also on the equilibrium between adsorption onto and desorption from the surfaces of dust grains. Thermal evaporation times are extremely long at the very low temperatures typical of interstellar grains; at 15 K, a rather weakly adsorbed molecule, with $\Delta H_{ads}/R = 1000$ K, doesn't evaporate in the lifetime of the universe. The 'average' unshielded interstellar photon flux**, $\phi_p \sim 2 \times 10^8$ cm^{-2} s^{-1} ($\lambda < 400$ nm), bombards each adsorption site on a grain surface with one photon every 2.5×10^6 s. If photodesorption occurs with any appreciable efficiency then it is the dominant evaporation mechanism from grain surfaces (Watson and Salpeter, 1972a). This paper gives the results of an experiment to determine the quantum efficiency for photodesorption.

Photodesorption has received little study either experimentally or theoretically, even though it was reported nearly forty years ago (Terenin and Solonitzin 1959). Most of the published experimental work[†] has been on the desorption of chemisorbed O and CO from metals and semiconductors; almost no work has been reported on physisorbed species or at low temperatures.

Harrison and Schoen (1967) have proposed the erosion of ice particles in Saturn's rings by photodesorption.[‡] Watson and Salpeter (1972a, b) have recently argued for efficient photodesorption by 100–200 nm photons and then calculated molecular abundances in the interstellar gas on that basis.

* Work partially supported by the National Aeronautics and Space Administration Grant NGL 05-003-272.
** The radiation field adopted is that of Werner and Salpeter (1969).
† See most recently: Lee (1972), Menzel *et al.* (1971), Baidyaroy *et al.* (1971), Peavey and Lichtman (1971), and Genequand (1971), and references contained therein. The most relevant experiment was never published (Reeves *et al.*, 1966).
‡ The word photodesorption will be used to cover all photon induced processes which remove the first layer of atoms or molecules from a surface by nonthermal mechanisms. Many other terms are used, e.g.: photoejection, photosputtering, photoevaporation, optical erosion, photon induced desorption, photodetachment, etc.

Greenberg and Van de Hulst (eds.), Interstellar Dust and Related Topics, 413–419.

2. The Experiment

A typical experiment to observe photodesorption consists of an adsorption substrate in a vacuum chamber, a source of illumination and a detector for measuring the increase in pressure caused by the additional gas desorbed under irradiation. The major difficulty in making quantitative studies is separating the true photodesorption from thermal desorption due to radiant heating of the sample and chamber by the large photon fluxes required for measurable pressure increases. Various means have been used in previous experiments to rule out thermal effects, with success that is in dispute (Genequand, 1971; Schubert and Böer, 1971). The design of the present experiment involved three strategems to reduce thermal interference: the sample chamber and substrate were made of fused quartz in order to be transparent to the radiation; the radiation was chopped at a high rate compared to thermal time constants; finally the spectral dependence of the desorption signal was observed.

The vacuum system was a bakeable (700 K) uhv system made of pyrex, quartz and metal, pumped by a combination of a Hg-diffusion pump (with two re-entrant liquid nitrogen traps) and a titanium getter pump; residual pressure was typically $< 2 \times 10^{-10}$ torr (1 torr $= 1.333 \times 10^2$ N m^{-2}). Sample gases were admitted with a metal seal variable leak valve at pressures between $1 - 100 \times 10^{-9}$ torr. The adsorption substrate was cooled with liquid nitrogen although some desorption was seen at room temperature. Depending on the vapor pressure of the sample gas at 77 K, the pressure during measurements was either the residual pressure or maintained near 1×10^{-8} torr.

The light source was a 1000 W Hg–Xe high pressure arc lamp and the beam was mechanically chopped at 37 Hz. The optics formed an image of the arc of approximate area 3 cm^2 on the substrate. The lamp output was filtered through a water cell (no photodesorption signal was observed between the water cutoff at $\sim$ 1000 nm and the

TABLE I

Photodesorption yield

$$\text{Yield} \equiv \frac{\text{molecules out}}{\text{photons in}}$$

Y	Molecule	Comment
$\sim 1 \times 10^{-5}$	CS$_2$	abs. 290–400 nm, < 220 nm
$\sim 1 \times 10^{-6}$	CO$_2$	no abs. > 175 nm
	O$_2$(O$_4$?)	only forbidden abs. > 200 nm, $\theta < 1$
	CO	only forbidden abs. > 200 nm, $\theta < 1$
10^{-7}–10^{-6}	C$_6$H$_6$	abs. near 260 nm
	C$_4$H$_{10}$	
	N$_2$	$\theta < 1$
	CH$_4$	$\theta < 1$, no absorption
	H$_2$O	only first layer desorbed, no absorption
	CH$_3$OH	only first layer desorbed
	NH$_3$	only first layer desorbed
$< 10^{-8}$	Cu$_2$O	experiment by M. Johnson, strong absorption < 250 nm
	CdS	experiment by M. Johnson, strong absorption < 310 nm

quartz cutoff at $\sim$ 3000 nm, but this region was not carefully studied) and spectral regions could be isolated by a series of nested long-wavelength-pass filters. The desorption efficiencies in Table I below are for the wavelength interval between the short wavelength cuton of the arc-quartz combination ($\sim$ 200 nm) and the cuton of a Schott WG-280 filter ($\sim$ 275 nm). The illumination of the entire cooled substrate in this band was measured calorimetrically to be 160 mW (2×10^{17} photons s^{-1}); the total power between the lamp cuton and the cutoff of the water cell was approximately 2 W. If the entire 2 W, modulated at 37 Hz, were absorbed at the surface of the quartz, then the temperature modulation at the chopping frequency would be 0.2 K. The true 'ac' temperature is much less than this because the quartz is transparent to the radiation, which has already been filtered through many thermal wave relaxation lengths of quartz.

The pressure was measured with a Varian UHV-12 Bayard-Alpert type ionization gauge; no calibration was made for the gases used which will introduce absolute errors of approximately a factor of two. The collector current was put into a fast feedback FET electrometer and then into a lock-in amplifier, whose d.c. output represents the pressure fluctuations, $|P_{ac}|$, synchronous with the chopped photon flux. In the case of CS_2, the large signal permitted direct observation of the time evolution of the pressure signal when the electrometer output was put into a signal averaging computer. By using short pulses of light (pressure bursts), the vacuum system response was determined; by using very long pulses, a further indication was available that the signal was photodesorption, rather than thermally induced desorption.

The use of an ion gauge to measure the pressure has a major drawback in that one cannot identify the chemical species responsible for the signal. Therefore, in the present experiment one cannot completely exclude the possibility of either contamination (which can partly be ruled out by sample techniques, wavelength dependence, consistency and other experimental variables) or chemical reactions (Moesta and Trappen, 1970). A mass spectrometer can be used but suffers from shot noise $\sim$ 100 times worse than an ion gauge (Redhead et al., 1968). The shot noise with an ion gauge at a dc pressure of 3×10^{-9} torr is 10^{-13} torr $Hz^{-\frac{1}{2}}$ allowing a useful sensitivity of a $\sim 10^{-14}$ torr.

The response of the vacuum system is governed by the differential equation

$$V \frac{dP}{dt} = -SP + Q,$$

where V is the volume, P the pressure, S the pumping speed (customary unit, s^{-1}) and Q the gas throughput (customary unit, torr l s^{-1} = 3.536×10^{19} mol s^{-1}). Q in this case is due to outgassing, responsible for the residual pressure; thermal desorption due to slow heating of the substrate; and the photodesorbed gas $Q'e^{-i\omega t}$, where $\omega = (2\pi) \times$ (chopping frequency). The expression for the pressure modulation synchronous with the chopped light (with the appropriate phase shift) is

$$|P_{ac}| = \frac{Q'}{V}(\tau^{-2} + \omega^2)^{-\frac{1}{2}},$$

where $\tau = V/S = 5.4$ ms, $\omega = 232$ rad s^{-1} and $V \cong 500$ cm^3, yielding

$$Q' = 5.3 \times 10^{21} \left(\frac{P_{ac}}{1 \text{ torr}} \right) \text{mol s}^{-1}.$$

Typical signal levels were of the magnitude $P_{ac} = 10^{-11}$ torr or $\sim 10^{11}$ mol s^{-1}. No nonlinearity of signal intensity with changes in light intensity was observed.

3. Results

Table I presents the major results of this investigation. The quantity Y is the yield, the number of molecules desorbed per photon incident ($\lambda < 275$ nm). It is quoted only to an order of magnitude for the various gases because systematic errors such as calibration, contamination, and in some cases, thermal desorption, are not totally eliminated. For CS_2, which received the most study, the error is more likely to be less than a factor of three. Some of the gases with low adsorption energies do not adsorb a complete monolayer (the degree of coverage, θ, is less than 1) at 77 K, and it is hard to estimate θ accurately, but it cannot be very low judging by the signal levels.

Desorption was observed for a number of materials which show little or no uv absorption in this region when in the gaseous state. This is consistent with previous studies (Moesta and Trappen, 1970; Hubbard et al., 1971; Breuer, 1973) which have observed efficient photocatalysis on surfaces with photons of longer wavelength than are capable of gas phase photolysis. The spectral dependences of the photodesorption yields have not been carefully analyzed although certain features are obvious in the data: that the desorption yield falls off markedly with increasing λ, but that some desorption does persist through most of the visible region for some gases. No sharp features were observed with the available resolution of ~ 10 nm in the integrated spectra taken with the nested filter set. The spectra have different shapes for the various molecules, as is expected of true photodesorption.

The values of Y may seem to be small at first glance, but one must realize that only photons that are absorbed in the first monolayer (perhaps a few layers) are likely to lead to desorption. A quantity that is more revealing is the cross section for photodesorption, $\sigma_{PD} = Y \sigma_m / \theta$, where σ_m is the area of a molecule adsorbed on a surface (typically 2×10^{-15} cm^2) at monolayer coverage. A yield of 10^{-5} corresponds to a rather large cross section of $\sim 2 \times 10^{-20}$ cm^2 ($\theta = 1$).

One can also define the relative quantum efficiency $\eta = \sigma_{PD}/\sigma_{abs}$, where σ_{abs} is the cross section for absorption of photons by the adsorbed layer. One can express η in terms of macroscopic quantities

$$\eta = \frac{Y}{Kd},$$

where K[cm^{-1}] is the attenuation coefficient and d is the thickness of the monolayer. The denominator is just the fractional number of photons, $F = \Delta I/I$, absorbed in passing through one layer. Photodesorption in wavelength regions for which the sample gas is transparent excludes the use of the gas phase value for the attenuation coefficient in the calculation of η. At present one cannot calculate the absorption spectrum of the adsorbed layer, in fact, very little is known about the excited electronic states of adsorbed molecules. (Data is just becoming available on vibrational spectra, see for example Heidberg *et al.*, 1971.) Photodesorption is a useful technique for gaining this knowledge, due to its easily adjustable energy resolution and sensitivity to broad as well as narrow spectral features; presuming that we can understand the ejection mechanism well enough to decouple it from the absorption.

Let $f(\Delta v)$ be the total oscillator strength for all transitions of an adsorbed molecule in some extended wavelength region Δv. Surface phonons broaden out individual transitions, so that a monolayer is optically thin, even at a line center, i.e. $\sigma_{abs}(v) < \sigma_m$. The average absorption cross section is then

$$\bar{\sigma}_{abs} = \frac{1}{\Delta v} \int \sigma(v)\,dv = \frac{1}{\Delta v} \frac{\pi e^2}{mc} f(\Delta v)$$

and $F = \bar{\sigma}_{abs}/\sigma_m$. For the observed region $F = 3.2 \times 10^{-2} f(\Delta v)$. We might expect $f(\Delta v)$ to be of the order of 10^{-3} corresponding to a mean attenuation coefficient $K \sim 10^3$ cm^{-1}, but K could be as high as 10^5 cm^{-1} which is typical of vacuum UV absorptions in gases and direct band gap absorptions in solids.

If we provisionally adopt $f(\Delta v) = 10^{-3}$ for CS_2 then the observed yield of 10^{-5} implies that $\eta \sim \frac{1}{3}$, i.e. that a substantial fraction of the photons absorbed lead to desorption of a molecule. Of course a higher f value would lower η proportionately.

Water, ammonia and methanol gave moderate desorption yields for the first layer adsorbed on the quartz, but the next layer adsorbed on top was desorbed much less strongly, if at all. A variety of explanations are possible and the present data does not allow differentiation.

Also given in Table I are two unpublished results of a related experiment by Michael Johnson, also at Berkeley. Thin films ($\sim$ 30 nm thick) of the ionic materials CdS and Cu_2O were given prolonged intense uv bombardments, $> 3 \times 10^{24}$ photons cm^{-2} beyond the band gap energy, at the end of which no decrease in thickness was apparent. The upper limits set to the rate of photodesorption of these ionically bonded materials are consistent with the qualitative discussion by Dexter (1964).

4. Discussion

There are several major categories of ejection mechanisms that can be responsible for photodesorption.

(1) For chemisorbed species, the photon can break the bond to the surface, allowing desorption of a neutral molecule (atom) or an ion, possibly by subsequent thermal evaporation. Photodesorption and electron induced desorption of CO from metals is thought to occur this way (Menzel *et al.*, 1971).

(2) An adsorbed molecule can undergo photodissociation or photoionization, with desorption of the fragments.

(3) The adsorbed molecule makes a transition to an excited electronic state and is desorbed before it deexcites (Watson and Salpeter, 1972a).

(4) The molecule goes up to an excited electronic state, returns to the ground electronic state, but in an excited vibration-rotation level. Some of this internal kinetic energy is converted to translational energy away from the surface by collision with surface atoms.

(5) The bonds in the surface layer of the substrate are excited and lead to desorption either due to changes in the adsorption potential or transfer of the excitation as in 4.

Mechanism 1 can be ignored when discussing physisorbed molecules. Mechanism 2 is not likely with the low energy photons, $E < 6$ eV in this experiment. Mechanism 3 requires that the photon be of shorter wavelength than that absorbed by the gaseous molecule; it can be responsible for some but not all the observed desorption. Mechanism 5 is possible and may be responsible for some of the long wave ($\lambda > 400$ nm) desorption, which was similar in spectral shape for many of the molecules, but rapid diffusion of the excitation will inhibit this mechanism. Mechanism 4 is flexible enough to be possible whether the absorption is redshifted (i.e. the excited state is more tightly bound to the surface than the ground state, perhaps by chemical forces) or is blueshifted, with respect to the gaseous absorption. The high desorption yield for CS_2 which is linear in the ground state and bent in its excited states, may be an indication that Mechanism 4 is responsible, but this could just be a selection effect due to the wavelength dependence of σ_{abs}. All mechanisms except 4 can desorb single atoms as well as molecules.

5. Conclusions

The main conclusion is that photodesorption is an efficient process, with an efficiency, η, which may be as high as order unity. In spectral regions of high absorption, the photodesorption yield Y may be as high as 10^{-2} but even a yield of 10^{-6} cannot be neglected. In a region of moderately high density ($n_H + 2n_{H_2} = 10^3$ cm^{-3}, $n_0 = 1$ cm^{-3}, $n_{mol} = 10^{-2}$ cm^{-3}, $T_{gas} = 50$ K) the flux of molecules, ϕ_m, is ~ 50 cm^{-2} s^{-1} which is down by a factor of 4×10^6 from the average unshielded interstellar radiation field. Thus, the photons can return molecules to the gas as fast as they collide with grains, preventing depletion of the gas, at least in regions of modest shielding. Photodesorption can likewise maintain an equilibrium abundance of $\sim 10^{-2}$ cm^{-3} in a dense completely molecular cloud (molecular formation has gone to completion). The column density of molecules through such a cloud is approximately $Y\phi_p L/\bar{v}$ where L is the mean free path of the photons and $\bar{v}$ is the mean velocity of the molecules and ϕ_p is the photon flux outside the cloud.

The equilibrium coverage of grain surfaces and the question of mantle growth depends not only on ϕ_m, but also on the rate of molecular formation on grain surfaces. If the fraction of molecules that remain on the surface during formation is δ, ϕ_a is the flux of heavy atoms and we assume (Watson and Salpeter, 1972a) that every heavy atom forms a molecule when it hits a grain, the criterion for mantle growth becomes $(\phi_m + \delta\phi_a) > Y\phi_p$. The conditions for which this is satisfied depends on the value of Y

and the amount of UV extinction. If δ is nearly 1, then the molecules we observe in the gas that were formed on grains were photodesorbed, except H_2 which can evaporate thermally. If photodesorption occurs by Mechanism 4, then it may indicate that δ is small, since molecules are probably formed in states of high vibrational and rotational excitation. Mantles that are formed in regions of high shielding, will be destroyed if the grain migrates to the surface of the cloud or if a star should form in the cloud. The interstellar flux ϕ_p with a yield of 10^{-5} will remove a 100 nm thick mantle in 10^6 yr, but the yield may be lower for ice mantles. In circumstellar regions, around newly formed stars, and in the solar system the photon flux is much higher but thermal evaporation also becomes important. Another interesting application occurs in circumstellar regions. The momentum transfer to the grain due to recoil when a molecule becomes photodesorbed is greater than the momentum of the photon if $Y > 10^{-4}$, so photodesorption increases the radiation pressure of uv photons.

Acknowledgements

Photodesorption is an interesting and neglected process which certainly deserves further study which will contribute to both astrophysics and surface chemistry.

The author owes a great deal of thanks to Prof. Charles Townes who directed this research. He also acknowledges many helpful discussions with H.D. Breuer, M. A. Johnson, W. Klemperer, G. A. Somorjai, W. D. Watson and the members of the Quantum Electronics and Astrophysics Group at Berkeley.

References

Baidyaroy, S., Bottoms, W. R., and Mark, P.: 1971, *Surface Sci.* **28**, 517.
Breuer, H. D.: 1973, this volume, p. 399.
Dexter, D. L.: 1964, *Nuovo Cimento* **32**, 90.
Genequand, P.: 1971, *Surface Sci.* **25**, 643.
Harrison, H. and Schoen, R. I.: 1967, *Science* **157**, 1175.
Heidberg, J., Zehme, S., Chen, C. F., and Hartmann, H.: 1971, *Ber. Bunsenges. physik. Chem.* **75**, 1009.
Hubbard, J. S., Hardy, J. P., and Horowitz, N.H.: 1971, *Proc. Nat. Acad. Sci. U.S.A.* **68**, 574.
Lee, T. J.: 1972, *J. Vac. Sci. Technol.* **9**, 257.
Menzel, D., Kronauer, P., and Jelend, W.: 1971, *Ber. Bunsenges. physik. Chem.* **75**, 1074.
Moesta, H. and Trappen, N.: 1970, *Naturwissenschaften* **57**, 38.
Peavey, J. and Lichtman, D.: 1971, *Surface Sci.* **27**, 649.
Redhead, P. A., Hobson, J. P., and Kornelson, E.V.: 1968, *The Physical Basis of Ultrahigh Vacuum*, Chapman and Hall, London, p. 342.
Reeves, R. R., Jr., Harteck, P., and Albers, E. W.: 1966, unpublished.
Schubert, R. and Böer, K. W.: 1971, *J. Phys. Chem. Solids* **32**, 77.
Terenin, A. and Solonitzin, Yu.: 1959, *Discussions Faraday Soc.* **28**, 28.
Watson, W. D. and Salpeter, E. E.: 1972a, *Astrophys. J.* **174**, 321.
Watson, W. D. and Salpeter, E. E.: 1972b, *Astrophys. J.* **175**, 659.
Werner, M. W. and Salpeter, E. E.: 1969, *Monthly Notices Roy. Astron. Soc.* **145**, 249.

PART VIII

DUST AND H II REGIONS

INTERFEROMETRIC STUDIES OF OH SOURCES ASSOCIATED
WITH H II REGIONS AND IR OBJECTS

R. D. DAVIES

University of Manchester, Nuffield Radio Astronomy Laboratories, Jodrell Bank, U.K.

Abstract. An interferometer with a 24 km baseline has been used to map components of OH sources to a precision of a few hundredths of a second of arc. The W3 OH source which is associated with an H II region appears to be rotating slowly. There is evidence for Zeeman splitting in the components of this source. The IR/OH sources NML Cygnus and VY CMa are found to be expanding objects with a large rate of mass outflow. Rotation is also evident in both these objects.

1. Introduction

Interstellar OH is found in two main classes of object in the Galaxy. The first is the 'normal' interstellar cloud (or at least the more dense part of the distribution of these clouds) which probably contains the bulk of the molecules in the Galaxy. These clouds are sometimes observed in emission at radio frequencies but are most readily detected in absorption against background radio sources. The gas (hydrogen) density in these clouds is typically 10 to 10^3 hydrogen atoms cm^{-3}. The second class of OH object is the compact region which has a high brightness temperature in the spectral line as a result of masering action within the cloud. These objects have diameters of $\sim 10^{16}$ cm and gas densities estimated to lie in the range 10^8 to 10^{12} cm^{-3}. The present paper is concerned with new measurements of the second class of OH object.

The OH masering sources may themselves be subdivided into distinct groups each with its own properties. One group consists of the main line (1665 and 1667 MHz) emitters which are associated with H II regions; the satellite lines are weak. This group is strongly circularly polarized. Further, for a given source the velocities of the components in any one line are different from those of the other lines.

Another group contains those OH sources which are associated with IR sources; they are generally strongest in the 1612 MHz satellite line of OH. This group includes such diverse objects as the Mira variables, the Orion IR nebula, NML Cygnus and VY Canis Majoris.

Both the 'normal' and the masering clouds contain significant amounts of dust. The 'normal' clouds show a visual absorption ranging from a few tenths of a magnitude (Davies and Matthews, 1972) to 6–8 magnitudes (Heiles, 1968). The OH masering objects associated with H II regions have been found to be IR emitters in most cases (Wynn-Williams *et al.*, 1973). The IR emission from both groups of OH masering objects is believed to come from circumstellar dust clouds.

Interferometric studies of the spatial structure of the OH masering objects have been made at a number of observatories with baselines ranging up to several thousand kilometres. These have shown that the various emission components in the spectrum come

Greenberg and Van de Hulst (eds.), Interstellar Dust and Related Topics, 423–432.

from separate regions with diameters typically a few hundredths or thousandths of a second of arc. The present observations have been taken with the 24 km baseline of the Mark II–Mark III interferometer at Jodrell Bank. They provide more definitive positions for the brighter OH masering sources than have been made so far.

2. The W3 OH Source

The W3 OH source lies at the edge of the H II region IC 1795. The OH source is within a few seconds of arc of a compact H II region visible only at centimetre wavelengths.

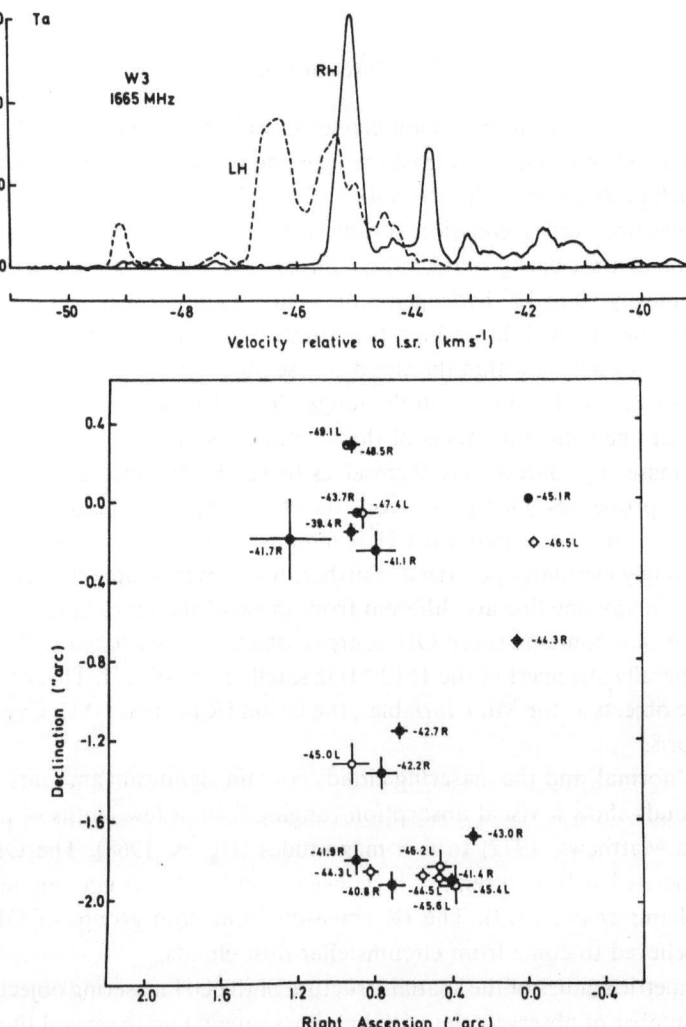

Fig. 1. The W3 OH source at 1665 MHz. LH and RH spectra taken with a single telescope are shown. The map gives the location of source components relative to the −45.1 km s⁻¹ RH component.

An IR source has also been found within the position uncertainty of the OH source. (Wynn-Williams *et al.*, 1973). The OH source, H II region and IR object could be concentric.

The most detailed map of the 1665 MHz OH source published so far (Cooper *et al.*, 1971) was obtained with an observing baseline of 127 km s⁻¹. It showed that 13 components were scattered over a region $\sim 2''$ in diameter. Very long baseline interferometry (Moran *et al.*, 1968) has shown that the components have diameters ranging down to 0.0045″.

New observations with the Mark II–Mark III interferometer have led to improved positions for 22 components. Although this number of components could not be conclusively identified in the RH and LH spectra alone, the interferometry provided the further means of separating the components in position as well as in frequency.

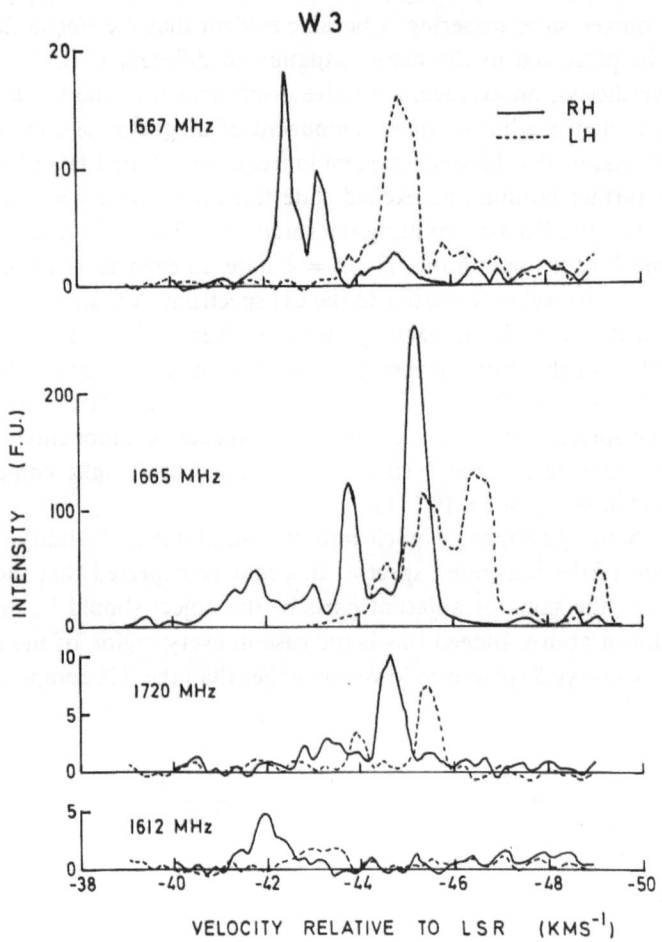

Fig. 2.   W3 OH source spectra in the 4 lines of the $^2\Pi_{\frac{3}{2}}$, $J = \frac{3}{2}$ state showing displacement of the LH and RH polarization which is interpreted as the result of the Zeeman effect. The 1612, 1667 and 1720 MHz spectra are taken from Rydbeck *et al.* (1970).

Figure 1 shows the LH and RH spectra and the location of the components relative to the -45.1 km s^{-1} RH component obtained in the interferometry. The relative position accuracy ranges from $\sim 0.01''$ for the brighter components such as -49.1 km s^{-1} LH to $\sim 0.08''$ for the fainter components such as -45.0 km s^{-1} LH.

The components shown in Figure 1 probably represent the total number that exist in W3. They are spread over a region $2.5'' \times 1.4''$ with the major axis at p.a. $= 174°$.

At first sight the map of W3 does not show any clear pattern either in the distribution of velocity or polarization. Before examining this point further, attention will be drawn to another feature of the spectra of W3 which will help in the interpretation of Figure 1.

As soon as circular polarization was discovered in OH sources it was noticed that the LH and RH spectra of W3 were displaced relative to one another in the way which might arise from Zeeman splitting (Davies *et al.*, 1966). However when it was found that these sources were masering it became evident that the *simple* Zeeman pattern would not be preserved in the maser situation in different parts of the interstellar clouds. Nevertheless, on average, a relative displacement of the LH and RH spectra would be expected if a line-of-sight component of magnetic field existed in the OH cloud. In the case of W_3 this displacement in frequency of the LH and RH spectra has persisted as further ground and excited state transitions have been measured giving added weight to the Zeeman splitting interpretation. Figure 2 gives the LH and RH spectra of the 4 transitions in the $^2\Pi_{\frac{3}{2}}$, $J = \frac{3}{2}$ state. In each case the RH spectrum is displaced to positive velocity relative to the LH spectrum. An estimate of the longitudinal component of the field in each spectrum is given in Table I. In each case it has been derived from the displacement between the mean velocity of the LH and RH spectra. The theoretical Zeeman split for the 1612 and 1720 MHz lines was calculated as the mean displacement of the 3 theoretically expected components weighted by the component intensities. Table I suggests that the line-of-sight component of the magnetic field in W3 is $\sim 5 \times 10^{-3}$ G.

Figure 1 should give more explicit information about the validity of the Zeeman interpretation of the integrated spectra. It would be expected that the LH and RH components in the same or adjacent parts of the object should be displaced in the sense mentioned above. Indeed this is the case in every region of the map – the LH component is always displaced to lower velocities than the RH component. The mean

TABLE I

Magnetic fields in W3 derived from Zeeman interpretation

Line	Theoretical Zeeman split (MHz G^{-1})	Observed split (kHz)	Magnetic field (10^{-3} G)
1612 MHz	1.31	5.8	5.8
1665	3.27	14.1	4.3
1667	1.96	10.0	5.1
1720	1.31	4.6	3.5

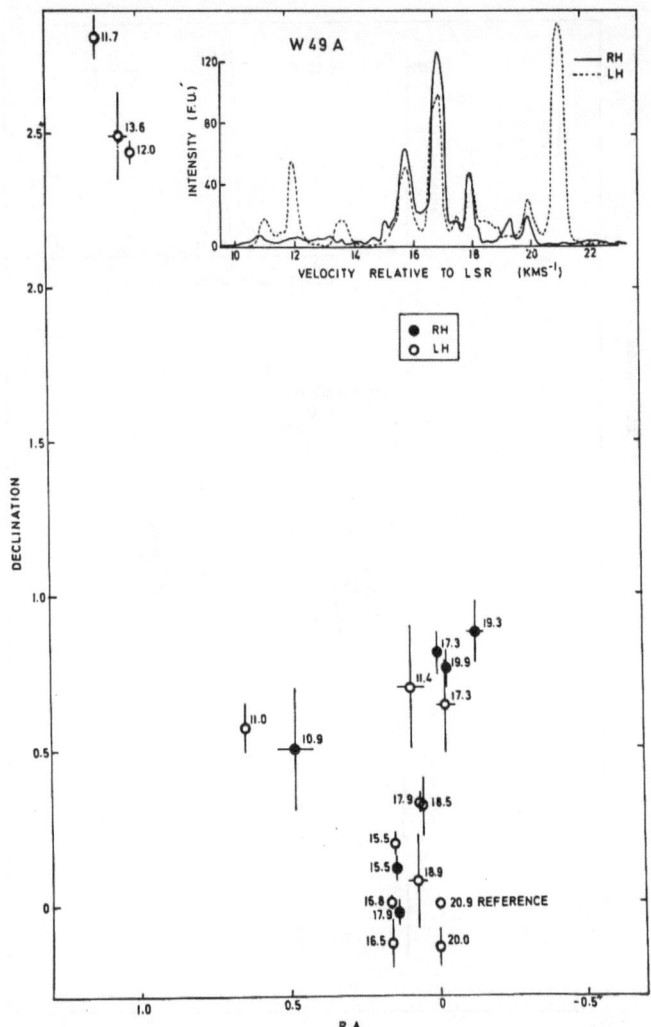

Fig. 3. The W49A OH source. This is the north preceding group of the W49 double OH source. The spectrum has been derived from the interferometric observations. Component positions in the map are plotted relative to the 20.9 km s^{-1} LH component.

displacement within the various groups of features is 15.4 kHz; this corresponds to a magnetic field of 4.7×10^{-3} G directed away from the observer.

The further consequence of the Zeeman interpretation of the data is that local velocities can be calculated over the object. The velocity increases from the northern to the southern end of the object with a gradient of 1.5 km s^{-1} per second of arc. This can most simply be interpreted as a rotation. The elongation of the W3 OH source is perpendicular to the inferred rotation axis, as would be expected for a flattened rotating object.

It is interesting to compare the magnetic field strength inferred from the present

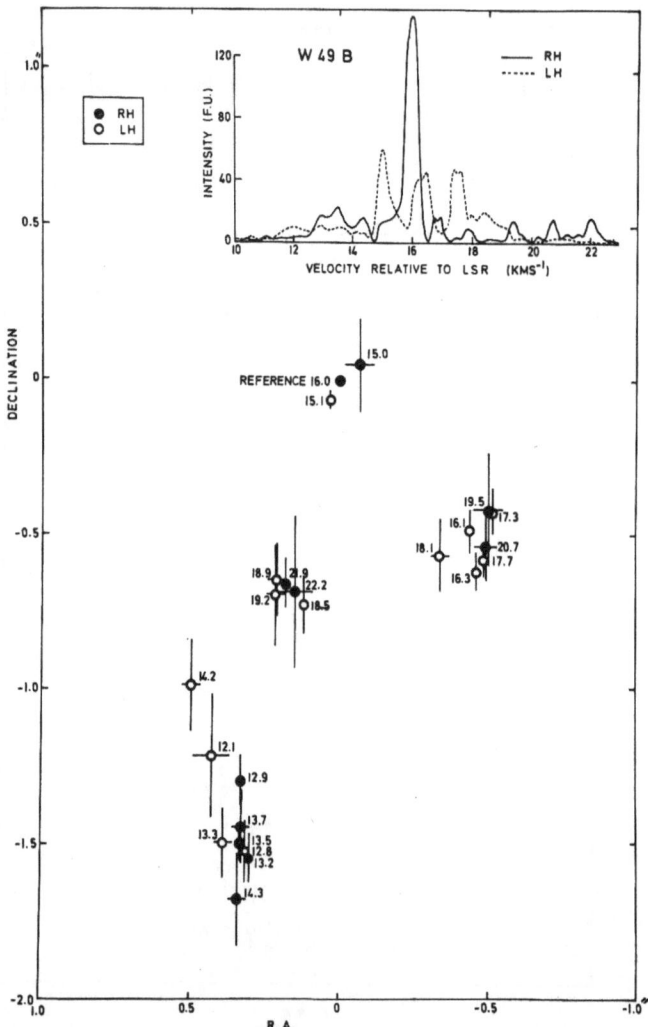

Fig. 4. The W49B source. The spectrum is derived from interferometer observations. Component positions on the map are plotted relative to the 16.0 km s^{-1} LH component.

measurements with that expected for a contracting interstellar cloud. We will take a field strength of 4×10^{-6} G to be typical of an interstellar cloud as determined from Zeeman effect measurements in neutral hydrogen absorption features. If a typical dimension of 10 pc is assumed then this field will be increased to ~ 1 G when the cloud contracts to a diameter of 2×10^{-2} pc, the size of the W3 OH source. Clearly the original cloud has shed away most of its magnetic flux in contracting to the size of an OH masering cloud. Also it can be readily shown that the OH cloud must lose most of its magnetic flux in collapsing to form a star. It would therefore seem that the loss of magnetic flux is a continuing phenomenon in the condensation of interstellar clouds to form stars.

3. The W49 OH Sources (1665 MHz)

The W49 region contains two main groups of OH sources and associated H II regions. The north preceding OH group (W49A) lies 8.5 s of RA and 69.5" of Declination away from the south following OH group. Since the components of each group overlap in velocity it is only possible to separate the spectra of the two groups by interferometric observations. Figures 3 and 4 show the two spectra derived from the present observations.

The maps of the W49A and W49B OH sources are plotted in Figures 3 and 4; the position error in Declination is a factor of approximately 3 times the error in RA because of the low declination of these objects. W49A itself appears to consist of two groupings – one small group of LH components with velocities in the range 11.7 to 13.6 km s^{-1} lies ∼ 2.5" away from the main concentration.

W49B could be a cluster of 4 groups with diameters in the range 0.1 to 0.5". Such an interpretation is suggested by the small diameter/separation ratio, by the separation into velocity groupings and by the fact that there is no systematic velocity gradient across the whole region. These individual groupings would be similar in size to W3, which at the distance of W49 (14 kpc) would have an angular diameter of 0.3".

Of the 6 suggested groupings in W49 two show some evidence for the Zeeman splitting which was evident in W3. These are the groupings of W49B at the relative positions ($\Delta\alpha = +0.2''$, $\Delta\delta = -0.7''$) and ($\Delta\alpha = -0.4''$, $\Delta\delta = -0.5''$). The longitudinal component of magnetic field in these two groupings is 5×10^{-3} G directed away from the observer in each case. If the other groupings of components contain magnetic fields of this magnitude they must be highly inclined to the line of sight.

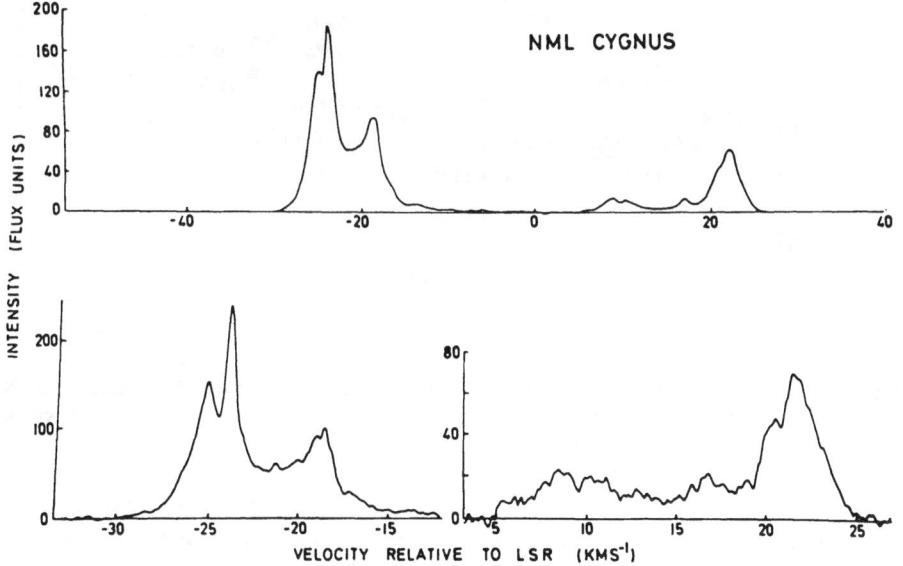

Fig. 5.   The 1612 MHz spectrum of NML Cygnus. Top, a frequency resolution of 4.4 kHz. Bottom, a frequency resolution of 1.1 kHz.

4. The OH/Infrared Objects NML Cygnus and VY Canis Majoris

The infrared objects NML Cygnus and VY Canis Majoris have strong OH maser emission in the 1612 MHz line. They are associated with stars of M spectral type (M3e-M6) which are probably supergiants. A characteristic of their 1612 MHz spectra is the two peaks separated by $\sim$ 40 km s^{-1}. The redshifted peak is at the velocity of the M-star.

NML Cyg and VY CMa have similar spatial structures as determined by the interferometric observations. NML Cyg only will be described here in detail because it is a circumpolar source at Jodrell Bank and allows accurate relative positions to be determined. Its spectrum is shown at two frequency resolutions in Figure 5. A map of the main groups of components is shown in Figure 6.

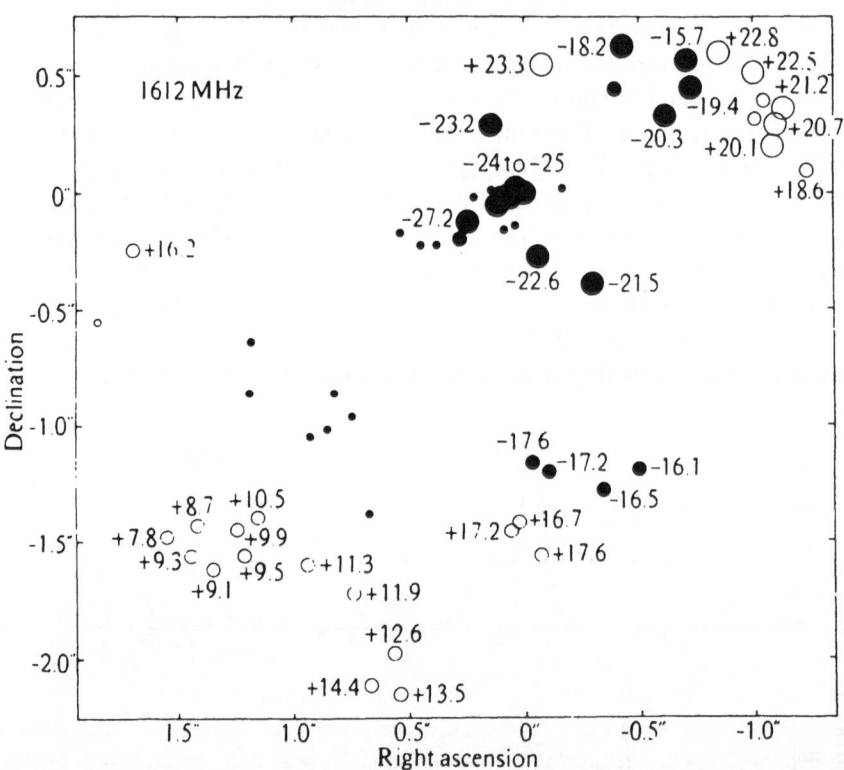

Fig. 6.   A map of the principal features or groups of features in the 1612 MHz spectrum of NML Cygnus. The positive velocity end of the spectrum is plotted as ○, and the negative velocity end as ●. The size of the symbols indicates the relative strength of the components.

The spatial structure of NML Cyg has a clearcut pattern; it is quite different from that found in the OH sources associated with H II regions such as W3 and W49. The components of NML Cyg lie in a region $3.3 \times 2.3''$ with its major axis at a position angle of 150°. The unique feature of the pattern in Figure 6 is that the components in the positive (receding) velocity part of the spectrum lie outside those in the negative (approach) velocity part.

The observations plotted in Figure 6 can be understood in terms of the emission from an expanding gas cloud which is also rotating. The expansion accounts for the fact that the highest approach velocities occur at the centre of the object. Gas at the outermost edge of the object is expanding from the centre in a plane orthogonal to the line of sight; it is expected to have a radial velocity equal to that of the central star. The gas expanding away from the star on the far side of the object is not seen in the OH or H_2O lines. It is proposed that the molecular line radiation on the far side is obscured by a high density, partially ionized, central cloud. Figure 7 illustrates the present model of NML Cyg and VY CMa.

The physical parameters of these objects can be estimated on certain simplifying assumptions (Davies *et al.*, 1972). At a distance of 500 pc the diameter of NML Cyg is 2.4×10^{16} cm. The gas density within this diameter is 2×10^7 cm^{-3} leading to an estimated gas mass of $0.12\,M_\odot$. If 10^{-3} to 10^{-4} of this gas is ionized it will be optically thick at the OH line frequency and obscure radiation from the far side of the object. The evidence of rotation comes from the observed gradient of the positive velocity gas; this gives a rotational velocity of 5 km s^{-1} at the outer edge which leads to a rotation period of 4200 yr. Another important parameter is the rate of mass outflow from the object. With the density and dimensions given above and an expansion velocity of 40 km s^{-1}, the rate of mass outflow is $3.5 \times 10^{-3}\,M_\odot$ yr^{-1}.

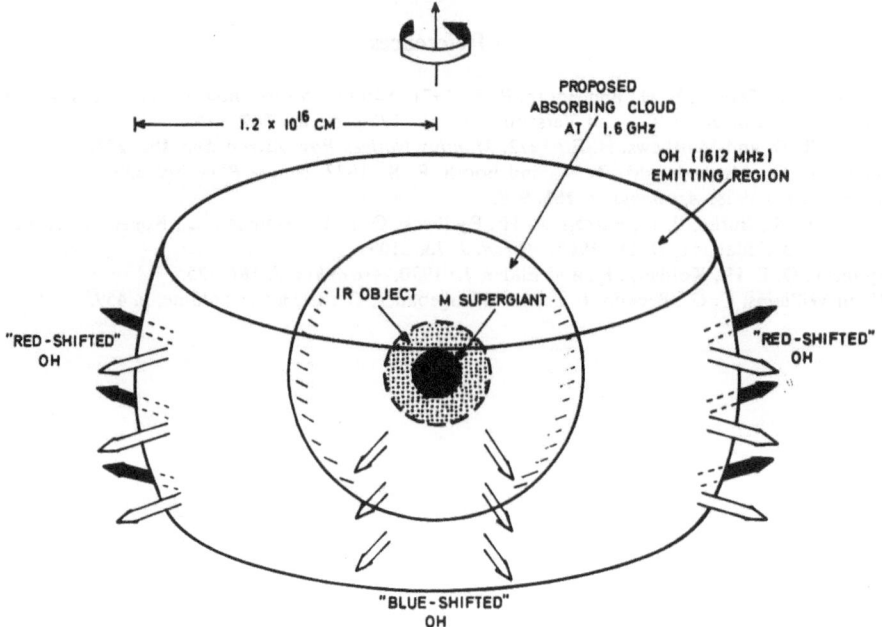

Fig. 7. The proposed model of NML Cygnus. The regions responsible for the redshifted and blue shifted features are indicated. The M supergiants has a radius of $\sim 2 \times 10^{14}$ cm and the IR object radius is $\sim 1.5 \times 10^{15}$ cm.

The above information provides substantial evidence that NML Cyg and VY CMa are in an early state of evolution. The cloud mass and rate of mass outflow is larger than has been found in any evolved star. It is most likely the remnant of the original cloud which condensed to form the M supergiant. Radiation pressure and thermal pressure from the absorbed stellar radiation are responsible for the present expansion of the gas and dust envelope. A further argument that these objects are remnants of the original cloud is their high angular momentum per unit mass – values far in excess of any evolved star.

The marked differences between the OH sources associated with infrared objects and those associated with H II regions are presumably attributable to the later stage of evolution of the gas cloud in objects like NML Cyg and VY CMa. In the H II region objects the gas cloud is presumably still contracting; it has a greater gas mass and a smaller velocity spread. In the 1612 MHz emitters the infall has been reversed to an outflow due to the action of the recently formed central star. In both stages large amounts of dust are present and become detectable in the infrared.

Acknowledgements

This paper describes work completed recently by the spectral line group at Jodrell Bank. Major contributions have been made by R. S. Booth, P. J. Harvey, M. R. W. Masheder, D. C. B. Whittet and A. Jane Wilson.

References

Cooper, A. J., Davies, R. D., and Booth, R. S.: 1971, *Monthly Notices Roy. Astron. Soc.* **152**, 383.
Davies, R. D., de Jager, G., and Verschuur, G. L.: 1966, *Nature* **209**, 974.
Davies, R. D. and Matthews, H. E.: 1972, *Monthly Notices Roy. Astron. Soc.* **156**, 253.
Davies, R. D., Masheder, M. R. W., and Booth, R. S., 1972, *Nature Phys. Sci.* **237**, 21.
Heiles, C. E.: 1968, *Astrophys. J.* **151**, 919.
Moran, J. M., Burke, B. F., Barrett, A. H., Rydbeck, O. E. H., Hansson, B., Rogers, A. E. E., Ball, J. E., and Cudaback, D. D.: 1968, *Astron. J.* **73**, S109.
Rydbeck, O. E. H., Kollberg, E., and Elldér, J.: 1970, *Astrophys. J.* **161**, 125.
Wynn-Williams, C. G., Becklin, E. E., and Neugebauer, G.: 1973, this volume, p. 459.

DUST, STAR, AND HYDROGEN DISTRIBUTIONS
IN THE ORION ASSOCIATION

SYUZO ISOBE

Tokyo Astronomical Observatory, University of Tokyo, Mitaka, Tokyo

Abstract. From the star counts on the Palomar Sky Survey prints, dust and star distributions in the Orion Association are studied. The Hydrogen distribution in the same region is obtained from an Hα photograph. It is found that a considerable part of the interstellar absorption on the line of sight is due to the circumstellar dust grains in the Orion Association.

1. Introduction

The age of the Orion Association is estimated to be 3×10^6 yr (Walker, 1969; Menon, 1958; Blaauw and Morgan, 1954). In this Association is the Orion Nebula whose age is said to be 10^4–10^6 yr (Parenago, 1953; Vandervoort, 1963; Sharpless, 1966). As was already shown by the present author (Isobe, 1971), in 3×10^6 yr a considerable part of the gas and dust surrounding early type stars should be blown out to the surrounding space. The Orion Association is relatively near to the Sun (Johnson and Hiltner, 1956) in the univrese and high enough in galactic latitude to study the apparent distributions of dust, gas, and stars without serious disturbances by the fore-ground stars and interstellar matter.

Menon (1958) obtained the distribution of neutral hydrogen in the Orion Association by 21 cm observations, and found that the neutral hydrogen atoms distribute with a shape of shell. O'Dell *et al.* (1967) and Henize *et al.* (1967) detected the near ultra-violet radiation in the Orion Association, which was identified as the scattered radiation by dust grains. However, Henry and Carruthers (1970) did not detect far ultra-violet radiation within the accuracy of their observations. These situations may show the fact that the scattering coefficients of dust grains at the far ultra-violet wavelengths are small (Nandy and Wickramasinghe, 1971). Photometric and spectroscopic observations of stars in the Orion Association have been made by many observers (e.g., Sharpless, 1952; Parenago, 1954; Straizus, 1963, 1964; Bartkus, 1964). Especially, Parenago (1954) presented the distributions of stars classified by spectral type and luminosity.

In this paper, we shall first study the distributions of dusts and stars from the star counts on the Palomar Sky Survey prints and the distribution of hydrogen atoms from an Hα photograph taken by the present author, and secondly compare these results with the previous results.

2. Hα Observations

An Hα photograph was taken by the Nikon F camera attached to the head of the 12″ telescope at the Okayama Astrophysical Observatory. The camera has an aperture of

Greenberg and Van de Hulst (eds.), Interstellar Dust and Related Topics, 433–444.

5 cm with a focal length 5 cm, which corresponds to a film scale of 70′ per milimeter. Figure 1 shows a reproduction of the photograph.

From the photograph, isodensity contour maps were obtained with the isodensito-

Fig. 1.   Hα photograph of the Orion Association. Up is north and left is east.

TABLE I

The data of Hα observations

Date	1969 January 17
Exposure time	100 minutes
Emulsion	Kodak 103a-E
Filter	Hoya R-64
Central wavelength	6563 A
Effective band width	297 A

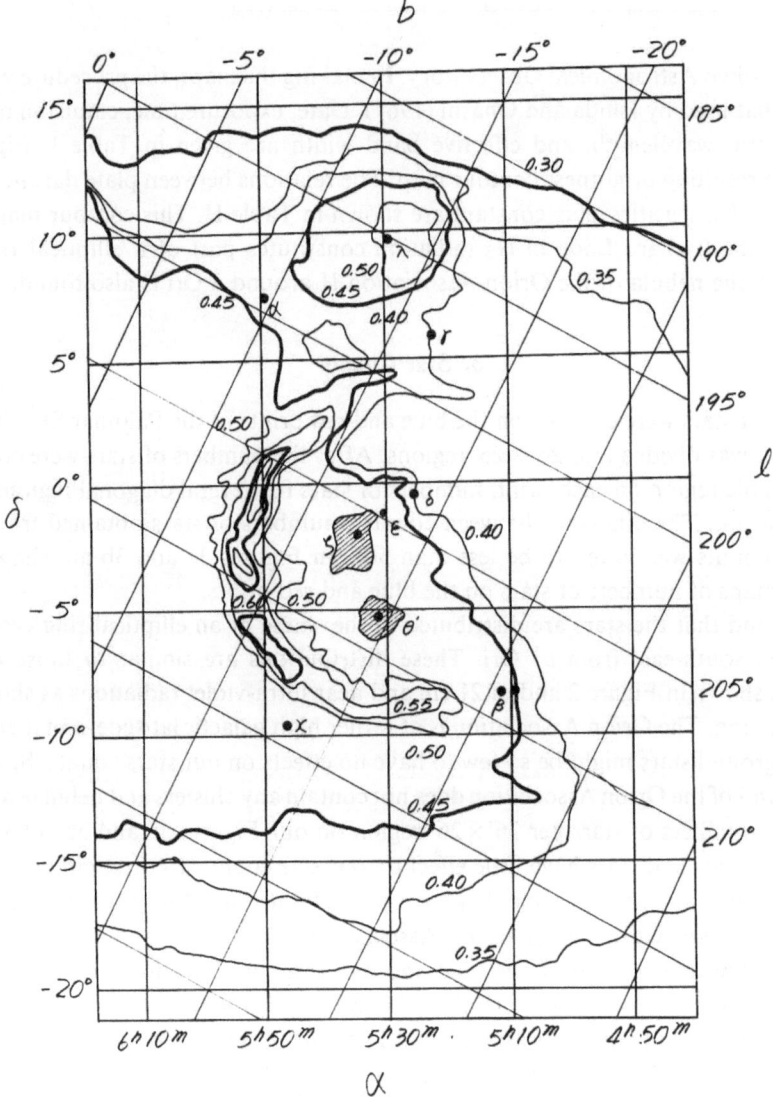

Fig. 2. Contour maps of Hα radiation in the Orion Association. The contour units are shown by plate density. The scale of all figures in this paper are same. The positions of the bright stars are shown.

TABLE II

The relations between plate darkness
and log (intensity + constant)

Plate density	log (intensity + constant)
0.40	0.84
0.45	1.11
0.50	1.19
0.55	1.32
0.60	1.39
0.65	1.47

meter of Tokyo Astronomical Observatory. In making the maps, the procedure was the same as that used by Ishida and Ohashi (1967). Date, exposure time, emulsion of film, filter, central wavelength, and effective band width are given in Table I. Figure 2 shows the resulting brightness contour map. The relations between plate darkness and logarithm of intensities plus constant are shown in Table II. This contour map indicates that the Barnard Loop in Hα radiation constitutes part of a elliptical ring. In this figure, the nebula of the Orion Association II around λ Ori is also found.

3. Star Counts

Numbers of stars were counted on the blue and red prints of the Palomar Sky Survey. Each print was divided into 26' × 26' regions. After the numbers of stars were counted for the whole region on each print, numbers of stars for several diagonal regions were counted again. The difference between counted numbers of stars obtained from two different counts was found to be less than 5%. In Figures 3a and 3b are shown the contour maps of numbers of stars on the blue and red prints.

It is found that the stars are distributed in the shape of an elliptical ring except in the region south-east from θ^1 Ori. These distributions are similar to those of Hα radiation shown in Figure 2 and of 21 cm and near ultra-violet radiations as shown in a later Section. The Orion Association is at rather high galactic latitude, and therefore, the background stars might be so few to have no effects on our stars counts. Since the fore-ground of the Orion Association does not contain any clusters and nebulae and the minimum numbers of stars per 26' × 26' region on our Figures 3a and 3b is less than 100, the fore-ground stars have little effect on our star counts. In these circumstances, it would be reasonable to assume that most of the faint stars observed in the direction of the Orion Association belong to this Association.

The limiting magnitudes of the blue and red original plates in Selected Area are 21.1 and 20.0 mag. respectively (Minkowski and Abell, 1963), but the corresponding ones for the blue and red prints are 20.0 and 17.5 mag., respectively, as deduced for the regions of the north equatorial pole and the north ecliptic pole (Tanabe and Mori, 1971). Since the 'Catalogue of Plates of the Palomar Sky Survey' shows that the observing conditions were similar for all plates and since the reproductions of the

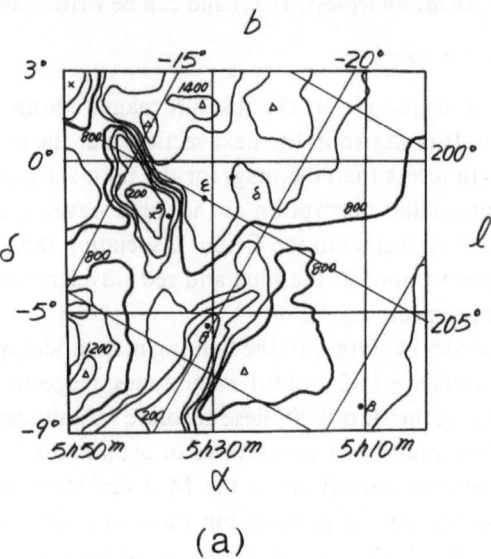

(a)

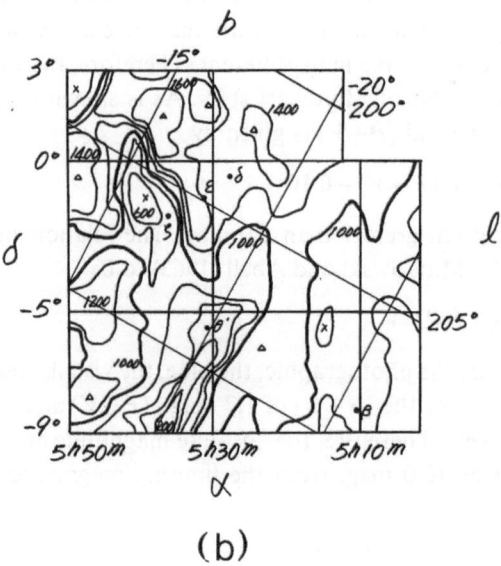

(b)

Fig. 3. (a) Contour maps of the number densities of stars in the Orion Association on the blue prints of Palomar Sky Survey. The numbers of stars in regions of 26′ × 26′ are shown. (b) Same as (a), but on the red prints.

paper prints were performed with care (Minkowski and Abell, 1963) limiting magnitudes of the blue and red prints in the region of Orion Association may be taken to be 20.0 and 17.5 mag. with little risk.

The distance modulus, $m_0 - M$, of the Orion Association is known to be 8.1

(Johnson and Hiltner, 1956; Sharpless, 1952) and can be written as

$$m_0 - M = m - M - A,$$ (1)

where m is the apparent magnitude, M the absolute magnitude and A the magnitude of interstellar absorption. It is shown in the next section that the interstellar absorption at the visual wavelength is less than 0.6 mag. for the faint stars in the Orion Association, and the mean interstellar absorption at the same wavelength is about 0.5 mag. for the O-, B-, A-, and F-type stars in the Orion Association (Sharpless, 1952). Therefore, the interstellar absorptions at the blue and red wavelengths are of the order of 0.6 mag. and 0.3 mag., respectively.

The absolute magnitudes of stars with the limiting magnitudes of 20.0 and 17.5 mag. on the blue and red prints are 11.3 and 9.1 magnitudes, respectively. Since the Orion Association is very young, the stars with these absolute magnitudes should be dwarf or pre-main sequence stars and have temperatures of around 3000 K. The mean color, $(B-V)$, and the bolometric correction of the M dwarf stars with temperatures of about 3000 K are not available at present, but those of stars with luminosity class I and III were obtained by Johnson (1964). From the comparisons of Johnson's (1964) Tables 3, 4, and 5 which give the mean colors and the bolometric corrections of stars of luminosity class III, I, and V respectively, the mean color of K-type stars with luminosity class III is only 0.1 mag. different with that of class V while the bolometric correction of the same stars is 0.5 mag. different. Therefore, the value of 1.7 mag. may be taken for the mean color of M5 dwarf stars. As is shown by Johnson (1964), the relation between $(B-V)$ and $(P-V)$ is given by

$$(B-V) = 1.19(P-V) - 0.10,$$ (2)

for the color, $B-V$, given greater than 1.0 mag. The relation between $(P-V)$ and $(P-R)$ is also given by Minkowski and Abell (1963) to be

$$(P-R) = 1.6(P-V),$$ (3)

where P, B, V, and R are the photographic, the blue, the visual, and the red magnitudes for a star, respectively. For the $(B-V)$ of 1.7 mag., $(P-V)$ and $(P-R)$ should be 1.5 and 2.4 mag. respectively. Therefore, the absolute magnitude at the visual wavelength is deduced to be 9.8 or 10.0 mag. from the limiting magnitudes of the blue or red prints, respectively.

Since the bolometric corrections of M-type stars were determined only from the observations of I, J, K, and L magnitudes by Johnson (1964), they may have errors of one magnitude or so because the intensity of molecular absorption band and the effect of blanketing are different from star to star. However, at the present stage we have no data available for these problems, and therefore we adopt -2.5–3.5 mag. for the bolometric correction of the star with temperatures of about 3000 K according to Johnson's (1964) Table 3. Therefore, the bolometric magnitudes of the faintest stars on the blue and red prints are to be 6.5–7.5 mag.

Hayashi (1961) and Ezer and Cameron (1967) calculated the paths of stellar evolutions on the HR diagram in the early phases of the gravitational contraction and obtained the curves of the constant age for stars on the HR diagram. According to their calculations, the stars with temperature of 3000 K should have the bolometric magnitude of 6.5 mag. at the age of 3×10^6 yr and situate at the pre-main sequence stage. These calculations are well consistent with our observational results.

It is concluded in this section that the stars with the limiting magnitude of 20.0 mag. on the blue prints are the same ones with the limiting magnitude of 17.5 mag. on the red prints if there is no interrtellar absorption in the region of the Orion Association.

4. Interstellar Extinction in the Direction of the Orion Association

We assume that the luminosity function is the same all over the Orion Association. In the absence of interstellar absorption, the apparent number of stars with a certain intrinsic luminosity in a unit area on the plane of sky is given by

$$N_0 = \alpha r_0^3, \tag{4}$$

while the apparent luminosity is

$$L_0 = \beta r_0^2, \tag{5}$$

where α and β are constants depending on the space density of stars and the intrinsic luminosity, respectively, and r_0 is the distance of the center of the Association. We denote the quantities associated with the red prints and the blue prints by the subscripts R and P respectively, and then we obtain

$$N_{R_0}/N_{P_0} = (L_{R_0}/L_{P_0})^{\frac{3}{2}}, \tag{6}$$

leading to

$$m_{P_0} - m_{R_0} = \tfrac{5}{3} \log(N_{R_0}/N_{P_0}), \tag{7}$$

where m_{R_0} and m_{P_0} are the respective apparent magnitudes in the case without interstellar absorption. In the presence of interstellar absorption on the line of sight, we denote each quantity shown in Equations (4), (5), (6), and (7) by omitting the subscript 0. In doing so, we have

$$m_P - m_R = \tfrac{5}{3} \log(N_R/N_P). \tag{8}$$

By taking the difference between expressions (7) and (8), it follows that

$$\begin{aligned} E_{P-R} &= (m_P - m_R) - (m_{P_0} - m_{R_0}) \\ &= \tfrac{5}{3}[\log(N_R/N_P) - \log(N_{R_0}/N_{P_0})]. \end{aligned} \tag{9}$$

From the results of Section 3, the stars observed on the blue prints should be the same as the stars observed on the red prints if the amounts of interstellar reddening are small. Therefore, we have adopted $N_{R_0}/N_{P_0} = 1.0$.

The amounts of reddening for the wavelength range from 4000 Å to 7000 Å are nearly inversely proportional to the wavelength in the normal reddening region and

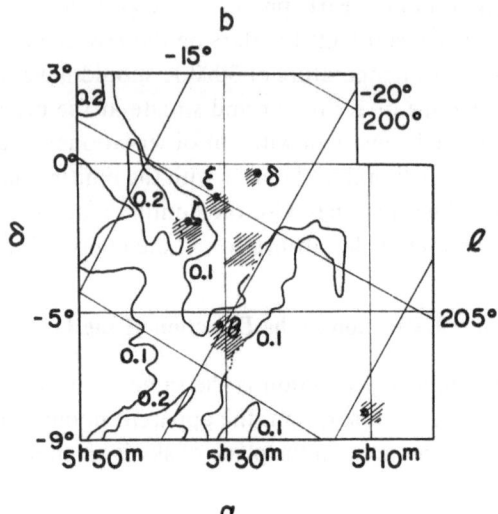

Fig. 4.   Contour maps of the color excess, E_{B-V}, obtained from the star counts. The contour units are shown by magnitude. The numbers of stars are not counted in the hatched region.

even in the abnormal reddening region. Therefore, we shall take the wavelengths 4200 Å, 4500 Å, 5500 Å, and 6500 Å for the effective wavelengths of P, B, V, and R respectively. Thus we have

$$E_{B-V} = E_{P-R}/2.0. \tag{10}$$

In Figure 4 are shown the contour maps of the color excess, E_{B-V}.

Figure 5 shows the mean interstellar absorption of stars for each region as quoted from Figure 1 of an earlier paper (Isobe, 1971). The region extending from north to south in the parts east of ζ Ori has in some parts a color excess, in E_{B-V}, larger than 0.3 mag. in Figures 4 and 5. However, the regions with a color excess, E_{B-V}, larger than 0.3 mag. in Figure 5 except the above region dont have such a large color excess, E_{B-V}, in Figure 4. Reddish (1967) showed that the amount of dust grains surrounding stars depends on the spectral types of stars and that there are few dust grains surrounding stars with spectral types later than G-type. Isobe (1971) showed that a considerable part of interstellar absorption for the O-, B-, and A-type stars in the direction of the Orion Association is caused by circumstellar grains. The color excess shown in Figure 4 is mainly due to the late type stars such as K- and M-type, while the color excess in Figure 5 is due to the early type stars such as O-, B-, A-, and F-type. Therefore, it may be concluded that the most part of interstellar absorption is caused by the circumstellar dust grains in the Orion Association except for the eastern part of ζ Ori.

5. Comparison of Contour Maps

The scale of the contour maps is the same. In Figure 6 is shown the contour map of the intensity of 21 cm radiation which is obtained from Figure 8 in van Woerden's work

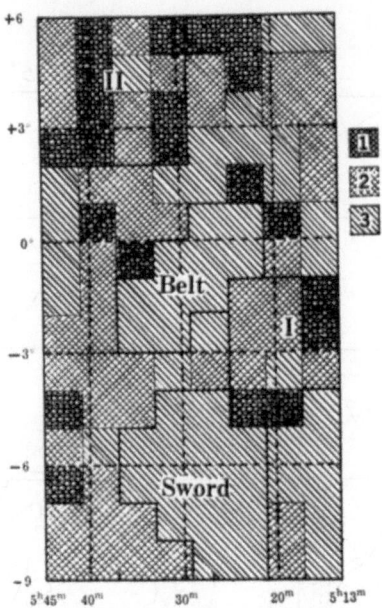

Fig. 5. Distribution of interstellar absorption obtained by Isobe (1971) from the observations of O-, B-, A-, and F-type stars. Hatched regions 1 show a mean color excess greater than 0.3 mag.; regions 2 show a mean color excess greater than 0.10 mag.; regions 3 show a mean color excess smaller than 0.10 mag.

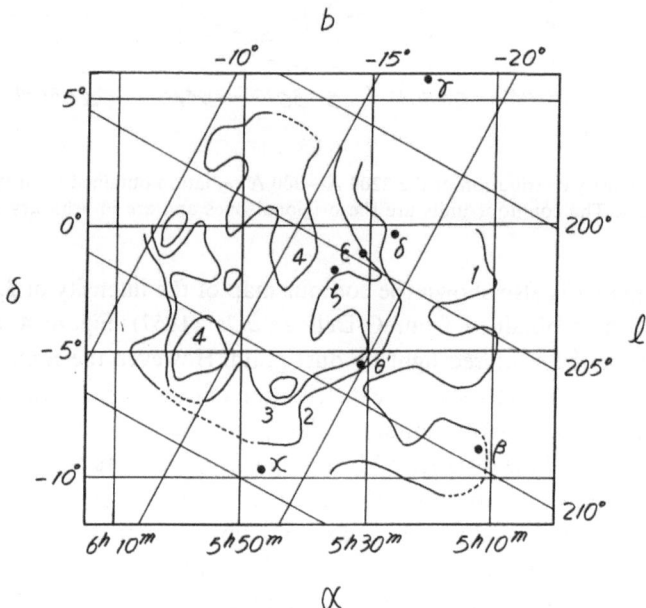

Fig. 6. Intensity distribution of the 21 cm radiation obtained from van Woerden's Figure 8. Large numbers of the contour unit correspond to larger values of the equivalent width.

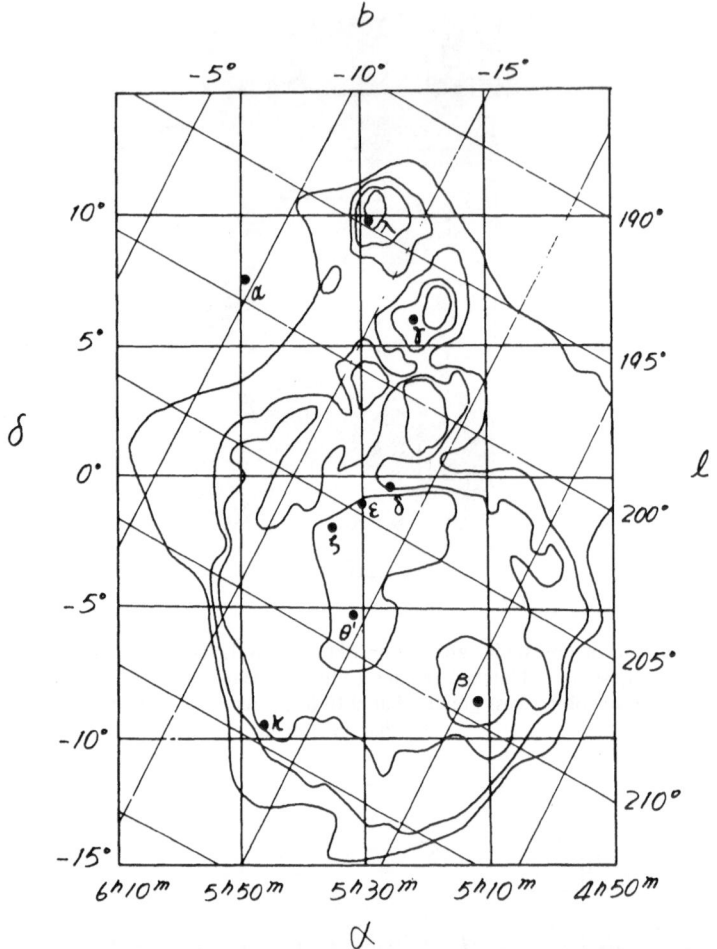

Fig. 7. The intensity distribution of the 2200 Å–4900 Å radiation obtained from the O'Dell *et al.*'s (1967) Figure 4. The countour units are the origional ones and are in arbitrary intensity units.

(1967). In Figure 7 is also shown the contour map of the intensity of 2200 Å–4900 Å radiation which is obtained from O'Dell *et al.*'s (1967) Figure 4 and which is considered to be the scattered light by dust grain. However, the contour map shown in Figure 7 does not directly show the distribution of dust grains because the observed intensity depends not only the amount of dust grains but also on the brightness of the illuminating stars and the distance of dust grains from the stars.

In all figures shown in this paper, we can easily recognize ring structures. Each ring has a marked ellipticity and the major axis of ellipse is approximately parallel to the galactic equator. It was suggested by Menon (1958) that the galactic magnetic field is the possible mechanism which can produce a distortion of an initially spherical cloud situated in the Galaxy. In Figures 3a and 3b, it is easily seen that the star density is small in the region from θ^1 Ori to the south-east direction. This feature is quite unique.

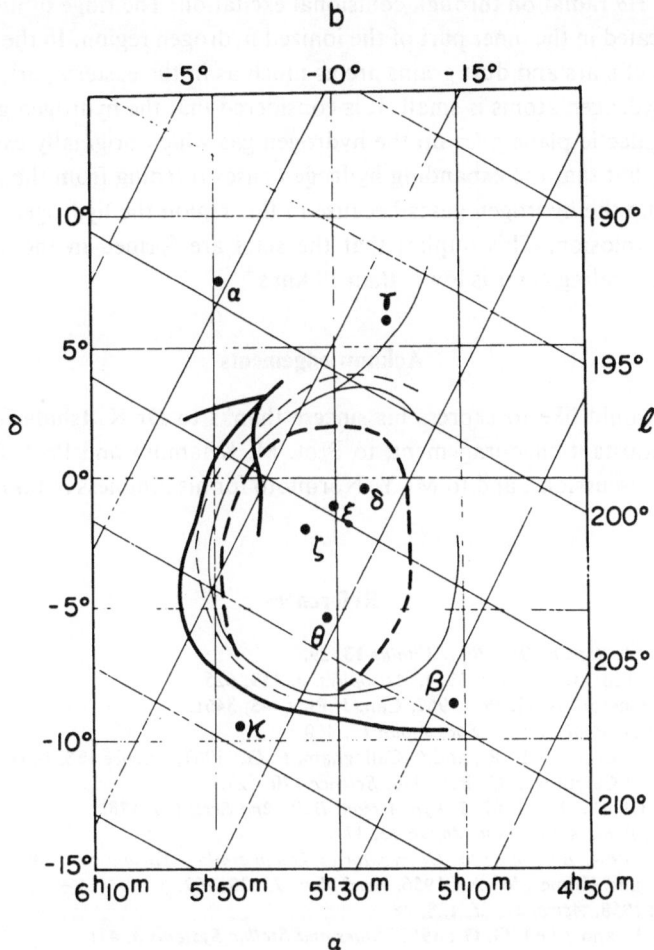

Fig. 8. Comparisons of the contour maps shown in this paper. Thick solid line shows the ridge positions of the 21 cm intensity; thin solid line shows the ridge positions of 2200 Å–4900 Å radiation; thick broken line shows the ridge positions of the star density; thin broken line shows the ridge positions of the Hα radiation.

As this region is also parallel to the galactic equator and would be parallel to the initial galactic magnetic line, the interaction of ionized gas and galactic magnetic field would be weak in the direction, and therefore the formation of ionized gas condensation is hard.

Figure 8 permits a comparison of the ridge positions of the contour maps. Those referring to the neutral hydrogen atoms (21 cm) are found to extend farthest east. According to Menon (1958), those atoms are also expanding with a velocity 10 km^{-1}. Figure 8 also reveals that the ionized hydrogen atoms (Hα) make a fine loop in the inner part of the neutral hydrogen region. 3×10^6 yr ago, a big explosion might have happened in the Orion Association. This explosion energy is considered to be transferred to the Hα radiation through collisional excitation. The ridge of number density of stars is located in the inner part of the ionized hydrogen region. In the western part, the amounts of stars and dust grains are as much as in the eastern part, although the amount of hydrogen atoms is small. It is considered that the hydrogen gas expanding toward the galactic plane piles up the hydrogen gas which originally existed in interstellar space, but that the expanding hydrogen gases receding from the galactic plane did not pile up the hydrogen gases because in this region the hydrogen gas density is low before explosion. This implies that the stars are formed in the stage that the velocity of expanding gases is lower than 10 km s^{-1}.

Acknowledgements

The author would like to express his sincere thanks to Dr K. Ishida for his helpful advice and constant encouragement, to Prof. M. Kitamura and Prof. B. Takase for their helpful discussions, and to Mr Y. Norimoto for his kindness in taking Hα photograph.

References

Bartkus, R.: 1964, *Astron. Obs. Biul. Vilnius* **13**, 29.
Blaauw, A. and Morgan, W. W.: 1954, *Astrophys. J.* **119**, 625.
Ezer, D. and Cameron, A. G. W.: 1967, Can. J. Phys. **45**, 3461.
Hayashi, C.: 1961, *Publ. Astron. Soc. Japan* **13**, 450.
Henize, K. G., Wackerling, L. R., and O'Callagham, F. G.: 1967, *Science* **155**, 1407.
Henry, R. C. and Carruthers, G. R.: 1970, *Science* **170**, 527.
Ishida, K. and Ohashi, M.: 1967, *Tokyo Astron. Bull.*, *2nd Ser.*, No. 178.
Isobe, S.: 1971, *Publ. Astron. Soc. Japan* **23**, 371.
Johnson, H. L.: 1964, *Boletin de los Observatorios Tonantzintla y Tacubaya* **3**, 305.
Johnson, H. L. and Hiltner, W. A.: 1956, *Astrophys. J.* **123**, 267.
Menon, T. K.: 1958, *Astrophys. J.* **128**, 28.
Minkowski, R. L. and Abell, G. O.: 1963, *Stars and Stellar Systems* **3**, 471.
Nandy, K. and Wickramasinghe, N. C.: 1971, *Monthly Notices Roy. Astron. Soc.* **154**, 255.
O'Dell, C. R., York, D. G., and Henize, K. G.: 1967, *Astrophys. J.* **150**, 835.
Parenago, P. P.: 1953, *Astron. Zh.* **30**, 249.
Parenago, P. P.: 1954, *Trudy. Gas. Astron. Inst. Sternberga* **25**, 1.
Reddish, V. C.: 1967, *Monthly Notices Roy Astron, Soc.* **135**, 251.
Sharpless, S.: 1952, *Astrophys. J.* **116**, 251.
Sharpless, S.: 1966, *Vistas in Astronomy* **8**, 127.
Straizus, B.: 1963, *Astron. Obs. Biul. Vilnius* **7**, 1.
Straizus, B.: 1964, *Astron. Obs. Biul. Vilnius* **9**, 4.
Tanabe, H. and Mori, K.: 1971, *Publ. Roy. Obs. Edinburgh* **8**, 173.
Vandervoort, P.: 1963, *Astrophys. J.* **138**, 294.
Van Woerden, H.: 1967, in H. van Woerden (ed.), 'Radio Astronomy and the Galactic System', *IAU Symp.* **31**, 3.
Walker, M. F.: 1969, *Astrophys. J.* **155**, 447.

DUST IN H II REGIONS

VAHE PETROSIAN

Institute for Plasma Research, Stanford University, Stanford, Calif., U.S.A.

Abstract. The observed far infrared flux of H II regions is interpreted as evidence for the presence of internal dust absorbing a fraction of Lyman continuum photons. The effect of this dust on the required number of ionizing photons and on the helium to hydrogen line intensity ratios are discussed and analytic formulae describing these effects are presented for simplified models.

1. Introduction

The purpose of this paper is to discuss the effects of the presence of dust in H II regions and in other gaseous nebulae on the ionization structures of these nebulae. I consider a spherically symmetric nebula with a single ionizing star located at its center, and present here analytic expressions describing two effects. First I shall discuss how the presence of dust alters the required number of ionizing photons. This effect depends on only one parameter, namely the total absorption optical depth τ_D due to dust throughout the nebula averaged over all frequencies beyond the Lyman limit. If n_D is the number density of dust and r_o is the observed radius of the nebula then

$$\tau_D = \int_0^{r_o} n_D \bar{\sigma}_D \, dr \tag{1}$$

and

$$\bar{\sigma}_D(r) = \int_{\nu_1}^{\infty} \sigma_D(\nu) N(r, \nu) \, d\nu \Big/ \int_{\nu_1}^{\infty} N(r, \nu) \, d\nu, \tag{2}$$

where $N(r, \nu)$ is the net flux of photons (per sec per frequency interval) flowing through a shell of radius r and ν_1 is the frequency of the Lyman limit.

The second effect I consider here is how the presence of dust alters the ratios of the Strömgren radii of other elements to that of hydrogen. In particular I shall consider the change in this ratio for helium which depends on the relative values of optical depths τ_D and τ'_D, the absorption optical depth for frequencies greater than $1.8 \, \nu_1$ ($1.8 \, h\nu_1 =$ ionization potential of He I).

$$\tau'_D = \int_0^{r'_o} n_D \bar{\sigma}'_D \, dr, \tag{3}$$

where r'_o is the helium Strömgren radius* and $\bar{\sigma}'_D$ is the average dust absorption cross

---

* In this paper I assume that $r'_0 \leq r_0$ which is true for most galactic nebulae where the sources of ionization are stars of effective temperature $< 6 \times 10^4$ K (Hummer and Seaton, 1964). I also neglect the presence of doubly ionized helium so that the results apply to regions of nebulae outside the He III zones.

Greenberg and Van de Hulst (eds.), Interstellar Dust and Related Topics, 445–452.
All Rights Reserved. Copyright © 1973 by the IAU.

section for frequencies above 1.8 v_1, defined as in Equation (2) with v_1 replaced by 1.8 v_1.

2. Evidence for Dust in H II Regions

Before describing these effects let me discuss briefly the evidence for the presence of dust in H II regions. In the first place, if the ratio of gas to dust in the H II regions (typically for bright H II regions the hydrogen column density is few times 10^{21} cm^{-2}) is the same as in the interstellar space, where there is on the average one magnitude visual extinction per 10^{21} cm^{-2} H-atoms in the line of sight, then we expect in H II regions a visual optical depth (scattering and absorption) of greater than one and even greater optical depth for ultraviolet photons. Another evidence for dust in H II regions is provided by the work of O'Dell and his collaborators (O'Dell and Hubbard, 1965; O'Dell et al., 1966; Krishna Swamy and O'Dell, 1967) who discovered that the diffuse nebular emission is more than that expected from atomic processes. They attribute the excess to scattering of stellar light by dust grains in the nebula. These and other similar observations could be interpreted as evidence for dust which only scatters stellar radiation.

However, there are now observations indicating absorption by dust. Münch and Persson (1971), from high resolution observations of the differential reddening of the Balmer lines in Orion nebula, conclude that the dust is mixed with the gas and therefore it is absorption by dust which is responsible for the reddening. They also find that high gas density regions correspond to high dust density regions. Because of this evidence I shall assume in this discussion that the ratio of dust to gas densities is independent of position in the nebula.

Finally, an indirect evidence for absorbing dust in H II regions is their large far infrared fluxes, observed by Harper and Low (1971), by Hoffmann et al. (1971), and by Sofier et al. (1972). This flux is presumably due to dust. The question that arises now is what heats up the dust. One heating source is the Lyman alpha (Lα) photons trapped in the H II regions. The expected infrared intensities due to Lα absorption are compared with the observed infrared intensities on Table I. The first column gives the infrared source numbers from Hoffman et al. (1971) and the second column gives their observed fluxes in the wavelength range 95 to 135 μ. Harper and Low (1971) have observed some of these sources in the wavelength range 45–750 μ. Their reported fluxes are shown on the third column, which are on the average three times larger than the fluxes in the second column. Consequently, I have estimated the total far infrared fluxes of those sources not observed by Harper and Low by multiplying their 85 to 135 μ fluxes by three. These numbers are shown in parenthesis. The fourth column gives the observed 5000 MHz fluxes of these sources using the identifications of Hoffmann et al. (1971). Finally, the fifth column gives the expected Lα fluxes calculated from the radio fluxes assuming that the nebulae are optically thick to Lα (case B). Comparison of column 5 and 3 shows that the expected infrared intensities due to absorption of Lα photons falls short of the observed intensities by factors ranging from 2 to 30. Consequently, I conclude that a fraction of stellar photons are

TABLE I

Source Hoffmann et al.	Other name	I_{IR} 85–135 μ erg cm^{-2} s^{-1}	I_{IR} total erg cm^{-2} s^{-1}	$S_{5000\,MH2}$ flux units	I_{L3} erg cm^{-2} s^{-1}
6	Orion A	5.3×10^{-6}	18.6×10^{-6}	340	2.3×10^{-6}
23		0.92	(2.8)	40	0.27
26	NGC6357	0.81	(2.4)	156	1.0
27	NGC6357	0.95	(2.8)	215	1.4
42		1.14	(3.4)	15	0.10
43		1.00	(3.0)	21	0.14
46	M8	0.51	(1.5)	100	0.66
48	W30	0.35	(1.0)	60	0.40
49	W31	0.71	(2.1)	60	0.40
50	W33	0.63	(1.9)	40	0.26
51		0.62	(1.9)	30	0.20
55	M17	3.6	7.3	550	3.6
56		0.55	(1.6)	20	0.13
57	W43	1.03	(3.0)	95	0.62
58	W49	1.14	3.1	50	0.33
59	W51a	0.36	1.2	34	0.22
60	W51b	1.95×10^{-6}	6.4×10^{-6}	100	0.66×10^{-6}

absorbed by dust, so that $\tau_D > 0$. The exact value of τ_D will depend on the fraction of infrared flux which is due to absorption by dust of Lyman continuum photons. This fraction depends on the spectrum of the central star and on the variation of albedo and total extinction of dust grains with frequency.

3. Effects of Dust

Let us now return to the calculation of the two effects described above. These calculations are similar to that described by Code (1973) and will not be discussed in detail here. In fact, in the present case the equations of the transfer of radiation are much simpler since I am interested only in photons which are absorbed completely in the H II region. As a result the scattering part of the cross section does not enter the calculations directly. The effect of scattering is to trap photons in the nebula and increase the effective optical depths (due to both gas and dust) to values larger than what one obtains from Equation (1). The effects of scattering can be neglected by replacing the line of sight absorption optical depths (Equation (1) with the *effective* absorption optical depths obtained from the integration along the photon paths.

(a) Effect of dust on the required number of ionizing photons: The net flux of ionizing photons crossing a spherical shell of radius r is given by

$$S(r) = \int_{v_1}^{\infty} N(r, v)\,dv, \tag{4}$$

with

$$S(r_o) = 0, \qquad S(O) \equiv S^*. \tag{5}$$

For a pure hydrogen nebula of density $n_H(r)$, the variation of $S(r)$ with the radial coordinates r is described by

$$\frac{dS(r)}{dr} = -n_D \sigma_D S(r) - 4\pi\alpha^{(2)} \int_0^r xn_H n_e r'^2 \, dr', \tag{6}$$

where n_e is the electron density, x is the fraction of hydrogen ionized and $\alpha^{(2)}$ is the recombination coefficient to excited levels of hydrogen.* The effect of dust is expressed by the first term on the right hand side of Equation (6). The rest of the terms in this equation are identical with the terms one obtains for a nondusty nebula (cf. Spitzer, 1968). In the absence of dust, Equation (6) gives the usual relationship between the Strömgren radius r_0 and the number of ionizing photons emitted by the central star,

$$S^*_{gas} = 4\pi\alpha^{(2)} \int_0^{r_o} xn_H n_e r^2 \, dr. \tag{7}$$

If we now introduce dust in this nebula, we will require a larger number of ionizing photons, since in addition to S^*_{gas} photons absorbed by the gas, photons are absorbed also by dust. In order to see how the presence of dust changes the required value of ionizing photons let us look at the solution of Equation (6).

$$S(r) = e^{-\tau}\left(S^* - \frac{4\pi\alpha^{(2)}}{\bar{\sigma}_D} \int_0^r \frac{xn_e n_H}{n_D} r'^2 e^{\tau'} \, d\tau'\right), \tag{8}$$

where $d\tau = n_D \bar{\sigma}_D \, dr$. At the boundary of the H II region $S(r_o) = 0$, so that we require

$$S^* = \frac{4\pi\alpha^{(2)}}{\bar{\sigma}_D} \int_0^{\tau_D} \frac{xn_e n_H}{n_D} r^2 e^{\tau} \, d\tau \tag{9}$$

ionizing photons for a dusty nebula compared to S^*_{gas} photons for a nebulae of the same size containing only gas. The presence of dust increases the required number of ionizing photons by a factor

$$\frac{S^*}{S^*_{gas}} = \int_0^{\tau_D} g(r)e^{\tau} \, d\tau \Big/ \int_0^{\tau_D} g(r) \, d\tau, \qquad g(r) = r^2 n_e xn_H/n_D. \tag{10}$$

For a uniform nebula ($n_H(r) = $ const.) we find

$$\frac{S^*}{S^*_{gas}} \equiv f(\tau_D) = \frac{3e^{\tau_D}}{\tau_D^3}\left[\tau_D^2 - 2\tau_D + 2(1 - e^{-\tau_D})\right]. \tag{11}$$

This result was derived also independently by J. Silk and G. Field. This and some other effects of dust are discussed in a joint paper (Petrosian et al., 1972). The inverse of the quantity $f(\tau_0)$, which is the fraction of ionizing radiation absorbed by the gas,

---

* For a hydrogen nebula, the electron density $n_e = xn_H$. For a nebula with both hydrogen and helium $n_e = xn_H + yn_{He}$, where y is the fraction of He ionized, Equation (6) is still valid to a good approximation. This is because recombination coefficients for hydrogen and helium are approximately equal (Burgess and Seaton, 1960) and because each helium ionization is followed by a recombination which injects into the nebula approximately one photon capable of ionizing hydrogen (Robbins, 1970).

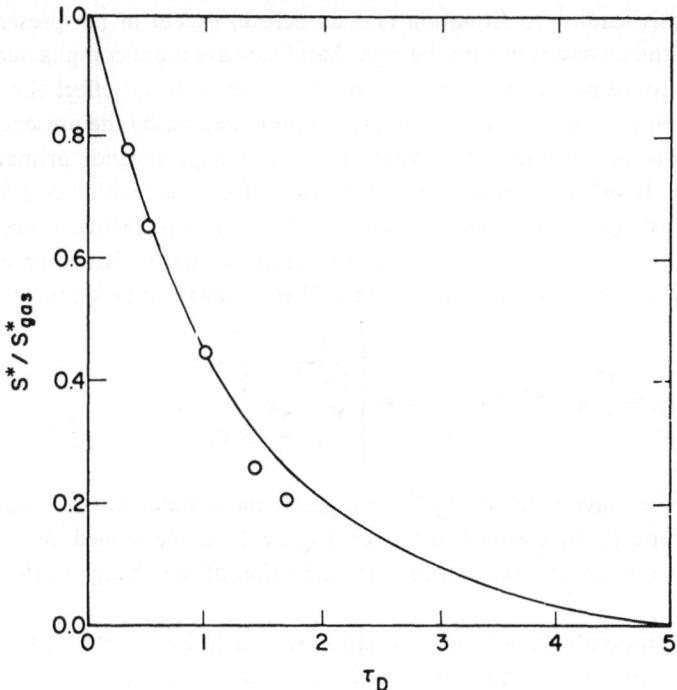

Fig. 1. Fraction of ionizing photons absorbed by gas vs the total effective absorption optical depth. The points are from the numerical results of Mathis (1971) which refer to the line of sight absorption optical depths.

is plotted vs τ_D (the effective absorption optical depth of dust) on Figure 1. As can be seen, even for optical depths of about unity, by neglecting the effect of dust one underestimates the required ionizing flux by a factor of about 2 to 3. For comparison with the expression (11) I have plotted on Figure 1 the points from numerical calculations of Mathis (1971). The discrepancies at large optical depths are due to scattering by dust which was included in Mathis' calculation. I have neglected the scattering by dust and have interpreted τ_D as the *effective* absorption optical depth (along the photon path) while points from Mathis' work refer to the absorption optical depth along the radius vector which is smaller than the effective optical depth when the albedo of dust is not equal to zero.

(b) Effect of dust on the ratio of helium to hydrogen Strömgren radii: In order to discuss this effect one must consider the fate of photons which can ionize only hydrogen (photons with $v_1 < v < 1.8\, v_1$) and photons which can ionize both hydrogen and helium ($v > 1.8\, v_1$) separately. I define the following quantities:

$$S_1(r) = \int_{v_1}^{1.8v_1} N(v,r)\,dv, \qquad S_2(r) = \int_{1.8v_1}^{\infty} N(v,r)\,dv, \qquad S_2(r_0') = 0,$$

$$= S_1 + S_2, \qquad \gamma^* = S_2^*/S^*, \qquad Y = n(\text{He})/n(\text{H}), \qquad R = (r_0'/r_0)^3 . \quad (12)$$

One can write differential equations similar to Equation (6) for S_1 and S_2 (and carry out the analysis leading to Equation (11) as before) except in the present case the integrands in the second terms on the right hand side are more compliacted functions of the radial coordinate. This complication, however, will not effect the changes in the ratio R of the volumes of the Strömgren spheres caused by the presence of dust.* For a nondusty nebula this ratio, which I will call R_{gas}, depends primarily on the quantity γ^*/Y. It will also depend on the fraction of photons with $v > 1.8v_1$ that are absorbed by hydrogen. For small values of $\gamma^* (\gamma^*/Y < 1)$ this fraction is negligible and $R_{gas} \approx \gamma^*/Y(1 + Y)$. For $\gamma^*/Y \gg 1$ this ratio tends to unity. Hummer and Seaton (1964) have discussed this problem in detail. Their results can be summarized by

$$
R_{gas} = \frac{\gamma^*}{Y} g(\bar{\gamma}, \gamma^*, Y), \qquad g = \begin{cases} \dfrac{1}{1+Y}, & \dfrac{\gamma^*}{Y} < 1 \\[2mm] \dfrac{Y}{\gamma^*}, & \dfrac{\gamma^*}{Y} \gg 1, \end{cases} \tag{13}
$$

where $\bar{\gamma}$ is the average value of S_2/S throughout the nebula. This relation has been shown schematically by the dashed line on Figure 2. As mentioned above, the exact shape of this curve is not essential for the derivation of the change in the value of R due to dust.

For a dusty but uniform nebula an equation similar to Equation (13) still holds, but in this case the quantity on the right hand side is no longer equal to the ratio of the volumes, instead we have

$$
\int_0^{\tau'_D} \tau^2 e^\tau \, d\tau / \int_0^{\tau_D} \tau^2 e^\tau \, d\tau = \frac{\gamma^*}{Y} g, \tag{14|}
$$

so that the ratio of the Strömgren volumes is

$$
R \equiv \left(\frac{\tau'_D}{\tau_D}\right)^3 = \frac{f(\tau_D)}{f(\tau'_D)} R_{gas}, \tag{15}
$$

where f is defined in Equation (9). In deriving Equation (15) I have assumed that $\bar{\sigma}_D = \bar{\sigma}'_D$ and that the function g is not changed by the presence of dust. For a given value of R_{gas} and τ_D, Equations (9) and (13), give the value of τ'_D and the ratio R. The ratio R is plotted on Figure (2 for various values of τ_D. As is evident, the effect of dust is to bring this ratio closer to unity.

4. Summary and Results

Now let us see how the effects I have discussed here will change the helium to hydrogen

---

* For comparison with observation (optical or radio line intensity ratios) we need the ratio of emission measures

$$
R' = \int_0^{r'_o} n_e n_{He\,II} r^2 \, dr / \int_0^{r_o} n_e n_{H\,II} r^2 \, dr .
$$

For a uniform nebula $R' = RY$.

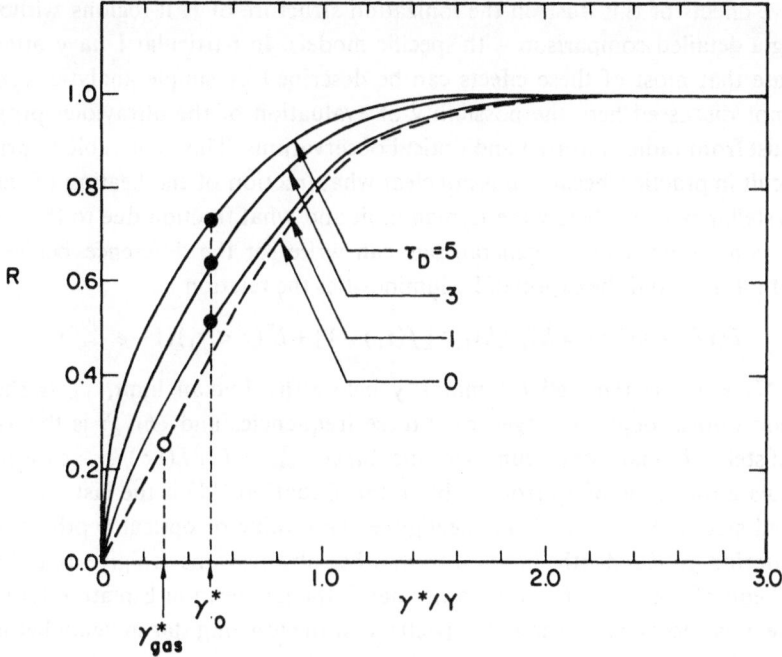

Fig. 2. The ratio of the values of Strömgren Spheres of helium to hydrogen versus the ratio of fraction of stellar ionizing photons capable of ionizing helium γ^* to helium to hydrogen abundance ratio Y. For details see Section 4.

line intensity ratios. Let us consider an H II region of observed size r_o, and an assumed abundance ratio Y. Neglecting the presence of dust one can calculate, from Equation (7), the required number of ionizing photons and find the spectral type of the ionizing star. From this and model atmospheric calculations (e.g. Auer and Mihalas, 1971) one finds the value of γ^*_{gas}. The dashed line on Figure 2 then gives the expected value of the ratio R_{gas} (the open circle on Figure 2) If there is absorbing dust of optical depth τ_D in the H II region we know that the required number of ionizing photons is increased by $f(\tau_D)$ (Equation (9)). This will require as an ionizing agent a new star (of larger effective temperature) which implies a new and in general a larger value for γ^*, say γ^*_0 (Figure 2). The correct value of the ratio R is then obtained by reading on Figure 2 for $\gamma^* = \gamma^*_0$ the value of the ratio from the appropriate curve (filled circles). These two effects increase the ratio R by a substantial factor, so that by neglecting dust, one underestimates the values of helium to hydrogen lines intensity ratios. This effect is more important for smaller values of γ^*/Y and tends to zero for $\gamma^*/Y > 1$, where both the dusty and nondusty models give a value near unity for the ratio R. It is believed that for bright H II regions the ratio R is near unity. Whether this is due to a large value of γ^*/Y or due to dust (or both) depends on various parameters which can vary from nebula to nebula.

In summary, I have interpreted the large far infrared flux observed from H II regions as emission by internal dust. My purpose here has been to point out the qualititive effects of this dust on the ionization structure of H II regions without attempting a detailed comparison with specific models. In particular I have attempted to indicate that most of these effects can be described by simple analytic formulae. I have not discussed here the possibility of evaluation of the ultraviolet properties of the dust from radio, infrared and optical observations. This is possible in principle but difficult in practice, because it is not clear what fraction of the heating of the dust is due to stellar photons below the Lyman limit and what fraction due to the ionizing photons considered here. In general, one can write for the difference between the observed infrared and the expected Lα luminosities the relation

$$L(\mathrm{I\,R}) - L(\mathrm{L}\alpha) = S_{\mathrm{gas}}^{*} \langle h\nu_{\mathrm{Lc}} \rangle [f(\tau_D) - 1] + L^{*}(\nu < \nu_1)(1 - e^{-\tau''_D}), \qquad (16)$$

where $L^{*}(\nu < \nu_1)$ is the stellar luminosity below the Lyman limit, τ''_D is the dust absorption optical depth averaged over these frequencies and $\langle h\nu_{\mathrm{Lc}} \rangle$ is the average energy of stellar Lyman continuum photons. Since $S_{\mathrm{gas}}^{*} = (\frac{3}{2}) L(\mathrm{L}\alpha)/h\nu_1$ (case B), one can calculate the value of τ_D from Table I and Equation (11) if the last term on the right hand side of Equation (16) is negligible. This value of optical depth combined with scattering optical depths inferred from other observations can give an indication of the albedo of the dust. If, on the other hand, the last term in Equation (16) is not negligible, it is necessary to know the spectrum of the ionizing star in detail before one can proceed with the above calculation.

Acknowledgements

This work was supported by National Aeronautics and Space Administration Grant Number NGR 05-020-510.

References

Burgess, A. and Seaton, M. J.: 1960, *Monthly Notices Roy. Astron. Soc.* **121**, 471.
Code, A.: 1973, this volume, p. 505.
Harper, D. A. and Low, F. J.: 1971, *Astrophys. J. Letters* **165**, L9.
Hoffmann, W. F., Frederick, C. L., and Emery, R. J.: 1971, *Astrophys. J. Letters* **170**, L89.
Hummer, D. G. and Seaton, M. J.: 1964, *Monthly Notices Roy. Astron. Soc.* **127**, 217.
Krishna Swamy, K. S. and O'Dell, C. R.: 1967, *Astrophys. J.* **147**, 529.
Mathis, J. S.: 1971, *Astrophys. J.* **167**, 261.
Auer, L. H. and Mihalas, D.: 1971, *Astrophys. J. Suppl.* **24**, 193.
Munch, G. and Persson, S. E.: 1971, *Astrophys. J.* **165**, 241.
O'Dell, C. R. and Hubbard, W. B.: 1965, *Astrophys. J.* **142**, 591.
O'Dell, C. R., Hubbard, W. B., and Peimbert, M.: 1966, *Astrophys. J.* **143**, 743.
Petrosian, V., Silk, J., and Field, G. B.: 1972, *Astrophys. J. Letters* **177**, L69.
Robbins, R. R.: 1970, *Astrophys. J.* **160**, 519.
Soifer, B. T., Pipher, J. D., and Houck, J. R.: 1972, paper presented at the IAU Symp. 52 on 'Interstellar Dust and Related Topics', not published.
Spitzer, L., Jr.: 1968, in *Diffuse Matter in Space*, Interscience Publishers, Division of J. Wiley and Sons, New York, p. 115.

LOW RESOLUTION SPECTRUM OF THE
ORION NEBULA IN THE 60–300 μ RANGE

C. D. SWIFT, F. C. WITTEBORN, E. F. ERICKSON, L. J. CAROFF,
G. C. AUGASON, A. J. MORD, L. W. KUNZ, and L. P. GIVER

Ames Research Center, NASA, Moffett Field, Calif. 94035, U.S.A.

Abstract. Airborne observations of the Trapezium region of the Orion Nebula in the 60–300 μ range have been made from 13.7 km altitude using a Michelson interferometer with the Rice University 12-in. Flying Infrared Telescope. Fourier analysis of five interferometer scans provided spectra with resolution ranging from 7 cm^{-1} to 20 cm^{-1}. These spectra were compared with lunar spectra taken with the same instrument at the same altitude to correct for instrumental and atmospheric effects. A weighted combination of these scans provides a low resolution spectrum. The radiation per unit frequency interval at 190 μ was found to be at least 40% of that measured between 75 and 90 μ. Neglecting possible features in the spectrum, its general shape is consistent with a blackbody near 70 K. The resolution is inadequate to resolve spectral lines.

Broadband observations of the Trapezium region of the Orion Nebula made by Low and Aumann (1970) revealed a large amount of radiation in the far infrared which was peaked near 70 μ. Their observations were made with a 31-cm open port telescope (Low *et al.*, 1970) on board a NASA Lear jet flown at an altitude of 13.7 km to eliminate most of the atmospheric water vapor absorption. To determine the spectral shape of the radiation we have added a Michelson interferometer to the airborne telescope to perform Fourier spectroscopy beyond 50 μ. The measurements described here result in a crude spectrum of a four arc minute diameter (estimated FWHM) portion of the Trapezium region of the Orion nebula from 60 to 300 μ.

A schematic diagram of the apparatus is shown in Figure 1. Light enters the Cassegrain telescope (Dall-Kirkham optics, Cervit primary with aluminum coating and aluminized silicon secondary) and then passes through a 1-mm white polyethylene window into an evacuated chamber which contains a Michelson interferometer developed by Augason and Young (1971) and since modified for airborne use. The interferometer employs a 12.5 μ mylar beam splitter with gold coated plane mirrors. The detector is a gallium-doped-germanium bolometer cooled to 2 K and filtered to transmit at wavelengths longer than 50 μ. Actually we found that there was some leakage at shorter wavelengths. This was measured in the laboratory and appropriate small corrections were applied to the spectra obtained from the Lear jet.

In order to cancel radiation emitted by the sky and other sources such as the telescope, the signal is modulated by oscillating the secondary mirror horizontally. This permits the detector to view either of two 4′ diameter fields or 'beams' separated by 12′ and located symmetrically to the right and left of the axis of the primary mirror. To remove any 'offset signal remaining due to residual asymmetries in the system, signals can be recorded with the object in each of the two beams. The AC signal produced by the oscillating secondary mirror is amplified by a preamplifier and subsequently by a phase-lock amplifier synchronized with the secondary mirror drive signal. The output

Greenberg and Van de Hulst (eds.), Interstellar Dust and Related Topics, 453–457.

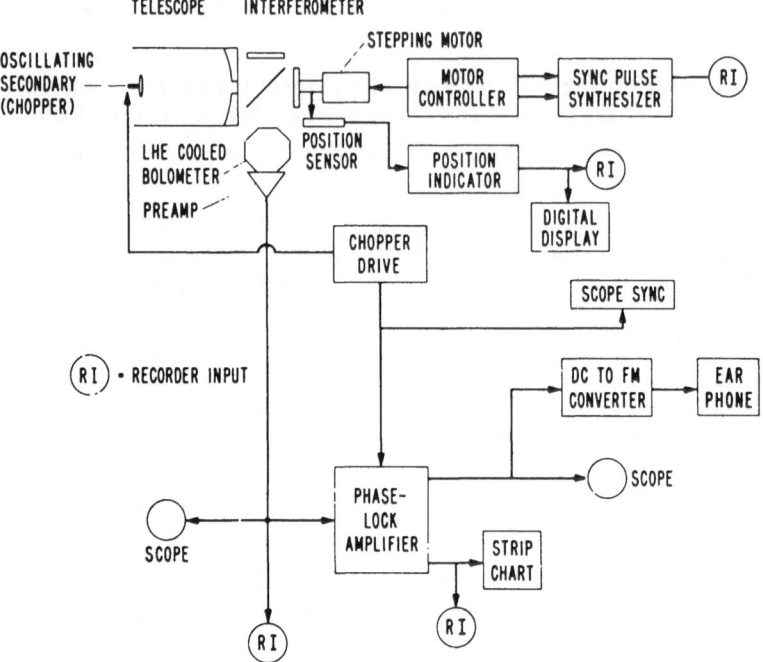

Fig. 1.   Interferometer electronics block diagram.

of the phase-lock amplifier is recorded as a function of the position of the movable interferometer mirror. This mirror is moved in 9.6 μ steps. When looking at a bright extended source such as the Moon, the interferometer may be stepped automatically every one-half second or less, and as many as 256 steps are taken in a single scan. When looking at fainter sources, the interferometer mirror is not moved until the signal level is clearly established in both the left and right beams. This process is slow, so that during a single flight there was time for only 30 to 70 steps when observing the Orion Nebula. The resulting interferograms are then Fourier analyzed to yield unapodized spectra.

The spectrum obtained from a single two minute, 256 step lunar scan is shown in Figure 2. Its resolution is approximately 2.5 cm^{-1}. The main sharp features are due to telluric water vapor absorption. The overall shape is determined by the filter cut-on near 200 cm^{-1}. the beam splitter function and the blackbody character of the lunar emission. An average of five spectra of the Trapezium region is shown in Figure 3. The resolution is 10 cm^{-1} and the line lengths indicate two standard deviations. To correct the Trapezium spectrum for telluric absorption and instrument effects we have divided the spectrum in Figure 3 by the lunar spectrum in Figure 4 which was degraded to the same resolution. The result is then multiplied by a 360 K balckbody function which we considered appropriate (Kopal, 1969) for the obliquely illuminated portions of the moon which we observed. The result appears in Figure 5. Since it is not possible to take both lunar and Orion spectra simultaneously and usually not even the same night,

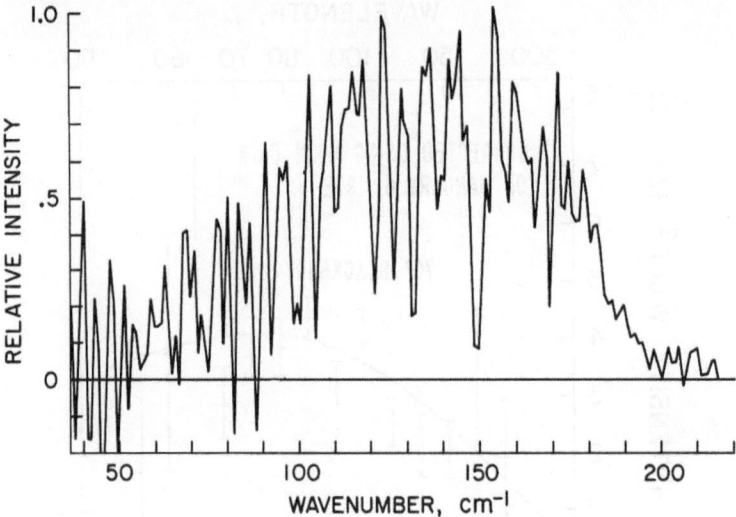

Fig. 2.   Moon spectrum not corrected for atmospheric absorption and instrument efficiency.

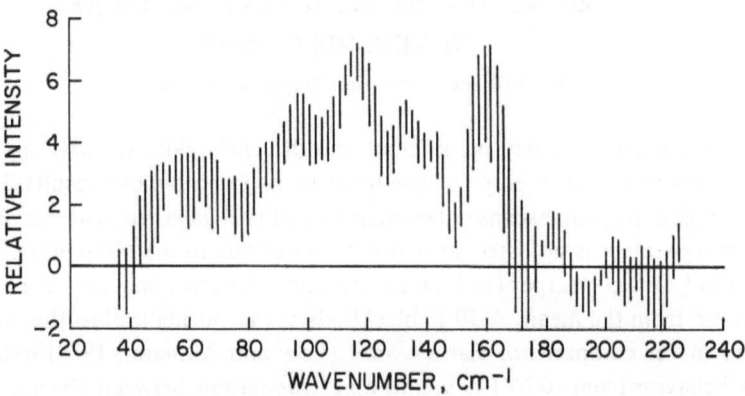

Fig. 3.   FIR spectrum of Orion Nebula-Trapezium region uncorrected for atmospheric absorption and instrument efficiency. The calculated resolution is 10 cm⁻¹.

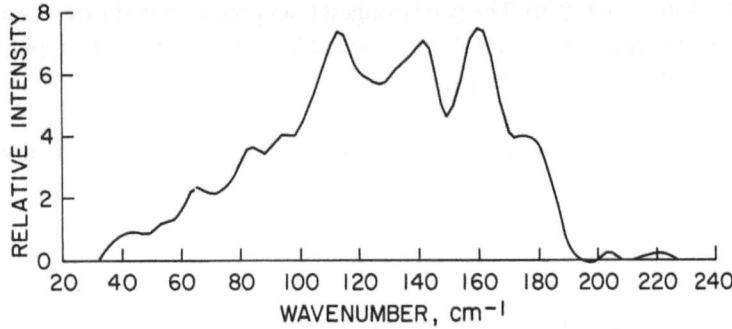

Fig. 4.   Degraded moon spectrum uncorrected for atmospheric absorption and instrument efficiency.

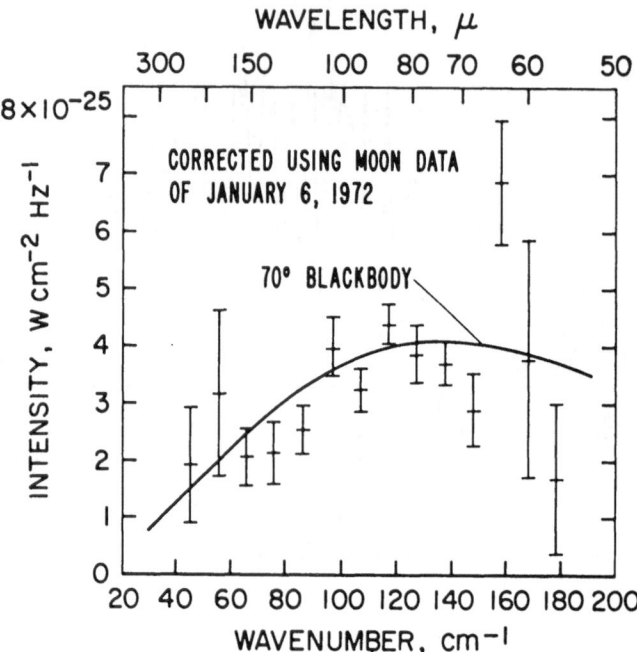

Fig. 5.   FIR spectrum Orion-Trapezium region.

the water vapor correction may be different on different nights. We used an average of five Orion scans and an average of five lunar scans to obtain our results. Use of a different average of five lunar scans obtained in August 1971 gave nearly identical results.

The intensity scale was adjusted so that the overall flux in the 60 tp 300 μ range is normalized to Low and Harper (1971) measurement. The error bars indicate ± 1 standard deviation from the mean. A 70 K blackbody curve, normalized to the same total flux, is shown for comparison. Earlier work (Low and Aumann, 1970) indicated a blackbody behavior from 10 to 100 μ, and an extrapolation between 100 μ and 1 mm suggested a $v^{3.5 \pm 0.5}$ behavior at longer wavelengths. Figure 5 shows that the blackbody shape extends out to 200 μ.

Resolution and signal to noise were insufficient to detect lines having intensities predicted by Petrosian (1970). The two strongest lines predicted in the observed spectral range are the O I line at 63 μ and the O III line at 88 μ. While there are apparent peaks in Figure 5 at these wavelengths, the large standard deviations (determined directly from the spread in the five measured spectra) precludes a positive identification. The instrument used in these measurements is capable of better resolution and with sufficient integration time should be capable of producing a more accurate spectrum.

Acknowledgements

We are indebted to F. J. Low and C. M. Gillespie not only for the use of the airborne telescope which they developed, but also for instruction in its use. We are also pleased to acknowledge the excellent flying of the NASA test pilots and the work of the ground crew and the Space Science technicians which made the measurements possible.

References

Augason, G. C. and Young, N.: 1971, in G. A. Vanasse, A. T. Stair and D. J. B. Baker (eds.), *Aspen International Conference on Fourier Spectroscopy*, Air Force Cambridge Research Laboratories, Bedford, Mass. AFCRL 71-0019, p. 281.

Kopal, Z.: 1969, *The Moon*, Reidel, Holland, p. 373–4.

Low, F. J. and Aumann, H. H.: 1970, *Astrophys. J. Letters* **162**, L79.

Low, F. J., Aumann, H. H., and Gillespie, C. M., Jr.: 1970, *Astronautics Aeronautics* **7**, 26.

Low, F. J. and Harper, D. A.: 1971, *Astrophys. J. Letters* **165**, L9.

Petrosian, V.: 1970, *Astrophys. J.* **159**, 833.

NEW INFRARED SOURCES ASSOCIATED WITH H II REGIONS

C. G. WYNN-WILLIAMS

California Institute of Technology, Pasadena, Calif., U.S.A.

and

E. E. BECKLIN and G. NEUGEBAUER

California Institute of Technology, Pasadena, Calif., U.S.A.

and

Hale Observatories, California Institute of Technology, and Carnegie Institution of Washington, D.C., U.S.A.

Abstract. Photometry and mapping in the wavelength range 1.65–20 μ of the H II regions W3 and NGC 7538 have led to the discovery of a dozen new infrared emission sources. The sources have flux densities ranging up to 10^{-23} W m^{-2} Hz^{-1} at 20 μ and diameters from < 3 to 40″. Some are associated with, and have similar brightness distributions to compact H II condensations observed at radio wavelengths; they have, however, 20-μ flux densities much greater than, and 2-μ flux densities much less than are predicted from the radio free-free spectra. It is concluded that in many cases there exist close spatial relationships between the hot dust and the ionized gas, and between infrared and OH/H$_2$O emission sources.

1. Introduction

We have been using the Mount Wilson 100-in. and Mount Palomar 200-in. telescopes to make broad-band observations of several galactic H II regions in the range 1.65 to 20 μ with as high as 5″ spatial resolution. The H II regions studied have generally been those which contain OH or H$_2$O maser sources and which have been shown by radio astronomers to contain compact condensations of ionized gas which may be connected with star formation processes. In this paper a brief account of two such sources, W3 and NGC 7538, is given. More detailed descriptions of these and other sources will shortly be published in the regular literature (Wynn-Williams *et al.*, 1972, 1973).

2. W3

W3 has several regions of interest at radio wavelengths. Most of the observations described here, however, relate to the W3 (continuum) source (G133.7 + 1.2), an area some 3′ in diameter, from which the bulk of the continuum radio emission originates at high frequencies.

Figure 1 shows the maps of the region at 2.2 and 20 μ compared with the 5-GHz map of the same area made with the Cambridge One Mile Radio Telescope (Wynn-Williams, 1971). The resolution of all these maps is in the range 5–10″. It can be seen that three of the compact H II condensations visible at 5 GHz have associated infrared sources.

The nature of the energy distribution of these sources (Figure 2) indicates that

Greenberg and Van de Hulst (eds.), Interstellar Dust and Related Topics, 459–463.

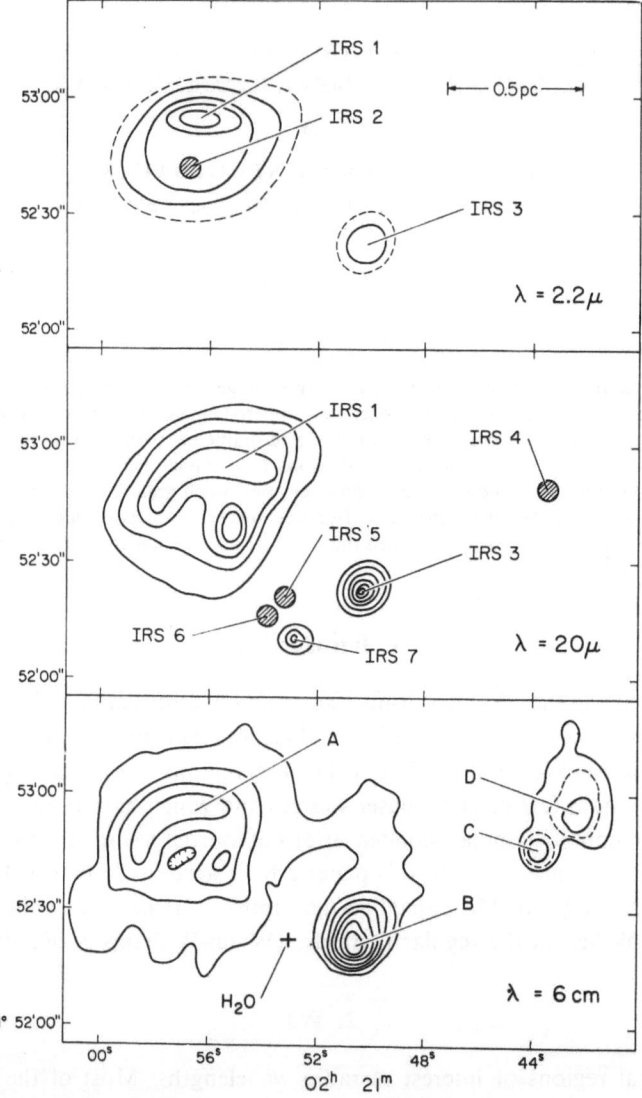

Fig. 1.   The W3 (continuum) region at 2.2-μ, 20-μ, and 6-cm wavelength. The radio continuum data
are taken from Wynn-Williams (1971) and have a resolution of 6.5″ × 7.4″; the position of the H_2O
maser source is as given by Hills *et al.* (1972). Hatched circles indicate unresolved sources and dashed
contours are half intervals. Sources to the west of $02^h21^m48^s$ are not shown at 2.2 μ because of the
presence of confusing field stars.

heated dust must be responsible for the emission in the 3- to 20-μ range. The close
correspondence between the 20-μ and 5-GHz maps indicates that the heated dust and
the ionized gas occupy the same regions of space. Observations at 10 and 20 μ indicate
a dust temperature of the order of 150 K, and that the infrared emission from the
sources can be caused by a collection of small particles with a total mass of only 10^{-2}

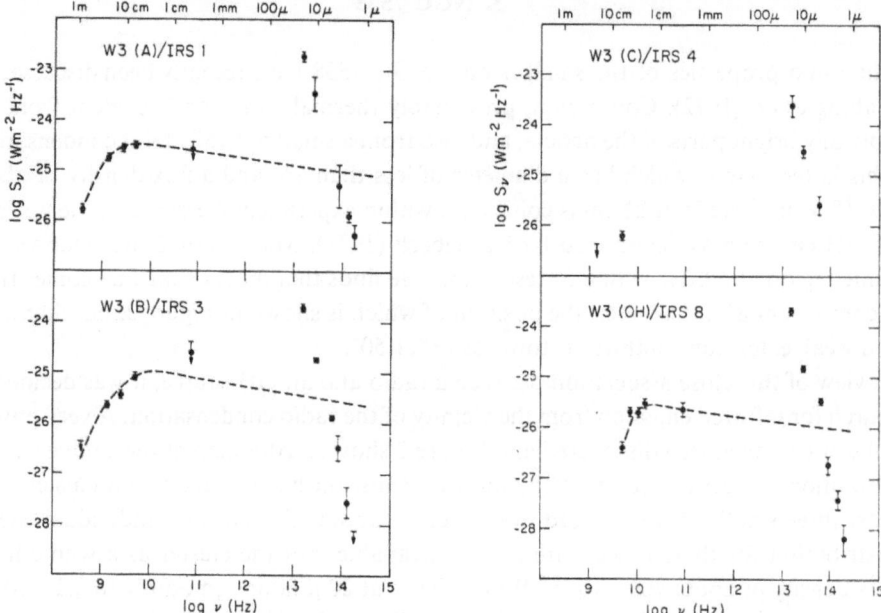

Fig. 2. The energy distributions of the four sources W3(A)/IRS1, W3(B)/IRS(3), W3(C)/IRS4, and W3(OH)/IRS8 at radio and infrared wavelengths. The dashed lines indicate the expected infrared thermal emission from the ionized hydrogen as predicted from the radio emission. References for the radio data are given in Wynn-Williams *et al.* (1972).

of the ionized gas. It is thus unnecessary to make any special assumptions about the nature or concentration of the dust mixed with the gas.

It can also be seen in Figure 2 that the 2-μ emission from the sources is well below that predicted from free-free and free-bound transitions in the ionized hydrogen plasma. There must therefore be large amounts of obscuring material associated with and in front of the ionized gas/hot dust region. These amounts vary, but reach 50 visual magnitudes in front of W3 (B).

There are several other infrared sources in Figure 1 as well as those already mentioned, which do not correspond to H II condensations. The magnitude and infrared color of one of these, IRS2, show that it is almost certainly a highly obscured O5 star, and the main source of excitation and energy in W3 (A)/IRS1. One of the others, IRS5, has a flux density of 500×10^{-26} W m^{-2} Hz^{-1} at 20 μ, but a diameter of less than 3″ (10^4 AU). Its luminosity (3×10^4 $L_\odot$) and temperature (300–400 K) lead to speculation that it may be a massive protostar. Its position agrees within 3″ with that of the H$_2$O maser source as determined by Hills *et al.* (1972).

Finally, two further infrared sources were discovered close to the W3(OH) source, some 12′ away from the region mapped in Figure 1. One of these, IRS8, whose spectrum is shown in Figure 2, is coincident with the OH, H$_2$O, and radio continuum positions, while the other, with a rather flatter energy distribution than that of IRS8, is not coincident with any known optical or radio feature.

3. NGC 7538

The radio properties of the small nebula NGC 7538 have recently been discussed by Habing *et al.* (1972). Continuum, presumably thermal, emission is evident from the optically bright parts of the nebula, and also from a small optically faint condensation. This latter source, which has a diameter of less than 15" and a flux density of about 10^{-26} Wm^{-2} Hz^{-1} at 21 cm is coincident within experimental error with the position of OH emission as determined by Hardebeck (1971). Martin (1972) has mapped the same region at 5 GHz with 6.5" resolution. He finds that most of the flux comes from a component 8" in diameter, the position of which is shown in Figure 3, but that there is a weak extension southwards towards 61°11′50″.

In view of this close association between a radio and an OH source, it was decided to search for infrared emission from the vicinity of the radio condensation. A very powerful source was indeed discovered and Figure 3 shows a 20-μ map of the source with 5" resolution. It can be seen that the infrared emission has its origin in a cluster of at least three small sources, spread over a region about 15" diameter. Individual energy distribution for these objects are not yet available, but the cluster as a whole has a flux density of about 700×10^{-26} Wm^{-2} Hz^{-1} at 20 μ falling smoothly to 0.2×10^{-26} Wm^{-2} Hz^{-1} at 1.65 μ in a manner fairly similar to the W3 objects shown in Figure 2.

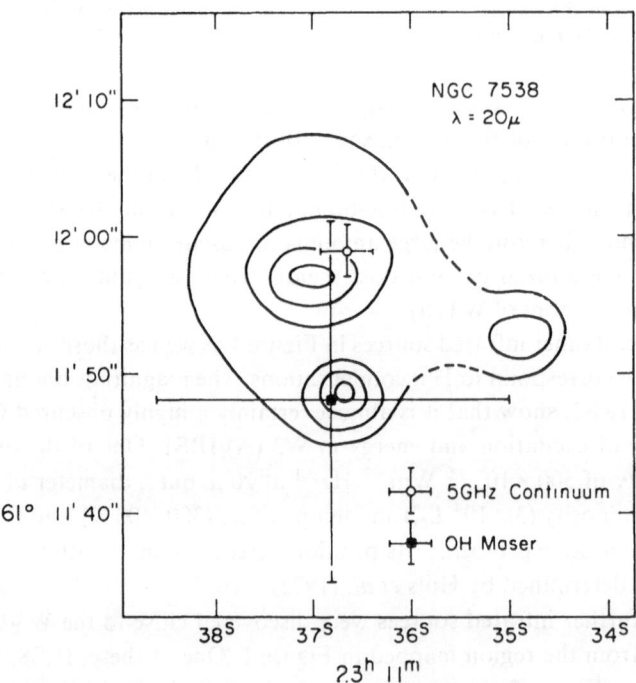

Fig. 3. The 20-μ emission from the vicinity of the OH source in NGC 7538. The crosses show the peaks of the 1720-MHz OH emission (Hardebeck, 1971) and of the 5-GHz thermal radio continuum emission (Martin, 1972).

Comparison with Martin's results would therefore indicate that, as in W3, some, but not all, of the infrared sources in NGC 7538 are coincident with H II condensations or microwave maser sources.

4. Conclusions

On the basis of the observations of W3, NGC 7538, and some other H II regions which are still being analyzed, it may be concluded that:

(1) In general H II condensations are infrared sources in the range 3–20 μ. The data are consistent with the emission having its origin in heated dust mixed with the ionized gas.

(2) The condensations are hidden optically by large quantities of obscuring matter, up to 50 visual magnitudes thick in one case.

(3) There exist additional infrared sources in the vicinity of the H II condensations. One of these may be a protostar with a luminosity of $3 \times 10^4 L_\odot$.

(4) There is a close connection between infrared sources and OH maser sources in H II regions.

Acknowledgements

This work could not have been done without the support of the Caltech infrared group; we especially thank J. Bennett, G. Forrester, and L. Roberts for their assistance. We further thank T. Hilgeman who made some of the early infrared measurements of W3 with us, and our night assistants E. Hancock and J. Corresco for their help. A. H. M. Martin kindly allowed us to use his radio data on NGC 7538 prior to publication. C.G.W-W. is grateful to the Commonwealth Fund for the receipt of a Harkness Fellowship and to Hale Observatories for guest observer's privileges. This work was supported in part by National Aeronautics and Space Administration Contracts NGL 05-002-007 and NGL 05-002-207.

References

Habing, H. J., Israel, F. P., and de Jong, T.: 1972, *Astron. Astrophys.* **17**, 329.
Hardebeck, E.: 1971, *Astrophys. J.* **170**, 281.
Hills, R., Janssen, M. A., Thornton, D. D., and Welch, W. J.: 1972, *Astrophys. J. Letters* **175**, L59.
Martin, A. H. M.: 1972, private communication.
Wynn-Williams, C. G.: 1971, *Monthly Notices Roy. Astron. Soc.* **151**, 397.
Wynn-Williams, C. G., Becklin, E. E., and Neugebauer, G.: 1972, *Monthly Notices Roy. Astron. Soc.* **160**, 1.
Wynn-Williams, C. G., Becklin, E. E., and Neugebauer, G.: 1974, *Astrophys. J.* (in press).

PHYSICAL STUDY OF A BRIGHT RIM
IN THE HORSEHEAD NEBULA

R. LOUISE and C. SAPIN

Observatoire de Marseille, 2, place Le Verrier, F 13004 Marseille, France

Abstract. The wellknown 'Horsehead' nebula is photographed with narrow interference filters centered respectively on Hα and [N II] ($\lambda = 6584$ Å). The [N II]/Hα ratio is deduced and its variation towards the Horsehead bright rim is studied. It is shown that the maximum value of the temperature is not reached on the rim itself but in a narrow zone located just behind it and ahead of the dark matter.

1. Introduction

Among the most intriguing features in H II regions are 'elephant trunk' structures, so designated because of their elongated form; they also are called bright rims (Pottasch, 1958). They are characterised by neutral dark matter intruding into the H II zone and are brighter than the rest of the nebula. An important theoretical work done by Pottasch shows that both electron density and temperature increase towards the bright rims. Up to now, only a few observations (Osterbrock, 1957; Louise, 1970) have been made concerning them.

Our purpose is to measure the variation of temperature from the normal H II zone towards the bright rim by observing the ratio of [N II ($\lambda = 6584$ Å) and Hα ($\lambda = 6563$ Å) emission lines. The wellknown and typical bright rim in IC 434, the Horsehead nebula, is chosen to be observed at the 120 cm telescope of the Haute Provence Observatory.

2. Observational Method

We may choose one of the observational methods described by Courtès *et al.* (1969). Unfortunately, the bright rims are generally small structures and high angular resolution observations appear to be the desirable.

The method we have adopted is based on the fact that if the [N II]/Hα ratio is known in one point (A) of the nebula it is possible to derive that ratio is at any point (B) by photographing the nebula through narrow interference filters ($\Delta\lambda = 10$ Å) centered respectively on $\lambda = 6584$ Å and Hα.
Indeed we have:

$$\frac{([N\ II]/H\alpha)_B}{([N\ II]/H\alpha)_A} = \frac{([N\ II])_B}{([N\ II])_A} \cdot \frac{(H\alpha)_A}{(H\alpha)_B}. \tag{1}$$

The above formula shows that the Hα and [N II] plates may be taken separately with an adequate exposure time. Furthermore it may be interpreted as the [N II]/Hα relative variation over the whole nebula compared to the [N II]/Hα of the point (A).

Greenberg and Van de Hulst (eds.), Interstellar Dust and Related Topics, 465–470.

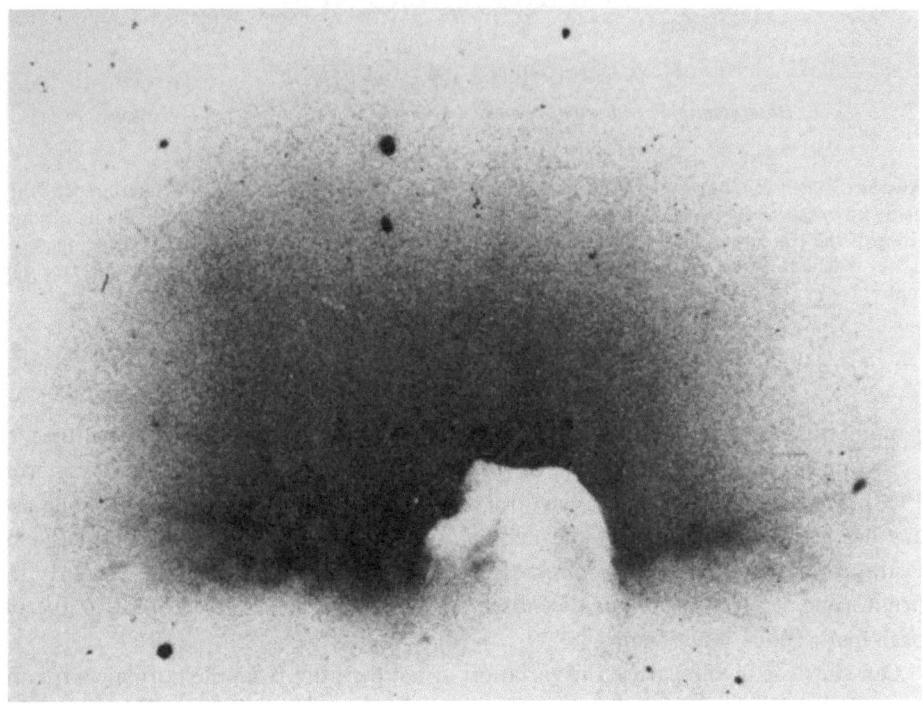

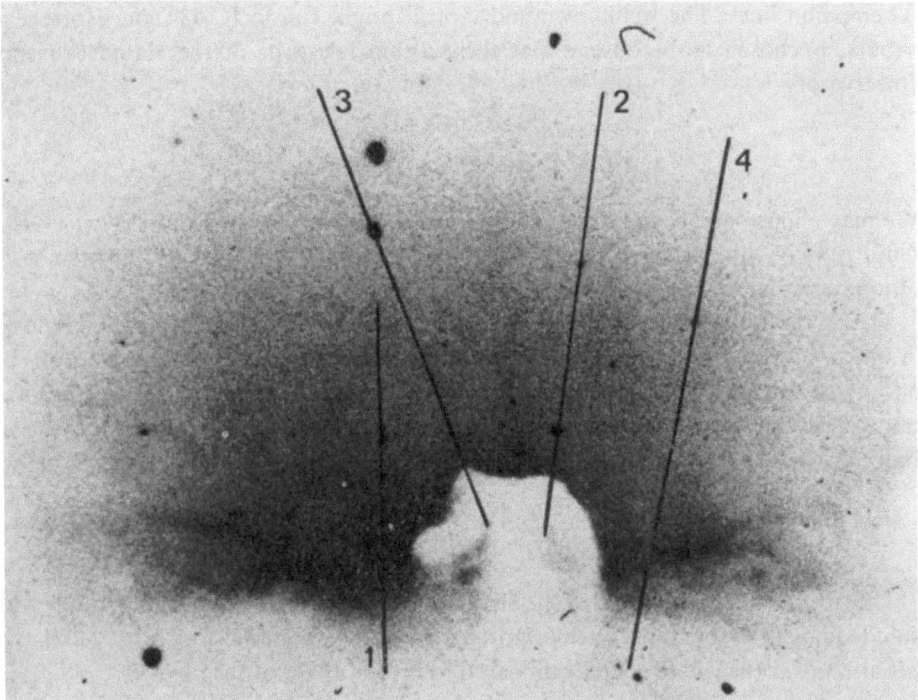

Fig. 1. Monochromatic exposures of the Horsehead nebula in Hα (top) and in [N ɪɪ] (bottom). Lines show the four microphotometer scans.

In this work the point (A) is chosen at the top of the bright rim photometric profile. On the other hand, it is not necessary to know the [N II]/Hα of the point (A) if we only want to study the variation of that ratio from the normal H II zone towards the bright rim.

3. Observation and Reduction

Plates with various exposure times were taken with the classical Focal Reducer (Courtès, 1960), without Fabry-Perot etalon, at the Newtonian focus of the 120 cm telescope of the Haute Provence Observatory. Figure 1 shows the image of the Horse-head nebula taken in Hα and [N II] monochromatic light respectively. The exposure times, 40 min for Hα and 1^h30 for [N II], are adopted because they allow to obtain practically the same optical density on the Kodak 103a-E plate. The Joyce and Loebl microphotometer is used to record the brightness along 4 directions as indicated by Figure 1. These directions are adopted in a rather arbitrary way owing to the positions

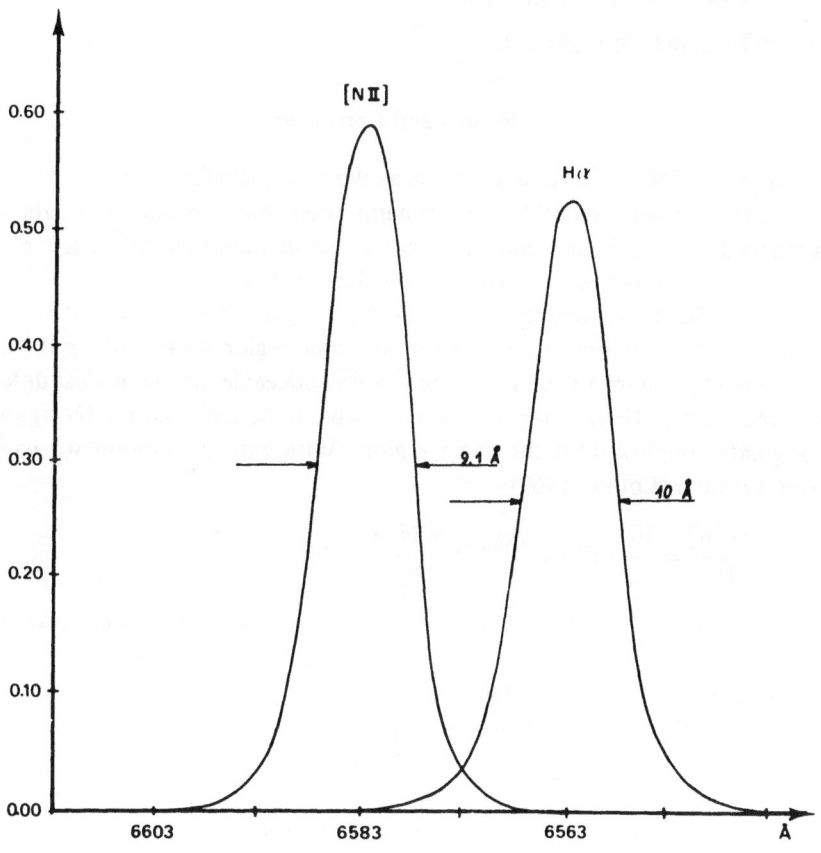

Transmission curve given by the constructor

Fig. 2. Transmission curves of the two filters.

of stars on the plate; but they must be as perpendicular as possible to the bright rim.

The photographic calibration is obtained by using the classical ETA photometric spectrograph of the Haute Provence Observatory.

The interference filters we used present a narrow bandwidth. A sample of their transmission curve is given in Figure 2. The Hα image is not in practice contaminated by [N II] light and vice versa. Although the bandwidth is narrow ($\Delta\lambda = 10$ Å) and the continuum emission spectrum of the nebula is weak, the observed intensity is always overestimated.

Indeed, we have:

$$I_{H\alpha}(\text{observed}) = I_{H\alpha}(\text{Line}) + \varepsilon(\text{continuum})$$
$$I_{[N\,II]}(\text{observed}) = I_{[N\,II]}(\text{Line}) + \varepsilon'(\text{continuum}).$$

ε and ε' are respectively the continuum intensity of the nebula admitted by Hα and [N II] interference filters. For $I_{H\alpha}$ (line) $\simeq 3\,I_{[N\,II]}$ (line) we have:

$$\left(\frac{I_{[N\,II]}}{I_{H\alpha}}\right)_{observed} = \left(\frac{I_{[N\,II]}}{I_{H\alpha}}\right)_{Line} + \Delta.$$

$\Delta = (\varepsilon' - \varepsilon/3)/I_{H\alpha}$ may be neglected.

4. Results and Comments

Figure 3 gives the [N II]/Hα variation towards the bright rim for 4 selected directions. This ratio first increases smoothly, then remains practically constant before the bright rim is reached. Its maximum value does not occur in the bright rim itself but in a narrow zone situated between the rim and the dark matter.

It is very difficult to interpret the observed [N II]/Hα ratio (Mein, 1968; Baudel, 1970) for these lines do not originate from the same region located along the line of sight. Fortunately Figure 1 shows that there is no noticeable morphological difference between the Hα and [H II] images. It is reasonable to assume that the [N II] and Hα lines originate roughly from the same region. With this assumption we have the following formula (Louise, 1969);

$$\frac{[N\,II]}{H\alpha} = \frac{10^4}{6.2} \cdot T_e^{\frac{1}{2}} 10^{-9500/T_e} \cdot \frac{N(N^+)}{N_e}.$$

$N(N^+)/N_e$ is the relative abundance of singly ionized nitrogen. It is not certain that this quantity remains constant within a given nebula because of the ionisation structure effect (Hummer and Seaton, 1963).

Consequently the observed [N II]/Hα variation must be interpreted as the variation of both $N(N^+)/N_e$ and T_e towards the bright rim. For many theoretical considerations T_e must increase towards the bright rim; so, if we assume $N(N^+)/N_e$ to be constant, we may interpret Figure 3 in terms of variation in T_e. Doing so, the figure shows that Te increases smoothly and/or remains constant until the bright rim is reached, and then suddenly increases behind it. There are good reasons to think that

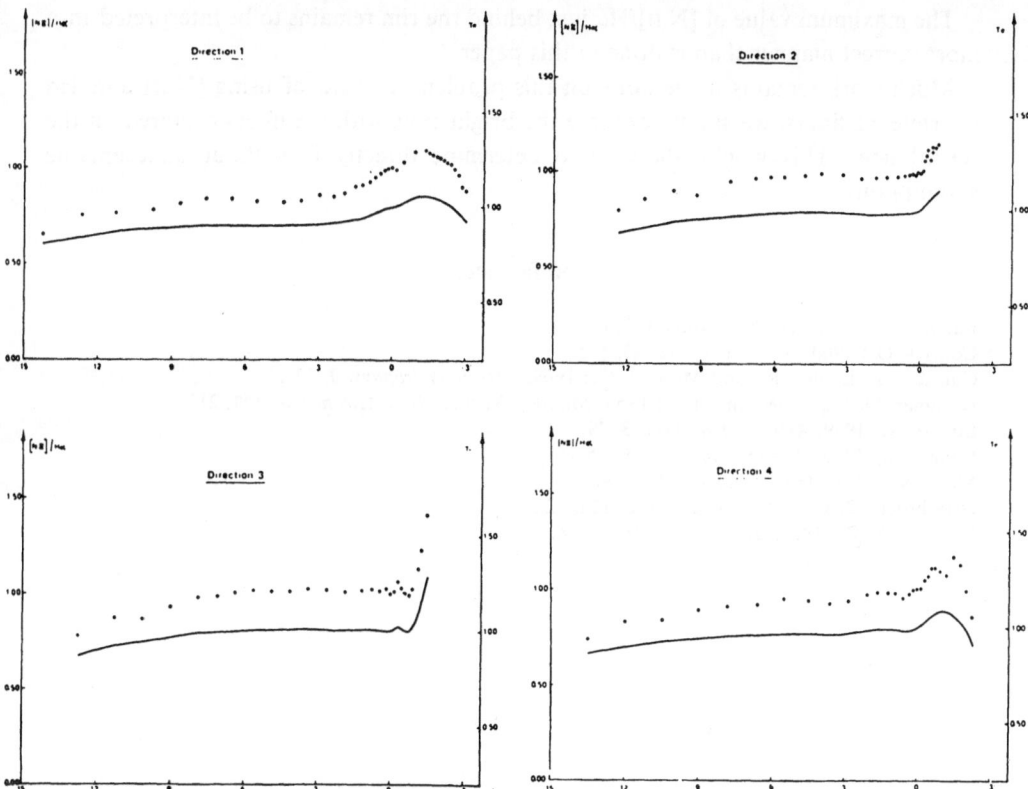

Fig. 3.   Values of the observed ratio [N II]/Hα (points, left scale) and of the derived electron temperature T_e (curve, right scale) along the four scan lines.

this effect is connected with the dissipation of energy of shock waves entering from the H II zone into the neutral material. The bright rim is certainly a very hot ionization front, but the hottest zone is located between it and the dark matter. This transition zone is very narrow, from 1' to 1.5 behind the rim.

5. Conclusion

The most important study of bright rims is to determine their physical parameters, especially their electron density and temperature. The observation of the [N II]/Hα ratio is a very interesting step; but it is not sufficient for a complete study concerning T_e variation towards the bright rim because two fundamental assumptions have to be made:

(a) Hα and [N II] originate from the same region;

(b) $N(N^+)/N_e$ remains constant.

These assumptions are not entirely justified by theoretical considerations but they allow us to obtain the first qualitative results about the electron temperature variation towards the bright rim.

The maximum value of [N II]/Hα just behind the rim remains to be interpreted in a more correct manner than is done in this paper.

Much work remains to be done on this problem. Instead of using [N II] and Hα interference filters, we should observe the bright rims with the filters centered on the [O III] lines. This would allow us to determine directly T_e without unacceptable assumptions.

References

Baudel, L.: 1970, *Astron. Astrophys.* **8**, 65.
Courtès, G.: 1960, *Ann. Astrophys.* **23**, 115.
Courtès, G., Louise, R., and Monnet, G.: 1969, *Astron. Astrophys.* **3**, 222.
Hummer, D. J. and Seaton, M. J.: 1963, *Monthly Notices Roy. Astron. Soc.* **127**, 217.
Louise, R.: 1969, *Astron. Astrophys.* **3**, 29.
Louise, R.: 1970, *Astron. Astrophys.* **5**, 35.
Mein, N.: 1968, *Ann. Astrophys.* **31**, 579.
Osterbrock, D. E.: 1957, *Astrophys. J.* **125**, 622.
Pottasch, S. R.: 1958, *Bull. Astron. Inst. Neth.* **14**, 29.

DUST IN THE LARGE MAGELLANIC CLOUD

HUGH M. JOHNSON*

Lockheed Missiles and Space Company, Palo Alto, Calif., U.S.A.

Abstract. A brief review of extinction effects is followed by a discussion of new Cerro Tololo Schmidt plates. They were exposed on parts of the LMC with filters which reveal the presence or absence of a continuum, such as one scattered on dust, in H II regions. The results are presented in comparison with the Orion nebula which was observed in the same way. The continuum is always relatively weak in those LMC H II regions which are significantly larger than Orion, but the continuum averages strong in the Orion-class and point-image nebulae.

1. Extinction Effects

Shapley (1951) made counts of extragalactic nebulae through the LMC and derived extinction of the order of nil to 0.27 mag. in several areas. From UBV photometry the maximum extinction affecting any part of the LMC in open regions is $A_V = 0.5$ mag. and 0.2–0.3 mag. on the average; in young associations, especially those embedded in gaseous nebulosity, the average rises to $A_V = 0.4$–0.5 mag. outside 30 Dor, and 1.3 mag. in 30 Dor (Thackeray, 1963; Bok, 1966). Visvanathan (1966) measured the polarization of 30 stars in the LMC and found values up to 0.068 mag. Brück *et al.* (1970) went on to conclude that the optical properties of dust grains in the Magellanic Clouds resemble those in the anticenter (Perseus) region of the Galaxy (in contrast with the Cygnus region.) Lastly, Hodge (1972) has catalogued 68 dark nebulae in the LMC 1'–21' in size. The survey he made was incomplete where silhouetting was weak or confused by H II regions.

2. Bright Continua

We expect that scattering on dust would produce reflection nebulae in the LMC, or that atomic continua or nonthermal synchrotron emission might be found (Johnson, 1971). Although emission nebulae are abundant in the LMC, it is not clear that reflection nebulae have been reported. Most photometric effort (Doherty *et al.*, 1956; Feast, 1964; and Dickel, 1965) has gone into measuring atomic-line intensities. Henize (1956) notes several spectral continua in emission nebulae of the LMC, but he remarks that "in general the presence of a continuum probably indicates the existence of a cluster or group of stars whose extent is about equal to that of the nebulosity."

* Visiting Astronomer, Cerro Tololo Inter-American Observatory, which is operated by the Association of Universities for Research in Astronomy, Inc., under contract with the National Science Foundation.

TABLE I

Curtis Schmidt plates of the LMC and Orion

1970 January (UT)	CTIO plate No.	Field center RA (1950) Dec.	Passband	Exposure (min)
10	5777	05ʰ35ᵐ0 −67°29′	N	60
10	78	05ʰ35ᵐ0 −67°29′	W	3
11	85	05 33.9 −05 24	N	120
11	86	05 33.9 −05 24	W	5
11	87	05 38.4 −69 58	W	3.6
11	88	05 38.4 −69 58	N	35.3
12	90	05 05.2 −68 07	N	120
12	91	05 05.2 −68 07	W	5
13	93	05 01.6 −70 34	N	120
13	94	05 01.6 −70 34	W	5

3. New Observations

We have taken plates of parts of the LMC and comparison plates of the Orion nebula with the Curtis Schmidt telescope of the Cerro Tololo Inter-American Observatory, as listed in Table I. All plates were exposed in a clear and moonless sky on filtered IIa-O emulsion. The 'N' filter restricted field size to $3°8$ sq, and the 'W' filter to $4°1$ sq. In both passbands the LMC fields overlap and cover a total area of about 38 sq deg.

4. The Wide and Narrow Passbands

The wide (W) band is bounded at the upper wavelength limit by the sensitivity of the emulsion, and at the lower limit by a Wratten 2B filter, giving 50 per cent of peak transmission near λ 4100 and λ 5000. The narrow (N) band is produced by a Baird-Atomic interference filter designed to peak at λ 4690 $\pm$ 5 in the $f/3.5$ beam of the camera. FWHM = 70Å $\pm$ 5Å, and transmission is less than 3% in the band wings below λ 4610 and above λ 4780. The measurements of the interference filter have been made in the spectrophotometer of the Kitt Peak National Observatory[*] with the aid of J. C. Golson, and in the spectrophotometer of the Hale Observatories by W. C. Miller. It has been checked to be uniform in its characteristics within a few per cent in a sample of different areas and to be stable over a period of more than a year. There are no leaks outside the specified passband in the range of sensitivity of IIa-O emulsion. The spectrophotometer data were the basis for an extrapolation to a more convergent beam on the theory of Lissberger and Wilcock (1959).

Exposures in the two passbands with the ratios of Table I (except plates 5787–88) nearly equalize the images of most stars. The inverse exposure-ratio exceeds FWHM (W-band)/FWHM(N-band) = 900 Å/70 Å by a factor of about 2 simply because of reciprocity failure of the emulsion. Near-equality of effective wavelengths practically

* The Kitt Peak National Observatory is operated by the Assocation of Universities for Research in Astronomy, Inc., under contract with the National Science Foundation.

eliminates effects of interstellar reddening. Stars such as Wolf-Rayet stars and high-excitation nebulae with He II λ 4686 emission have unequal images, stronger in the N band, as Reddish (1968) first demonstrated with a similar pair of passbands. However, low-excitation nebulae have no He II λ 4686 emission, and their line spectrum contributes little to the N band. Orion and the LMC nebulae which have been classified (Dickel *et al.*, 1964) are all of fairly low and nearly equal excitation class (3-4).

TABLE II

Schmidt telescope response to three compositions of light

Category	Composition	Predicted ratio of total light in N band to total light in W band
(a)	Low-excitation lines (Orion *less* continuum)	0.005
(b)	Orion nebula (continuum *plus* low-excitation lines)	0.5
(c)	Pure continuum	1.0

Table II compares three categories of interest for the study of the LMC. For the Orion nebula we have estimated a mean intensity ratio of λ 4690 continuum to Hβ from O'Dell and Hubbard's (1965) data, and we have summed the intensities of lines relative to Hβ as reported in Johnson (1968), each line weighted by filter transmissivity. About 40% of the light in the W band and 80% of the light in the N band is continuum. Structural details of the Orion nebula are remarkably similar in the N and W images, but the W image is appreciably denser, as predicted from the material compiled for Table II. We see that it should be fairly easy to distinguish categories (a), (b), and (c) of Table II among nebulae in the LMC. The Orion-nebula N and W images would be about 20″ in diameter if that nebula were placed in the LMC, so it is resolved.

5. Conclusions

We have inspected the plates directly, and by blinking them in the Lick Observatory machine, and by superposing positive-copy films of the W plates on the N plates. The blink machine was made available by courtesy of R. P. Kraft, and the film copies were made by M. M. Hanna in the photographic laboratory of the Kitt Peak National Observatory. Not all plate areas received equal attention, so this is not a homogeneous survey but rather a partial sampling of Henize (1956) emission nebulae. The sensitivity of the present combination of exposures and passbands to emission nebulae is not so great as Henize achieved with a narrow band around Hα. Table III identifies the inspected nebulae and gives the dimensions and relative intensity of Hα estimated by Henize on a scale of 1-5 (T = trace < 1). Point images on Henize's plates have no tabulated dimensions. The Hα intensity becomes a surface-brightness estimate for extended nebulae. The new-data column labeled N/W indicates the subjective impression of the relative strength of the N and W images on a scale of 0-10, with the following meaning in terms of Table-II categories: (a) should correspond to $N/W = 0$ and (c)

TABLE III
LMC nebulae examined for blue continuum

Henize-N	Diam.(")	Hα	N/W	Henize-N	Diam.(")	Hα	N/W
4A	53 × 50	5	2	34C		2	10
8	85 × 116	5	0	38	64 × 52	5	0
11B	262 × 212	5	0	44B	67 × 57	5	2
11C	191 × 158	5	0	44C	47 × 67	5	2
11E	119 × 127	4	0	44D	91 × 74	5	1
11F	199 × 97	4	0	49	68 × 82	5	3
16		3	6	59	152 × 174	4	1
18		4	6	59B	117 × 118	4	0
19		T	9	63A	40 × 40	5	3
22		5	9	77A	30 × 32	5	2
23A	82 × 107	3	1	79A	78 × 75	5	2
24		2	10	79B	21 × 20	5	8
25		2	10	79C, D, E	180 × 154	3	2
26ᵃ	27 × 30	4	10	82		5	10
27ᵃ	48 × 53	3	10	83	340 × 297	3	1
28		2	10	85	23 × 23	2	6
29		2	10	87	19 × 14	4	10
30A	30 × 41	4	0	88	36 × 21	2	10
30B	25 × 17	3	6	89		4	3
32	16 × 19	2	10	90ᵃ	25 × 27	4	10
33	81 × 69	4	0	91A, B	32 × 48	4-5	2
34A		3	10	93ᵃ	41 × 26	1	10
34B		2	10	95ᵃ	16 × 13	1	9
96		2	10	138	391 × 363	1	0
97		3	9	157A	922 × 1063	5	2
98		2	10	182		3	2
99		1	10	183		2	10
101		1	10	186A		3	5
102		2	10	186B	38 × 36	3	10
105A	160 × 270	4	2	186Cᵃ	27 × 28	2	10
106		1	10	187		1	10
109ᵃ	24 × 29	1	10	188		2	10
111		1	10	190	114 × 133	3	2
112	48 × 38	4	2	192		2	10
113A	21 × 29	5	1	193A	29 × 25	5	6
113C	81 × 65	4	1	193Bᵃ	32 × 36	2	10
113D	65 × 73	4	1	193Cᵃ	38 × 52	2	10
113F	49 × 44	4	1	193Dᵃ	29 × 35	T	10
115	38 × 37	1	10	193Eᵃ	29 × 35	T	10
117	28 × 27	4	10	194ᵃ	16 × 19	3	8
118	26 × 27	4	5	195A	34 × 27	4	6
119	800 × 932	3	1	195B	46 × 32	4	1
120	520 × 365	3	1				

ᵃ N and W images appear to be stellar.

to N/W = 10 (equal intensities). The plates of Orion (b) are estimated at N/W = 7. Thus the subjective scale is no doubt related nonlinearly to real intensity ratios and is multiplied by 10.

An interesting difficulty is that 12 non-stellar Hα images of the Henize (1956) catalog appeared to be stellar in N and W. These are superscripted 'a' in Table III. Only three of them are noted by Henize with any continuum around Hα. Although the small scale of Henize's finding charts admit the possibility of misidentifications, most of the 12 images which are discrepant cannot be so explained. It is possible then that a blue stellar continuum, or a concentrated nebular continuum, sometimes leads to $N/W = 10$ while the weaker, extended nebulosity around the star or central concentration is absent because of underexposure. The estimate $N/W = 0$ may also result from under-exposure of the N image.

Disregarding the 12 asterisked observations, we find that all nebulae with point images show $N/W \geqslant 2$, and they prefer the value $N/W = 10$. No definite case of $N/W > 10$ was found among Henize's nebulae, but the presence of He II λ 4686 cannot be distinguished from strong continuum in cases of high N/W values.

Among 'Orion-class' nebulae, i.e. resolved with geometrical mean diameters in the range 20"–50", N/W takes on all values in the range 0–10. The conclusion is that nebulae with a ratio of continuum to monochromatic-line emissions like Orion's, and variants, are present in the LMC.

Among large nebuale with geometrical mean diameters > 50", $N/W \leqslant 2$. These nebulae therefore exhibit a relatively weak continuum. It is found from the data of Feast (1964), Dickel et al. (1964), and Dickel (1965) that they have rms electron densities much less than that of Orion. However, within Orion O'Dell and Hubbard (1965) determined that the ratio of continuum to monochromatic line emission at Hβ grew as the borders of the nebula were approached, so it may not be simply said that relatively less scattering takes place where the electron density is low.

These conclusions about the continua of the LMC nebulae may be translated into the dust/gas density ratio in so far as the scattering of starlight on dust indeed accounts for the continua in the nebulae. The degree to which dust is responsible for continua in emission nebulae is a complicated subject in itself, and atomic continua are often taken to be dominant (e.g. Pottasch, 1965).

Three nebulae in Table III are of special interest: N 49, N 63A, and N 157A (30 Dor), which are among four LMC objects known to have a nonthermal radiofrequency spectrum (Mathewson and Healy, 1964). But their optical behavior in N/W is not notably different from the other nebulae observed in this work. The N-image details of 30 Dor imitate the W-image details closely, as in the images of Orion, but N/W is considerably weaker than it is in Orion. The fourth nonthermal nebula, N 132D, is in a very dense field of stars, which prevents satisfactory examination.

Acknowledgement

This work has been done under the Lockheed Independent Research Program.

References

Bok, B. J.: 1966, *Ann. Rev. Astron. Astrophys.* **4**, 95.
Brück, M. T., Lawrence, L. C., Nandy, K. N., Thackeray, A. D., and Wood, R.: 1970, *Nature* **225**, 532.
Dickel, H. R.: 1965, *Astrophys. J.* **141**, 1306.
Dickel, H. R., Aller, L. H., and Faulkner, D. J.: 1964, in F. J. Kerr and A. W. Rodgers (eds.), *The Galaxy and the Magellanic Clouds*, Australian Acad. Sci., Canberra, p. 294.
Doherty, L., Henize, K. G., and Aller, L. H.: 1956, *Astrophys. J. Suppl.* **2**, 345.
Henize, K. G.: 1956, *Astrophys. J. Suppl.* **2**, 315.
Hodge, P. W.: 1972, *Publ. Astron. Soc. Pacific* **84**, 365.
Johnson, H. M.: 1971, in A. B. Muller (ed.), *The Magellanic Clouds*, D. Reidel Publ. Co., Dordrecht-Holland, p. 95.
Johnson, H. M.: 1968, in B. M. Middlehurst and L. H. Aller (eds.), *Nebulae and Interstellar Matter*, University of Chicago Press, Chicago, p. 63.
Lissberger, P. H. and Wilcock, W. L.: 1959, *J. Opt. Soc. Am.* **49**, 126.
Mathewson, D. S. and Healey, J. R.: 1964, in F. J. Kerr and A. W. Rodgers (eds.), *The Galaxy and the Magellanic Clouds*, Australian Acad. Sci., Canberra, p. 283.
O'Dell, C. R. and Hubbard, W. B.: 1965, *Astrophys. J.* **142**, 591.
Pottasch, S. R.: *Vistas in Astronomy* **6**, 149.
Reddish, V. C.: 1968, *Observatory* **88**, 139.
Shapley, H.: 1951, *Proc. Nat. Acad. Sci.* **37**, 136.
Thackeray, A. D.: 1963, *Adv. Astron. Astrophys.* **2**, 264.
Visvanathan, N.: 1966, *Monthly Notices Roy. Astron. Soc.* **132**, 423.

THE MIXING OF NEUTRAL GAS AND DUST
WITH THE IONIZED GAS OF H II REGIONS

M. C. LORTET-ZUCKERMANN

Département d'Astronomie Fondamentale, Observatoire de Meudon, 92190 Meudon, France

Abstract. Two arguments are presented in support of the now well-established fact of the tight mixing of neutral gas and dust with the ionized gas of H II regions.

1. Theoretical Point of View

(1) Take a blue star and compute the mass of gas which it is able to ionize. This mass is inversely proportional to the density of the ionized gas

$$\mathfrak{M}_{H\,II} \propto N_e^{-1}.$$

The results are given in Table I, where the mass of ionized gas is expressed in units of the mass of the exciting star; the computation is made for a density $N_e = 100$ and using the rate of ultraviolet photons given by Rubin (1968). Now, if clumping is present, for instance, if the true density is rather 10^3 or 10^4 (may be 10^5), all the figures in the last column should be divided by ten or one hundred (may be one thousand) and it is seen that the star is able to ionize but a very small mass of gas. This result holds even when we take into account the fact that the rate of ultraviolet photons given by Rubin may be underestimated by a factor as large as 8 (Davidson and Terzian, 1969; Churchwell and Walmsley, 1973; Chopinet *et al.*, 1972) and so may

TABLE I

Mass of ionized gas in a medium where $N_e = 100$, compared to the mass of the exciting star. The radius of the Strömgren's sphere has been taken from Rubin (1968) and may be underestimated (see text)

Sp	T_{eff}	$\mathfrak{M}_*/\mathfrak{M}_\odot$	$\mathfrak{M}_{H\,II}/\mathfrak{M}_*$
O4	50	58.4	11.1
O5	45	33.7	7.7
O6	40	24.5	3.6
O7	38	22.1	2.3
O8	36	19.7	1.2
O9	35	18.3	0.90
O9.5	33	16.1	0.34
B0	31	14.8	0.088
B0.5	29	14.2	0.023

Greenberg and Van de Hulst (eds.), Interstellar Dust and Related Topics, 477–481.

be the mass of ionized gas given in Table I. The question is *where is the initial mass out of which the star was formed gone*? and the answer is: it is still neutral, around or inside the nebula.

(2) The second argument is related to the origin of the high degree of clumping, which has been found in *all* the H II regions investigated carefully enough. Aperture synthesis continuum radio maps (see for instance Wynn-Williams, 1971) analysis of the recombination line data (Hjellming and Gordon, 1971: M 17 and Ori A; Andrews *et al.*, 1971, five H II regions; Gordon and Wallace, 1971: W 49) and optical determinations of density (see for instance Deharveng-Baudel, 1972a) show that the actual density of the ionized gas may be 10 to 10^3 higher than the rms density obtained from a few arc minutes resolution radio maps and that steep density gradients are present. How are these inhomogeneities and gradients formed and *maintained*? The only explanation at hand at the moment and which does work is that neutral gas is present in high density clumps, inside the whole H II region.

2. Observational Results

The existence of large H I–H II complexes has been established beyond doubt in a number of cases: often, the mass of neutral gas is much larger than that of the ionized part (Table II).

A common case is that in which the ionized gas visible in the optical wavelengths (Hα) constitutes but a very unimportant fraction of the total mass of ionized gas. As

TABLE II

Association of H I and H II

Region	$\mathfrak{M}_{H I}/\mathfrak{M}$	$\mathfrak{M}_{H II}/\mathfrak{M}$	Notes	Ref. H II, H I
S–86 NGC 6820, 23	1 E 3	$\simeq$ 1.8 E 3	Assumed distance 1.65 kpc	Gordon *et al.* (1968) from Felli and Churchwell (1972)
S–131 IC 1396	2 E 4	$\leqslant$ 7 E 3	$\mathfrak{M}. \simeq$ 1 E 3	Pottasch (1965), Simonson (1968) (stars) Simonson (1973)
S–222 IC 1579	$\simeq$ 80	0.22	Reflex. neb.	Riegel (1967), Chopinet *et al.* (1972) from Terzian and Pankonin (1972)
S–125 IC 5146	6.7 E 2	2.0 E 1	cluster	Riegel (1967)
S–184 NGC 281	1.6 E 4	8.8 E 2	cluster	Riegel (1967)
Orion	7 E 4	8	Ori A	Schraml and Mezger (1969) Gordon (1970)
Cyg X	6.2 E 4		assumed distance 2 kpc	Cutcheon and Shuter (1970)
W 58 S–99 and 100 K 3–50	1.0 E 5 to 2.5 E 5	1.4 E 2		Felli and M. Fossi (1970) Bridle and Kesteven (1970)

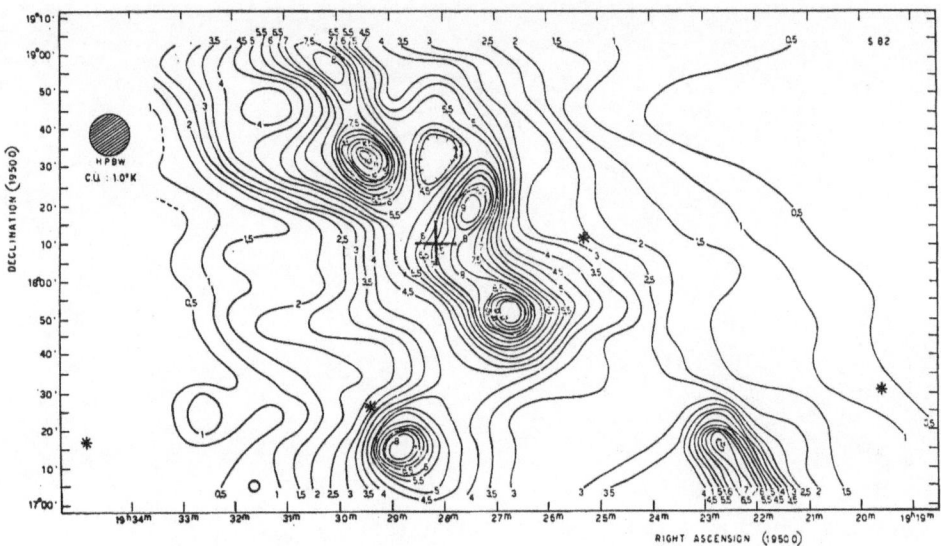

Fig. 1.   From Felli and Churchwell, 1972: Radio isophotes at 1400 Mhz in the region of Sharpless 82. The optical region is indicated by a cross.

an example, Figure 1 shows a continuum radio map of the region of Sharpless 82 (taken from Felli and Churchwell, 1972); the optical part is indicated by the cross. The optical aspect is seen in Figure 2: it is a fairly small and rather symmetrical nebula, north of which a smaller reflection nebula DG 159 is visible.

As a rule, the optical aspect may be largely determined by the presence of a large mass of neutral gas and associated dust, as is probably the case for the strange claw-like shaped nebula Sharpless 157 (see Figure 1a in Chopinet and Lortet-Zuckermann 1972a). This nebula contains a bright Hα-knot, designated Sh2-157 A, which has been studied by these authors and Deharveng-Baudel (1972b) and which shows the following properties:

– it is excited from the inside by a blue O9–B0 star,

– the rms density, deduced from poor spatial resolution radio observations (half power beam width 2.8, Aikman, 1968) is about $120 \, \text{cm}^{-3}$ and the corresponding mass of ionized hydrogen is 10 solar masses.

– the true density, deduced from the S II lines intensity ratio 6717/6731 ranges from 500 to $3500 \, \text{cm}^{-3}$.

– thus the actual mass of ionized gas is about one solar mass, that is about 1/20 of the mass of the star.

– Finally, the most unexpected result is that the brightness and density distributions seem to be anticorrelated in this object: the ionized gas density has a minimum *inside* the bright bar east of the central star and increases where the brightness begins to fall down, on each side.

The interpretation is that the knot of ionized gas is associated with a large mass of

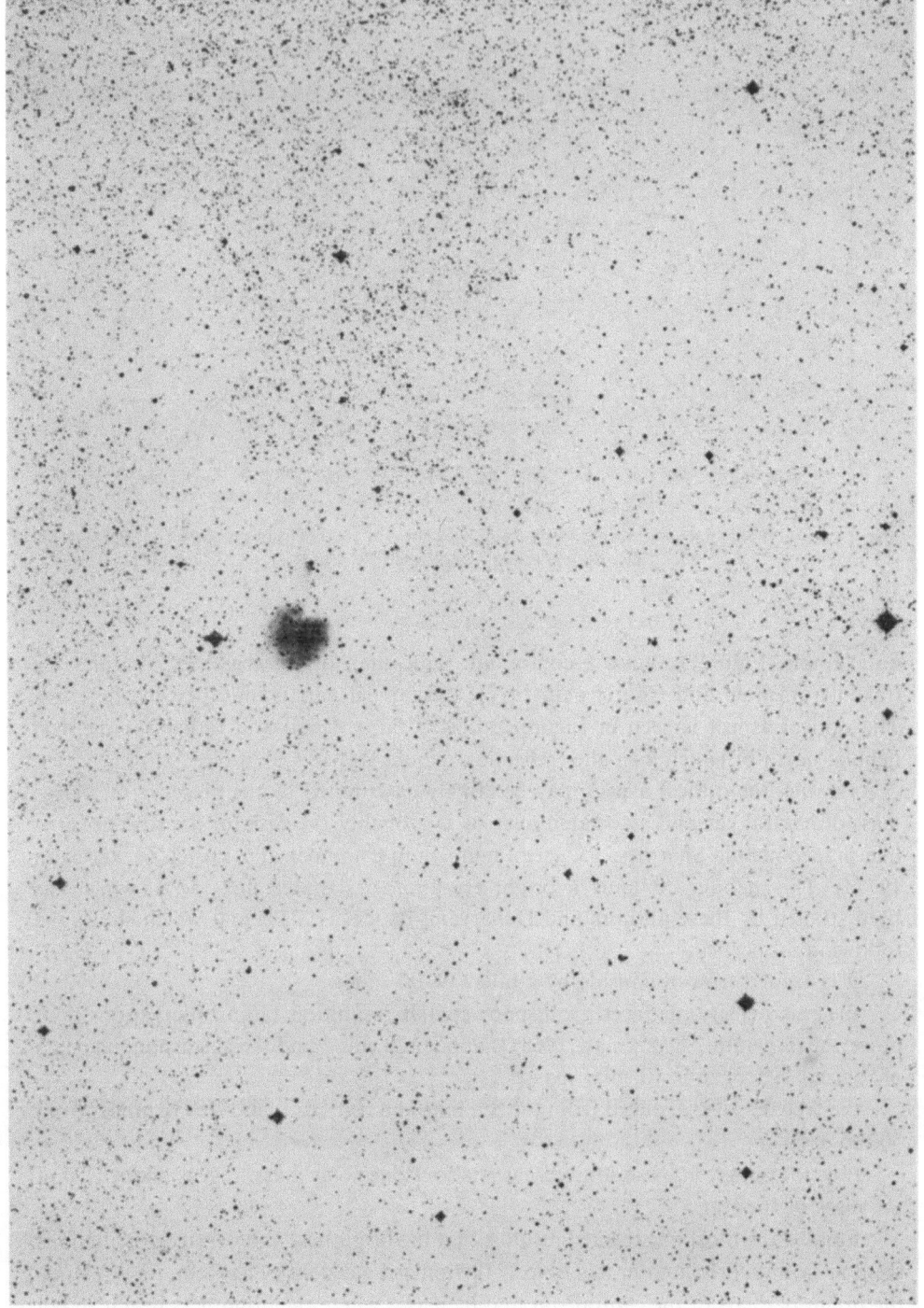

Fig. 2. Optical aspect of Sharpless 82 on the red print of the Palomar Observatory Sky Survey. North is at the top, east at the left (as in Figure 1). The scale is one degree represented by 154 mm.

neutral gas and dust a fraction of which is inside the knot (the remaining fraction probably being around the knot).

Similar bright bars are being studied in other nebulae, as for instance in Sharpless 206 (Chopinet and Lortet-Zuckermann, 1972b): they are certainly always associated with the existence of a large amount of neutral gas and dust, and this is suggested in the specific case of Sharpless 206 by the presence of the neutral oxygen forbidden line at 6300 Å in the brightest bars.

3. Conclusion

The most important points to be now investigated are to determine the exact amount of neutral gas and dust associated with H II regions, in connection with the problem of infrared emission and molecular formation and emission. New results such as those currently obtained by infrared observations (comparisons of the radio and infrared sources dimensions) or on the high positive charge of grains in the transition region between neutral and ionized regions (Flower, 1972) may be of considerable significance.

References

Aikman, G. C.; 1968, unpublished M. S. Thesis, University of Toronto.
Andrews, M. H., Hjellming, R. M., and Churchwell, E.: 1971, *Astrophys. J.* **167**, 245.
Bridle, A. H. and Kesteven, M. J. L.: 1970, *Astron. J.* **75**, 902.
Chopinet, M., Georgelin, Y. M., and Lortet-Zuckermann, M. C.: 1972, in preparation.
Chopinet, M. and Lortet-Zuckermann, M. C.: 1972a, *Astron. Astrophys.* **18**, 373.
Chopinet, M. and Lortet-Zuckermann, M. C.: 1972b, in preparation.
Churchwell, E. and Walmsley, C. M.: 1973, *Astron. Astrophys.*, **23**. 117.
Cutcheon, Mc W. H. and Schuter, W. L. H.: 1970, *Astron. J.* **75**, 910.
Davidson, K. and Terzian, Y.: 1969, *Nature* **221**, 729.
Deharveng-Baudel, L.: 1972a, Communication at the 18th colloquium, held in Liège, June 1972, '*Planetary Nebulae*', in press.
Deharveng-Baudel, L.: 1972b, in preparation.
Felli, M. and Churchwell, E.: 1972, *Astron. Astrophys. Suppl.* **5**, 369.
Felli, M. and Monsignori Fossi, B. C.: 1970, *Astron. J.* **75**, 1033.
Flower, D.; 1972, Communication at the 18th colloquium, held in Liège, June 1972, '*Planetary Nebulae*', in press.
Gordon, C. P.; 1970, *Astron. J.* **75**, 914.
Gordon, C. P., Howard, W. E., III, and Westerhout, G.: 1968, *Astrophys. J.* **154**, 103.
Gordon, M. A. and Wallace, D. C.: 1971, *Astrophys. J.* **167**, 235.
Hjellming, R. M. and Gordon, M. A.: 1971, *Astrophys. J.* **164**, 47.
Pottasch, S. R.: 1965, *Vistas in Astronomy* **6**, 149.
Riegel, K. W.: 1967, *Astrophys. J.* **148**, 87.
Rubin, R. H.: 1968, *Astrophys. J.* **154**, 391.
Sharpless, S.: 1959, *Astrophys. Suppl.* **4**, 257.
Simonson, S. C., III: 1968, *Astrophys. J.* **154**, 923.
Simonson, S. C., III: 1973, this volume, p. 227.
Terzian, Y. and Pankonin, V.: 1972, *Astron. J.* **77**, 115.
Wynn-Williams, C.: 1971, *Monthly Notices Roy. Astron. Soc.* **151**, 397.

PART IX

CIRCUMSTELLAR DUST

CIRCUMSTELLAR DUST

CIRCUMSTELLAR INFRARED EMISSION

I: *The Circumstellar Origin of Interstellar Dust*

NEVILLE J. WOOLF

School of Physics and Astronomy, University of Minnesota, Minneapolis, Minn., U.S.A.

Abstract. Infrared astronomy has in the past decade emerged from being a part-time occupation of a few astronomers. Three major subdivisions of research have become apparent, solar system, galactic and extragalactic studies. In each of these fields infrared studies have made unique contributions. Planets emit the bulk of their radiation in the infrared, and infrared studies are essential to study planetary thermal problems. Many extragalactic objects have been found to emit astonishingly large fluxes in the infrared.

In galactic astronomy the current major contribution of infrared studies has been to act as a bridge between two separate disciplines, stellar astronomy, and studies of the interstellar medium. Infrared studies have proved invaluable for studying star birth and star death. Both of these phases had previously seemed mysterious and invisible. And indeed they were not visible, because they occurred shrouded in dust that blocked transmission of visible rays. However, the dust that is merely opaque in the visible, is self-luminous in the infrared, and so in the midst of this optical darkness there has appeared a great infrared light.

At this time, we have progressed further with the study of star death than of star birth. The ejected matter from dying stars carries the dust shroud with it into space, and so the gas and dust become part of the interstellar medium. This process is clearly significant for understanding the composition and origin of interstellar dust.

Because star death and birth are embedded in dust, there has developed a separate interest in explaining the physical processes at work in these dust clouds. This study explains processes of optical circumstellar absorption lines, intrinsic polarization of cool star light, and stellar molecular masers.

Perhaps what these two paragraphs have just said is that our conceptual scheme of separating stellar astronomy and interstellar astronomy still acts as such a division that the infrared astronomer needs to present different aspects of this one topic, circumstellar infrared emission to different audiences. Such an opportunity has been given to the author in that he has been asked to give within a few weeks two talks. The first of these reviews is being presented at IAU Symposium # 52 on Interstellar Dust and Related Topics. The second is being given at the summer meeting of the Astronomical Society of the Pacific which has a symposium on Circumstellar Dust.

The two reviews have been made complementary. The first of these is primarily an observational study. It shows the infrared observations of stellar and interstellar dust, and in a qualitative way shows that one gives rise to the other. The second review is theoretical and attempts to place the first study in its theoretical context. It deals almost exclusively with the stellar and circumstellar parts of the topic. Together they present one man's view of *Circumstellar Infrared Emission*.

The literature relevant to this topic is voluminous. There have been false leads, dead ends, and irrelevant detail. This review has attempted to follow a thread through this detail, and to expose the skeleton of a scheme for understanding the processes at work. Such a review is intrinsically more dangerous, more likely to become obsolete than a comprehensive one. However, by carrying the seeds of its own destruction it seems to offer a greater opportunity for the growth of astronomy.

1. Theories of the Origin of Interstellar Dust

There are two and a half hypotheses about the origin of interstellar dust. The first is that it condenses in interstellar space, the second that it condenses in stellar envelopes and atmospheres, and the half hypothesis combines these by making condensation

Greenberg and Van de Hulst (eds.), Interstellar Dust and Related Topics, 485–504.

nuclei from circumstellar dust, but having the bulk of the material as more volatile materials that could condense in interstellar space.

This review presents arguments in favor of the second hypothesis. First, it collects the infrared evidence for the composition of interstellar dust from studies of emission and absorption spectra of diffuse matter. Then it discusses the observations of circumstellar matter around evolved objects, where dust is clearly not a remnant of the protostellar cloud. Some brief comments, for completeness, show how the circumstellar clouds around young objects compare with these. Finally, there is a discussion of the nature and origin of interstellar dust. Many of the observations discussed here are presented in other contexts in a comprehensive review article by Neugebauer *et al.* (1971).

2. Emission and Absorption Spectra of Diffuse matter

This past year Gillett has constructed a scanning spectrometer with resolution 0.1 μ for the spectral range 3-14 μ. This sensitive device employs a copper doped germanium photoconductive detector, and a cold, circular, variable wavelength pass-band filter wheel. Figures 1-6 show observations made by UCSD and University of Minnesota observers with this spectrometer attached to a 60 in telescope at Mt. Lemmon Infrared Observatory.

Figure 1 from Gillett and Stein shows the emission spectrum of the trapezium region of the Orion Nebula. No bright stars were admitted by the spectrometer entrance aperture. Note that the emssion peaks at 9.7 μ declines rapidly to short wavelengths, and slowly to long wavelengths. Photometric results, e.g. by Ney and Allen (1969) show that there is then a further rise into the 20 μ region. Because the emission decreases rapidly to wavelengths shorter than 9.7 μ, it is apparent that no other

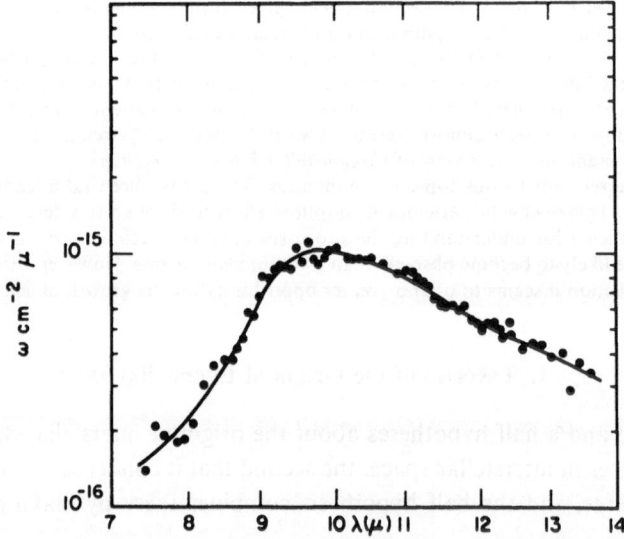

Fig. 1.   Spectrum of the trapezium region of the Orion Nebula. The feature is silicate emission.

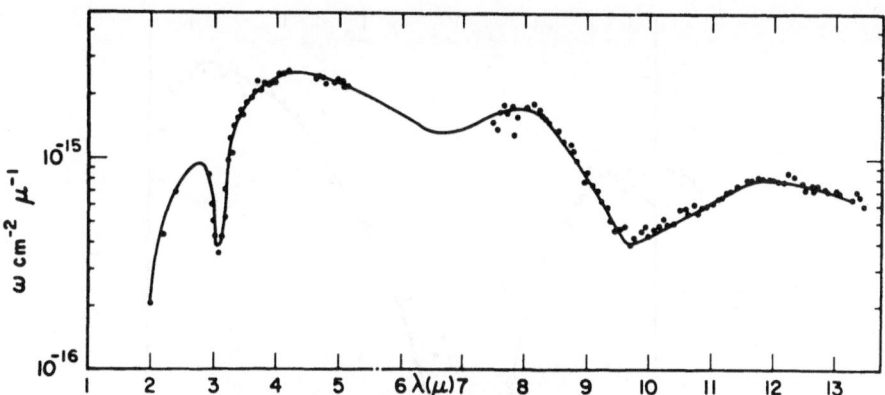

Fig. 2. Spectrum of the Becklin-Neugebauer star with wings embedded in the Kleinmann-Low infrared nebula in Orion. The 3.1 μ band is identified as ice, and the 9.7 μ feature is silicates.

source of emission has become significant in the 7–8 μ spectral band. If there is more than one solid material present here, and one produces the 9.7 μ feature, and the second material produces a warmer, black-body like spectrum, as one might expect for graphite or iron, then there is not very much of this second material present here.

Figure 2 from Gillett, Forrest and Cohen shows the Becklin-Neugebauer 'star with wings' that is in or behind the Kleinmann-Low infrared nebula in Orion. Additional observations with different size diaphragms show that out to 12 μ almost all of the energy comes from the starlike object. Note the two strong absorptions in this spectrum, one peaking at 9.7 μ, and the other at 3.1 μ. The 9.7 μ feature looks like an inversion of Figure 1, though the peak appears a little sharper.

The 3.1 μ feature appears at the same wavelength as a feature seen in CIT + 40°448 (NML Cygni) and some stars, but not seen in others, e.g. the B star VI Cygni #12 that suffers about 10 magnitudes of interstellar extinction (Danielson *et al.*, 1965; Knacke *et al.*, 1969; Stein *et al.*, 1969b; Low *et al.*, 1970). It seems probable that this feature is due to interstellar ice. A second band, weaker in equivalent width by a factor 3 should be present at 12.5 μ, or possibly resolved into two features at 11.5 μ and 13.5 μ. However, because of the greater width of this feature, the band should be so shallow that its absence is perhaps still compatible with the ice identification. The amount of ice needed to produce the 3.1 μ band is $\sim 6 \times 10^{-5}$ gm cm^{-2}.

If the 9.7 μ feature is identified as a silicate absorption band, and a band strength of 3000 cm$^2 \cdot$ gm^{-1} at the peak is assumed, then 4×10^{-4} gm cm^{-2} of silicate are in the line of sight. However, Hackwell (1971b) found that a carbonaceous chondrite, composed of $\sim 70\%$ silicate, and with absorption spectrum resembling Figure 2 required about 3 times as much matter to produce this absorption, thus the amount of silicate may be as high as 10^{-3} gm cm^{-2}. There is 6–20 times as much silicate as ice present.

Figure 3 from Woolf and Gillett shows the spectrum of the infrared core at the galactic center observed with a 22″ diam diaphragm. Note that peak absorption also occurs here at 9.7 μ, and the optical depth is $\tau \approx 4$. Earlier broadband observations by Low *et al.* (1969), and Hackwell *et al.* (1970), had suggested $\tau \approx 0.5$ both in this

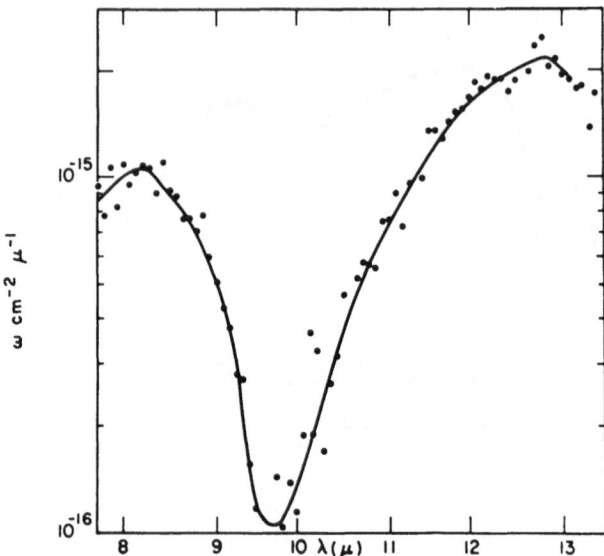

Fig. 3.   Spectrum of the IR core of the galactic center.

band, and a 20 μ band. The new observation seems quite certain, being a mean of four separate scans, and preliminary observations at an earlier time had also indicated a large optical depth. Apparently the broadband photometry washed out the feature by a large factor.

A very weak possible absorption appears near 12.5 μ. If this is an ice band, then at most $\tau \approx 0.2$. This value is probably consistent with the ratio of silicate to ice indicated by the Orion point source.

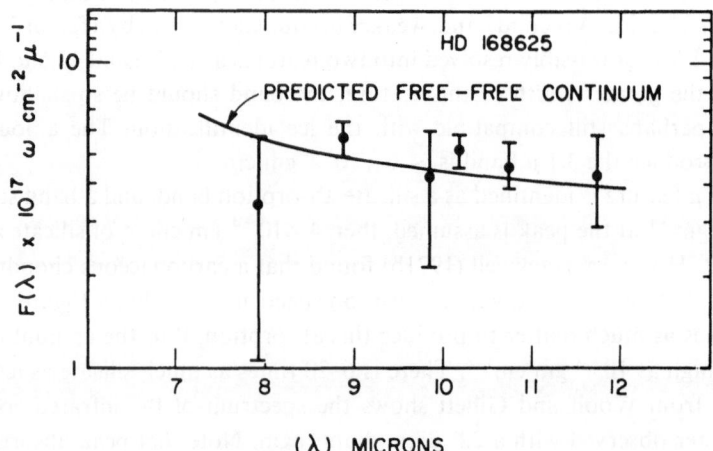

Fig. 4.   Points in the spectrum of HD 168625 B8Ia$^+$. This star suffers about 5.7 mag. visual interstellar extinction.

Stein and Gillett (1971) made broadband 10 μ region observations of VI Cygni #12 to search for an interstellar silicate absorption. They concluded that $0.1 > \tau$. If, however, we make some correction as with the galactic center for washing out of the feature, then more probably $0.3 > \tau$. Thus the ratio of visual extinction for this object to silicate absorption at the peak is greater than 30:1. For the Orion point source 1–2 μ spectral distribution implies a visual extinction close to 70 mag., giving its ratio of visual/IR silicate absorption 50:1. However, if we assume that the extinction to galactic center stars of 25 ± 5 mag. (Becklin and Neugebauer, 1968) also applies to the IR core, we have a ratio of about 7:1. Clearly if the line to the galactic center is typical, then the silicate absorption of the Orion object and VI Cygni #12 is weak. Or if these two objects are normal, then the galactic center core has much local extinction, with a total visual extinction of about 200 mag. Such absorption is compatible with radio molecular line studies.

In order to dinstinguish between these possibilities, Gillett and Woolf have observed a few points in the spectrum of a different reddened star HD 168625, B8Ia$^+$. This star has a slightly elevated 10 μ continuum, probably because of free-free emission. It has an otherwise very similar companion, with a visual extinction of 5^{m}7 (Hackwell *et al.*, 1970). Both are seen through the dense dust cloud at the edge of M17, giving us a good indication that this extinction is not circumstellar. Figure 4 shows six points in the spectrum of this rather faint object. The uncertainties are very high, but for this object $0.3 > \tau_{9.7}$. Thus the ratio of visual to silicate absorption is >20. This shows that the value obtained in Orion is likely to be correct, and that the extinction of the object at the galactic center is vastly greater than that found for the stars in the galactic bulge.

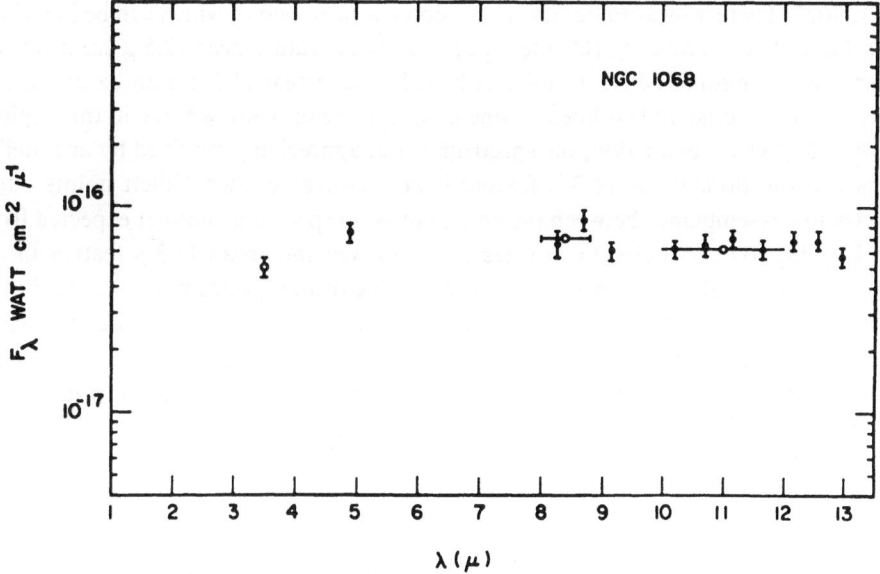

Fig. 5. Points in the spectrum of the Seyfert galaxy NGC 1068.

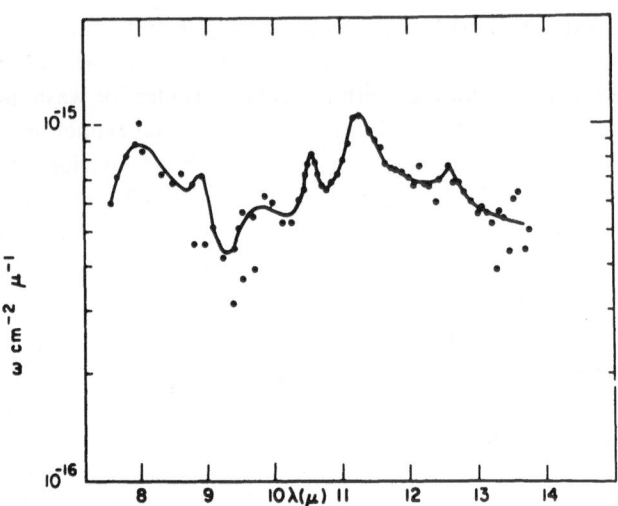

Fig. 6. The spectrum of the planetary nebula NGC 7027.

Figure 5 from Stein, Gillett and Merrill is a similar set of spaced points for the Seyfert galaxy nucleus NGC 1068. This spectrum appears featureless. If the emitting
material is a solid, then either it is remarkably optically thick, or it does not have the
same spectral characteristics as Figures 1–3. Emission from a conducting or semi-
conducting solid such as iron or graphite could cause an appearance like this. Or a
non-thermal process could be responsible.

Figure 6 from Gillett, Forrest and Stein shows the spectrum of the planetary nebula
NGC 7027. There appear to be sharp emission features, two of which can be identified
with lines of A III and S IV. (Delmer *et al.*, 1967). A feature near 12.5 μ seems at too
short a wavelength to be Ne II 12.8 μ. A broad feature near 11.3 μ is probably a blend
of two or three unidentified lines. A line of Cl I is the only known line in this region.
With 3."2 of visual extinction, the spectrum is not appreciably modified by interstellar
silicate absorption. If the 11.3 μ feature is *not* resolvable, then Gillett points out a
remarkable resemblance between the entire emission spectrum and that expected from
hot $MgCO_3$ solid particles. Gillett has also observed the broad 11.3 μ feature in the
planetary nebula BD + 30°3639. The general shape of the spectrum and depression in
the 9.5 μ region repeats earlier observations by Gillett *et al.* (1967). If enough features
can be identified to leave a smooth continuum unexplained, then again as in NGC
1068 iron or carbon particles could be responsible. However, because protons would
combine with carbon, iron would be more likely here.

3. Solar System Condensates

In the past few years, meteorite research has in general concluded that many meteorites
condensed directly from the vapor phase. Some of the more curious boundaries be-

tween crystalline iron-nickel and silicate droplets can then be explained. In this model, at some phases iron condenses as a metal from the gas, while at other times it condenses in silicates. At lower temperatures it sometimes forms magnetite Fe_3O_4, and condenses with silicates of unusual structure, and with ring carbon compounds to form carbonaceous chondrites.

The presence of metallic iron is a puzzle. Arrhenius and Alfvén suggest that grain growth is limited by the sticking coefficient of materials, and that neutrals stick better than ions. Iron would then be a condensate of phases where a gas is mildly ionized. However, regardless of the explanation, processes that occurred in the solar nebula should occur elsewhere, and both silicates and iron should be found in astrophysical environments.

The evidence for direct condensation of silicates is important, because the spectra of these materials show two bands near 10 and 20 μ, similar to those found for the Orion trapezium region, and the galactic center. These features have also been found in diffuse matter in the solar system. Comet Bennet was observed photometrically by Maas et al. (1970), and showed a 10 μ peak. Hackwell (1971a) studied the 10 μ peak spectroscopically and found it to have the same energy distribution as the supergiant μ Cephei, or essentially that of Figure 1. Comet Bennet also showed a weaker continuum of black body character that was probably iron since in Comet Ikeya Seki, Westphal and Becklin (1966) found a similar continuum existed to color temperatures greater than 1000 K.

The current basis of the silicate identification will be discussed in Paper II. Originally the identification as a solid emitter was made from some elaborate arguments. The circumstantial evidence given here is used here as adequate evidence for the identification of the 9.7 μ feature as silicates.

4. Dust Around Cool Evolved Stars

Cool evolved stars fit into one of three broad categories, M, S and C. The chemical compositions for these types probably have a decreasing oxygen to carbon ratio through the sequence with a near equality at the boundary between S and C. It also seems probable that both oxygen and carbon are depleted in a number of carbon stars with the beneficiary being nitrogen.

Hackwell has obtained photometric observations of M, S and C stars (Hackwell, 1971b). Figures 7 and 8 show that M and S stars show a double humped feature with strong emission near 10 μ and 20 μ, whereas C stars show a single feature near 10 μ. The subtle differences between the different M and S stars may imply slightly different materials being present, or that in some cases the material is optically thick.

Gillett et al. (1968) published spectra of one S star and a few M stars all of which had emission excesses closely resembling Figure 1 (see also Woolf and Ney, 1969). An example of a cool star in which the dust seems optically thick, R Aqr, has been published by Stein et al. (1969).

The emission around carbon stars has been spectroscopically studied by Hackwell

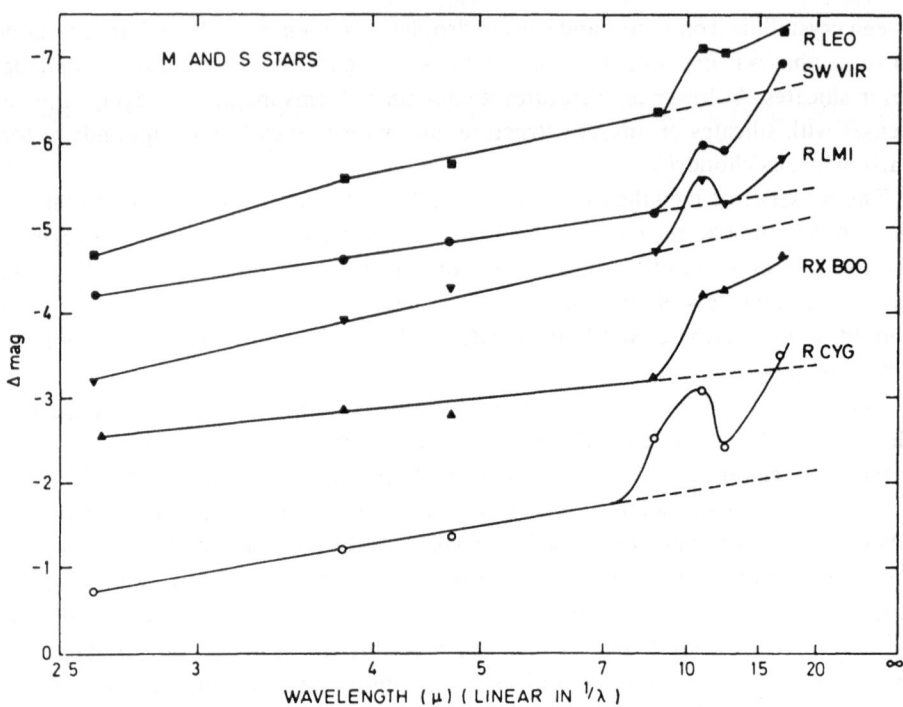

Fig. 7. Photometric measures of M and S stars.

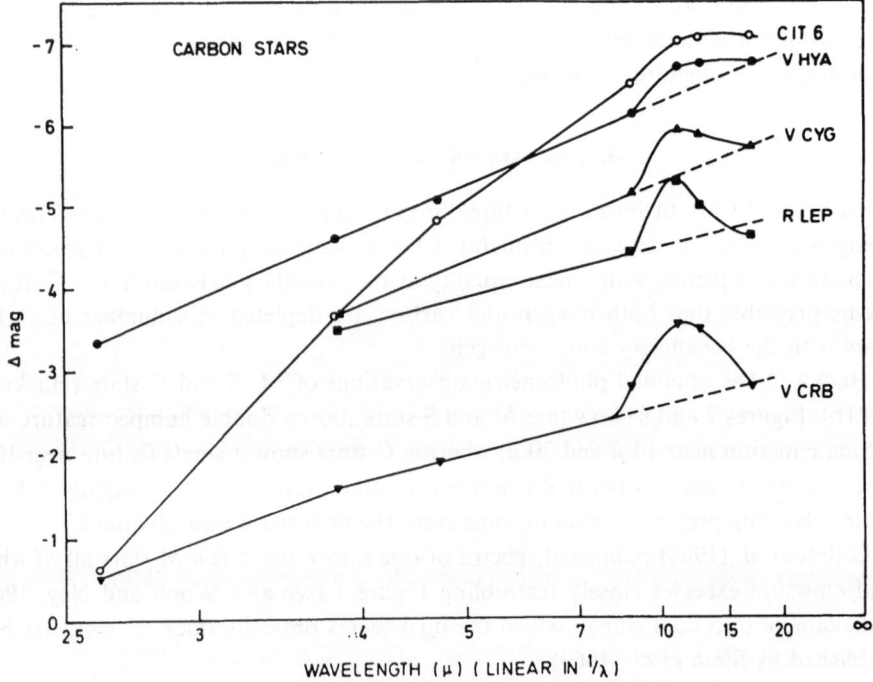

Fig. 8. Photometric measures of carbon stars.

in $+30°219$ (CIT6) and V Hya. The excess seen in Figure 9 is tentatively ascribed to Si_3N_4. If carbon is not very abundant in the first place, and what there is condenses at high temperatures as graphite, then a mixture is left with a little C and O as CO, and a great deal of nitrogen. On further cooling, the next condensate would be iron, and after this one of the first condensates with obvious spectral features would be Si_3N_4. It is not clear whether carbon stars are producing carbon and/or iron dust. The absence of clearcut spectral features makes identification very difficult. However, the temperature implied by the long wavelength continua of most of the carbon stars in Figure 8 are close to 2000 K, and this is just about the temperature of condensing carbon.

For cool stars of normal composition, the variation of the silicate feature with spectral type has been well studied. It seems to arise in all stars later than a certain spectral type, the type being luminosity class sensitive. This boundary is at G0Ia[+], G8Ia, M1Iab, M5Ib and M6III. Figures 10, 11 and 12 from Gehrz (1972) give examples of stars with and without silicate excesses. Some of the observations are from Humphreys *et al.* (1971).

There are a few exceptions of stars earlier and later than these boundaries that do not obey the rule stated above, thus VV Cephei with an M2I a component does not show an excess. On the other hand, the SRa variable RY UMa, M2–3III shows a substantial peak.

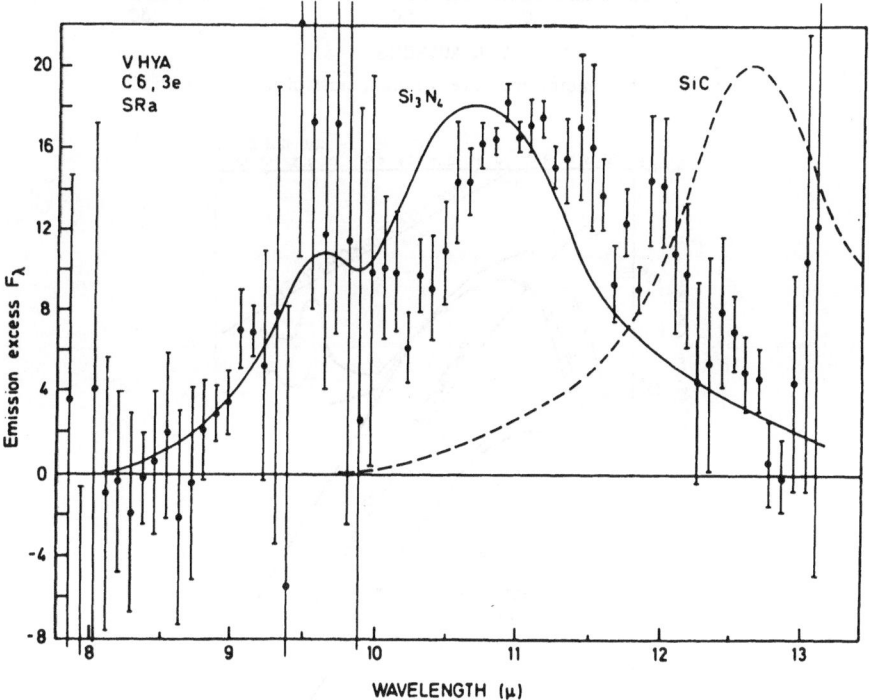

Fig. 9. Average excess above a black body for the carbon star V Hya.

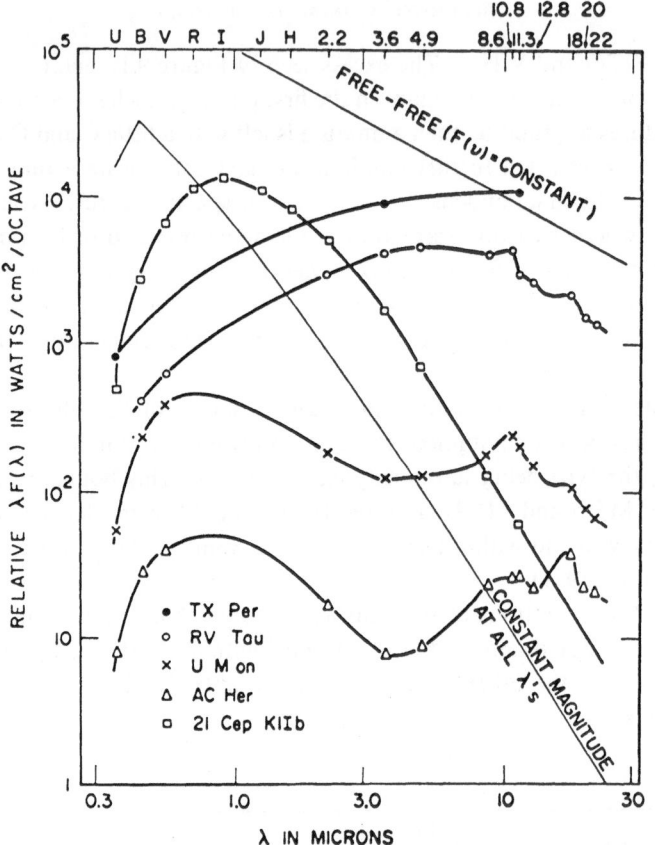

Fig. 10. Photometric measures of giant stars of luminosity Ib.

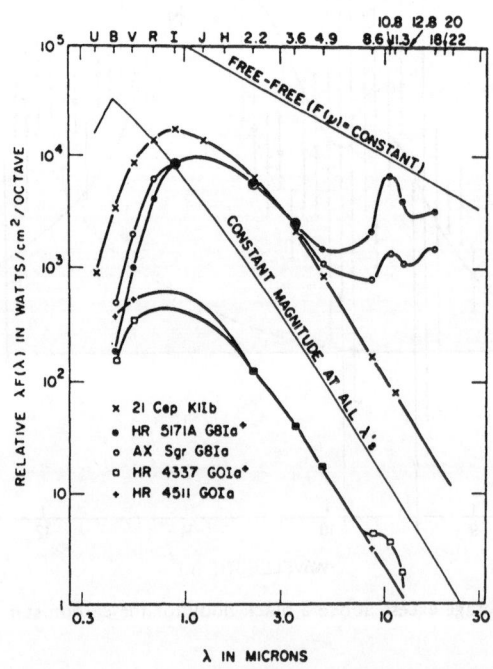

Fig. 11. Photometric measures of G supergiants.

Woolf and Pepin (1972) observed five Mira stars of early spectral type, period about 200 days, and high space motion. All these stars showed excess emission. However, the observations were interpreted as showing dust condensing in a more compact, high gas density shell than for more normal Mira stars. It was inferred that stars in globular clusters perhaps as heavy element rich as Messier 3 would show the phenomenon of circumstellar emission.

It should be noted that thus far the infrared emitting cloud has only been resolved on the sky for the one carbon star CIT + 10216 (Toombs *et al.*, 1972). However, circumstellar gas has been resolved a few seconds away from α Herculis by Deutsch (1956), and this star shows a weak silicate double feature (Figure 10).

A number of stars were found in the CIT 2 μ sky survey (Neugebauer and Leighton, 1969) to have apparently very low temperature continua. Some of these appear to be cool M or S Mira variables showing silicate emission peaks. One star + 40°448 (NML Cygni) is particularly mysterious. It has a near IR spectrum resembling an M giant rather than a supergiant; however, radio maser observations show outflowing gas to have velocities more characteristic of supergiants. There is a 3.1 μ band in the spectrum possibly indicating interstellar ice absorption. At 9.7 μ there is neither a peak nor a trough. However, interstellar absorption could be exactly compensating for circumstellar emission. If the star is indeed a relatively normal giant, then it is remarkably close, and suffers extra-ordinarily large absorption.

Both NML Cygni and the M5Ia star VY CMa are believed to be evolved. Davies *et al.* (1972) interpret some OH circumstellar features as showing rotating circumstellar clouds for these two stars. Such rotation, if confirmed, would imply such large angular momentum that one would need to assume these were protostars. However, Woolf (1971) explained optical polarization of such stars by asymmetric envelopes. The observations of Davies *et al.* follow the pattern of radial velocities predicted by Woolf, with a core apparently moving towards the observer, and a halo at the star's velocity. Similarly the asymmetric expansion predicted by Woolf would also lead to gas motions that Davies *et al.* would interpret as rotation. For more details see part two, Section 6. The main argument that these are indeed evolved stars is simply a probabilistic one. Time scales of protostar phases are very short compared with thermonuclear burning phases.

5. Dust Around Warm Evolved Stars

G and K type Ia and Ia$^+$ supergiants have silicate emission that appears to be an extension of the phenomenon that occurs with lower luminosity cool stars. Spectra of some of these stars show no optical emission lines, confirming that one is not seeing interstellar emission as in Figure 1.

Some pulsating stars that would be ineligible for silicate emission from their spectral type and luminosity class do have such emission. Most noteworthy are the RV Tauri stars. Figure 13 from Gehrz (1972) shows such stars in which dust is unmistakably present. Thus U Mon clearly shows a silicate double hump. The carbon

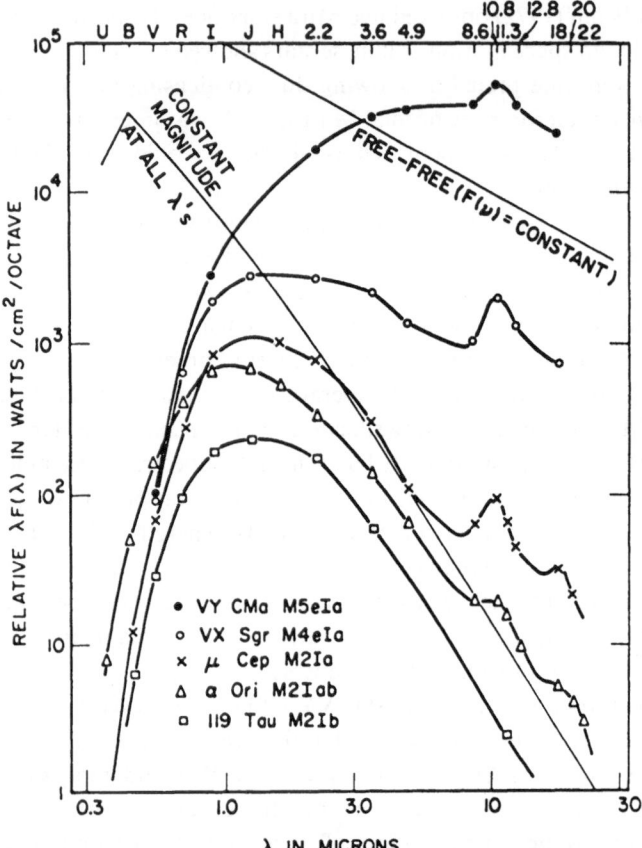

Fig. 12.   Photometric measures of M supergiants.

star AC Her shows a spectral distribution that is unique, and the condensate here is unknown.

Other RV Tauri stars show continua that slowly rise to long wavelengths, with little or no spectral evidence near 10 μ of particular solids. It seems more reasonable to assume these are optically thick dust shells rather than that a new class of gaseous continuum should be evoked.

Two carbon rich hydrogen deficient supergiants are known to have excess emission. These are the variable stars R Cr B and RY Sgr. The sudden decline in light of these stars and slower recovery, together with apparent absence of early extinction of chromospheric optical emission lines is good evidence that dust is formed here. The spectra (Stein *et al.*, 1969a; Lee and Feast, 1969) show a smooth black body like continuum temperature ~ 800 K and more detailed observations in the 10 and 20 μ range confirm the smoothness there. For stars like this, the near absence of hydrogen permits carbon to condense at a fairly high temperature, and this is probably the emitting material that is observed.

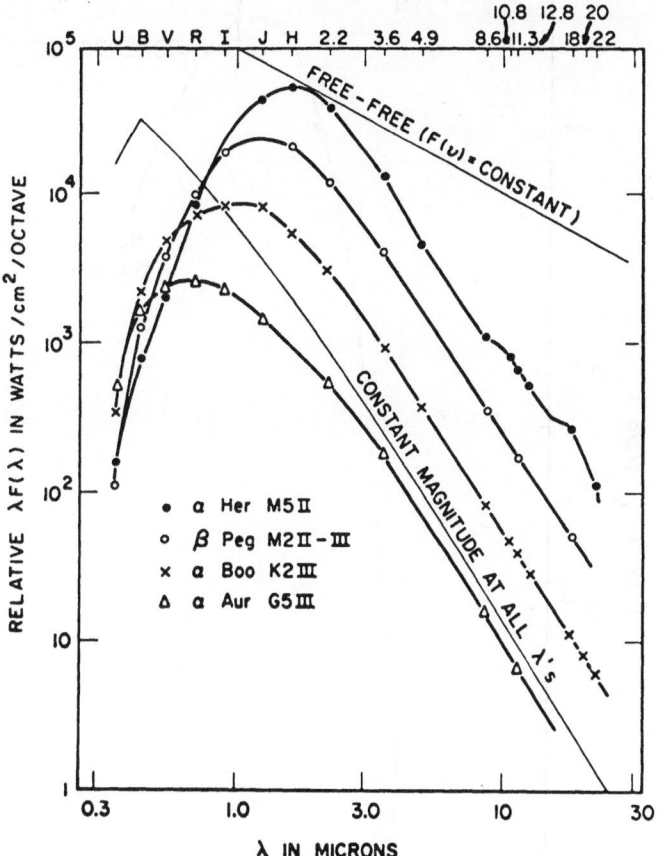

Fig. 13. Photometric measures of RV Tauri stars.

Two somewhat warmer luminous stars are known to also have black body like circumstellar emission. These are 89 Her, F2Ia (Gillett *et al.*, 1970) and the hydrogen deficient A star binary υ Sgr (Lee and Nariai, 1967). For both these stars there seem to be weak emission features at 10 and 20 μ superposed on the black body, indicating that some dielectric solid material is formed. However, there is no certainty for these two stars that the black body ≈ 800 K feature is due to solids. Other luminous F supergiants such as ε Aur, F0Ia and ι Sco F2Ia show no excess emission from 3–10 μ. Even a star such as ρ Cas, F8Ia⁺ that is known to eject matter, currently shows no circumstellar emission.

For 89 Her, the discussion of NGC 7027 and of solar system condensates suggests iron as a more likely condensate than carbon. For υ Sgr both would seem possible.

6. Dust Around Hot Evolved Stars

Hot stars with circumstellar matter generate excess infrared radiation from free-free

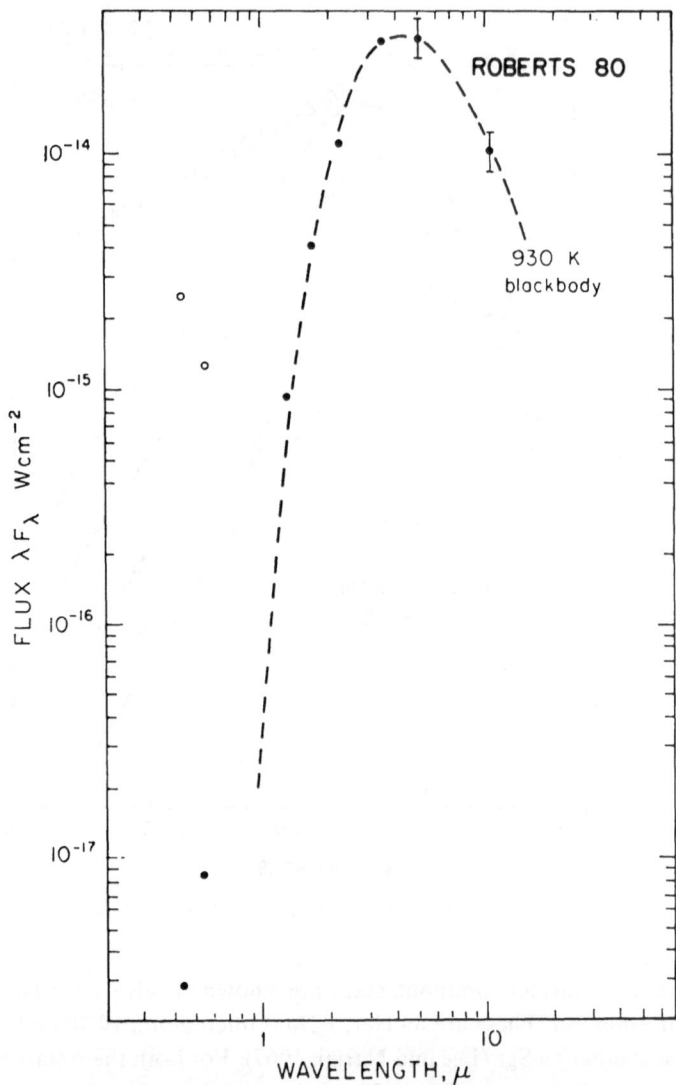

Fig. 14.   Photometric measures of the WR star Roberts 80.

and free-bound transition in the matter ionized by the star's light, and/or collisions. However, in some stars there may be excess infrared radiation attributiable to dust. In these cases extreme care must be exercised in insisting that dust is present. Cool black body like emission could also be produced by a dense lightly ionized gas, by electrons scattering off neutral particles (Dyck, private communication).

Two possible cases of time variable dust emission from Be stars were mentioned by Woolf *et al.* (1971). A far better case has been made by Allen *et al.* (1972) for the Wolf-Rayet star Roberts 80, Figure 14. Here a substantial fraction of the star's energy appears to come out as a black body continuum of $T \approx 930$ K.

Other cases of excess infrared continua in Be, P Cygni and WR stars seem more consistent with free-free emission, and in these cases the IR emission can be shown to be consistent with predictions from the optical emission lines.

A number of planetary nebulae show both a cool continuum and some emission lines (Gillett *et al.*, 1972). The emission is spectrally unlike that found in cool M, S or C stars, therefore if it is dust, it would seem to be manufactured *in situ*, perhaps in small condensations embedded in the nebula like raisins in a pudding. There is a similarity between these observations, and observations of H II region in the 100 μ region, (Hoffmann *et al.*, 1971; Harper and Low, 1971) where there seems little possibility of evoking a gaseous continuum. Thus we shall assume that the planetary nebulae condense iron dust as was discussed. Further observations of the semi-sharp 11.3 μ feature reported for NGC 7027 may imply a need to modify this conclusion.

Novae appear to condense dust some weeks after their outburst (Kleinmann, 1972). This is being discussed in another paper at this meeting. The only comment appropriate here is that iron would seem the likely first condensate. Presumably supernovae may also be expected to follow this behavior, though no infrared observations of supernovae have yet been reported, and the time from outburst before dust condenses may be many months, or even years.

7. Evolved Stars; Summary and Conclusions

All of the objects discussed in the sections on evolved stars are known spectroscopically to be ejecting matter. The condensates will become interstellar dust. Therefore, it is of interest to ask which condensates are likely to arise from which stars. Table I below attempts to summarize the discussion above.

8. Solids Around Young Stars and Protostars

When a star is surrounded by circumstellar dust, this is subject to radiation pressure. The dust flows through the gas imparting momentum to it, and typical outflow velocities of 10–100 km s^{-1} are generated. Thus at such speeds the dust moves out a distance of 1 pc in less than 10^5 yr. This naive calculation fails if there is a net inward pressure on the star. This pressure could arise in one of two ways. Condensation of a protostar occurs with the outer layers lagging behind the inner layers. This lag is comparable with the free fall time of the outer layers, and in an initial condensation of density $N_H \approx 10^2$, this phase may last a few million years. In the second case the star may be in a gas flow. For example, if an O star is ionizing a dense cloud of dust and gas from the outside, then the ionized gas will flow back past the star. Dust embedded in the gas would then be brought close to the star.

Both of these processes are probably significant for young stars. The second process probably explains the emission of the trapezium region of the Orion Nebula. Similar spectral features are found in several stars in the Orion nebula (Ney, private communication). They are also seen in the O star Herschel 36 embedded in a dust cloud in M8,

TABLE I

Solids produced by evolved stars

Material	Types of star
Silicates	M and S stars
	G and K supergiants
	RV Tauri stars (some)
	Solar Nebula (meteorites)
	Comets
Iron	Solar Nebula (meteorites)
	A and F unusual supergiants[b]
	Planetary nebulae[b]
	Be P Cygni and WR stars (some)[b]
	Novae and supernovae[b]
Carbon	R CrB stars
	Hot H deficient stars[a]
	C stars[b]
Si_3N_4	C stars[a]
Unknown dielectrics	RV Tauri stars (some)
	A and F unusual supergiants
	Planetary Nebulae[a]

[a] Spectral feature observed, identification in doubt.
[b] Black body like spectra, with negligible evidence for identification.

in CoD 12403 in the same nebula, but not in the O5 star 9 Sgr which is in gas of density 10^2 now. Nor is it seen in S Mon, another early O star mainly free of gas.

It can be predicted in such cases that for a given luminosity star and velocity of flow, the dust stays out to a distance proportional to (gas density)$^{-\frac{1}{4}}$. The temperature of the gas is independent of the luminosity of the star, and is proportional to (gas density)$^{\frac{1}{4}}$. However, if the gas is ionized by UV radiation, it will be optically thick for Lα photons, which will then be absorbed by the dust (Krishna Swamy and O'Dell, 1967). In such cases the dust will be hotter than simple theory predicts. Other such stars with silicate dust around include the two T Tauri stars that flared up, V1057 Cygni (Cohen and Woolf, 1971) and FU Ori (Cohen, private communication).

Stars that probably still have some condensation occurring are the T Tauri stars, particularly the stars in comet shaped nebulae. Incoming matter will bring angular momentum. This will be ejected, with least possible accompanying mass from the outer edge of a disc. The core's light will mainly escape perpendicular to the plane of rotation and illuminate nearby condensations of dust. Since the light escapes along two co-axial opposed cones, this will explain cometary nebulae such as those accompanying R Mon and R Cr A (Mendoza, 1968) or LK Hα101 and T Tauri itself. Since the rotating disks are likely to have a very high density, spectral features from the hot dust should not be apparent. Figure 15 from Gehrz (1972) shows examples of such objects.

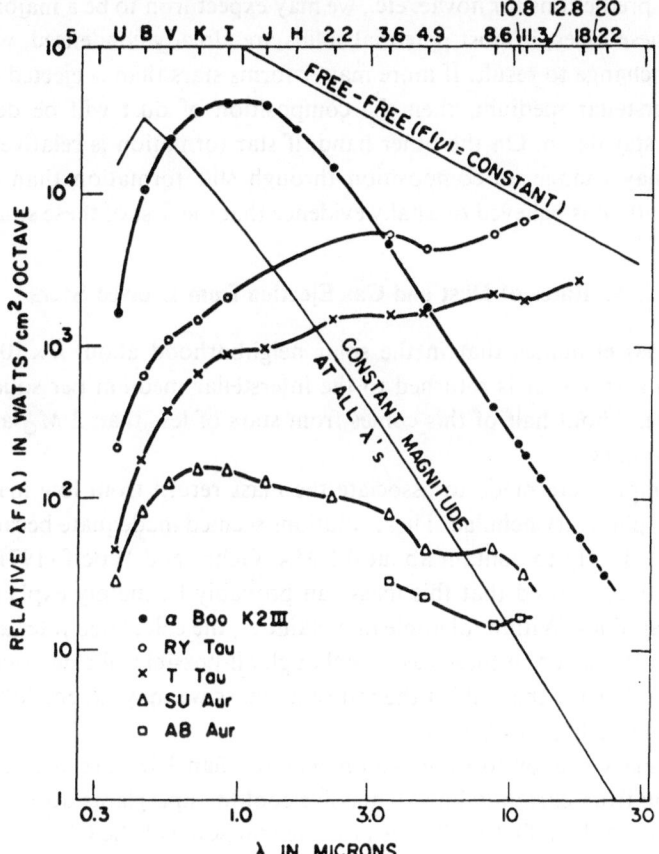

Fig. 15. Photometric measures of T Tauri stars.

Young stars will form from dust and gas as it occurs in space. Some of the matter will fail to condense, and will be returned unchanged to space. Other matter may get close to the star, have molecular bands broken, and perhaps remade. Under such circumstances it is possible that silicates would become iron or gas, or any combination of change between the three might take place. However, it seems that these changes will occur in relatively high density cool places, and so most material that can form non-volatile solids will leave protostars as solids.

Sometimes events may be explosive as in η Carina where matter started moving out in 1843. In the late last century the object changed from a gaseous absorption spectrum to an emission spectrum and slowly dimmed. However, Westphal and Neugebauer (1969) showed that the dimming was probably the result of dust formation, since the IR luminosity now is comparable with the optical luminosity at light maximum. Ney has found the spectrum to resemble a cool black body with a weak silicate feature superposed. The angular structure is resolved on the sky. Here it seems there may have been a substantial conversion from one kind of solid matter into another. By analogy with predictions for novae, etc., we may expect iron to be a major condensate.

In all of these cases, if most potential solids are already condensed, we should not expect much change to result. If more matter forms stars than is ejected directly back into the interstellar medium, then the composition of dust will be determined by processes of star death. On the other hand, if star formation is relatively inefficient, more dust may change its composition through star formation than through star death. Currently it is believed on shaky evidence that the first of these situations holds.

9. Rates of Dust and Gas Ejection from Evolved Stars

Deutsch (1968) estimates that in the solar neighborhood about $4 \times 10^{-10} \, M_\odot$ per year of dead star matter is returned to the interstellar medium per square parsec of galactic plane. About half of this comes from stars of less than $2 \, M_\odot$ and half from more massive stars.

Early attempts were made to associate the mass return from low mass stars with ejection from planetary nebulae. These solutions seemed inadequate because planetary nebulae seemed only to contain about $0.1 \, M_\odot$. Gehrz and Woolf (1971) and Woolf and Pepin (1972) showed that this mass can probably be mainly explained as ejecta from Mira variables. Within tolerable uncertainties, the calculated rates agree with the observations. It seemed in these cases as though all possible silicates were condensed in the ejecta. Thus for this half of the returning mass we may expect 90% of solids to be silicates and perhaps 10% iron.

Studies of mass ejection from stars more massive than $2 \, M_\odot$ are in a less satisfactory state. Ejection from currently known massive cool M supergiants does not seem to be adequate, perhaps by a factor 10. It is possible that some of the CIT objects represent cool remnants of massive stars, and perhaps these are a significant fraction of objects that return matter. Some carbon stars represent the death of moderately massive stars, and so do some S stars. However, carbon variables are several times as rare as M variables. Thus again it would seem likely that these would only represent perhaps 10% of the ejecta of massive stars.

Then one is left with the possibility that the remainder, less than 80% of the return from high mass stars, comes off as winds from Be and related stars, or in supernova explosions. In such cases the solid ejecta is probably mainly iron.

If we attempt to total these we find the composition of solids entering interstellar space as silicates > 50%, carbon > 5%, iron < 45%, silicon nitride > 1.5%. Such numbers only hint at the results of the detailed study necessary.

10. The Composition of Interstellar Dust

Current estimates of star formation rates and of star deaths are the same, implying that the fraction of the Galaxy as interstellar medium is not changing rapidly with time. However, both of these numbers are small compared with the current mass of the interstellar medium, and imply change only on a cosmological time scale. Thus although we know a little about the way dust now enters the interstellar medium, we do not know whether this entering matter is similar to the matter now there.

It seems highly unlikely that the interstellar medium is primordial. Heavy element abundances seem if anything higher than in the Sun. Therefore if the medium started as hydrogen, with perhaps helium, it has been processed through being a star at least once. Our possible doubts about the current composition of dust arise because we do not know what these early stars were, or how they died. We only know that they emitted copious amounts of heavy elements. Fortunately, the discussion we have just been through about ejecta from stars suggests that in all such cases solids do condense, and the main uncertainty is the fraction of the matter ejected as silicates, and as iron. Here the observations of the Orion nebula in Figure 1 are of paramount importance. Silicates are obviously present. There is no sign of a hot black body continuum to signify the presence of iron or carbon. The Orion nebula is presumably a typical sample of the interstellar medium.

The next question is, are there mantles on the solid matter? Here we have observations of two dusty regions, one in the Orion Nebula, the other in the galactic center. Water ice may be present, but only represents about 10% by mass of the silicate matter in the line of sight to these objects. The long searches by Knacke et al. for the 3.1 ice band show that though ice mantles may exist in some places, they are a rarity.

Finally therefore we can try to estimate the amount of interstellar silicate. From the observation of the Orion star, and the negative evidence of VI Cygni #12 and HD 168625, the ratio of interstellar visual extinction to silicates 9.7 μ extinction is $\approx 50:1$. Thus we find that there are 6×10^{-26} to 2×10^{-25} g of silicates per magnitude of interstellar absorption. Such absorption is probably associated with 2.5 to 5×10^{-3} g of interstellar gas. This implies a mass ratio of silicates to gas of between 1:125 to 1:800. These ratios must be decreased, by perhaps 10% to allow for ice, and at least a further 10% to allow for carbon, iron, silicon nitride etc. to obtain the ratio of total solids to gas.

If we assume an actual ratio of 1:250 then most silicon, iron, aluminum, magnesium, and calcium would be locked into the dust. Some problems of curiously low cosmic abundances of these materials in the interstellar gas would then have been explained. Again, factor of two uncertainties in cosmic compositions make the appropriate gas to dust ratios uncertain. However, we may now invert the abundance argument, and suggest that most matter condensable into nonvolatile silicates exists in that form.

It is up to us to then use the interstellar extinction curve to infer what these cosmic abundances are. Some difficulties of interpretation will arise because cosmic ray and low energy proton bombardment fracture bonds, and make the optical properties of the material anomalous. However, almost regardless of these problems it seems that the interstellar dust needs to be about 1% of the gas by mass, and this does seem to be possible if the bulk of the dust is silicates that condensed in cool stellar atmospheres.

Acknowledgements

This article could not have been written without many discussions with almost all active infrared observers. However, special acknowledgements go to Dr F. C. Gillett without whose work this paper would have had a different form, and some erroneous conclusions. The research has been supported by the National Science Foundation under grants GP-27983 and GP-32772.

References

Allen, D. A., Harvey, P. M.. and Swings, J. P.: 1972, *Astron. Astrophys.*, **20**, 333.
Becklin, E. E. and Westphal, J. A.: 1966, *Astrophys. J.* **145**, 445.
Becklin, E. E. and Neugebauer, G.: 1968, *Astrophys. J.* **151**, 145.
Cohen, M. and Woolf, N. J.: 1971, *Astrophys. J.* **169**, 543.
Danielson, R. E., Woolf, N. J., and Gaustad, J. E.: 1965, *Astrophys. J.* **141**, 116.
Davies, R. D., Masheder, M. R., and Booth, R. S.: 1972, *Nature* **237**, 21.
Delmer, T. N., Gould, R. J., and Ramsay, W.: 1967, *Astrophys. J.* **147**, 495.
Deutsch, A. J.: 1956, *Astrophys. J.* **123**, 210.
Deutsch, A. J.: 1968, in M. Hack (ed.), *Mass Loss from Stars*, Reidel Publ. Co., Dordrecht, p. 1.
Gehrz, R. D.: 1972, Thesis, University of Minnesota, Minneapolis.
Gehrz, R. D.: 1971, *Bull. Am. Astron. Soc.* **3**, 454.
Gehrz, R. D. and Woolf, N. J.: 1971, *Astrophys. J.* **165**, 285.
Gillett, F. C., Low, F. J., and Stein, W. A.: 1967, *Astrophys. J.* **149**, L97.
Sillett, F. C., Low, F. J., and Stein, W. A.: 1968, *Astrophys. J.* **154**, 677.
Gillett, F. C., Hyland, A. R., and Stein, W. A.: 1970, *Astrophys. J.* **161**, L219.
Gillett, F. C., Merrill, K. M., and Stein, W. A.: 1972, *Astrophys. J.* **172**, 367.
Hackwell, J. A.: 1971a, *Observatory* **91**, 37.
Hackwell, J. A.: 1971b, Thesis, London University College.
Hackwell, J. A., Gehrz, R. D., and Woolf, N. J.: 1970, *Nature* **227**, 822.
Harper, D. A. and Low, F. J.: 1971, *Astrophys. J.* **165**, L9.
Hoffmann, W. F., Frederick, C. L., and Emery, R. J.: 1971, *Astrophys. J.* **170**, L89.
Humphreys, R. M., Strecker, D. W., and Ney, E. P.: 1971, *Astrophys. J.* **172**, 75.
Kleinmann, S. L.: 1972, Thesis, Rice University.
Knacke, R. F., Cudaback, D., and Gaustad, J. E.: 1969, *Astrophys. J.* **158**, 151.
Krishna Swamy, K. S. and O'Dell, R. C.: 1967, *Astrophys. J.* **147**, 529.
Lee, T. A. and Nariai, K.: 1967, *Astrophys. J.* **149**, L93.
Lee, T. A. and Feast, M. W.: 1969, *Astrophys. J.* **157**, L173.
Low, F. J., Kleinmann, D. E., Forbes, F. F., and Aumann, H. H.: 1969, *Astrophys. J.* **157**, L97.
Low, F. J., Johnson, H. L., Kleinmann, D. E., Latham, A. S., and Geisel, S. L.: 1970, *Astrophys. J.* **160**, 531.
Maas, R. W., Ney, E. P., and Woolf, N. J.: 1970, *Astrophys. J.* **160**, L101.
Mendoza, E. E.: 1968, *Astrophys. J.* **151**, 977.
Neugebauer, G. and Leighton, R. B.: 1969, *Two Micron Sky Survey: A Preliminary Catalog*, NASA, Washington, D.C.
Neugebauer, G., Becklin, E. E., and Hyland, A. R.: 1971, *Ann. Rev. Astron. Astrophys.* **9**, 67–102.
Ney, E. P. and Allen, D. A.: 1969, *Astrophys. J.* **155**, L193.
Stein, W. A., Gaustad, J. E., Gillett, F. C., and Knacke, R. F.: 1969a, *Astrophys. J.* **155**, L3.
Stein, W. A., Gaustad, J. E., Sillett, F. C., and Knacke, R. F.: 1969b, *Astrophys. J.* **155**, L177.
Stein, W. A. and Gillett, F. C.: 1971, *Nature* **233**, 72.
Toombs, R. I., Becklin, E. E., Frogel, J. A., Law, S. K., Porter, F. C., and Westphal, J. A.: 1972, *Astrophys. J.* **173**, L71.
Westphal, J. A. and Neugebauer, G.: 1969, *Astrophys. J.* **156**, L45.
Woolf, N. J.: 1971, *Mem. Soc. Roy. Sci. Liège* **6**, III, 205.
Woolf, N. J. and Ney, E. P.: 1969, *Astrophys. J.* **155**, L181.
Woolf, N. J., Strittmatter, P. S., and Stein, W. A.: 1970, *Astron. Astrophys.* **9**, 252.
Woolf, N. J. and Pepin, T. J.: 1972, unpublished.

RADIATIVE TRANSFER IN CIRCUMSTELLAR DUST CLOUDS

A. D. CODE

Washburn Observatory, University of Wisconsin, Wis., U.S.A.

Abstract. Radiative transfer methods are used to show how and to what extent the usual techniques relating color excess to optical depth through circumstellar material are not applicable.

1. Introduction

The manner in which the observed flux from a star is modified by interstellar grains depends upon the spatial distribution as well as the optical properties of the grains. It is the purpose of this paper to describe the radiation transfer for several simple models in which the distinction between interstellar extinction and interstellar absorption becomes important.

If the interstellar grains are distributed along the path between the source and the observer in such a way that only an infinitesimal fraction of the scattered radiation appears in the field of view, the stellar flux is simply reduced by the extinction optical depth exponentially. The derivation of interstellar extinction curves determined by taking the ratio of a reddened and an unreddened star assumes this to be the case. If, however, scattering is produced by a nearby cloud, the extinction curve can be significantly modified. Collins and Code (1965) discussed this case for isotropic conservative scattering and Capriotti (1967) applied these considerations to the Balmer decrement in diffuse nebulae. In a series of papers by Mathis (1970, 1971, 1972) the problem of internal dust in gaseous nebulae has been investigated by an iterative numerical technique for a variety of geometries and albedos. More recently the wavelength dependence of interstellar extinction has been extended into the infrared (Johnson, 1968) and the vacuum ultraviolet (Stecher, 1959; Bless and Savage, 1972). In these spectral regions the extinction is highly non-linear and therefore provides important information on the nature of the interstellar grains. In these investigations significant variations in the derived extinction curves were found from star to star and region to region. Hallam (1959) was the first to show that variations in the law of reddening were correlated with the presence of surrounding diffuse nebulae. The most anomalous case so far documented is for the Trapezium, θ Orionis (c.f. Carruthers, 1969). It is important to separate the variations in extinction caused by differences in the nature of the interstellar grains from those produced by multiple sacttering in a nearby cloud. In their classical discussion of diffuse galactic light, Henyey and Greenstein (1941) emphasized the fact that multiple scattering rather than the precise geometry was the dominant feature in determining the modification of the stellar radiation produced by interstellar grains.

A discussion of radiative transfer in an interstellar or circumstellar dust cloud re-

Greenberg and Van de Hulst (eds.), Interstellar Dust and Related Topics, 505–515.

quires knowledge of both the extinction cross-section and the albedo of the grains. To appreciate the importance of these considerations for a circumstellar dust cloud one need only consider the dramatic difference between the two limiting cases. If the surrounding dust cloud were spherically symmetric then in the case of pure scattering, where the albedo is unity, all the stellar radiation would escape and the luminosity and spectral distribution of the star would be unchanged. Unless we could resolve the circumstellar shell we would not know it was there. On the other hand, if the grains were pure absorbers, namely an albedo of zero, the radiation at a particular wavelength would be reduced by $e^{-\tau_\lambda}$. The total luminosity would remain the same; however, the absorbed radiation would reappear as thermal radiation in the infrared provided all energy transport is radiative.

The general characteristics of the interstellar extinction curve consist of a toe in the infrared, a nearly linear increase from 8000 Å to 4000 Å, a shoulder between 4000 Å and 3000 Å, a strong peak around 2200 Å followed by a minimum near 1600 Å, and a rise to shorter wavelengths. The albedo of the interstellar grains is uncertain. Code (1971) has discussed evidence for a high albedo shortward of the 2200 Å bump. Witt and Lillie (1972) have derived an albedo curve in the spectral region from 4250 Å to 1400 Å from measures of the diffuse galactic light utilizing OAO-2 data. The outstanding feature of their results is a pronounced decrease in albedo around 2200 Å, which is coincident with the extinction curve bump, followed by a rapid rise in albedo shortward of 2000 Å. This is a difficult observational problem and the detailed albedo curve is probably quite uncertain although the decrease in albedo near 2200 Å is consistent with the data. A significant decrease in albedo at 2200 Å is also consistent with the identification of this feature with the presence of graphite particles by Stecher (1969), Bless and Savage (1972), Gilra (1971), and Wickramasinghe and Nandy (1971). Gilra (1972) has argued on rather general grounds that whatever constituent is responsible for the bump, the feature is due to an absorption process. We shall assume this to be true and examine the consequences.

In the discussion to follow we shall assume that the extinction curve is the Johnson (1968) Cygnus curve joined to the Bless-Savage (1972) 'average' ultraviolet curve and that the albedo is represented by the Witt-Lillie (1972) results in the ultraviolet and is of the order of 0.6 in the visual (Van de Hulst and de Jong, 1969). Figure 1 shows the adopted extinction curve for a $B - V$ color excess of 1 mag. and the assumed variation of albedo that shall be adopted for illustrative purposes. It is the purpose of this paper to indicate the effect such particles would have on the observed stellar flux for several simple models.

2. Plane Parallel Slab

We consider first a star illuminating a plane parallel slab of total optical thickness, τ_λ, and albedo $\tilde{\omega}$. The total energy reflected plus the direct stellar radiation in the direction (μ, φ) follows from the physical interpretation of the X-function (Van de Hulst, 1948; Chandrasekhar, 1950) and is in fact just $X(\mu)$ times the stellar flux in the direction

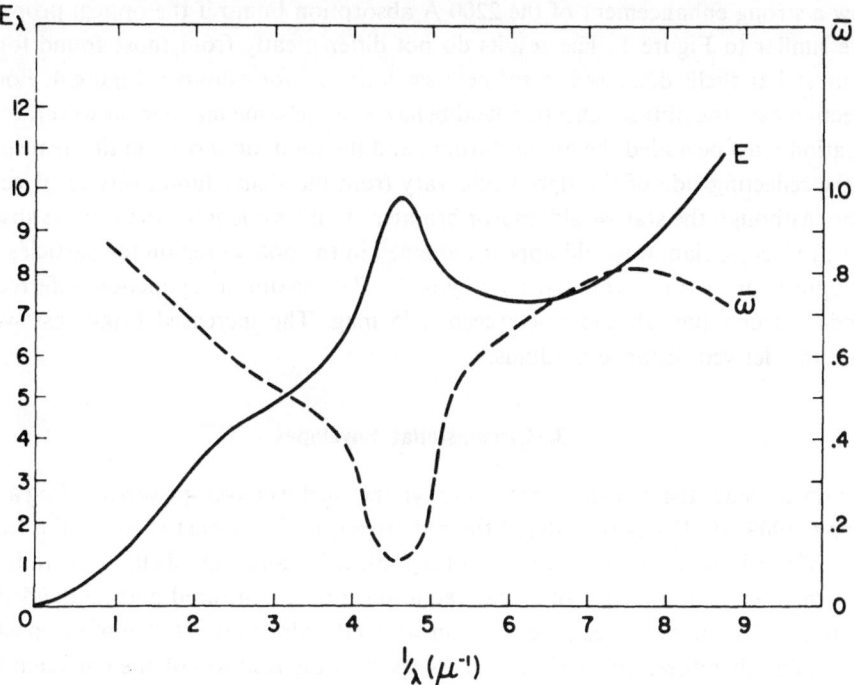

Fig. 1.   The adopted extinction and albedo curves. The solid line shows the extinction in magnitude for a $B - V$ color excess of 1 mag. plotted against wavenumber in μ^{-1}. The dashed curve represents the assumed variation of albedo as a function of wavenumber as given by the right hand coordinate.

$I(\mu, \varphi)$. Similarly the total diffusely transmitted energy plus the direct transmitted energy in the direction (μ, φ) is proportional to $Y(\mu)$. Thus

$$E(\varphi, \phi)_R = X(\mu)F^*(\mu, \varphi)$$
$$E(\mu, \varphi)_T = Y(\mu)F^*(\mu, \varphi), \tag{1}$$

where $F^*(\mu, \varphi)$ is the stellar flux that would be observed in the direction (μ, φ) if the plane plarallel slab were not there. If the field of view is restricted so that less and less of the surface of the slab is seen, the transmitted radiation approaches just that of the direct transmitted radiation $F^*e^{-\tau/\mu}$. One can expect cases, however, where most of the diffusely reflected or transmitted radiation will be observed. For example, virtually all the reflected energy, in the direction (μ, φ) would be observed in a field of view of 1' for a star located 100 astronomical units in front of the slab if its distance were the order of 1 kpc.

    In the case where the star is observed through the slab, two qualitative features are clear. Due to the addition of the diffusely transmitted light the extinction derived by the usual techniques would be very much smaller than if only the direct transmitted radiation is observed. Secondly, any absorption feature would be enhanced since for low albedos the diffuse radiation is small. The extinction curve resulting from the transmission of light through a slab with a visual optical depth of unity, for example, would show a strong enhancement of the 2200 Å absorption bump if the optical properties were similar to Figure 1. The results do not differ greatly from those found for the circumstellar shells discussed in the next section; see, for example, Figure 4. For the reflection case the diffuse reflection field behaves in the same manner; however, to this radiation must be added the direct starlight and the total luminosity in the hemisphere on the reflecting side of the slab would vary from the stellar luminosity to twice this value. Although the star would appear brighter at all wavelengths than in the absence of the reflecting slab, it would appear reddened in the optical region for particles with the optical properties exhibited by Figure 1. The maximum extinction inferred by standard techniques should not exceed 0.75 mag. The increased brightness would affect the derived distance modulus.

3. Circumstellar Envelopes

We now consider the case of a spherically symmetric dust envelope surrounding a star. Huang (1969, 1971) has considered the radiative transfer in dust shells in the framework of the Eddington approximation for both thick and thin shells. His study was concerned with the transfer of stellar radiation into the infrared and he divided the radiation field into two spectral regions in which the albedo was taken to be constant. His results, therefore, give only the gross spectral distribution of the emergent flux. Furthermore, the variation of albedo with wavelength is an important feature of the problem. Huang's technique has been extended by Apruzese (1972) to consider three or more spectral regions, which then makes it possible to determine the spectral distribution by shifting the third wavelength band across the spectrum. Some of Apruzese's results will be noted later. Mathis (1972) has also carried out numerical integrations for a few cases of spherical dust envelopes for the stellar case to which we shall refer.

Let us first consider the radiation transfer problem in a spherical shell in the simplest approximation which still preserves the basic features of the problem, namely multiple non-conservative scattering. If κ represents the mass absorption coefficient and σ the mass scattering coefficient, then the extinction optical depth is given by

$$d\tau = -(\kappa + \sigma)\varrho \, dx \tag{2}$$

and the albedo by

$$\bar{\omega} = \frac{\sigma}{\kappa + \sigma}. \tag{3}$$

We shall simply divide the radiation into a forward and a backward stream (Schwarzschild approximation). In this case the equation of radiative transfer becomes simply

$$\frac{dI_+}{d\tau} = I_+ - \frac{\bar{\omega}(1+g)}{2} I_+ - \frac{\bar{\omega}(1-g)}{2} I_- \tag{4}$$

$$-\frac{dI_-}{d\tau} = I_- - \frac{\bar{\omega}(1-g)}{2} I_+ - \frac{\bar{\omega}(1+g)}{2} I_-, \tag{5}$$

where g is the Henyey-Greenstein phase function, which in this approximation is

$$g = \frac{\sigma_+ - \sigma_-}{\sigma}, \tag{6}$$

where σ_+, σ_- are the scattering cross-sections in the forward and backward direction, and σ their sum. The appropriate boundary conditions for a spherical shell are that there be no incident radiation on the outside boundary where $\tau = 0$. That is

$$I_-(0) = 0. \tag{7}$$

If the radius of the inner surface of the envelope is large with respect to the stellar radius, then the radiation incident on the inner boundary is equal to the stellar intensity, I^*, plus the radiation scattered from the inner boundary. Thus at $\tau = \tau_1$

$$I_+(\tau_1) = I_* + I_-(\tau_1). \tag{8}$$

The solution of Equations (4) and (5) under the boundary conditions (7) and (8) is straightforward. The ratio of the emergent luminosity to that of the stellar luminosity is

$$\frac{L}{L_*} = \frac{I_+(0)}{I_*} = \frac{2}{(1+\zeta)e^{\xi\tau_1} + (1-\zeta)e^{-\xi\tau_1}}, \tag{9}$$

where

$$\zeta = \sqrt{(1-\bar{\omega})/(1-\bar{\omega}g)} \tag{10}$$

and

$$\xi = \sqrt{(1-\bar{\omega})(1-\bar{\omega}g)}. \tag{11}$$

For isotropic scattering $g = 0$ and for a completely forward scattering phase function $g = 1$. In the limit $g = 1$ the solution is exact, since with spherical symmetry there would be only a radial stream of radiation. For $g = 0$ the approximation is still very good, differing from numerical calculations by less than 10% in most cases. The dash-dot curve in Figure 2 shows a plot of L/L^* as a function of extinction optical depth and albedo determined from Equation (9) for isotropic scattering. Numerical results obtained by Mathis (1972) and Apruzese (1972) are indicated by the open circles and open triangles respectively. For an albedo of one, all the radiation escapes and $L/L^* = 1$. For an albedo of zero, we have the case of pure absorption and Equation (9) becomes simply $e^{-\tau}$, shown by the lower envelope. Even for albedos very near to unity the multiple scattering at high optical depths increases the effective absorption optical depth. In the two stream approximation given by Equation (9) the ratio of the inner radius to the outer radius does not enter, as is clear from the geometry of the two stream approximation. Mathis' results are for a ratio of radii of 0.2 and he finds essentially the same value for any smaller ratio.

A somewhat closer approximation is obtained by noting that in the exact solutions for a plane parallel atmosphere the ratio of transmitted intensities for two different

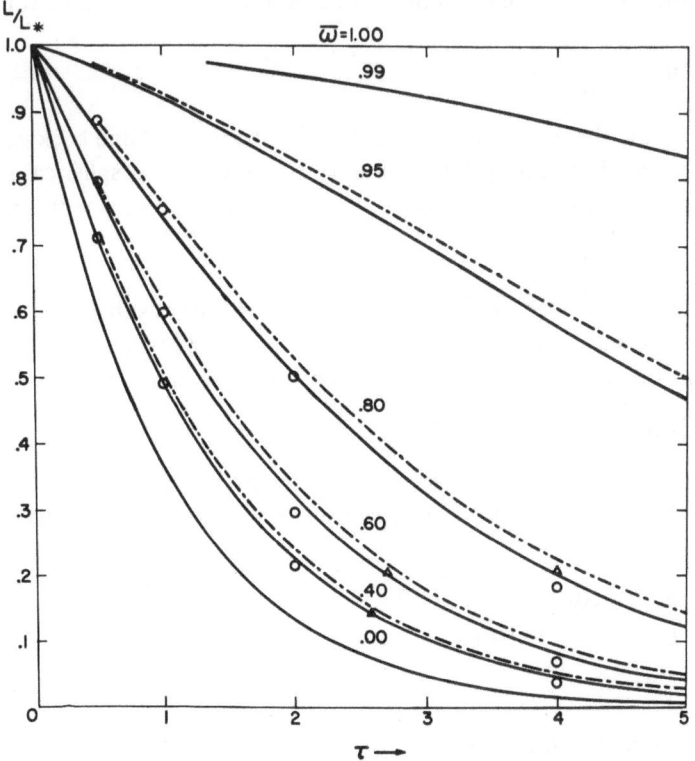

Fig. 2.   The ratio of emergent luminosity to the stellar luminosity as a function of extinction optical depth and albedo for $g = 0$. The solid lines give the results obtained from Equation (12), while the dash-dot lines are for the two stream approximation given by Equation (9). Each set of curves is labeled by its appropriate albedo, $\bar{\omega}$. The open circles are from the results by Mathis (1972) for a ratio of radii of 0.2. The open triangles are from calculations by Apruzese (1972) for a ratio of radii of 0.001. Mathis quotes accuracies better than 1% for $\tau < 1$, 3% for $\tau = 2$, and 10% for $\tau = 4$.

albedos is not very sensitive to the angle of incidence. The equation

$$\frac{L}{L_*} = e^{-\tau} + (1 - e^{-\tau}) \frac{J(\bar{\omega})}{J(1)}, \tag{12}$$

where

$$\frac{J(\bar{\omega})}{J(1)} = \frac{Y(\bar{\omega}, 1) - e^{-\tau_1}}{Y(1, 1) - e^{-\tau_1}} \tag{13}$$

is in excellent agreement with detailed numerical calculations. Here $Y(\bar{\omega}, 1)$ refers to the Y-function for an albedo $\bar{\omega}$, at normal incidence, $\mu = 1$. The result of applying Equation (12) is shown as the solid curve in Figure 2.

If the ratio of radii is near unity the ratio of luminosities is given by the thin shell case which is the same as a plane parallel atmosphere with the appropriate boundary condition. This can be solved exactly by iteration using the X, Y functions and is

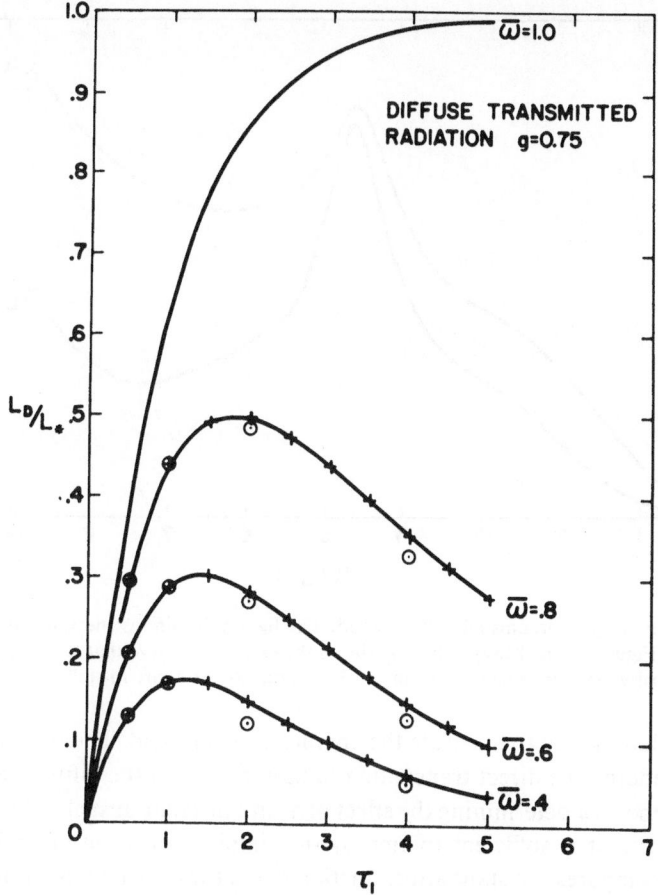

Fig. 3. Diffuse transmitted luminosity for $g = 0.75$. The solid lines are from Equation (9) where the direct transmitted component, $e^{-\tau}$, has been subtracted. The open circles are from the numerical calculations by Mathis (1972). The albedo for each curve is indicated.

approximately

$$\frac{L}{L_*} = \frac{F^{\circ}_{\text{trans}}/F_*}{1 - F^{\circ}_{\text{ref}}/F_*}, \tag{14}$$

where F°_{trans} and F°_{ref} are the total transmitted and reflected fluxes for a plane parallel atmosphere with normal incident flux. The ratios of luminosities given by Equation (14) are less than those found for the case of large curvature (small ratios of inner to outer radius). This is because large angle scattering traverses a larger optical depth in a plane parallel atmosphere than in a curved atmosphere. The luminosity curve given by (14) has the same shape and qualitatively the results insofar as they affect the extinction curve are the same. Figure 3 shows a comparison of the results obtained for $g = 0.75$ from Equation (9) with the calculations by Mathis. The agreement is excellent. The quantity plotted in Figure 3 is the ratio of the diffuse transmitted luminosity

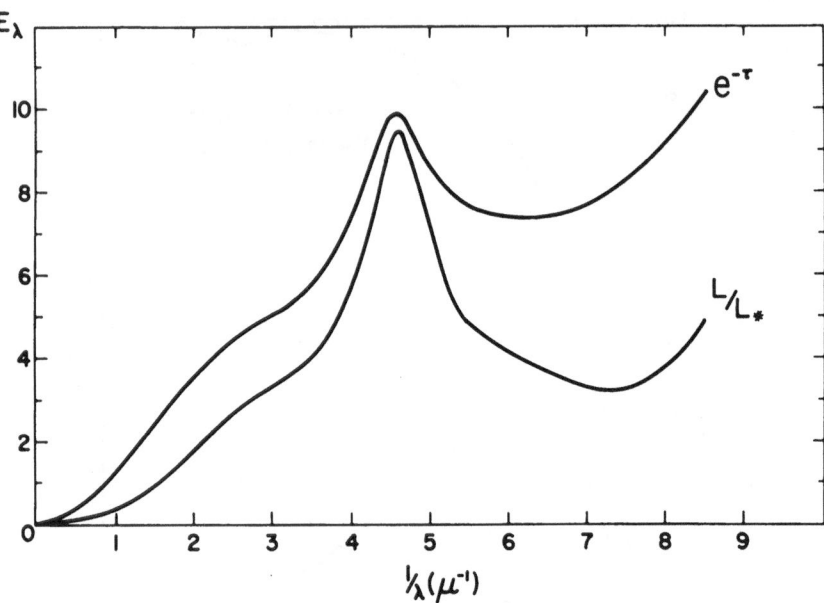

Fig. 4. Extinction by a circumstellar dust cloud. The light solid line is the same as the extinction curve shown in Figure 1. The heavy solid line shows the extinction curve that would be derived if the luminosity were modified according to Equation (9) for a circumstellar envelope.

to the stellar luminosity to illustrate the character of this field. The total luminosity is obtained by adding the direct transmitted radiation, $e^{-\tau}$, to the diffuse field.

For the purpose of determining the effect of a circumstellar dust cloud on the derived extinction curve, it is sufficient to employ the simple expression given by Equation (9). Figure 4 compares the standard extinction curve for a visual absorption of 3 mag. ($E_{B-V} = 1.0$) with the extinction curve found for a circumstellar shell of the same optical depth if the albedo varies as in Figure 1. Note the enhanced 2200 Å peak and the low apparent extinction in the 1500 Å region.

It is appropriate at this point to comment on the anomalous extinction curve found for θ Orionis (Carruthers, 1969; Bless and Savage, 1972). In these measurements the possibility of including significant scattered light is very real since the field of view included all the components of θ^1 and θ^2 Orionis as well as a significant part of the nebula. They found that the extinction in the 1500 Å region is about the same as in the visual. This is true of the results plotted in Figure 4. The bump at 2200 Å, however, is found observationally to be very small in contrast to the enhancement shown in Figure 4. If the 2200 Å feature is due to absorption, all the transfer models considered in this paper yield an enhancement of the bump and the explanation of the anomalous reddening for θ Orionis must result from an actual difference in the nature of the grain composition or size distribution. The possibility that contributions from an atomic continuum and reflected flux, however, still deserves consideration. The Trapezium is a very complicated system.

Figure 5 shows a comparison of the color excesses obtained with the UBV system if

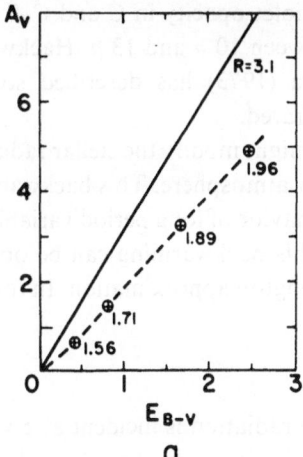
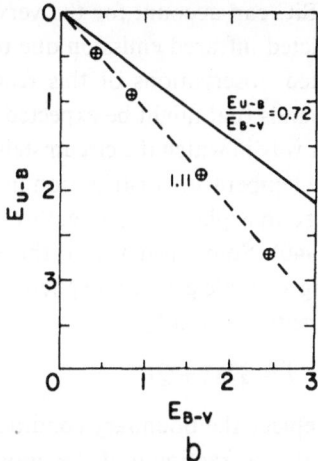

Fig. 5. (a) Total visual absorption vs $B - V$ color excess for the average extinction curve shown in Figure 1, solid curve, and for a circumstellar envelope, dashed curve. The ratio of total to selective absorption is indicated along the curve. (b) $U - B$ color excess vs $B - V$ color excess is plotted for average extinction curve and circumstellar envelope.

the extinction is circumstellar. The ratio of total to selective absorption is smaller, increasing with optical depth, and the ratio E_{U-B}/E_{B-V} is greater. The moral to be derived from these curves is obvious. Inferences as to the total optical depth from color measurements should be accepted only with reservation if there is reason to suspect that the dust is near the star.

4. Discussion

The presence of a circumstellar shell may modify the observed stellar radiation in several ways. It may be that a part of the radiant energy is converted into kinetic energy or into expansion of the dust cloud. In this case the total radiant energy is not conserved in passing through the dust shell. If the luminosity remains constant we can expect the radiation in the ultraviolet and visual to be redistributed in the infrared as a result of the absorption of grains and their thermal radiation. Apruzese (1972) has attempted to construct a model to describe the infrared star HD 45677. Some of the properties of this Be star are described by Swings and Allen (1971). The spectral energy distribution consists of a continuum in the visual, characteristic of a reddened B star, and a much stronger broad maximum in the $5-10~\mu$ region. The infrared flux is about 4 times that of the visual. Apruzese finds that a middle B star in the center of a spherical dust cloud can quantitatively account for the observed energy distribution. For large curvature he finds that the observations require an optical depth in the visual of 3.97, 2.68, or 2.12 for albedos of 0.8, 0.6, and 0.4 respectively. His calculations also give the temperature distribution of the grains in the dust envelope. For a given temperature distribution it would be possible to compute the infrared emission for particles with known optical constants. Gilra and Code (1971) have suggested that circumstellar clouds of SiC can account for the very strong violet opacity in C and C–S stars and have predicted infrared emission due to SiC between 10 μ and 13 μ. Hackwell (1972) has reported observations of this feature. Gilra (1973) has described some other emission bands that might be expected in the infrared.

Another way in which the circumstellar cloud might modify the stellar radiation is to change the temperature distribution in the stellar atmosphere. This backwarming was first invoked to explain certain features in the behavior of long period variable stars by Merrill (1940). Some insight into the effect of this backwarming can be obtained by considering a simple grey atmosphere in the Eddington approximation. In this case the mean intensity is given by

$$J = \tfrac{3}{4}F(\tau + Q).\tag{15}$$

We then replace the boundary condition that no radiation is incident at $\tau = 0$ by the condition that a fraction α of the outward intensity is returned to the stellar surface. That is, at $\tau = 0$,

$$I_+ = I_0\tag{16}$$

and

$$I_- = \alpha I_0 ;\tag{17}$$

then

$$J(0) = \tfrac{1}{2}I_0(1 + \alpha)\tag{18}$$

and

$$F(0) = I_0(1 - \alpha).\tag{19}$$

Solving (15) for Q we find that

$$J = \tfrac{3}{4}F\left(\tau + \frac{2}{3}\frac{(1 + \alpha)}{(1 - \alpha)}\right)\tag{20}$$

or the temperature distribution is

$$T^4 = \tfrac{3}{4}T_e^4\left(\tau + \frac{2}{3}\frac{(1 + \alpha)}{(1 - \alpha)}\right).\tag{21}$$

The temperature at small optical depths is increased by the radiation scattered or reemitted from the shell towards the stellar surface. For example, the temperature at an optical depth of unity is increased by about 15% if $\alpha = 0.5$. Buerger (1972) has recently computed a series of model atmospheres in which incident radiation is included. He finds effects of the same order as given by the simple grey atmosphere. It would be worthwhile carrying out more detailed calculations to determine the spectral distribution and the strength of particular atomic lines to see if the change in the stellar spectrum produced by circumstellar backwarming should be observable.

In summary, we have tried to indicate the manner in which the presence of nearby dust clouds can influence the observed flux from stars. The discussion indicates that for stars with significant infrared excesses neither the nature of the embedded star nor the optical depth of the dust cloud can be determined by the usual techniques employing color excesses and ratios of total to selective absorption. The errors can be large. Finally, it should be remarked that features considered in this paper are also applicable to stellar systems such as galaxies (Code *et al.*, 1972).

References

Apruzese, J.: 1972, private communication.
Bless, R. C. and Savage, B. D.: 1972, *Astrophys. J.* **171**, 293.
Buerger, P. F.: 1972, *Astrophys. J.* **177**, 567.
Capriotti, E. R.: 1967, *Astrophys. J.* **148**, 318.
Carruthers, G. R.: 1969, *Astrophys. J. Letters* **157**, L113.
Chandrasekhar, S.: 1950, in *Radiative Transfer*, Oxford Press, p. 211.
Code, A. D.: 1971, in B. T. Lynds (ed.), *Dark Nebulae, Globules and Protostars*, University of Arizona Press, Tucson, p. 101.
Code, A. D., Welch, G. A., and Page, T. L.: 1972, in A. D. Code (ed.), *The Scientific Results from the Orbiting Astronomical Observatory*, NASA SP-310, p. 559.
Collins, G. W., II and Code, A. D.: 1965, *Astrophys. J.* **142**, 1576.
Gilra, D. P.: 1971, *Nature* **229**, 237.
Gilra, D. P.: 1972, in A. D. Code (ed.), *The Scientific Results from the Orbiting Astronomical Observatory*, NASA SP-310, p. 295.
Gilra, D. P.: 1973, this volume, p. 517.
Gilra, D. P. and Code, A. D.: 1971, *Bull. Am. Astron. Soc.* **3**, 379.
Hackwell, J. A.: 1972, preprint.
Hallam, K. L.: 1959, Thesis, University of Wisconsin.
Henyey, L. G. and Greenstein, J. L.: 1941, *Astrophys. J.* **93**, 70.
Huang, Su-Shu: 1969, *Astrophys. J.* **157**, 843.
Huang, Su-Shu: 1971, *Astrophys. J.* **164**, 91.
Johnson, H. L.: 1968, in B. M. Middlehurst and L. H. Aller (eds.), *Nebulae and Interstellar Matter*, University of Chicago Press, Chicago, p. 167.
Mathis, J. S.: 1970, *Astrophys. J.* **159**, 263.
Mathis, J. S.: 1971, *Astrophys. J.* **167**, 261.
Mathis, J. S.: 1972, *Astrophys. J.* **176**, 651.
Merrill, P. W.: 1940, *Spectra of Long-Period Variable Stars*, University of Chicago Press, Chicago.
Stecher, T. P.: 1969, *Astrophys. J. Letters* **157**, L125.
Swings, J. P. and Allen, D. A.: 1971, *Astrophys. J. Letters* **167**, L41.
Van de Hulst, H. C.: 1948, *Astrophys. J.* **107**, 220.
Van de Hulst, H. C. and de Jong, T.: 1969, *Physica* **41**, 151.
Wickramasinghe, N. C. and Nandy, K.: 1971, *Nature* **229**, 81.
Witt, A. N. and Lillie, C. F.: 1972, in A. D. Code (ed.), *The Scientific Results from the Orbiting Astronomical Observatory*, NASA SP-310, p. 199.

dust clouds can influence the observed flux from Sgr A*. The discussion implies that for stars with no obscuration, the particles neither die nor at the embedded sun nor the peak absorption of the dust cloud can be determined by the usual techniques employed. Impede the process and value of total to free-free absorption. These must be be larger. Finally, it should be reiterated that for the conclusion in this paper are also applicable to stellar systems such as galaxies [Oke et al., 1974].

References

Aarseth, S., 1971, private communication.
Bland, R. C. and Jeans, R. L., Astrophys. J. Lett. **175**, L97.
Burbidge, E. M., 1970, *Astrophys. J.* **160**, 469.
Chandrasekhar, S., 1960, *Principles of Stellar Dynamics*, Dover Publ.
Chandrasekhar, S., 1969, *Astrophys. J.* **159**, L13, L61.
Chandrasekhar, S., 1960, in *Radiative Transfer*, Oxford Publ. Press, p. 390.
Colla, A. G. 1971, in P. V. Blanco (ed.), *IAU Symposium No. 44*, D. Reidel Publ. Comp., Dordrecht, p. 318.
Fricke, K., Gold, R. A., and Page, T. L., 1968, *Astron. Astrophys.*
Galactic Astronomical Observatory, NASA SP-314, p. 594.
Gebbie, C. W. F. and Glass, A. P., 1969, *Astrophys. J.* **157**, 389.
Gillett, F. C. 1971, *Astrophys. J.* **170**, 617.
Hall, D. N. B. 1971, in L. E. C. and Low, J. Astron. J., Space Sci. Rev.
Harwit, M., 1973, *Astrophys. J.* **186**, 344.
Hoyle, F. 1951, *Astrophys. J.* **117**.
Greenstein, J. L., Hoyle, F. 1972, *Astrophys. J.* **188**.
Low, F. J. and Aumann, H. H., 1970, *Astrophys. J. Lett.* **162**, L79.
Hoyle, F. 1953, *Astrophys. J.* **118**.
Hoyle, F. and Wickramasinghe, N. C., 1970, *Nature* **227**, 51.
Hoyle, F. and Wickramasinghe, N. C., 1969, *Nature* **223**, 459.
Hoyle, F. and Wickramasinghe, N. C., 1962, *Mon. Not. Roy. Astron. Soc.* **124**, 417.
Johnson, H. L., 1966, in K. N. Sandulak and Early, Astr. Space Sci. Structure and Evolution of Stars, Univ. of Chicago Press, Chicago, p. 141.
Lightman, A. P. 1969, *Astrophys. J.* **157**.
Matthews, T. A., 1971, *Astrophys. J.* **157**.
Mathis, J. S., 1970, *Astrophys. J.* **176**.
Mathis, J. S., 1972, *Astrophys. J.* **176**, 651.
Merrill, P. W., 1960, *Spectra of Long-Period Variable Stars*, University of Chicago Press, Chicago.
Sciama, D. W., 1960, *Astrophys. J.*, L. Terrell, 1912.
Sanger, P. E. and Allen, D. A., 1974, *Astrophys. J. Lett.*, L97.
Van de Hulst, H. C., 1957, *Light Scattering by Small Particles*, Wiley, New York.
Van de Hulst, H. C., Rossberg, T., 1969, *Nature* **41**, 138.
Wickramasinghe, N. C. and Guillaume, S., 1965, *Nature* **207**, 366.
Woolf, C. N. and Ney, E. P., 1967, *IAU Symp. No. 75*, in M. Hack (ed.), Circumstellar Atmospheres of Stars, D. Reidel Publ. Comp., Dordrecht, p. 105.

DUST PARTICLES AND MOLECULES
IN THE EXTENDED ATMOSPHERES OF CARBON STARS

DAYA P. GILRA

Space Astronomy Laboratory, Washburn Observatory, University of Wisconsin,
Madison, Wis., U.S.A.

Abstract. It is shown that the absorption due to a circumstellar shell containing solid silicon carbide particles can very nicely explain the observed strong violet opacity in stars in which the carbon to oxygen ratio is > 1. It has been shown by Friedemann and Gilman that solid SiC particles can form in the cooler outer layers of such stars. Thermal re-emission from SiC particles is predicted to be in the 10–13 μ region and recent infrared observations by Hackwell show an emission band in this region, thereby strongly supporting the SiC suggestion. It is also shown that the opacity due to C_3 pseudo-continuum is not adequate to explain the observed violet opacity.

It is suggested that the vibrational bands of C_3 and SiC_2 molecules should be among the major opacity sources in the infrared spectra of the late N-type carbon stars and some of the observed bands may be, at least in part, due to these molecules. The frequencies of their isotopic species have been calculated and attempts should be made to observe these bands.

Thus the atmosphere of a late N-star should be pictured as containing, probably on its outskirts, solid carbon particles. There is some kind of smoke veil around the star causing a reddening by absorption of the ultraviolet.

A veiling effect by smoke has been occasionally envisaged to interpret various astronomical phenomena, even in the case of novae. A late N-star would be a striking example. The smoke veil would vary in variable N-stars.

B. Rosen and P. Swings (1953)

It is well known that in the spectra of late N-type carbon stars the intensity drops very rapidly at about 4400 Å, this rapid drop continuing throughout the ultraviolet. In fact, due to this problem, so far, spectra of very few late N-stars have been photographed shortward of about 4000 Å (Swings *et al.*, 1953, to be precise, spectra of only four stars have been obtained at 4000 Å and of only one star shortward of 3800 Å). More than 40 yr ago Shane (1928) noted that the could photograph the spectrum of Y CVn, a star one can see with the naked eye, in the 4800 Å region with a small quartz slitless spectrograph in *one second*, whereas an exposure of *5 hours* with the same instrument failed to show any light shortward of 3900 Å.

The late N-type stars also show the 4050 Å bands of C_3 and the blue-green Merrill-Sanford Bands of SiC_2. It was shown by Swings *et al.* (1953) that the intensities of the C_3 and the Merrill-Sanford bands are related to the violet drop in the sense that the stronger the bands, the more pronounced is the violet opacity. They also pointed out that the star U Hya sometimes shows the SiC_2 bands, the C_3 bands and strong violet opacity and sometimes all three are absent. On the basis of these observations Swings *et al.* were led to conclude that the source or sources of the violet opacity are related to the formation of the polyatomic molecules SiC_2 and C_3. (It should be pointed out that

Greenberg and Van de Hulst (eds.), Interstellar Dust and Related Topics, 517–528.

SiC$_2$ as the carrier of the Merrill-Sanford bands was not identified until 1956, Kleman 1956.) Rosen and Swings (1953) and Swings (1953) suggested that circumstellar solid carbon particles may be responsible for the strong violet opacity observed in these stars.

In 1954 McKellar and Richardson (1955) compared their observations of carbon stars Y CVn and U Hya in the blue-violet with the experimental C$_3$ continuum – it is now believed to be pseudo-continuum (Brewer and Engelke 1962) – and found good agreement. It was shown by Feast (1955) that a similar rapid decrease in intensity at about 4400 Å occurs in some S and C–S stars and suggested that the agent is the same in the two cases. He argued that the source cannot be solid carbon particles as suggested by Rosen and Swings (1953) and Swings (1953) because they are not likely to form in S stars. In a subsequent paper, however, he pointed out (Feast, 1957) that there are apparent differences in the energy distributions in this spectral region in the two cases, the late N-stars on one hand and the S and C-S stars on the other, and the sources of opacity may be different.

Recently Stephenson and Ross (1970) have shown that this phenomenon, the sudden decrease in intensity at about 4400 Å, is a characteristic of quite a few S and C–S stars. They favor Feast's original suggestion (Feast, 1955) that the source of opacity in late N-stars, and S and C-S stars is the same and argue that C$_3$ cannot be the source for S stars.

Recent observational studies by Catchpole and Feast (1971) and by Greene (quoted by Slettebak, 1972) show that all S–C and C–S stars show strong violet opacity. Greene (quoted by Slettebak (1972), see also Stephenson (1965), and Catchpole and Feast (1971)) has drawn attention to one star, Case 621, in particular. This star has neither ZrO bands which are characteristic of S stars, nor C$_2$ bands which are characteristic of carbon stars, but has strong violet opacity and shows evidence of SiC$_2$ bands. It also shows CaCl bands and a very strong Li λ 6707 line. Greene suggests that in the atmosphere of Case 621 the carbon to oxygen ratio is one.

Two points must be noted about this violet opacity effect: its 'sudden' appearance at about 4400 Å and the continued rapid increase throughout the ultraviolet. The main problem to be investigated in this paper is the identification of the source or sources of this dramatic opacity effect in the stars in which the carbon to oxygen ratio is approximately equal to or greater than one. This work has been done in collaboration with Code and the preliminary results, with the predictions, were presented at the Amherst Meeting of the American Astronomical Society (Gilra and Code, 1971), and a detailed paper is in preparation (Gilra and Code, 1972). The other problem to be discussed is the nature of the vibrational bands of C$_3$ and SiC$_2$ in the infrared with the possibility that information about the isotopes of carbon and silicon may be obtained from high resolution observations of carbon stars in this spectral region.

1. The Violet Opacity

It has been shown by Friedemann (1969a, b) and Gilman (1969) that solid silicon carbide particles can form in the cooler outer layers of the stars in which the carbon to oxygen ratio is approximately equal to or greater than one. SiC_2 gas is a major vaporization product of solid SiC such as C_3 is that of graphite (Weltner and McLeod, 1964b). Bands of SiC_2 are seen in the spectra of late N-type carbon stars. A very important point is that depending upon the polytype, temperature, and concentration of impurities, the fundamental absorption edge of SiC lies between about 2.2 eV and 3.3 eV (Choyke, 1969).

Since solid SiC particles can form in the atmospheres of all the stars in which strong violet opacity is observed, let us consider a circumstellar shell containing solid SiC particles. We consider particles of cubic SiC. The imaginary part of the refractive index, k, has been calculated from the absorption coefficient given by Patrick and Choyke (1969) for two cases of cubic SiC: pure, and with nitrogen impurity. Mie calculations have been made for both the cases for a gaussian size distribution with a mean radius of 0.105 μ and a standard deviation of 0.06 μ. (Friedemann (1969b) has calculated the mean radius to be about 0.1 μ.) The resulting albedoes and the extinction cross-sections between 5400 Å and 3600 Å are shown in Figure 1. The curve for albedo 'with nitrogen' is for SiC particles having nitrogen impurity and the albedo curve 'without nitrogen' is for pure SiC. The extinction cross-sections are identical in the two cases. The point to note is the sudden decrease in albedo for both the cases at about 4500 Å and its continued decrease throughout the ultraviolet. The absorption cross-section for 'pure' SiC particles increases by about a factor of 30 between 5200 Å and 3600 Å. (The 'band' in the albedo curve for particles with nitrogen impurity arises because there is an absorption band at about 4000 Å in the bulk crystals with the nitrogen impurity.)

With these cross-sections and albedoes the ratio of the emergent flux to the incident flux, $F(O)/F(\tau)$, was computed using the theory of radiation transfer described by Code (1973). This ratio is plotted against wavelength in Figure 2 for two values of

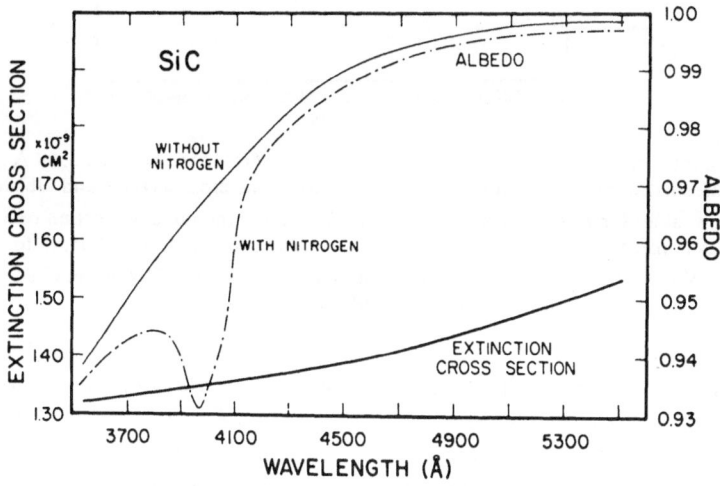

Fig. 1.   The albedoes and extinction cross-sections between 5500 Å and 3500 Å for two cases of SiC particles: pure ('without nitrogen'), and with nitrogen impurity ('with nitrogen'). The extinction cross-sections are identical in the two cases.

the optical depth for each of the two cases: pure SiC particles, and SiC particles
with nitrogen impurity. The values for τ given in the figure are $(\sqrt{3})^{-1}$ times the
extinction optical depth at λ 5220; for example, the case $\tau = 15$ implies that the
extinction optical depth for that case at λ 5220 Å is $15 \times \sqrt{3}$ ($= 26$). Also plotted in
this figure is the observed absorption continuum for Y CVn as given by McKellar and
Richardson (1955). The values were measured from their Figure 4 at 50 Å each and
a constant vertical shift has been made to match with the theoretical curves.

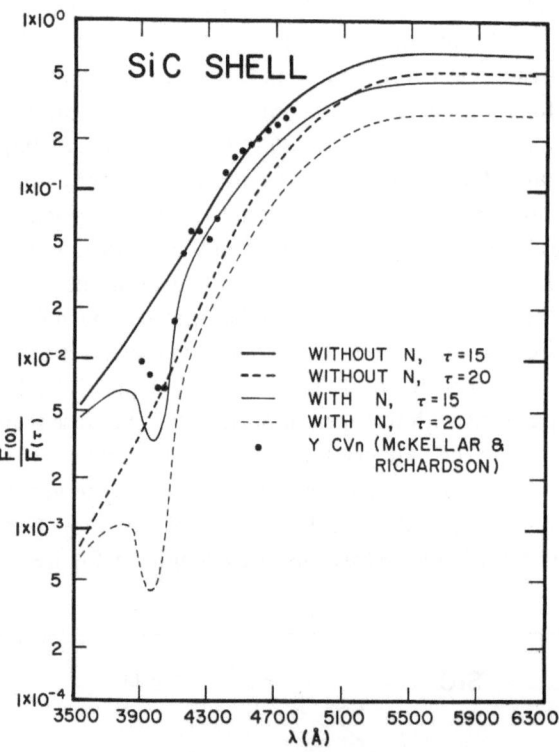

Fig. 2.   The ratio of the emergent flux to the incident flux, $F(O)/F(\tau)$, between 6300 Å and 3500 Å,
through a circumstellar shell containing SiC particles using the cross-sections and albedoes given in
Figure 1. The values for τ given in the figure are $(\sqrt{3})^{-1}$ times the radial extinction optical depth at
λ 5220: for example, the case $\tau = 15$ implies that the radial extinction optical depth for that case at
λ 5220 is $15 \times \sqrt{3}$ ($= 26$). The dots give the absorption continuum of Y CVn as obtained by McKellar
and Richardson (1955).

Let us first consider the difficulties with the identification of the C_3 pseudo-continuum (McKellar and Richardson, 1955; Brewer and Engelke, 1962) as the source of
the observed absorption continuum. The C_3 identification was based on the apparent
similarity between the observed continuum and the experimental C_3 continuum, with
the stellar absorption features at 4300–4350 Å and at about 4000 Å significantly aiding
the identification. McKellar and Richardson mention in their paper that the observed

feature at 4300–4350 Å is not attributable to the λ 4383 2,0 sequence of the Swan bands of C_2 nor to the λ 4315 CH band. They do not mention that the strong red degraded SiC_2 band at λ 4352, which can be seen very clearly in the spectra reproduced by them, can be a significant contributor to this feature. It seems the whole feature is due to these and other (λ 4261 band of SiC_2, etc.) molecular bands in this spectral region. The photoelectric scans by Fay and Honeycutt (1972) of W Ori, a star which shows much stronger SiC_2 bands and much stronger violet opacity than Y CVn (Swings et al., 1953), show this very well. The broad stellar absorption feature at λ 4000 also seems to be due to various discrete molecular absorptions in this region (for a list of absorption features in this region, see Swings et al., 1953). Again, the observations of Fay and Honeycutt support this conclusion. As regards the main absorption continuum, the observations of McKellar and Richardson extended down to only λ 3900. Recent wide band photometry of several carbon stars by Mendoza and Johnson (1965) and the narrow band photometry extending down to λ 3300 of four carbon stars, in particular of Y CVn, by Johnson, Mitchell and Latham (1967) show that the rapid decrease in intensity continues throughout the ultraviolet. For example, the narrow band photometry by Johnson et al. (1967) of Y CVn shows that the stellar flux decreases by a factor of one hundred and fifty between 4020 Å and 3530 Å. On the other hand, the experimental absorption coefficient of C_3 in this spectral region is decreasing (Brewer and Engelke, 1962). We can thus conclude that C_3 is not the major source of opacity in this spectral region in the spectra of late N-type carbon stars. (The nature of the observed stellar SiC_2 and C_3 bands in the optical region and their implications on the opacity problem and the work of Shajn and Struve (1947) will be discussed in detail elsewhere (Gilra and Code, 1972)).

We now compare the observations with our theoretical results for a circumstellar shell of SiC particles as shown in Figure 2. As can be seen, the agreement is very good for both the cases of SiC particles. The extinction optical depth at 5220 Å is about 30. (We are not suggesting that the absorption band at 4000 Å in the theoretical curves for SiC particles with nitrogen impurity is responsible for the observed feature.) The agreement between our theoretical curves and the observations of Feast (1957) of AM Cen, an SC star, is also very good. From their photoelectric observations, Fay and Honeycutt (1972) also find support for our suggestion.

Similar calculations have been performed for circumstellar graphite particles. The absorption coefficient of graphite increases throughout the ultraviolet but there is no 'suddenness' in the increase. Therefore the conclusion is that circumstellar graphite particles do not contribute much to the sudden opacity increase in the blue-violet. It should be possible to calculate an upper limit for the mass in circumstellar graphite particles.

We can obtain the mass in SiC particles in the entire circumstellar shell from the values given in Figures 1 and 2. It comes out to be of the order of 10^{-9} solar masses. The mass thus derived will be an upper limit because the molecular absorptions and the possible contribution from circumstellar graphite particles have not been subtracted from the observations. The crystalline structure of circumstellar SiC particles, the impurities they might have and their temperature also affect the calculations for the mass. But the primary aim here is to identify the major source of this dramatic opacity effect and we can say with reasonable confidence that it is circumstellar SiC particles. Recent infrared observations, as discussed below, by Hackwell (1972) confirm this suggestion.

There is another observational test of the SiC hypothesis. The energy the SiC grains are absorbing in the blue-violet will be re-emitted in the infrared. Based on the calculations of Gilra (1972a), it was predicted by Gilra and Code (1971) that this thermal re-emission should be in the 10–13 μ region, a region in which ground-based observations can be made. The theory is discussed elsewhere (Gilra, 1972a, b) and the main results are given below. For about a tenth of a micron radius particles the *shape* of the particles is the most important parameter, the size is not important and we can use the Rayleigh approximation. Depending upon the shape, the emission band(s) should appear between about 10.2 μ and 12.8 μ. A distribution of *shapes* will make a broad emission feature between about 10.2 and 12.8 μ. The shape is not a significant factor in the absorption in the blue-violet, so the optical depth derived from this region will be, so to speak, redistributed throughout this broad feature in the 10–13 μ region and the optical depth may not be high at any given wavelength in this infrared band. The temperature of the particles depends upon their distance from the star; therefore, the distance may be of some importance.

There is a slight problem, however. The Si–C stretching frequency, v_1'', of SiC_2 molecules is at 11.75 μ (see the discussion below). In late N–type stars there should be a strong absorption band at 11.75 μ which will somewhat 'fill in' the predicted emission. Observations with high spectroscopic resolution may be required to separate the two effects.

During this symposium Hackwell showed me his scans of two carbon stars, CIT6 and V Hya, in the 10 μ region (Hackwell, 1972). There is a strong emission feature present between about 10 μ and 13 μ almost similar to what the theoretical calculations (Gilra, 1972a, b) show. There is an indication of an absorption feature at 11.75 μ which may be the SiC_2 band. His photometry of carbon stars (Hackwell, 1972) in the infrared also shows emission at 11 μ. An emission feature at 11 μ is also present in the photometry of R CMi, a C–S star.

On the basis of all the theoretical and observational evidence, it seems the existence of circumstellar SiC particles should be considered well-established.

2. C_3 and SiC_2 Bands in the Infrared

Strong vibrational bands of C_3 and SiC_2 should appear in the spectra of late N-type carbon stars in the infrared. A brief discussion of these bands with special emphasis on the possibility of detecting and measuring the isotopic bands is given. To the best of my knowledge there has been no such discussion in the astronomical literature (cf. Spinrad and Wing, 1969; Vardya, 1970), even though most of the information has been available since 1964 (Weltner and McLeod, 1964a, b).

2.1. C_3 BANDS

For C_3 the asymmetric stretching frequency, v_3'', is 2040 cm^{-1} (Weltner and McLeod, 1964a) and the symmetric stretching frequency, v_1'', is 1224.5 cm^{-1} (Merer, 1967; Weltner and McLeod 1964a). Since the value 1224.5 cm^{-1} for v_1'' is somewhat smaller than the value 1240 cm^{-1} used by Weltner and McLeod, I have recalculated the force constants of C_3 in the ground state from formulas given by Herzberg (1945). The values are:

$$k_{11} = 10.20 \times 10^5 \text{ dyn cm}^{-1},$$
$$k_{12} = +0.397 \times 10^5 \text{ dyn cm}^{-1}.$$

(The values calculated by Weltner and McLeod (1964a) are 10.34×10^5 dyn cm^{-1} and $+0.542 \times 10^5$ dyn cm^{-1} respectively.) With these force constants one can calculate (Bartunek and Barker, 1935) the corresponding frequencies for the isotopic specicis as was done for v_3 by Weltner and McLeod (1964a) who found very good agreement between the theoretical and experimental values. The values are given in Table I. Also given in Table I are the values of $v_1 + v_3$. The anharmonicity constants x_{13} are not known but should be of the order of 5 to 10 cm^{-1}. Thus the values $v_1 + v_3$ should be within about 5 to 10 cm^{-1} of the frequencies of the combination band ($v_1 + v_3$). C_3 is a linear symmetric molecule (Gausset et al., 1965) and therefore v_3 and ($v_1 + v_3$) are infrared active whereas v_1 is not (Herzberg, 1945). However, the species $C^{12}-C^{12}-C^{13}$ and C^{13} $C^{13}-C^{12}$ are not symmetrical and v_1 for them is infrared active (Herzberg, 1945). Weltner and McLeod (1964a) in their experimental work on the infrared spectrum of C^{13}-substituted C_3 observed all the six v_3 bands. Apparently their observations were not extended to the 1200 cm^{-1} region so the $C^{12}-C^{12}-C^{13}$ and $C^{13}-C^{13}-C^{12}$ v_1 bands at 1200 cm^{-1} seem to have remained undetected.

We note from this table that in the spectra of late N-type carbon stars a strong band at 2040 cm^{-1} ($= 4.90 \mu$) due to $C^{12}-C^{12}-C^{12}$ should be observed. Many of these stars show relatively high abundance of C^{13}, but the interpretation is not easy (Fujita, 1970). In these stars all the six bands extending to 1960 cm^{-1} ($= 5.1 \mu$) may be observed and it should be possible to derive, perhaps less ambiguously, the ratio of C^{12}/C^{13}. Low resolution infrared spectra of carbon stars CIT 6, CIT 13, and T Cnc

TABLE I

Isotopic bands of C_3

C–C–C	v_1 (cm^{-1})	v_3 (cm^{-1})	$v_1 + v_3$ (cm^{-1})
12–12–12	1224.5	2040	3264.5
12–12–13	1200	2027	3227
13–12–13	1176	2013	3189
12–13–12	1224.5	1987	3211.5
13–13–12	1200	1974	3174
13–13–13	1176	1960	3136

obtained by Gaustad *et al.* (1969) show an absorption feature at about 4.9 μ. For T Cnc a 'broad depression' at 5 μ has been observed. They have attributed these absorptions in CIT 6, CIT 13 and T Cnc to the fundamental vibrational band of CN. I would like to suggest that C_3 is a significant contributor to these features.

Some support for this suggestion comes from the broad band photometric observations. Figure 3 shows a color-color plot for 11 carbon stars common in the photometric observations of Mendoza and Johnson (1965) and Gillett *et al.* (1971). The $U-B$ color is from Mendoza and Johnson and the infrared color [4.9 μ]−[8.4 μ] is from Gillett *et al.* The horizontal arrows for two stars, the upper point is for T Cnc and the lower point for V CrB, mean that their U magnitudes are fainter than 22 (the V magnitudes are 9.05 and 9.33 respectively). (Incidentally, the values of $U-B$ show very clearly the extreme faintness of the late N-type carbon stars in the ultra-violet.) Figure 3 shows an apparent correlation which can be interpreted to mean that the stronger absorption in the ultraviolet goes with the stronger absorption in the 5 μ region. Since the presence of the 4050 Å C_3 bands is related to the opacity in the ultraviolet (Swings *et al.*, 1953), even though we have shown earlier that C_3 is not a major source of this opacity, it can be inferred that the C_3 bands may be a significant contributor to the opacity in the 5 μ region. However, it is necessary to make observations with high spectroscopic resolution to definitely identify the source(s) of opacity in the 5 μ region.

In the spectra of the three carbon stars, CIT 6, CIT 13 and T Cnc Gaustad *et al.*

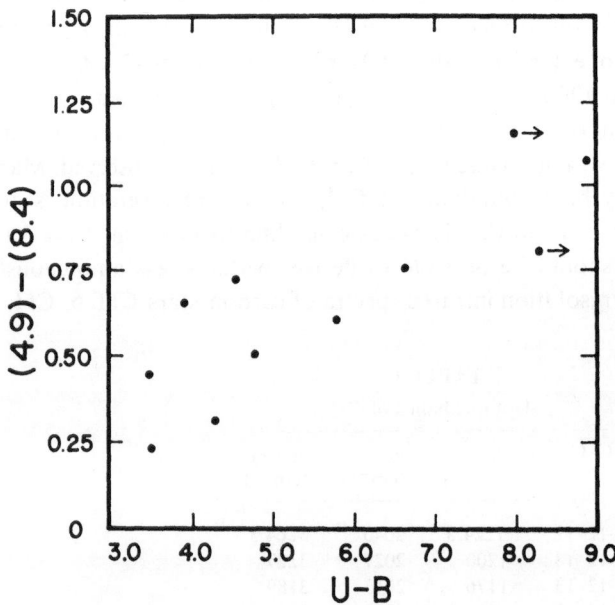

Fig. 3.   Infrared color [4.9 μ]−[8.4 μ] (Gillett *et al.*, 1971) against $U-B$ (Mendoza and Johnson, 1965) for 11 carbon stars common in their observational programs. The horizontal arrows for two stars mean that the U magnitudes of these stars are fainter than 22.

(1969) observed, a strong absorption feature is present at 3 μ which they suggested may be due to the v_3 fundamental of either HCN or C_2H_2. The existence of this feature was confirmed by Low et al. (1970) who remarked that the carbon stars as a group exhibit a broad absorption at about 3250 cm^{-1}, "with the coolest carbon stars having the strongest absorption." They pointed out that the intensity at 3250 cm^{-1} for T Lyrae is only about 20 percent of the interpolated continuum.

As indicated in Table I the $(v_1 + v_3)$ combination band of C_3 occurs at about 3265 cm^{-1}. The photometry of Mendoza and Johnson (1965) for T Lyrae gives V = 8.18 and U > 22, that is, the star shows extremely strong violet opacity. On the basis of a similar argument as made earlier it is suggested that the combination band $(v_1 + v_3)$ of C_3 may be a significant contributor to the observed feature at about 3250 cm^{-1}. Once again, high resolution observations are required to identify the source(s) definitely.

The value of the bending frequency, v_2 is 63.1 cm^{-1} for C_3 and is quite a low value (Gausset et al. 1965). Several combination bands involving v_2, $2v_2$, etc. should be observable in the spectra of the late N-type carbon stars. Also, the carbon stars which show relatively high abundance of C^{13} should show the C^{13}–C^{13}–C^{12} and C^{12}–C^{12}–C^{13} bands at 1200 cm^{-1}.

2.2. SiC$_2$ BANDS

Weltner and McLeod (1964b) from their matrix infrared observations obtained the values of v_1 and v_3 fundamentals of SiC$_2$ as 853 cm^{-1} and 1742 cm^{-1} respectively. Verma and Nagaraj (1971) have obtained 134 cm^{-1} for the v_2 fundamental. SiC$_2$ is a linear asymmetric molecule (Si–C–C) in the ground state and all the fundamentals are infrared active. The following stretching force constants for SiC$_2$ in the ground state were calculated by Weltner and McLeod (1964b):

$$k(\text{Si–C}) = 7.43 \times 10^5 \text{ dyn cm}^{-1},$$
$$k(\text{C–C}) = 7.98 \times 10^5 \text{ dyn cm}^{-1}.$$

By assuming that k_{12} is zero, that is how the force constants given above were calculated, I have calculated from the formulas of Bartunek and Barker (1935) the frequencies for the v_1 and v_3 fundamentals for various isotopic species of silicon and carbon (Table II). (In the light of the work of Weltner and McLeod (1964b), and Verma and Nagaraj (1971), the force constants derived by Yamashita (1967) for the ground state are wrong and the 'identification' of the isotopic SiC$_2$ (Yamashita and Utsumi, 1968) is not correct.)

The v_3 fundamental lies in a spectral region which is not accessible from the ground. It should be possible to observe the absorption due to the v_1 fundamental. A search should be made to detect the isotopic bands; we have as yet no information on the isotopic abundances of silicon. However, the differences are not large and the observations may not be easy to interpret. The observations of V Hya by Hackwell (1972) indicate an absorption feature at 11.75 μ; it may be the SiC$_2$ v_1 fundamental band.

Quite a few overtone and combination bands are infrared active and a search for

TABLE II

Isotopic bands of SiC_2

Si–C–C	v_1 (cm^{-1})	v_3 (cm^{-1})
28–12–12	853	1742
28–12–13	835	1725
28–13–12	850	1694
28–13–13	833	1677
29–12–12	846	1740
29–12–13	829	1724
29–13–12	844	1693
29–13–13	827	1675
30–12–12	841	1739
30–12–13	823	1722
30–13–12	838	1691
30–13–13	821	1674

them should be made, especially because, as in the case of C_3, the bending frequency is low (v_2 is 134 cm^{-1}, Verma and Nagaraj (1971)).

3. Concluding Remarks

(1) The first part of this paper describes the work of Gilra and Code (1971, 1972) on the identification of the source of the violet opacity in the stars in which the carbon to oxygen ratio is approximately equal to or greater than one. The major source was identified as circumstellar solid silicon carbide particles, with circumstellar graphite particles and C_3 molecules being possible minor contributors. There is strong support for this identification from the infrared observations. Thus the 'picture' envisaged by Rosen and Swings (1953), as quoted at the beginning of this paper, is essentially correct, we would just add solid SiC particles with the carbon particles. The behavior of U Hya, as described earlier, is also understood. This identification also confirms the suggestions of Feast (1955) and Stephenson and Ross (1970) that the opacity source should be the same in the late N-type carbon stars and, SC and CS stars.

Detailed observations both in the 10 μ region and in the blue violet are needed for a good understanding of the atmospheric structure of these stars. High resolution observations of some SC and CS stars, in particular, Case 621, which I have identified as IRC 00404, should be made in the 10 μ region.

(2) In the second part a brief discussion of the vibrational bands of C_3 and SiC_2 is given. It is pointed out that C_3 and SiC_2 should be among the major opacity sources in the infrared spectra of late N-type carbon stars. The frequencies of their isotopic species have been calculated and attempts should be made to observe them. From the infrared observations of the C_3 and SiC_2 bands in the spectra of late N-type stars it should be possible

(i) to derive the number densities of the C_3 and SiC_2 molecules,

(ii) to obtain information about the exciting conditions in the regions of the atmosphere in which these molecules exist, and

(iii) to derive the isotopic composition of these molecules. These results will provide us with very valuable information about the atmospheric structure of the late N-type carbon stars.

Acknowledgements

I would like to thank Dr R. D. Verma, Dr S. Nagaraj and Dr John Hackwell for communicating their results in advance of publication. This research has been supported in part by NASA grant NGL 50-002-013.

References

Bartunek, P. F. and Barker, E. F.: 1935, *Phys. Rev.* **48**, 516.
Brewer, L. and Engelke, J. L.: 1962, *J. Chem. Phys.* **36**, 992.
Catchpole, R. M. and Feast, M. W.: 1971, *Monthly Notices Roy. Astron. Soc.* **154**, 197.
Choyke, W. J.: 1969, *Materials Research Bulletin* **4**, S141.
Code, A. D.: 1973, this volume, p. 505.
Fay, T. and Honeycutt, R. K.: 1972, *Astron. J.* **77**, 29.
Feast, M. W.: 1955, *Mém. Soc. Roy. Sci. Liège* **15**, 280.
Feast, M. W.: 1957, *Mém. Soc. Roy. Sci. Liège* **18**, 301.
Friedemann, Chr.: 1969a, *Physica* **41**, 139.
Friedemann, Chr.: 1969b, *Astron. Nachr.* **291**, 177.
Fujita, Y.: 1970, *Interpretation of Spectra and Atmospheric Structure in Cool Stars*, University Park Press, Baltimore.
Gausset, L., Herzberg, G., Lagerquist, A., and Rosen, B.: 1965, *Astrophys. J.* **142**, 45.
Gaustad, J. E., Gillett, F. C., Knacke, R. F., and Stein, W. A.: 1969, *Astrophys. J.* **158**, 613.
Gillett, F. C., Merrill, K. M., and Stein, W. A.: 1971, *Astrophys. J.* **164**, 83.
Gilman, R. C.: 1969, *Astrophys. J. Letters* **155**, L185.
Gilra, D. P.: 1972a, in A. D. Code (ed.), *The Scientific Results from the Orbiting Astronomical Observatory*, NASA-SP 310, p. 295.
Gilra, D. P.: 1972b, unpublished thesis, University of Wisconsin.
Gilra, D. P. and Code, A. D.: 1971, *Bull. Am. Astron. Soc.* **3**, 379.
Gilra, D. P. and Code, A. D.: 1972, in prepration.
Hackwell, J. A.: 1972, *Astron. Astrophys.* **21**, 239.
Herzberg, G.: 1945, *Infrared and Raman Spectra of Polyatomic Molecules*, D. van Nostrand Co., Princeton.
Johnson, H. L., Mitchell, R. I., and Latham, A. S.: 1967, *Comm. Lunar Planetary Lab.* **6**, 85.
Kleman, B.: 1956, *Astrophys. J.* **123**, 162.
Low, F. J., Johnson, H. L., Kleinman, D. E., Latham, A. S., and Geisel, S. L.: 1970, *Astrophys. J.* **160**, 531.
McKellar, A. and Richardson, E. H.: 1955, *Mém. Soc. Roy. Sci. Liège* **15**, 256.
Mendoza, E. E. and Johnson, H. L.: 1965, *Astrophys. J.* **141**, 165.
Merer, A. J.: 1967, *Can. J. Phys.* **45**, 4103.
Patrick, L. and Choyke, W. J.: 1969, *Phys. Rev.* **186**, 775.
Rosen, B. and Swings, P.: 1953, *Ann. Astrophys.* **16**, 82.
Shajn, G. and Struve, D.: 1947, *Astrophys. J.* **106**, 86.
Shane, C. D.: 1928, *Lick Obs. Bull.* **13**, 123.
Slettebak, A.: 1972, *Bull. Am. Astron. Soc.* **4**, 152.
Spinrad, H. and Wing, R. F.: 1969, *Ann. Rev. Astron. Astrophys.* **7**, 249.
Stephenson, C. B.: 1965, *Astrophys. J.* **142**, 712.

Stephenson, C. B. and Ross, H. E.: 1970, *Astron. J.* **75**, 321.
Swings, P.: 1953, *Ann. Astrophys.* **16**, 287.
Swings, P., McKellar, A., and Rao, K. N.: 1953, *Monthly Notices Roy. Astron. Soc.* **113**, 571.
Vardya, M. S.: 1970, *Ann. Rev. Astron. Astrophys.* **8**, 87.
Verma, R. D. and Nagaraj, S.: 1971, private communication.
Weltner, W. and McLeod, D.: 1964a, *J. Chem. Phys.* **40**, 1305.
Weltner, W. and McLeod, D.: 1964b, *J. Chem. Phys.* **41**, 235.
Yamashita, Y.: 1967, *Publ. Dominion Astrophys. Obs.* **13**, 67.
Yamashita, Y. and Utsumi, K.: 1968, *Publ. Astron. Soc. Japan* **20**, 73.

INFRARED OBSERVATIONS OF STARS
IN THE ASSOCIATION CYG OB2

K. VOELCKER* and H. ELSÄSSER

*Max-Planck-Institut für Astronomie und Landessternwarte, 69 Heidelberg 1,
Königstuhl, W. Germany*

Abstract. About 80 stars of the association Cyg OB2 have been observed in *J*, *H*, *K* and *L*. These measurements have been made in search for infrared circumstellar emission in this highly reddened association.

1. Introduction

The most highly reddened association Cyg OB2 in the northern sky, in which Schulte (1956, 1958) has found a large number of OB stars, has been investigated thoroughly by Reddish *et al.* (1966). They used a photographic UBV-photometry with limiting magnitudes of 16^m5 in *B* and 15^m7 in *V*. These authors concluded that the high visual absorption (up to about 10^m) is caused mainly by dust clouds associated with the individual stars. They come to the result that about 60% of the total association mass is in circumstellar shells. The age of the association is estimated to be less than 10^6 yr. There exist so far observations in the infrared only for a few very bright members of the association. We observed approximately 80 stars of different visual magnitudes from Reddish's catalogue in the infrared in order to look for circumstellar emission.

2. Observations

Our measurements were made in the four broadband filters *J*, *H*, *K* and *L* in the spectral range from 1.2μ to 3.5μ with the 50″ telescope at Kitt Peak. In addition several stars also were observed at 5μ and 10μ, but they were too faint for detection. The observed stars were selected according to the following criteria:

(1) OB stars identified spectroscopically by Schulte (1956, 1958) with large color-excesses E_{B-V}.

(2) other stars unambiguously of early spectral type (earlier than B3) taken from the $U-B/B-V$ diagram, also with large color-excesses.

(3) several stars later than A0 and stars of different spectral types with small color-excesses.

Figure 1 shows the $U-B/B-V$ diagram of the observed early type stars with color-excesses roughly between 1^m0 and 2^m8 and the normal reddening line. Schulte's spectral classification and UBV spectral types are in good agreement.

* Visiting astronomer, Kitt Peak National Observatory, which is operated by the Association of Universities for Research in Astronomy, Inc., under contract with the National Science Foundation.

Greenberg and Van de Hulst (eds.), Interstellar Dust and Related Topics, 529–533.

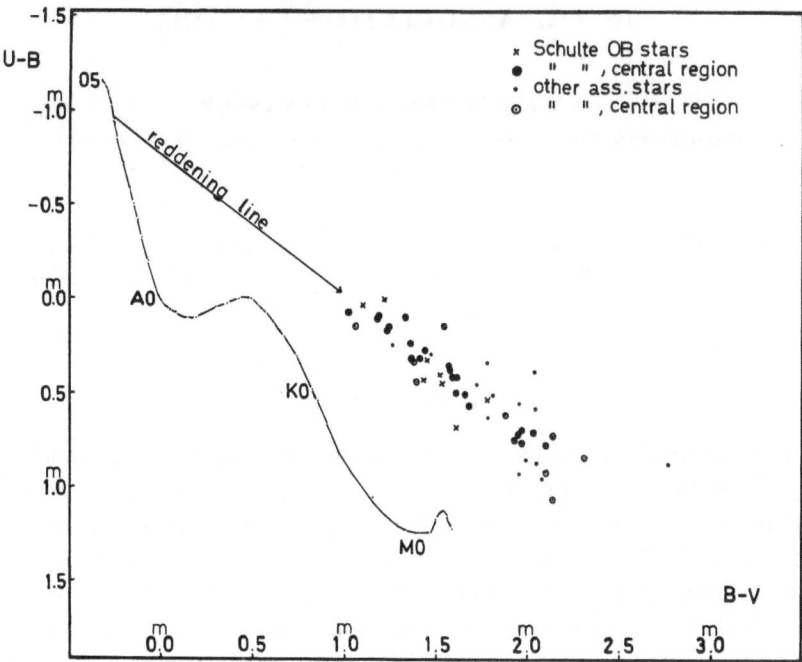

Fig. 1. $U - B/B - V$ diagram for most of the observed stars. The different symbols mark Schulte's OB stars and other association stars inside and outside the central part of the association. The unreddened main-sequence and the reddening line also are shown.

The question now arises whether or not these high absorption values are produced inside the association itself. From Reddish's two-color diagram for all his observed stars (more than 1200) we obtain a minimal value $A_V \approx 2$ mag. for the foreground absorption. The catalogue of early type stars by Ikhsanov (1959) yields a crude estimate of 3 mag. visual absorption within a distance of 2.1 kpc, the adopted distance of the association Cyg OB2. Field 30 of Neckel's catalogue (Neckel, 1967) gives $A_V = 2\overset{m}{.}6$ within 2 kpc. As an average we adopt an absorption A_V of about $2\overset{m}{.}5$ in the line of sight to the association.

Most of the stars observed by us lie in the central region of the association where the visual absorption and OB star density are highest. In Figure 2 the lines of high constant absorption in this inner part are shown. It clearly stands out against the surroundings. The major axis of these ellipsoids is coincident with the axis of the ellipsoid of highest star density. This also confirms the fact that most of the extinction takes place within the association.

The indicated values for A_V are based on $R \equiv A_V/E_{B-V} = 3.25$. R has been obtained by extrapolating $(1/\lambda \to 0)$ our infrared data of about 22 stars with spectral types deduced from objective prism classification. Figure 3 shows our extinction curve vs wavelength. Not a single star indicates a value of R significantly different from 3.0 in contrast to the large values of R found by Strom et al. (1972) in the young cluster NGC 2264.

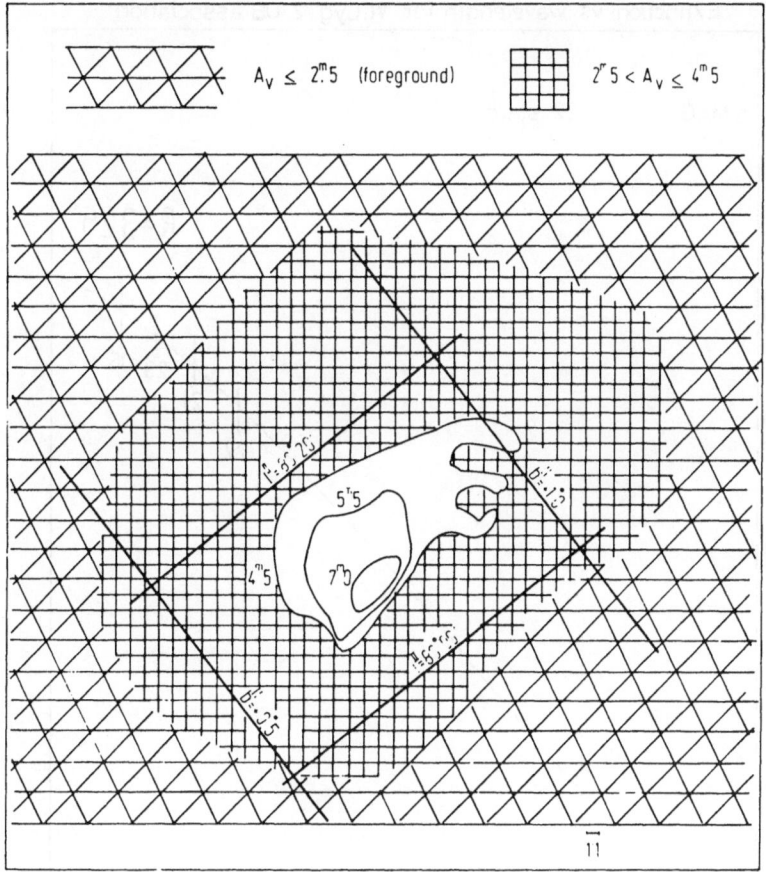

Fig. 2.   Lines of constant high absorption in the central part of the association Cyg OB2, where most of Schulte's OB stars are concentrated. Foreground absorption A_V amounts to approximately $2\overset{m}{.}5$.

Figure 4 represents the $V - L/B - V$ diagram for 35 of the observed stars for which L measurements were possible. The colors all can be explained by extinction according to the known interstellar law. There is no infrared excess (including even the extremely reddened OB star Schulte No. 12) that would indicate circumstellar shell emission.

For only one star (Reddish-catalogue No. 662) with $V = 15\overset{m}{.}6$ there could be found an excess in K (2.2 μ) of about $3\overset{m}{.}5$. Unfortunately this measurement is not very reliable because it is near the limiting magnitude in K; the other infrared colors have not been measured.

3. Conclusion

Our conclusion therefore is that the stars brighter than $V = 15\overset{m}{.}6$ have no circumstellar shells within the usual temperature range from 700 to 1000 K. We are inclined to interpret the correlation found by Reddish, namely that the absolutely brighter

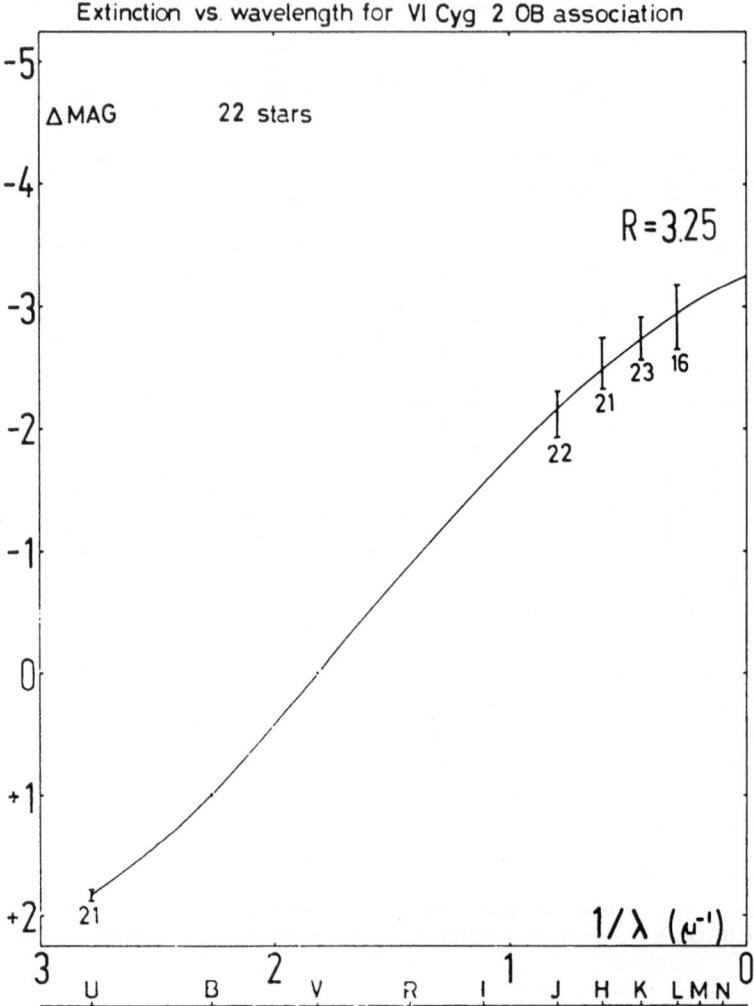

Fig. 3.   Mean normalized extinction curve vs 1/λ. The numbers of stars with known spectral types measured in different colors are given below the error bars.

stars have larger color-excesses E_{B-V}, rather by the fact that these stars are located in the densest regions of the dust cloud out of which they have been formed and not by the existence of circumstellar shells with variable optical depth. It might however be that these bright stars already have lost their original thick shells that could have formed according to current models of pre-mainsequence evolution (Larson, 1969, 1972). Then the question arises whether or not circumstellar emission might be found for the stars fainter than $V = 15\overset{m}{.}7$.

On sensitized IN plates ($\lambda_{eff} \approx 0.8\ \mu$) taken of this association at Heidelberg we were able to identify approximately 20 very faint ($V > 16\overset{m}{.}0$) and extremely red objects with $B - V \geqslant 2\overset{m}{.}5$, which need further photoelectric observations at longer wavelengths.

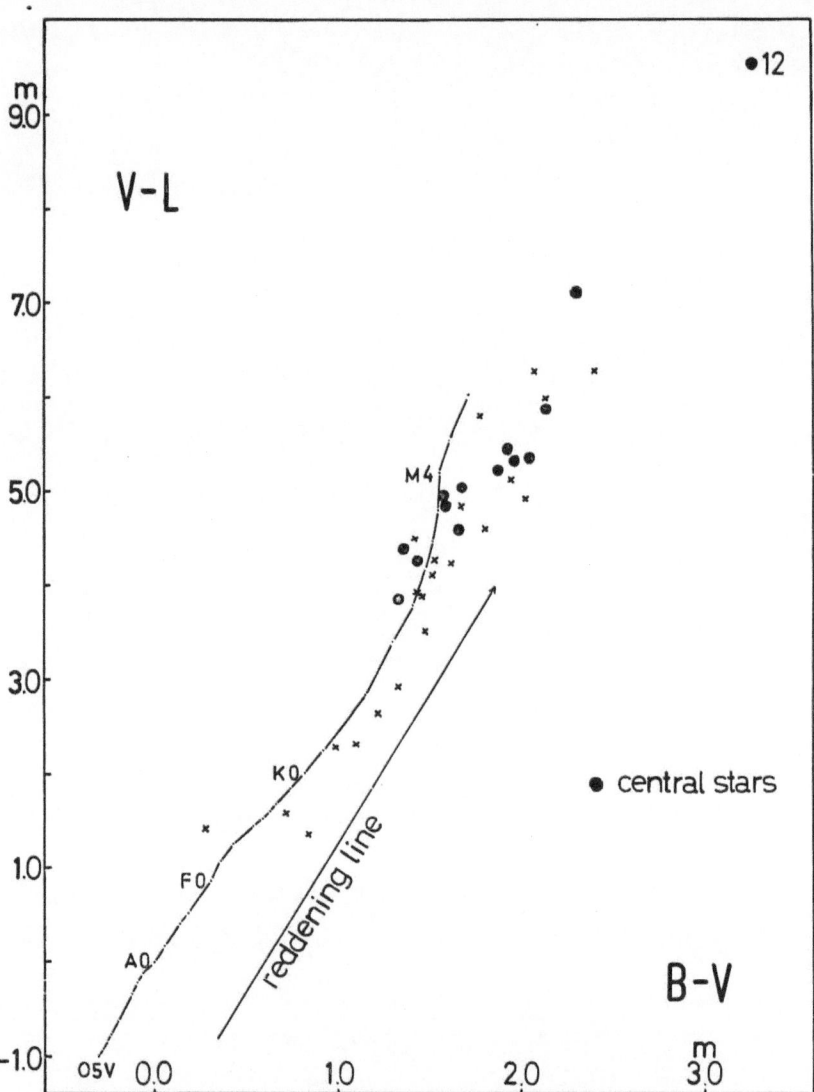

Fig. 4. $V-L/B-V$ diagram for the stars inside (●) and outside the central part. The unreddened main-sequence and the reddening line are shown. Not a single star (including the extremely reddened OB star Schulte No. 12) has a significant L-excess.

References

Ikhsanov, R. N.: 1959, *Crimean Comm.* **21**, 229 and 257.

Larson, B. R.: 1969, *Monthly Notices Roy. Astron. Soc.* **145**, 271; 1972, *Monthly Notices Roy. Astron. Soc.* **157**, 121.

Neckel, Th.: 1967, *Veröff. d. Landessternwarte Heidelberg* **19**.

Reddish, V. C., Lawrence, L. C., and Pratt, N. M.: 1966. *Publ. Obs. Edinburgh* **5**, 111.

Schulte, D. H.: 1956, *Astrophys. J.* **123**, 250; **124**, 530.

Schulte, D. H.: 1958, *Astrophys. J.* **128**, 41.

Strom, S. E., Strom, K. M., Brooke, A. L., Bregman, J., and Yost, J.: 1972, *Astrophys. J.* **171**, 267.

PART X

MISCELLANEOUS

A STATISTICAL MODEL FOR THE CLOUD STRUCTURE
OF THE INTERSTELLAR MEDIUM

L. G. TAFF and M. P. SAVEDOFF

*C. E. Kenneth Mees Observatory, Dept. of Physics and Astronomy, University of Rochester, Rochester,
N.Y., U.S.A.*

Abstract. We have formulated and studied the properties of a mathematical model intended to span
between the hydrodynamical and discontinuous regimes. Model parameters and predictions permit
matching the observed mass spectrum of those interstellar clouds with masses between 10^3 and 10^6 $M\odot$.
The central element of the theory is the specification of P $(m, m'; \mu, \mu', \dots \mu'')$, the conditional probability that once a collision has occurred between clouds of mass m, m' the result is clouds of mass
$\mu, \mu', \dots \mu''$. For stable solutions to exist we require that the total mass in the system be conserved.
Such simple models as total coalescence, geometric overlap and partition statistics were considered for
P as well as several probabilities based on possible physical conditions that might prevail in cloud-
cloud collisions. One immediate numerical result of these models is that nearly total coalescence must
obtain in actual cloud-cloud collisions before one can build up a non-infinitesimal concentration of
large mass couds. Field and coworkers (1965, 1968) indicate that total coalescence would produce a
minimum in the mass spectrum and that as the position of the minimum went to infinity the spectrum
itself more closely approximates a power law curve with index $-\frac{3}{2}$. In none of our calculations was
anything but strict monotonic decrease observed and under the same physical assumptions as Field
the curve we obtain (analytically) has no minimum, independent of the largest mass in the system.
Both of these spectra are flatter than recent observational evidence indicate. True equilibrium solutions
exist in our formulation with an e-folding time $\simeq 10^7$ yr for objects such as these. We are continuing
work on the detailed time evolution of interstellar clouds in particular and formulating other astronomically interesting applications of the general theory.

1. Introduction

There are many physical, astronomical and chemical processes of interest wherein
there is a reaction of the form $X + Y \rightarrow A + B + \cdots + C$ where the products of the reaction are not necessarily of the same generic type as the reactants. To describe the time
evolution of the processes with particular application to the cloud structure of the
interstellar medium (ISM), a mathematical formulation of such processes has been
created in a very general way. This formulation and applications to the ISM will be
presented here. The nature of the mathematical structure will be indicated without
proof in Section 3. The results of some of our calculations as pertains to the mass
distribution of clouds is contained in the last section. Further possible applications
of the formalism to astronomical problems are molecular formation in the (interstellar
medium) ISM, the mass structure of the asteroid belt and sporadic meteors, nucleo-
genesis and the production of large energy fluxes via stellar collisions in galactic nuclei.

2. Mathematical Formulation

We now consider the coalescence and disruption of clouds in the ISM. Since we are
primarily interested in deriving the mass spectrum the velocity parameter has been

Greenberg and Van de Hulst (eds.), Interstellar Dust and Related Topics, 537–543.
All Rights Reserved. Copyright © 1973 by the IAU.

averaged over. The simplest geometrical situation we invision for cloud-cloud collisions will have at most three fragments and further generalizations will not be included here. For complete details see Taff (1973). When the fraction of space occupied by the clouds is small compared to the total available volume only binary collisions will be of primary importance. We assume this to be the case. We shall formulate the problem in a discrete language for simplicity. Thus, we introduce for convenience the assumption that all cloud masses are integral multiples of some unit mass m, i.e. if the jth cloud (or cloud of type j) has mass m_j then

$$m_j = j \cdot m. \tag{1}$$

In the discussion of the equilibrium problem for the mass distribution we are concerned with closed systems and allow no unbalanced sources or sinks. This implies large objects cannot leave the system as in the original work by Field and coworkers (Field and Saslaw, 1965; Field and Hutchins, 1968). Finite computing machine memories also limit the largest mass, m_n, which we can consider in detail.

The specification of a collision between clouds of mass i,j involves two steps. First is the number of such collisions occurring per unit time per unit volume. We shall symbolize this as $<\sigma v>_{ij} N_i(t) N_j(t)$. $N_i(t)$ is the number of clouds of type i per unit volume at time t and $<\sigma v>_{ij}$ is the velocity averaged collison rate between clouds i and j. The time dependence will not always be explicitly exhibited. Once a collision has occurred we need the conditional probability that it will yield fragments of mass p, q, r in accordance with

$$m_i + m_j = m_p + m_q + m_r. \tag{2}$$

We denote this probability by $P(i, j; p, q, r)$. In any particular application of this statistical mechanical formulation the *mechanics* lies in the specification of the P's. To determine the differential equation governing the time evolution of $N_k(t)$ we need the production and destruction rates for clouds of type k. The processes are:

(I) Input to N_k.
    (a) When one reactant is a k.
        (i) Only one is.
        (ii) Both are.
    (b) Neither of the reactants is a k
        (i) The reactants are different from each other.
        (ii) The reactants are identical.

(II) Output from N_k.
    (a) One reactant is a k but none of the products is.
    (b) Both of the reactants are k's.
        (i) None of the products is a k.
        (ii) One of the products is a k.

Processes not listed above (apart from redistribution discussed below and spontaneous input-output terms) produce no net change in N. Let us write $<\sigma v>_{ij}$ as A_{ij}.

Summing all of the above terms we have apart from the matter of masses $>n$ the equation

$$\frac{dN_k}{dt} = (\tfrac{1}{2}) \sum_{i=1}^{n} \sum_{j=1}^{n} A_{ij} N_i N_j H^k_{ij} \tag{3}$$

with

$$H^k_{ij} = H^k_{ji} = \sum_{p=0}^{2n} (P(i,j;p,k,i+j-p-k)+P(i,j;i+j-p-k,p,k)+ \\ + P(i,j;k,i+j-p-k,p))-\delta_{ik}-\delta_{jk} \\ = p^k_{ij} - \delta_{ik} - \delta_{jk}, \tag{4}$$

where, as an example, the *Ibi* term can be seen to be equivalent to

$$(\tfrac{1}{2}) \sum_{i=1}^{n} \sum_{j=1}^{n} A_{ij} N_i N_j (1-\delta_{ij})(1-\delta_{ik})(1-\delta_{jk}).$$
$$\sum_{p,q,r=0}^{2n} P(i,j;p,q,r)\delta_{i+j,\,p+q+r}(\delta_{pk}+\delta_{qk}+\delta_{rk}). \tag{5}$$

The massive clouds formed from clouds with masses less than or equal to n are assumed to disrupt because of gravitational instability and then be redistributed into clouds of smaller mass. Although a spectrum of redistribution is more probable then redistribution into clouds of one particular mass we shall follow the procedure of Field and coworkers, based on a suggestion due to Oort (1954), and assume all of this mass return as clouds of unit mass. H^1_{ij} will now be assumed to be corrected for this.

We have postulated a closed system for the equilibrium discussion and must not only take into account redistribution but insure that the total mass is conserved, i.e.

$$\sum_{k=1}^{n} k\,dN_k/dt = 0, \tag{6}$$

which, in turn, is a condition on the type of probability function P that we are using since it implies that

$$\sum_{k=1}^{2n} kH^k_{ij} = 0, \quad \text{for all} \quad i \text{ and } j. \tag{7}$$

This constraint will be called the k sum. For simple probability functions one can show that this is the same statement as the one that P is a true probability. The mathematical implications of the k sum are such to insure the existence of a mathematically nontrivial solution for the equilibrium problem and its stability relative to perturbations.

Once the set of H matrices, which are real, symmetric, non-definite and non-pairwise commuting*, along with $N_k(0)$ are specified the time evolution of the system can

* This means the commutator
$$\sum_{j=1}^{n} (H'_{ij}H^s_{jk}-H^s_{kj}H'_{ji}) \neq 0, \quad r \neq s.$$

be studied. We shall instead study the equilibrium problem here to indicate the kind
of result the time-dependent problem will yield. The statement of the equilibrium
problem is the solution of

$$\sum_{i=1}^{n} \sum_{j=1}^{n} A_{ij} N_i N_j H_{ij}^k = 0, \quad k = 1,2,...,n, \tag{8}$$

with

$$\sum_{k=1}^{n} k N_k = M/(mV) = \text{constant}, \tag{9}$$

where M is the total mass of clouds contained in the fixed volume V.

3. Mathematical Properties

The above problem can be solved analytically only for the non-physical case of $n = 2$,
for all n, if there is total coalescence and A_{ij} is independent of i, j (Taff and Savedoff,
1972). Fortunately, the theorems below are not n dependent. The proofs can be found
in Taff (1973). The $A_{ij} H_{ij}^k$ will be assumed to have the general properties listed above.

Theorem 1. For a given set of inital conditions $N_i(0)$ there is one and only continuous
solution to Equations (3). This is true with or without the k sum. (Equation (9)).

Theorem 2. If the $A_{ij} H_{ij}^k$ approach, uniformly, a nonzero constant as t goes to infinity
there is a non-trivial solution of Equations (8) subject to the k sum.

Theorem 3. Under the time dependence assumptions of Theorem 2 if a solution vector
N exists to Equations (8) and (9) such that each of its components N_i is real, bounded
below by 0 and above by $M/(miV)$ then it is unique.

The stability of the equilibrium solution relative to perturbations depends on the
eigenvalues of the matrix H given by

$$\mathbf{H}_{ki} = 2 \sum_{j=1}^{n} A_{ij} H_{ij}^k N_j^0 \tag{10}$$

and we have introduced the superscript zero to indicate the equilibrium solution. The
properties of H are given by Theorem 4.

Theorem 4 The matrix H defined by Equation (10) when N^0 is the equilibrium solution
vector has the following properties:
   (i) $\det(H) = 0$
   (ii) The rank of H is $n-1$ and it therefore possesses only one zero eigenvalue whose
eigenvector is N^0.
   (iii) If λ is an eigenvalue of H then $\text{Re}(\lambda) < 0$ unless $\lambda = 0$.
With these characteristics for H one may prove Theorem 5.

Theorem 5. Every solution $N(t, N(0))$ of Equations (3) and (9) approaches N^0 as t approaches infinity. Thus the solution N^0 is absolutely stable relative to perturbations and N^0 is independent of $N(0)$.

4. Cloud Models

It is known that under the assumptions

(i) $p_{ij}^k = \delta_{i+j,\,k}$ (ie. total coalescence)

(ii) $n = \infty$

(iii) A_{ij} is independent of i, j.

that

$$N_k = \text{constant} \cdot k^{-3/2} \tag{11}$$

independent of the time dependence (independence) or type of redistribution (Taff and Savedoff, 1972). If one modifies the third assumption to one of geometric cross-section the mass spectrum is still a power law with index $-\frac{5}{3}$ (Taff, 1973). These mass distributions are lower limits to those of more realistic models. For simplicity let us assume all clouds are spheres with the same mean interior density. Then

$$A_{ij} = (3m/4\pi\varrho)^{2/3}(i^{1/3}+j^{1/3})^2\langle v\rangle_{ij}, \tag{12}$$

because gravitation is negligible. There are two reasonable forms for $\langle v\rangle_{ij}$. The first is that it is uncorrelated with i, j and the second that there is kinetic energy equipartition (KEE) in which case $\langle v\rangle_{ij} \propto ((i+j)/ij)^{\frac{1}{2}}$.

There are two simple geometrical models for $P(i, j; p, q, r)$ which we have considered. In the first when two clouds of mass i, j ($j>i$) collide the result is (for impact parameters greater than that of internal tangency $= c_{\text{int}}$) the part of i swept out by j

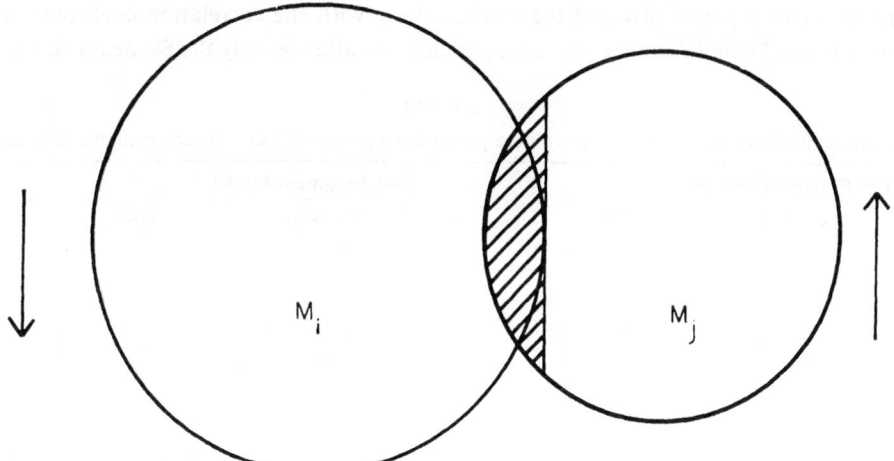

Fig. 1.   The two fragment model ($j > i$). The shaded part of i sticks to j and the remainder goes free. Arrows indicate directions of relative velocity.

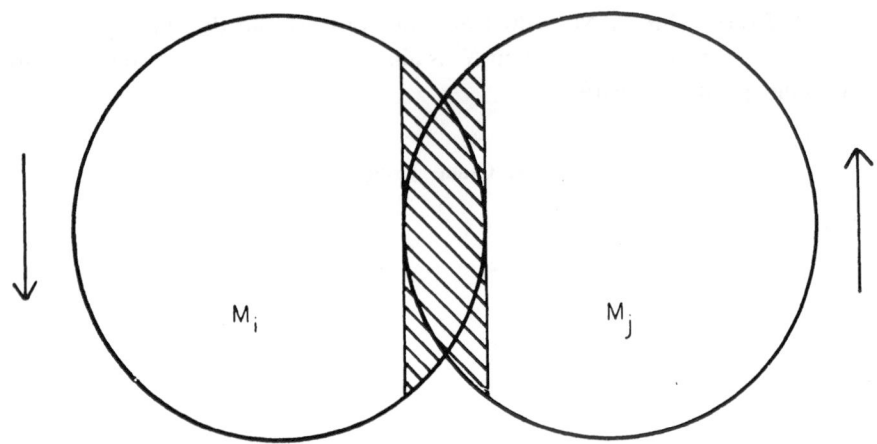

Fig. 2.   The two fragment model ($i = j$). The shaded parts of i and j coalesce and both remaining
outer pieces go free. Arrows indicate directions of relative velocity.

sticks to j, the remainder of i (the 'ear') going free. When $i = j$ one obtains three
fragments, the central region of coalescence and the two ears. See Figures 1 and 2.
In the remaining case of $j > i$ if the impact parameter is less than or equal to c_{int} we
assume total coalescence. It is clear that this is the next most sophisticated model
favoring growth as compared to total coalescence. A less favorable assumption to
mass accretion would be to always allow three fragments (two ears plus the central
region of coalescence) for impact parameters larger than c_{int}, the case of collisions
closer than c_{int} remains unchanged.

   We have numerically solved these two models (ie. solved Equations (8) and (9)) for
various values of n. Both of the above mentioned possibilities for the velocity dependence were included. In every case the resulting values for the N_k were least
squares fit to a power law and the results, along with the correlation coeficients are
contained in Table I. The correlation coefficient, or alternatively the Student t statistic,

TABLE I

Results of fit of theoretical mass spectrum to a power law $x_k = a \cdot k^{-p}$ and the correlation coefficient r

Three Fragment Model				Two Fragment Model				
n	No KEE		KEE		n	No KEE		KEE	
	p	r	p	r		p	r	p	r
3	4.53	0.96	4.33	0.96	5	2.83	0.99	2.61	0.99
5	5.51	0.97	5.35	0.97	10	2.45	0.98	2.25	0.98
10	7.40	0.97	7.30	0.97	20	2.11	0.98	1.92	0.97
20	9.51	0.98	9.47	0.98	40	1.82	0.97	1.65	0.97

can be used as a measure of the signifiance of the fit. It is clear that as n approaches
infinity in the three fragment model the index of the power law approaches $-\infty$. In
the case of the two fragment model the power law index is approximated by

$$p(n) = 1.68 + 12.62 \left(\frac{\ln(n)}{n^{3/2}} \right) - 30.82 \left(\frac{\ln(n)}{n^{3/2}} \right)^2 \tag{13a}$$

for no KEE and by

$$p(n) = 1.51 + 12.31 \left(\frac{\ln(n)}{n^{3/2}} \right) - 30.83 \left(\frac{\ln(n)}{n^{3/2}} \right)^2 \tag{13b}$$

when there is KEE. Thus there is little difference between these and the total coalescence model with geometric cross-section with/without KEE. Complete details can be found in Taff (1973).

References

Field, G. B. and Saslaw, W. C.: 1965, *Astrophys. J.* **142**, 568.
Field, G. B. and Hutchins, J.: 1968, *Astrophys. J.* **153**, 737.
Oort, J. H.: 1954, *Bull. Astron. Inst. Neth.* **12**, 177.
Taff, L. G.: 1973, unpublished Ph.D. thesis, University of Rochester.
Taff, L. G. and Savedoff, M. P.: 1972, *Monthly Notices Roy. Astron. Soc.* **160**, 89.

INTERSTELLAR GRAINS IN MUSEUMS?

A. G. W. CAMERON

Belfer Graduate School of Science, Yeshiva University, New York, N.Y., U.S.A.

and

Goddard Institute for Space Studies, NASA, New York, N.Y., U.S.A.

Abstract. It is argued that carbonaceous chondrites, particularly of type I, are probably collections of interstellar grains which have been mildly transformed through exposure to higher than normal temperatures, resulting in a loss of volatile materials.

During this conference there have been many discussions of methods for determining the properties of interstellar grains by means of electromagnetic observations and by investigation of the properties of possible laboratory prototypes. It has not been realized that it may be possible to study interstellar grains directly within the laboratory.

Theoretical investigations of the star formation process indicate that the collapse of an interstellar cloud is likely to take place under very low temperature conditions. The calculated temperature in the interior of a collapsing gas cloud is about 10 K (Larson, 1969). The temperature in the gas starts to rise only when the gas density becomes high enough so that the gas becomes opaque to the transmission of its own radiation, and the compression goes over from an isothermal stage to an adiabatic stage. M. R. Pine and I have recently constructed models of the primitive solar nebula which may have resulted from such an interstellar collapse process (Cameron and Pine, 1973). I believe that these models are sufficiently general in indicating that the temperature will not rise high enough to evaporate completely the interstellar grains, contained within the gas, beyond about one or two astronomical units from the central spin axis of the primitive solar nebula (Cameron, 1973). The important aspect of this conclusion is that small bodies formed at very much larger distances from the central spin axis are likely to be composed of collections of interstellar grains whose properties have been transformed by varying degrees. Naturally, if we are interested in the laboratory examination of interstellar grains, we must seek material which has suffered the least rise in temperature during the adiabatic compression of the interstellar gases.

Ideally, it would be nice to be able to examine a piece of a comet, which has probably formed at very great distances from the central spin axis of the primitive solar nebula, and in which the temperature has always stayed very low. Unfortunately, it is unlikely that we will obtain any sample of cometary material for some time to come, and therefore we must seek the next most primitive material.

This material is presumably carbonaceous chondritic material. Studies of elemental abundances in such meteorites, as well as of the ratios of the oxygen isotopes, tend to show that carbonaceous chondrites have been accumulated in the primitive solar nebula at a temperature in the vicinity of 350 K. Large amounts of water and carbon

Greenberg and Van de Hulst (eds.), Interstellar Dust and Related Topics, 545–547.

compounds have been retained, although it is obvious that most of the condensed ice has disappeared, along with ammonia and methane, with only the more complex carbon compounds and water of crystallization being retained at this temperature. The more refractory materials, silicates and metallic oxides, have probably been retained with very little change of properties.

One of the most common substances contained in carbonaceous chondrites is magnetite. Huffman and Stapp (1973) has already pointed out to us that magnetite has some interesting polarization properties which suggest interstellar grain material. It is also interesting that the small magnetite grains tend to have remanent magnetism. Even if this were not the case, the magnetite grains would be interesting as a possible subset of the interstellar grains because the larger ones would be ferromagnetic, and the smaller ones would be superparamagnetic. Thus those interstellar grains which have magnetite cores may play a large role in the polarization of interstellar starlight.

It is likely that interstellar grains are nucleated in cool stellar atmospheres, being condensed and expelled by radiation pressure, or possibly precipitated in rapidly expanding stellar atmospheres such as in nova explosions. Studies of the remanent magnetism of magnetite grains in carbonaceous chondrites tend to indicate that the magnetism was acquired in the presence of a field having a strength of 0.1 to 1 G (Brecher, 1972). Such fields naturally exist in the vicinities of stellar surfaces, but the precipitating substance is likely to be metallic iron rather than magnetite. However, I have been informed by Aviva Brecher, who has carried out a number of these studies, that in her opinion the remanent magnetism would be retained if a magnetized iron grain is oxidized to magnetite.

Thus it appears to me to be natural that the nucleation of some of the interstellar grains may provide magnetic particles, upon which other materials can condense in interstellar space. These particles should be alignable by relatively weak interstellar magnetic fields, and they will be part of any collapsing interstellar cloud which forms stars and planetary systems. Those grains which remain rather far out in any primitive gaseous nebula which is formed, will assemble into material resembling carbonaceous chondrites, with only a loss of the more volatile materials, and without a loss of magnetic properties. Of course, I refer here only to the matrix material in types II or III carbonaceous chondrites, and not to chondrules or inclusions. All of the material in type I carbonaceous chondrites should qualify.

Thus it is possible that meteorite collections in our museums may be very valuable for research on interstellar grains. Thin sections of the matrix material in carbonaceous chondrites cannot be expected to reproduce precisely the optical properties of interstellar grains, because there has been some loss of material by heating, and possibly some mineral changes in the remaining material, but some of the properties should undoubtedly resemble those of interstellar grains if this hypothesis is correct, and if so, then the materials responsible for these properties could be identified with somewhat greater certainty than is done at the present time. I therefore recommend that investigations of the properties of carbonaceous chondrites be carried out with this hypothesis in mind.

Acknowledgements

This research has been supported in part by grants from the National Science Foundation and the National Aeronautics and Space Administration.

References

Brecher, A.: 1972, 'On the Primordial Condensation and Accretion Environment and the Remanent Magnetism of Meteorites, in *Proc. IAU Symposium on the Evolutionary and Physical Properties of Meteoroids*, in press.
Cameron, A. G. W.: 1973, *Icarus* **18**, 407–450.
Cameron, A. G. W. and Pine, M. R.: 1973, *Icarus* **18**, 377–406.
Huffman, D. R. and Stapp, J. L.: 1973, this volume, p. 297.
Larson, R. B.: 1969, *Monthly Notices Roy. Astron. Soc.* **145**, 271–295.

DUST AND STAR FORMATION

D. McNALLY

University of London Observatory, Mill Hill Park, London NW7 2QS, England

Abstract. The view that dust is essential to star formation is challenged on the ground that other interstellar constituents can provide more rapid cooling. From the evidence of stellar minimum masses it is suggested that self absorption of the radiation emitted by the coolant H_2 is the dominant mechanism leading to the heating of a collapsing fragment. It is however shown that extensive dust and molecule formation may take place during star formation and that a natural explanation for the 4 terrestrial planets in the neighbourhood of the Sun is then provided.

1. Introduction

The presence of interstellar dust is usually considered to be a necessary prerequisite for the initiation of star formation. The basis for this hypothesis is the observation that young stars and regions of great obscuration seem to be associated spatially. However, this observation, as was pointed out by Ambartsumian (1958) could also be interpreted on the basis that dust formed during the star formation process. It is the purpose of this paper to examine the role that is played by dust in promoting or hindering star formation. Unfortunately the role of dust will be different according to the model of star formation chosen. For example in a model of star formation based on cosmological considerations such as that proposed by Layzer (1964), dust probably plays a minor role. The same is true of the random accretion model proposed by McCrea (1960) though it is probable that the presence of dust may be important at a late stage in this model. The classical model of star formation through collapse under gravity is probably the most sensitively dependent on the assumed dust content of interstellar space and, as a result of recent numerical studies, is the model most susceptible to definitive analysis. Although the physical state of the interstellar gas is ill defined particularly in regions of high density, the numerical calculations that already exist underline those parts of the collapse where the physical state of the gas plays an important role. However, the importance of dust in this model may only relate to the opacity of the gas and in some circumstances the role of dust may be simply to provide a catalytic surface for the formation of molecular species.

2. Models for Star Formation

Many different models for star formation have been proposed – collapse under gravity, association of small cloudlets, cosmological condensation, ejection from galactic nuclei (see McNally, 1971). These theories have attempted to formulate a broad framework within which processes leading to star formation could operate. Most of

Greenberg and Van de Hulst (eds.), Interstellar Dust and Related Topics, 549–558.

the theories have concentrated on dynamical considerations to the exclusion of any considerations of whether the state of the material out of which stars would form could support the dynamics proposed. In consequence such theories of star formation, for all their apparent universality, could not give detailed models of star formation.

Recently attention has been paid to the state of the interstellar gas. The work of Hayashi and his collaborators (e.g. see Hattori *et al.*, 1969) on the rates of likely interstellar cooling and heating processes has shown very clearly the influence that the state of the interstellar gas is likely to have. Field *et al.* (1969) have also shown the influence of the balance of heating and cooling mechanisms on the structure of the interstellar gas. Their work has shown clearly that unless the state of the interstellar gas is considered, it is misleading to consider the dynamics of star formation in isolation. It would be equally misleading to give undue weight to considerations based on the state of the gas alone. Unfortunately the simultaneous consideration of dynamics and the state of the interstellar material creates an intractable algebraic formulation of the equations of motion.

In one particular model of star formation – namely that of collapse under gravity – some progress has been made in the simultaneous consideration of the state of the interstellar gas and its dynamics. Gravitational collapse is based simply on the result that for a given gas density and temperature (neglecting rotational and magnetic forces) there is a critical mass for an isolated gas cloud above which gravitational forces dominate all other forces – in this case gas pressure. By maintaining a situation in which gravitational forces dominate pressure forces, a gas cloud will collapse. A simple way of maintaining such collapse is by efficient cooling of the gas cloud so that the kinetic temperature and so gas pressure, remain low and gravity is allowed to dominate.

Such a simple picture has considerable appeal for gravitation is a long range associative force. But the simple model has to face formidable difficulties. Since gravity is a weak force, a gas cloud of great mass ($\gtrsim 1000\ M_\odot$) is required to produce collapse under 'average' interstellar conditions. The problem of forming stellar objects whose masses lie in the range 0.1–$100\ M_\odot$ must then be faced. If the interstellar gas cloud is rotating, centrifugal forces have been considered to halt collapse after the cloud density has increased by a factor of 10 (a further factor of 10^{22} being required before stellar densities are achieved). These problems have been considered so formidable that other models of star formation have been sought. Three dimensional calculations now being carried out by the author (unpublished) suggest that the rotational constraint is not as severe as had been expected. Nevertheless despite these difficulties, it is useful to consider collapse under gravity since it can be described effectively in mathematical terms in a form suitable for numerical solution. Several authors (Disney *et al.*, 1969; Hunter, 1969; Larson, 1969; Penston, 1969) have tackled the problem of the collapse of gas clouds from the interstellar state for the case of spherical symmetry with no magnetic field or rotation, while Bodenheimer (1969), Hayashi (1970), and Larson (1969) have considered the collapse of more condensed objects of stellar mass to the proto-stellar state (i.e. the appearance of the object in the Hertzsprung-Russell diagram). While a consensus on how a spherically symmetrical gas cloud will collapse has been achieved, there is no such consensus on proto-stellar evolution and there is no real linkage of the cloud collapse problems with the evolution of proto-stellar objects. This state of affairs is not surprising. Leaving aside problems of a purely technical kind in the numerology, the physical description of all but the initial phases of any collapsing model is wholly inadequate. This is not merely a problem of available resources but the phenomena dominant throughout the range of densities and temperatures encountered during collapse are indifferently determined where they are known at all. Those workers who have tried to investigate the interaction of atomic, molecular and other processes with the dynamics of collapse have the double handicap of a difficult numerical problem and an essentially undefined physical situation. In this context wide divergences between authors are to be expected.

Nevertheless, the numerical studies have indicated an essential dynamical feature of gravitational collapse which had not been stressed sufficiently in the past. This is best demonstrated by consideration of free fall collapse (i.e. gravity is the only force acting). Motion under gravity alone takes a characteristic form. Changes in density take place slowly at first but take place at a rate which is accelerating. This means that rapid changes of density take place towards the end of collapse. In a gravitationally dominated situation collapse proceeds slowly at first and ultimately increases very rapidly. This phenomena is illustrated in Figure 1, for a slightly more realistic situation than free fall. Gas pressure and energy loss can be considered in a system in which the gas pressure is not zero provided the velocity distribution is assumed to be linear wave (i.e. the velocity increases linearly from centre to surface of the collapsing cloud).

In linear wave flow the initial density distribution is maintained so that ρ/ρ_i refers to any point in the gas cloud so long as the same relative position at all times is considered. The model used to determine Figure 1. is one in which a linear wave flow collapse from equilibrium is initiated either by a sudden initial acceleration or cooling and which attains a final radiating equilibrium state. From Figure 1, it is clear that 75% of the collapse time must elapse before the density increases by an order of magnitude. However, an increase in density by three orders of magnitude take place in the next 15% of the collapse time and a very rapid increase of density takes place in the very last stage before the final equilibrium configuration is reached. Such characteristic behaviour has an important consequence. Disney et al. (1969) investigated the collapse of a gas cloud when cooled by some physically realistic process (within wide limits it does not matter what the precise process is). They found that in the denser hotter central part of the cloud cooling was rapid while it was slow in the cooler less dense region near the boundary. As a consequence collapse under gravity was scarcely resisted by gas pressure in the cloud centre whereas collapse in the outer parts of the cloud proceeded rather slowly because gas pressure and gravitational forces remained of similar order. Such a situation capitalised on the characteristic nature of gravitational collapse to produce a region of high density near the cloud centre while the bulk of the gas cloud did not vary in density by a significant factor from the initial density. Such a situation gives separation of a collapsing region of small mass from a more

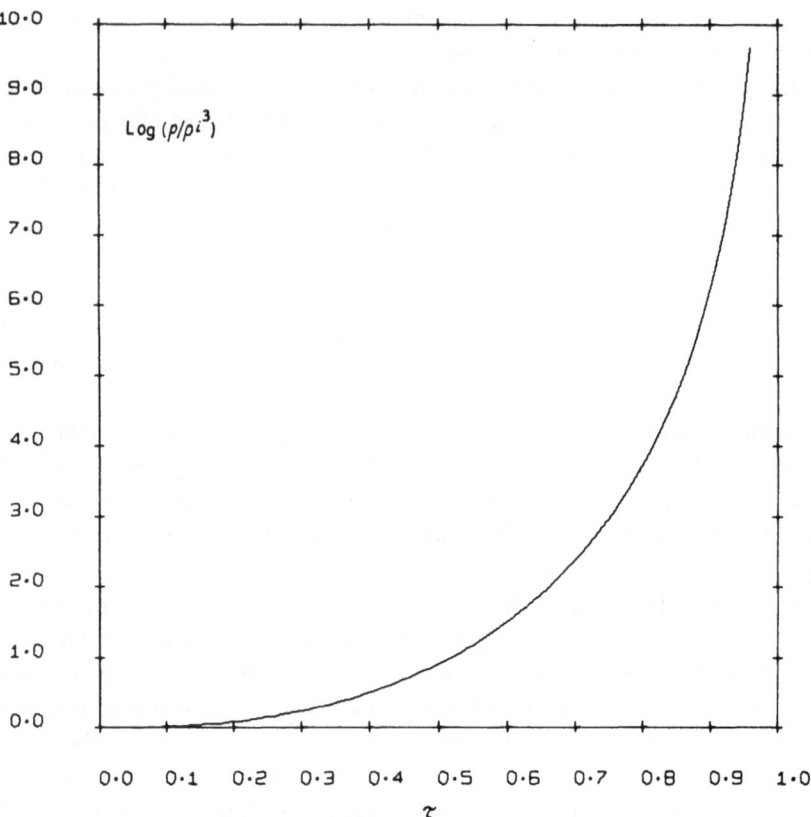

Fig. 1. The variation of density (ρ – with respect to an initial density ρ_i) with time (τ – relative to the collapse time for the system considered) in the case of a gas cloud collapsing with linear wave flow.

massive parent cloud. However, in the context of the present paper it is the possibility of the existence of a linking region of great density and temperature variation connecting the high density and hot collapsing fragment and the low density cool surrounding gas that is of importance. If the collapsing fragment is attaining stellar densities a density variation more than 10^{20} may be expected in the linking region accompanied by a temperature variation by at least a factor of 10^3. If such a region exists surrounding a collapsing fragment the consequences for molecule and dust formation may be considerable.

Disney *et al.* (1969) also drew attention to the possibility that opacity in the interstellar medium may well inhibit the rate at which heat can be removed from the interstellar gas. Opacity is contributed by the general opacity of the dust component and self-absorption by the coolants themselves. They showed that opacity effects would not become apparent until the stage of rapid density increase was attained. They therefore assumed a sudden transition from a cooling to an adiabatic regime in the collapsing fragment. More detailed calculations showed that the rate of the transition was unimportant since the rate of density increase was rapid. They found for adiabatic collapse that the kinetic temperature of the gas increased and that eventually pressure forces would exceed the gravitational. In such a situation re-expansion would occur. To maintain collapse they proposed that the ionisation of atomic hydrogen would be a sink for the heat of compression. However, for a given value of the opacity a minimum collapsing mass was required in order to raise the temperature of the gas to a value where collisional ionisation was important. It is of some interest that the value of this mass depended both on the opacity of the interstellar gas and the properties of atomic hydrogen. The limiting mass M_c is given by

$$M_c = 1.4 \times 10^{29} \alpha T_i T_c, \tag{1}$$

where α is related to the opacity of the gas, T_i is the temperature of the gas before the effect of opacity operates and T_c is the temperature at which pressure and gravitational forces balance. Clearly for small values of M_c, small values of α, T_i and T_c are required. In general T_i can be maintained at about 10 K and T_c is fixed by the temperature at which hydrogen becomes collisionally ionised ($\sim 10^4$ K). The value of α is therefore critical. If α is locally large then stars of large mass would be expected whereas if α were locally small then stars of low mass would be expected. A similar argument applies if dissociation of H_2 is used as the sink of compressional heating. It is essential therefore to know what component of the interstellar medium contributes most to α at the time when radiative cooling becomes inefficient in removing compressional heating.

Two properties of the interstellar gas are therefore important at two stages in the star formation process namely the cooling of the interstellar gas to initiate and maintain the early collapse stage and the opacity of the interstellar gas in retarding that collapse. Interstellar dust clearly contributes both to the cooling of the interstellar gas and to its opacity. But is the dust the dominating component of the interstellar medium? The contributions to cooling and opacity of the interstellar gas will be considered in the next two sections.

3. The Cooling of the Interstellar Gas

The interstellar gas can be cooled by a number of its constituents. The contribution to the cooling of O, H_2, atomic ions (in particular C^+, Si^+, Fe^+, Mg^+) and dust are compared in Figure 2. It is clear from an inspection of Figure 2 that there are several ways in which the interstellar gas could rid itself of energy through radiative processes. It might be argued that interstellar O, C, Fe, Si, Mg could be depleted by being components of the dust. This would still leave H_2 as a more effective coolant than the dust. Dust is the least efficient of the proposed coolants by several orders of magnitude so that almost total depletion of the interstellar gas of all atomic species other than hydrogen, in addition to impossibility of formation of molecular hydrogen, would be required before dust could be considered as an effective coolant. Therefore interstellar dust cannot be considered to play a dominant role as a coolant and therefore plays little part in initiating collapse through differential cooling of the interstellar gas.

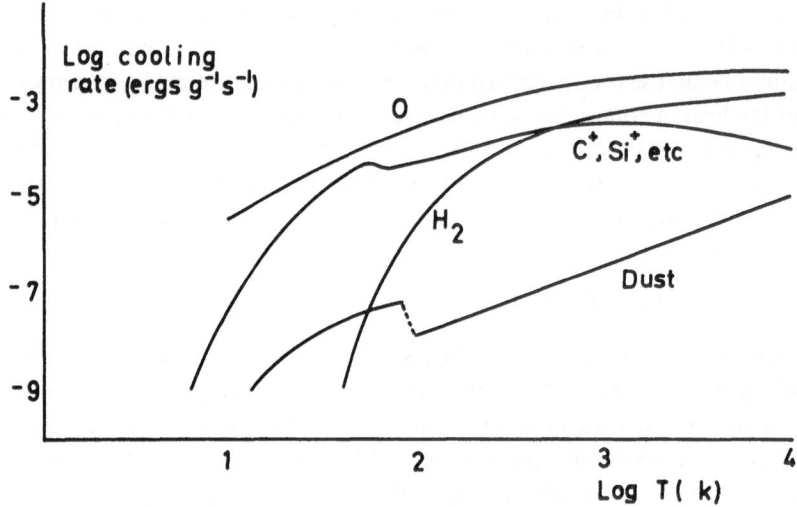

Fig. 2.   A comparison of interstellar cooling rates (adapted from Disney *et al.*, 1969).

4. Opacity of the Interstellar Gas

The opacity of the interstellar gas is conventially considered in terms of the amount of extinction caused by the dust. While the dust will contribute to the absorption of the radiation emitted by coolants, the self-absorption of this radiation by the coolants must also be considered. Such self-absorption plays an important part when considering opacity at the wavelength of the cooling radiation (in the infra-red).

Hattori *et al.* (1969) have considered in some detail the cross sections for self-absorption by some of the interstellar coolants considered in the last section. The absorption cross section for dust has been adapted from the study of interstellar opacity by Gaustad (1963). For the purpose of illustration the quantity α of Equation (1) has been evaluated from the cross section $\sigma(a)$ for the species a.

$$\alpha^{-1} = \varrho l(a) = M_H n(H) l(a), \tag{2}$$

where ρ is the gas density, M_H is the mass of a hydrogen atom, $n(H)$ is the number density of H atoms and $l(a)$ is a characteristic length defined by:

$$l(a) = \frac{1}{n(a)\sigma(a)} = \frac{1}{n(H)A(a)\sigma(a)}, \tag{3}$$

where $A(a)$ is the abundance of species a. From Equations (2) and (3)

$$\alpha = A(a)\sigma(a)/M_H. \tag{4}$$

Values of α for C^+, S^+, O, H_2 and dust are given in Table I.

It is clear from Table I that radiation from H_2 is least readily absorbed while that from O is most easily self-absorbed. This means that at an early stage in the collapse O, C^+, and Si^+ cease to be effective coolants. Only H_2 remains free of self-absorption

TABLE I

The relative opacity of interstellar constituents evaluated at 100 K

a	$A(a)$	$\sigma(a)^*$ (cm^2)	α
C$^+$	3×10^{-4}	1.1×10^{-17}	2.0×10^3
Si$^+$	3×10^{-5}	1.7×10^{-17}	3.1×10^2
O	7×10^{-3}	8.8×10^{-18}	3.7×10^4
H$_2$	10^{-1}–10^{-3}	6.6×10^{-25}	$4.0 \times (10^{-2} - 10^{-4})$
dust	3×10^{-13}	2×10^{-11}	3.6

* varies at $T^{-\frac{1}{2}}$ where T is the temperature

until high densities are attained. If dust can absorb the 28 μ radiation from H$_2$ ($J = 2 \rightarrow 0$ rotational transition) then dust could play the dominating role in interstellar opacity. On the basis of current concepts of the size distribution of interstellar dust particles it is unlikely that dust will be an efficient absorber of the 28 μ radiation of H$_2$. Consequently it would appear that the opacity which should be used in Equation (1) is that for self-absorption by H$_2$. Taking a value of 10^{-2} for α and setting $T_i = 10$ K, $T_c = 10^4$ K a minimum stellar mass of 1.4×10^{32} g is predicted. Since the minimum stellar mass is of this order it is tempting to propose that it is the self-absorption of H$_2$ which plays a dominating role in determining the minimum stellar mass. If an interstellar situation could be envisaged where neither dust nor H$_2$ was present, then only stars of large mass would be able to form assuming the presence of the necessary coolants.

The above suggestions require detailed elaboration in a model which takes account of the cooling mechanisms available and the sources of interstellar opacity. Such a model requires a detailed solution of the equation of radiative transfer within the collapsing cloud and so far this has not been attempted. The full number of physical processes contributing to the interstellar cooling, heating and opacity may not have been fully enumerated and the detailed nature of the interstellar grains are not known. Consequently any calculation based on a partial description of the state of the interstellar gas can only give an indication of a possible path of collapse.

5. The Formation of Dust

If dust does not play a significant role in cooling the interstellar gas or contribute to the opacity of the gas with respect to limitation of cooling rate it may legitimately be asked – of what use is dust in the star formation process? It is clear that H$_2$ is the most useful coolant for the interstellar gas and that it is self-absorption by H$_2$ which effectively limits its use as a coolant. A short answer would be that dust is at best of very secondary importance to star formation. However, since a detailed model of star formation is not available such a statement must be tempered with caution – the more particularly as it seems that H$_2$ is most efficiently produced by reactions at grain surfaces (McNally, 1972).

The role of dust may therefore lie in the production of H$_2$ and other molecules.

If this is the case it poses the awkward question of why the production of a simple diatomic molecule should require the presence of multiatomic dust particles. However, star formation at the present epoch may well be accompanied by formation of dust and molecular species.

The region linking a collapsing fragment with the remainder of the parent cloud is one which has a great variation of density and a smaller but still significant variation of temperature. Density variation by up to a factor of 10^{20} and a temperature variation by a factor of 10^3 could be anticipated. At the densities and temperatures (10 H atoms cm^{-3}, 100 K) conventionally assumed for interstellar gas clouds reaction mechanisms are slow. However, the greater the gas density the faster the reaction rates. In Figure 3 is compared the variation of the rate (see McNally, 1972) of the reaction

$$C + H \rightarrow CH$$

at a grain surface with the free fall times at the same density. For densities lower than 5×10^{11} H atoms cm^{-3} the chemical reaction is slower than the dynamical time scale. For higher densities the rate of the chemical reaction will be more rapid than the dynamical time scale. The example chosen here is of no significance being merely used to illustrate the point that at high densities chemical reaction times will become shorter than dynamical time scales. Consequently at some stage in the evolution of a collapsing gas cloud to the stellar state there will be a situation in the region surrounding the collapsing fragment when chemical reactions will become so fast that a great variety of reactions will be able to proceed to completion and further reactions of greater complexity initiated. The type of molecule produced will depend on the conditions of density and temperature but certainly complex molecular species

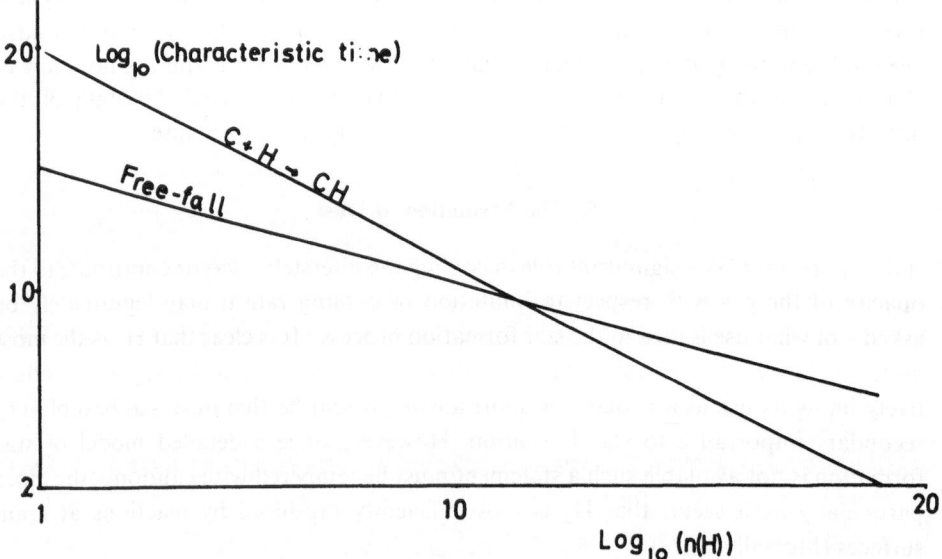

Fig. 3.   A comparison of dynamical (free fall) and chemical (C + H → CH) time scales.

and the growth and production of dust particles could be envisaged. If some initial ionisation of the linking region can decay subsequently a further range of reactions could be initiated. The chemical processes are likely to be very complex, but their rapidity will greatly alter the composition of the region linking the collapsing fragment and the remainder of the parent cloud.

On the basis of the relative opacities in Table I it would seem more plausible that extensive dust and star formation would be associated with stars of small mass. If the opacity is dominated by the self-absorption characteristics of H_2, the collapse of the central core will proceed to higher densities before the second phase of star formation takes place, initiated either by the ionisation of H or the dissociation of H_2. The existence of the terrestrial planets close to the Sun could be cited as evidence for dust and molecule production in the denser part of the linking region surrounding the collapsing fragment. However, detailed calculations are required to lift this proposal from the realm of speculation. Herbig (1970) has suggested that it is the particulate ejecta from planetary systems that maintains the interstellar dust. The recent radio and infra-red observations (see Mezger, 1971) of objects considered to be proto-stellar show evidence for both molecule and dust formation.

6. Conclusion

It has been shown that for star formation by collapse under gravity, interstellar dust is not likely to play a major role. The molecule H_2 has been shown to play a much more important role than dust. Ambartsumian's interpretation of the observation that dust and young stars are found in close association as showing that star formation and production of dust take place at the same time is given support. Furthermore it has been argued that extensive molecule formation must also be associated with the formation of stars particularly stars of low mass. The formation of the solid terrestrial planets in the neighbourhood of the Sun therefore receives a natural explanation in that it is just in the region nearest the forming Sun that dust and molecule production were most rapid.

The conventional view that dust is essential to star formation finds little support from the argument presented here. But it must be remembered that until more is known of interstellar dust, some of the numerical values used here could be considerably upset.

References

Ambartsumian, V. A.: 1958, in R. Stoop (ed.), *La structure et l'évolution de l'universe*, Brussels Institut International de Physique Solvay, p. 241.
Bodenheimer, P.: 1968, *Astrophys. J.* **153**, 483.
Disney, M. J., McNally, D., and Wright, A. E.: 1969, *Monthly Notices Roy. Astron. Soc.* **146**, 123.
Field, G. B., Goldsmith, D. W. and Habing, H. J.: 1969, *Astrophys. J.* **155**, L149.
Gaustad, J. E.: 1963, *Astrophys. J.* **138**, 1050.
Hayashi, C.: 1970, *Mem. Soc. Roy. Sci. Liège* **19**, 127.
Hattori, T., Nakano, T., and Hayashi, C.: 1969, *Prog. Theor. Phys. Japan* **42**, 781.

Herbig, G. H.: 1970, *Mem. Soc. Roy. Soc. Liège* **19**, 13.
Hunter, J. H. Jr.: 1969, *Monthly Notices Roy. Astron. Soc.* **142**, 473.
Larson, R. B.: 1969, *Monthly Notices Roy. Astron. Soc.* **145**, 271.
Layzer, D.: 1964, *Ann. Rev. Astron. Astrophys.* **2**, 341.
McCrea, W. H.: 1960, *Proc. Roy. Soc.* **A256**, 245.
McNally, D.: 1971, *Rep. Prog. Phys.* **34**, 71.
McNally, D.: 1972, *Adv. At. Mol. Phys.* **8**, 1.
Mezger, P. M.: 1971, in C. de Jager (ed.), *Highlights of Astronomy*, Vol. 2, D. Reidel Publishing
    Co., Dordrecht, p. 366.
Penston, M. V.: 1969, *Monthly Notices Roy. Astron. Soc.* **145**, 457.

ROCKET MEASUREMENTS OF THE
GALACTIC BACKGROUND AT 100 μ

JUDITH L. PIPHER*

Dept. of Physics and Astronmy, University of Rochester, Rochester, N.Y., U.S.A.

Abstract. Measurements of the diffuse background radiation in the 85–115 μ band are presented, as observed from altitudes close to 190 km with a rocket-borne, liquid helium cooled telescope. Evidence is given for detection of the galactic background due to thermal grain emission at galactic latitudes of 5–35°, and at a galactic longitude of ∼ 163°. At small latitudes, the background intensity is measured to be ∼ 9×10^{-11} W cm^{-2} sr^{-1}; the average number density of grains derived is consistent with the optically determined measure. The 100 μ data is compared with 20 μ data taken on the same flight, in order to draw some conclusions about the grain emissivity and temperature.

1. Introduction

Recent observations (Houck *et al.*, 1971; Hoffmann *et al.*, 1971; Soifer *et al.*, 1973; Harper and Low, 1971; Low and Aumann, 1970; Soifer, 1972). indicate that the galactic center, H II regions and dust clouds are strong infrared sources with peak intensity at 100 μ. Although the mechanism for such emission is not perfectly under-stood (partially because there is a lack of far infrared data with good spectral resolu-tion), absorption of ultraviolet and optical radiation by dust grains and subsequent re-emission in the infrared has been suggested as a possible source for all of the above objects. It has been recognized for some time that a diffuse galactic background should be observable in the far infrared due to the re-emission of galactic interstellar grains. Stein (1966) first made rough predictions of the expected intensity. The tem-perature that the grains assume and the wavelength of peak emission depend on the heating mechanism and the grain characteristics. Most models, however, predict that the galactic grain emission should peak in the 70–300 μ range. The observed intensity will depend on the column density of grains along the line of sight, and will vary relatively slowly in brightness gradient. Since atmospheric constraints at wavelengths proximate to 100 μ make differential chopping necessary even at balloon altitudes, it is possible only at rocket altitudes and higher to make wide field observations of diffuse galactic grain emission.

We report here observations of a diffuse galactic background, in the 85–115 μ band, and give evidence that thermal grain emission is the most likely source of the radiation. The observations were made with a liquid helium cooled telescope carried above the atmosphere on Aerobee 170 rockets launched from White Sands. The observations reported here took place at about 01:32 MST on 1970 December 2, and some sub-stantiation of these results were obtained from a flight at 21:53 on July 16, 1971. Pre-

---

* This work was done while the author was at Cornell University.

Greenberg and Van de Hulst (eds.), Interstellar Dust and Related Topics, 559–566.
All Rights Reserved. Copyright © 1973 by the IAU.

liminary results of the July flight were previously (Houck *et al.*, 1971; Soifer *et al.*, 1973) reported as were more detailed accounts of the December flight (Pipher *et al.*, 1971; Soifer *et al.*, 1971; Pipher, 1971).

2. Observations

The cryogenic telescope used for the observations is described by Pipher *et al.* (1971) cited above. Although the main emphasis here will be on the 100 μ (85–115 μ) observations, reference will be made to the results from some of the other detectors flown.

On this flight, the telescope first scanned along the galactic plane, and then scanned perpendicular to the plane at an approximate galactic longitude of 163°. Because there were technical difficulties on the scan along the galactic plane, only the scan

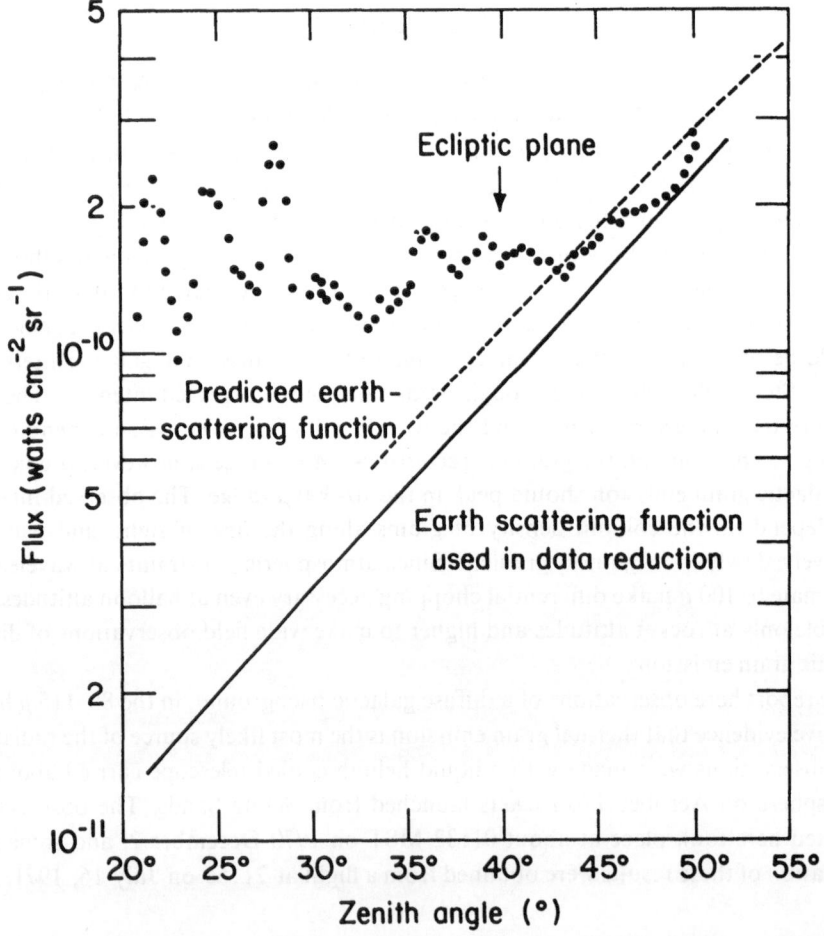

Fig. 1. Signal level observed by 85–115 μ detector as a function of zenith angle for a scan at $l^{II} = 163°$, and b^{II} ranging from 5°–35°. Predicted horizon shine (earth scattering function) and function used in data reduction are shown.

perpendicular to the plane will be discussed here, for galactic latitudes ranging from 5° to 35°. Because of an attitude control system (ACS) failure 225 s into the flight, the path was passed over only once. The raw data for this scan are presented in Figure 1 as a function of zenith angle. At zenith angles greater than 42°, the contribution from scattered earthshine predominates over celestial signals. In order to obtain the celestial flux, the scattered earthshine component is subtracted from the raw data by the method outlined by Soifer *et al.* (1971) and Pipher (1971). Because the signal to noise ratio is large for these measurements, the major source of error is the absolute detector calibration, which is described in detail by Pipher *et al.* (1971). On the 1971 flight (Houck *et al.*, 1971), 100μ observations of the galactic center agreed well with previous studies from balloon altitudes (Hoffmann *et al.*, 1971), giving added confidence in the calibration techniques. Another source of error is the subtraction of the scattered earthshine. At galactic latitudes less than 30° this error is thought to be minimal. We estimate our subtracted intensity levels to be accurate to $\pm 50\%$ close to the galactic plane. Figure 2 contains a linear plot of the difference data as a function of both galactic and ecliptic latitude. As can be seen, the mean intensity decreases with increasing latitude, and peaks as the ecliptic plane is crossed. As well, discrete sources,

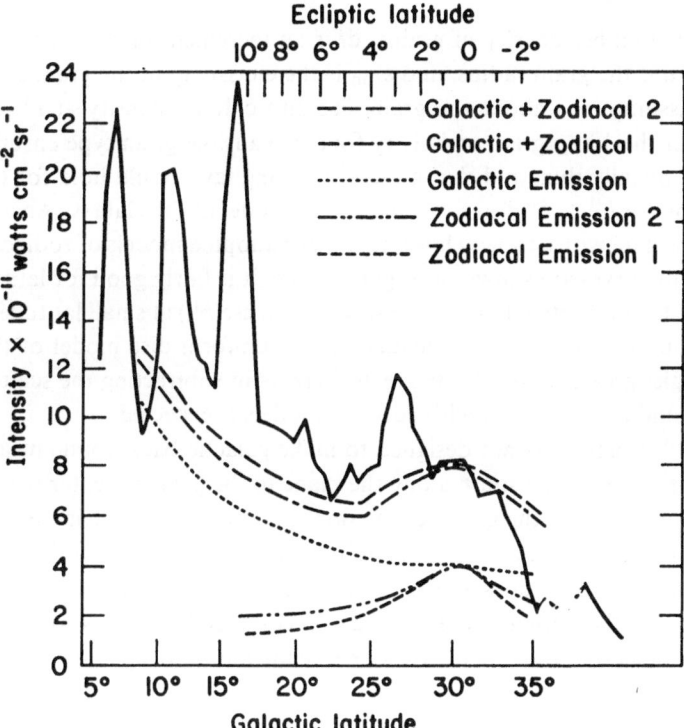

Fig. 2. Difference signal between raw data of Figure 1 and assumed horizon shine, as a function of galactic latitude, ecliptic latitude and zenith angle. A $\csc(b^{II})$ galactic emission and two zodiacal emission models are scaled so that their sum best fits the 100μ average background data. Discrete sources (corresponding to dust clouds, an H II region and nebulosity) are also evident.

which can be identified with dark clouds, reflection nebulae (Pipher *et al.*, 1971), and an H II region (Soifer *et al.*, 1973; Pipher *et al.*, 1971) are superimposed on the background signal; these will be discussed in a separate article.

Unfortunately, it is not an unambiguous exercise to separate the galactic and zodiacal dus emission in the 100 μ data shown in Figure 2. Soifer *et al.* (1971) have measured the thermal emission from zodiacal particles at 5–6 μ, 12–14 μ and 16–22 μ on this rocket flight. They find the spectrum of measured radiation in these bandwidths appropriate to a dilute 280 K blackbody. At these shorter wavelengths, the zodiacal emission has fallen in intensity by a factor of two to three by the time an ecliptic latitude of 10° is reached. For want of a more plausible model for the 100 μ zodiacal component, similar scaling laws with ecliptic latitude are assumed. As discussed in the next section, galactic grain emission seems to be the most plausible explanation for the radiation that decreases in intensity with increasing galactic latitude, at $l^{II} \sim 163°$. If the distribution of grains is approximately uniform, the radiation at 100 μ should be optically thin. The total optical depth along the line of sight can be written as

$$\tau_{100} = \int n \, dl \cdot \pi a^2 \cdot \varepsilon_{100}, \tag{1}$$

where n is the number density of grains, dl is an increment of path length along the line of sight, a is the grain radius, and ε_{100} is the emissivity of the grain at 100 μ. For reasonable assumptions about the grain size and column density at $l^{II} \sim 163°$, the small value of the 100 μ grain emissivity for any plausible grain type ensures that the radiation is optically thin, and that a $\csc(b^{II})$ scaling law should hold for the galactic grain emission for $b^{II} \geqslant 5°$ if one assumes a disc model of the Galaxy. An attempt was made to fit the data with a $\csc(b^{II})$ law and the two adopted models of zodiacal emission and the best fit curves are shown on Figure 2. The fit is fairly good for latitudes larger than 10°, but the intensity falls off more slowly than $\csc(b^{II})$ at small latitudes. This discrepancy probably reflects the inadequacies of a uniform disc model of the Galaxy. At zenith angles greater than 42° the greatest errors in subtracting the scattered earthshine occur, and a discrepancy with adopted models is expected.

The July 1971 flight was not designed to make galactic background measurements. The telescope first scanned south along the plane to the galactic center region. During this portion of the flight, the galactic emission is confused with the increasing horizon shine as the telescope points at increasingly larger zenith angles. In the region of the galactic center, not only does the horizon shine confuse observations of the background, but also the strong and extended sources at 100 μ make observations of the background difficult. However, on a scan perpendicular to the plane, an ACS pointing error resulted in a crossing of the plane some 2–3° away from the galactic center, beyond the extended 100 μ source (Houck *et al.*, 1971). As a consequence, the increasing intensity depicted at that time in Figure 3 is identified with thermal emission from grains. The detector field of view was $1\frac{1}{4}$°, so that the galactic disc at 20 kpc would be smaller than the beam size. If the grain density typical of the solar neighbor-

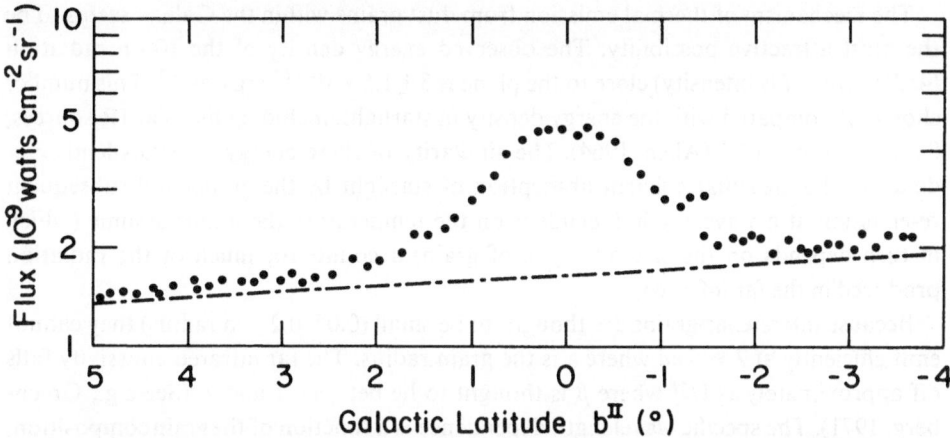

Fig. 3. Signal level observed by 85–115 μ as a function of galactic latitude for a scan at $l^{II} \sim 2°5$, and b^{II} ranging from $-5° - +5°$. The horizon shine predominates over celestial signals except close to $b^{II} = 0°$.

hood persists throughout the Galaxy within an order of magnitude, the total optical depth over this path length would still be less than unity. The intensity as the plane was crossed is some 30 times the value observed at $l^{II} \sim 163°$ and $b^{II} = 5°$. This intensity, within reasonable error, corroborates the December 1970 observation reported here.

3. Possible Mechanisms for the 100 μ Galactic Emission

There are several possible mechanisms for production of the 100 μ galactic emission. Thermal radiation from dust grains appears to be the most likely source.

By extrapolation from the radio data, several authors (Lequeux, 1970; Partridge and Peebles, 1967) have shown that synchrotron radiation, Inverse Compton radiation and free-free emission are not expected to be strong sources of 100 μ radiation, and in fact are all at least an order of magnitude down in strength from the observed signal in the galactic anticenter direction close to the plane. Synchrotron emission in the far infrared as the primary energy (Cavaliere *et al.*, 1970) requires moderately high magnetic fields; normal synchrotron emission not only would require high magnetic fields (Burbridge and Stein, 1970) but also a low frequency galactic absorption mechanism to reconcile the radio data with the 100 μ data. The density of relativisitic electrons required by an Inverse Compton mechanism is too high.

Fine structure line emission in the Galaxy is expected to be of limited concern in the 85–115 μ band (Petrosian, 1970). The only possible candidate, the 88.16 μ line of O III, is predicted to be strong only in H II regions. Hence, this mechanism is not considered likely.

Molecular line radiation in the 85–115 μ band, or resonance radiation from oscillators trapped in grains are possible sources of the 100 μ galactic background. The latter possibility is the more likely; however, without observations with increased spectral resolution, a specific model will not be examined in detail.

The mechanism of thermal emission from dust grains within the Galaxy seems to be the most attractive possibility. The observed energy density of the 100 μ radiation ($4\pi I/c$, where I is intensity) close to the plane is $3 \pm 1.5 \times 10^{-13}$ ergs cm^{-3}. This number should be compared with the energy density in starlight, including the near IR sources, 8×10^{-13} ergs cm^{-3} (Allen, 1964). The similarity of these energy densities lends credence to the idea that efficient absorption of starlight by the grains and subsequent re-emission at a wavelength dependent on the temperature the grains assume (which in turn depends on the size and type of grain) accounts for much of the radiation produced in the far infrared.

Because interstellar grains are thought to be small (0.05–0.2 μ in radius) they cannot emit efficiently at $\lambda \gg 2\pi a$ where a is the grain radius. The far infrared emissivity falls off approximately as $1/\lambda^\beta$ where β is thought to lie between 1 and 2. (See e.g., Greenberg, 1971). The specific wavelength dependence is a function of the grain composition, and the interstellar medium may very well contain a range of grain sizes, shapes and compositions. The 'free space' temperatures that the grains can attain are given by the radiative equilibrium of the grains with their environment

$$\int_0^\infty R(\lambda)\varepsilon(\lambda)\,d\lambda = \int_0^\infty \varepsilon(\lambda)B(\lambda, T_g)\,d\lambda, \tag{2}$$

where T_g is the grain temperature, $\varepsilon(\lambda)$ is the emissivity of the grain, radius a, at wavelength λ, $B(\lambda, T_g)$ is the Planck distribution of temperature of the grain, $R(\lambda)$ is the radiation field of the environment. The two integrals in Equation (2) generally refer to different wavelength regions. For grains in free space, or in clouds of moderate opacity, the ultraviolet optical and near infrared wavelengths dominate the left hand side of the equation, while the low grain temperatures attained imply the right hand integral covers far infrared wavelengths. If F is the flux of energy, in W cm^{-2}, emitted from a grain surface in the 85–115 μ bandwidth, then under the assumption of isotropic emission and a single grain size, the intensity observed at earth above the atmosphere along a line of sight is

$$I_{100} = F \cdot a^2 \cdot \int n\,dl\,(W\,cm^{-2}\,sr^{-1}). \tag{3}$$

Because the actual parameters describing galactic grains are quite uncertain, several extreme models of grains are considered here. F was calculated for these models, using free space grain temperatures and infrared emissivities appropriate to the grain, and assuming the stellar radiation field as the sole source of grain heating. A single representative grain size was adopted for each model. The important grain size ranges have been given by Greenberg (1971) as

$$a_{ice} = 0.05 - 0.3\,\mu$$
$$a_{graphite} = 0.05 - 0.1\,\mu$$
$$a_{silicate} = 0.05 - 0.1\,\mu$$
$$a_{core} = 0.05\,\mu, a_{grain} = 0.1 - 0.2\,\mu \text{ for core-mantle grains.}$$

TABLE I

Grain parameters

Grain type	T_g (K)	a (μ)	$\int n \, dl$ (cm^{-2})	$\bar{n}$ (cm^{-3})	Source of grain parameters
pure graphite	43	0.05	4×10^{10}	1.3×10^{-11}	Werner and Salpeter (1969)
dirty ice	15	0.2	2×10^{9}	7×10^{-13}	Stein (1966)
core-mantle	17	0.05, 0.15	2×10^{9}	7×10^{-13}	Greenberg (1971) and Werner and Salpeter (1969)

The value of I_{100} is 9×10^{-11} W cm^{-2} sr^{-1}, and corresponds to a total path length along the line of sight of roughly 1 kpc. On the assumption of a uniform distribution of grains along this path, the average number density of grains was calculated. These values are tabulated along with the relevant grain parameters in Table I. Silicate grains are not included, because the far infrared emissivities for such grains are very uncertain (Kushna Swamy, 1971): it is expected that the column and number densities derived for the silicate grains would be larger than those for the ice grains, because the grain temperature expected (~ 9 K) shifts much of the radiation out of the detector bandwidth.

Both the dirty and core-mantle grain models give reasonable agreement with the optically determined measure of the average number density of 2×10^{-13} cm^{-3} (Allen, 1964). Impure graphite grains, and graphite grains with moderate numbers of resonances near 100μ (see e.g., Werner and Salpeter, 1969) could reduce the column density of grains required to produce the observed 100μ intensity, both because the grain temperature would be reduced, shifting the peak wavelength of emission closer to the acceptance band of the detector system, and because the resonance might fall within the band. Although such possibilities are interesting, detailed calculations do not appear to be warranted until further spectral information on the radiation becomes available.

4. Discussion

It appears that pure graphite grains are excluded on the grounds that too large a grain density is required to give the appropriate 100μ intensity. In fact, graphite grains sufficiently pure to achieve a free space temperature of 43 ° K are not expected to exist in space. These grains are excluded on another ground as well. On the same flight, a 16–23 μ detector was also flown. Using the grain column density from 100μ data, an intensity of 2.4×10^{-11} W cm^{-2} sr^{-1} is predicted in this band due to thermal emission from pure graphite grains. The observed intensity, with the horizon-shine subtracted out, was at this time in the flight, 1.3×10^{-11} W cm^{-2} sr^{-1}. At first glance, these values seem fortuitously close; however, Soifer et al. (1971) have presented very convincing evidence that most if not all of this signal is due to zodiacal emission. As a consequence, we can conclude that the 16–23 μ observation supports the conjecture that pure graphite grains are ruled out. The 16–23 μ observation, however, is compatible with less pure graphite grains of free space temperatures of 33 K (Greenberg,

1971); however, the column density of grains required by the 100 μ observations is still rather high ($\sim 1 \times 10^{10}$ cm^{-2}), and it is concluded tentatively that such grains are only marginally allowed by the observations.

On the other hand, core-mantle grains and dirty ice grains, attaining lower free-space temperatures, are attractive not only because of the more reasonable column density required to explain the observation, but also because an upper limit of 1.4×10^{-10} W cm^{-2} sr^{-1} to the galactic emission observed at this time by a detector sensitive at 230–330 μ, does not conflict with this conclusion.

We conclude that grain temperatures of 15- $\leq$ 33 K are allowed by the observations for the grain types and sizes indicated. It is not a particularly restrictive conclusion; graphite grains with impurities and resonances can achieve temperatures within this range (Werner and Salpeter, 1969), as well as dirty ice and core-mantle grains. Wide-band observations at 200 μ and 40 μ would do much to restrict the allowable grain types and temperatures. Narrow band measurements are required to define the role that impurity resonances play in grains.

Acknowledgements

The author would like to thank B. T. Soifer, M. Harwit and J. R. Houck for helpful discussions. She held a National Research Council of Canada Graduate Fellowship during the time the experimental work was executed.

The research reported on was sponsored by AFCRL contract F 19628-70-C-0128, and KPNO rocket branch supplied excellent support.

References

Allen, C. W.: 1964, *Astrophysical Quantities*, University of London, The Athlone Press, 2nd ed.
Burbidge, G. R. and Stein, W.: 1970, *Astrophys. J.* **160**, 573.
Cavaliere, A., Morrison, P., and Pacini, F.: 1970, *Astrophys. J.* **162**, L133.
Greenberg, J. M.: 1971, *Astron. Astrophys.* **12**, 240.
Harper, D. A. and Low, F. J.: 1971, *Astrophys. J.* **165**, L9.
Hoffmann, W. F., Frederick, C. L., and Emery, R. J.: 1971, *Astrophys. J.* **164**, L23.
Houck, J. R., Soifer, B. T., Pipher, J. L., and Harwit, M.: 1971, *Astrophys. J.* **169**, L31.
Krishna Swamy, K. S.: 1971, *Astrophys. J.* **163**, 63.
Lequeux, J.: 1970, *Astrophys. J.* **159**, 459.
Low, F. J. and Aumann, H. H.: 1970, *Astrophys. J.* **162**, L79.
Partridge, R. B. and Peebles, P. J. E.: 1967, *Astrophys. J.* **148**, 377.
Petrosian, V.: 1970 *Astrophys. J.* **159**, 883.
Pipher, J. L.: 1971, Ph.D. Thesis.
Pipher, J. L., Houck, J. R., Harwit, M., and Jones, B. W.: 1971, *Nature* **231**, 375.
Soifer, B. T.: 1972, Ph.D. Thesis.
Soifer, B. T., Houck, J. R., and Harwit, M.: 1971, *Astrophys. J.* **168**, L73.
Soifer, B. T., Pipher, J. L., and Houck, J. R.: 1973, to be published in *Astrophys. J.*
Stein, W.: 1966, *Astrophys. J.* **144**, 318.
Werner, M. W. and Salpeter, E. E.: 1969, *Monthly Notices Roy. Astron. Soc.* **145**, 249.

CONCLUDING REMARKS

A QUICK REVIEW OF INTERSTELLAR GRAINS

H. C. VAN DE HULST

Sterrewacht, Leiden, The Netherlands

Abstract. This paper reviews the results of the Symposium on Interstellar Grains. It draws some historical lines and emphasizes the questions that require further study.

1. Introduction

This paper summarizes the main points made during the oral presentation of a symposium summary talk. I had the advantage of an evening's preparation and of the fresh impressions from listening to over 90 communications. But the texts, tables and figures of those communications were not yet available to me. For that reason references to individual contributions presented at the Symposium have generally been omitted.

2. Recommended Reading

During this meeting I have heard many references to 'original suggestions' made 1–5 yr ago. Some of these references were correct; others were not, but simply reflected the time at which the authors first became acquainted with interstellar matter. I wish to emphasize that some of the roots of the subject go back deeply in time. A somewhat arbitrary selection of significant early treatments is given in the following list (Table I).

3. Observations

About 25 yr ago a stationary point seemed to have been reached, in which no further relevant observations could be obtained and the remaining task seemed to be interpretation. Yet, history decided otherwise. In 1949 interstellar polarization was discovered by accident and a decade later the far UV and IR regions of the spectrum became

TABLE I

Recommended reading

Eddington (1923)	temperature of gas and dust
Oort (1932)	mass density near galactic plane
Lindblad (1935)	growth of grains
Greenstein (1937)	extinction and radiation pressure
Strömgren (1939)	H II-regions
Spitzer (1948)	charge on grains
Morgan *et al.* (1953)	spiral structure
Van de Hulst *et al.* (1954)	

Greenberg and Van de Hulst (eds.), Interstellar Dust and Related Topics, 569–573.

accessible. Altogether, the observations have greatly grown in accuracy, quantity and diversity. Figure 1 summarizes the present situation. The *extinction* curve, then limited to the range 1–3 on the λ^{-1} scale, is now well observed over the range 0–9. The blunt peak near 4.5 (2200 Å), the dip near 6, and the continued rise from 7 to 9, have been added to the familiar features of the extinction curve. *Linear polarization* has been well observed in the classical range 1–3. Somewhere in this range it shows a maximum. The first extensions of these measurements to the UV were reported by Stecher and to the IR by Okuda. They did not yet convince me. Polarization anomalies near the 4430 band were also reported, but the evidence still appears to be conflicting. The few reported observations of interstellar *circular polarization* are a strong appetizer for more. It would be premature to tell now what they mean.

The albedo, or ratio of scattering to extinction efficiency of individual grains, remains a key quantity in the interpretation. Unfortunately, its reliable observation is quite difficult. The situation appears to be that the old monopoly, set by the Henyey-Greenstein (1939) paper, has been taken over by a new monopoly of Lillie and Witt. If the alleged dip observed near $\lambda^{-1} = 4.5$ (see Figure 1) is correct, the discussions of a few days ago indicate that this means incredibly strange particles. Some held that

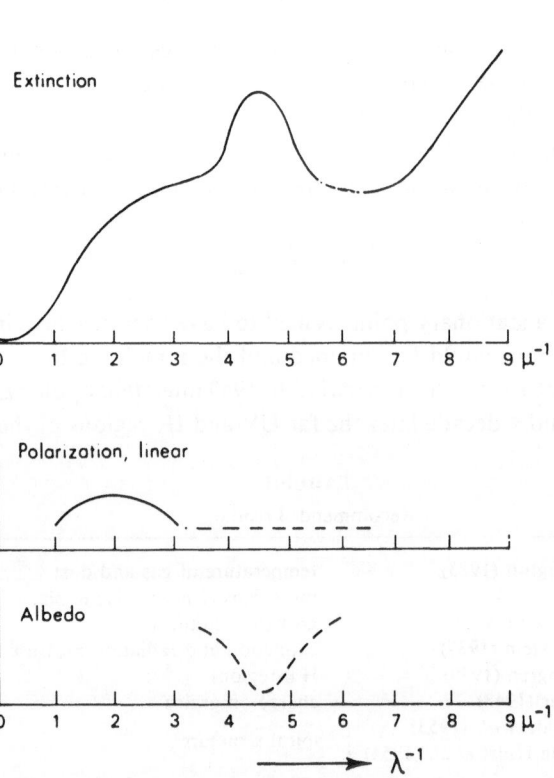

Fig. 1.   Rough sketches of curves showing the main observational data from which the nature of the interstellar grains has to be inferred.

nature might well be ahead of our ideas. I feel, more conservatively, that the observations may be allright, but that the interpretation, which was so far based on a very simple model calculation by Van de Hulst and De Jong (1969), would merit a reexamination. An added problem, recognized by all authors, is that determination of the albedo is strongly linked with determination of the asymmetry factor of the scattering diagram.

Finally, impressive new data on the *IR emission* by grains in the range 10 μ to 100 μ were presented by several authors.

4. Interpretation

The leading question is: 'do grains exist' and the classic answer stil holds: "presumably yes, for it is virtually impossible to think up another agent which can produce so much extinction without the dynamical effects of its mass becoming noticeable."

Next, we have to ask what determines the shape of the extinction curve: is it mainly a matter of size or mainly a matter of composition? This dilemma is shown by Figure 2. If size effects predominate, for grains with a refractive index m that does not vary strongly with wavelength, the theoretical curve links the extinction to the ratio of circumference to wavelength, $2\pi a/\lambda$. When only the curve between $\lambda^{-1} = 1$ and 3 was available, it seemed natural to match this to part of the theoretical curve and thus fix the size (Figure 2a). This classical explanation also meant that grains in their size distribution should be like men: a range of sizes, yet rather uniform in the predominant size. This consequence is a little awkward in view of the large variations that may exist in circumstances leading to grain growth. The alternative to the classical explanation is that at least certain features of the curve are caused by a fixed substance and bound to a fixed λ like an absorption line in the spectrum. The emphasis then falls on composition rather than size (Figure 2b). Presumably the correct explanation will be a combination of these extremes but we are still far from a complete theory.

Serkowski has shown that the polarization curves lead to a more uniform curve if plotted against $\lambda/\lambda_{max\ pol.}$. This is one way to partially eliminate the size effects and the

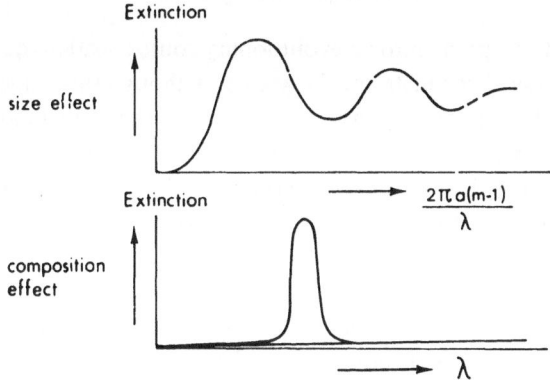

Fig. 2.   Contrast between two extreme approaches to interpret the extinction curve.

suggestion made in the discussion, that the extinction curve should be treated similarly, is worth pursuing.

About *shapes* even less can be said than about sizes. The extinction curves for most assumed shapes cannot be found in practice from theory, but may be obtained from microwave model measurements (Greenberg *et al.*, 1961). A constant warning sign should be up against papers in which conclusions are drawn from simplified formulae beyond the range of their validity.

5. Distribution

In the old (pre-1940) times it was customary to distinguish between the general interstellar medium and the dark clouds. This distinction about vanished when it became clear that the 'general extinction' could well be the combined effect of all clouds along the line of sight.

In recent years the insight has emerged that grains must somehow fit into the evolutionary sequence leading to spiral arms and other details of galactic structure. This makes careful distinctions necessary in interpreting the observations and this was the aim of many papers presented at this Symposium. From large to small objects we may distinguish:

External galaxies
Spiral arms / Galactic centre
Local dependence on cosec *b*
Dark nebuale / Reflection nebulae
Circumstellar clouds
Protostars / Novae.

To this we should add the comment that external galaxies and the galactic centre entered hardly at all in our discussions. These subjects, e.g., the beautiful results on molecular lines at the galactic centre, were left to other symposia and in this respect we followed a long tradition to treat interstellar extinction as a rather local subject.

6. Theory and Experiment

Any attempt to put the grains into an evolutionary context leads to questions regarding the physics and physical chemistry of the grains. Although this subject again is about 30 yr old, very little is known with certainty. The number of bold suggestions has multiplied, the number of firm facts has not. I feel that only a careful theoretical approach, combined with judiciously interpreted laboratory experiments, can help. And the observation, or non-observation, of some molecules can be a big help in ruling out certain alternatives.

Let me mention a few questions under this heading.

Smoke or dust ? All interpretations agree that the grains are formed from the gas phase, presumably on condensation nuclei, like smoke, and not broken up from a larger body like dust.

How hostile is the environment ? Our approach in the forties (Oort and Van de Hulst, 1946; Van de Hulst, 1949) assumed a rather gentle environment in which soft particles of 'dirty ice' could and should grow. There are at least three changes which all point to a more hostile environment: the presence of low-energy cosmic rays (LECR), a pervading UV continuum at 10–13 eV, and finally, the fact that some grains are known to be present even in H II-regions. This opens the possibility, pursued by many authors, that only hard grains or hard cores of grains remain.

Where does nucleation occur ? We now think that nuclei spontaneously formed in space as poly-atomic molecules, do not have a chance to survive because of LECR. (De Jong and Kamijo, 1973). Two possibilities remain. One is that condensation nuclei leave a star like smoke leaves a candle flame. The other is that non-volatile cores survive the formation of stars and H II-regions in one spiral arm and start the formation of new dust grains upon arrival in the next density wave.

How large are the accomodation and sticking coefficients ? Good guesses have been made since many decades but the number of really relevant experiments is extremely limited. For instance, the process of formation of H_2 molecules from H-atoms collecting with a cold surface, has only recently been the subject of quantitative interpretation (Marenco et al., 1971).

Alignment theory is another favorite subject, absolutely needed in order to interpret the polarization data. It is a tricky problem, but perhaps not so beset with fundamental physical uncertainties as some of the others mentioned. It has been well discussed at this symposium.

Dynamics of grains in grain-gas mixtures has a variety of consequences in processes on the largest scale (galactic structure) and on the smallest scale (star formation). It falls near the borderline of the subject matter discussed in this Symposium.

References

De Jong, T. and Kamijo, F.: 1973, *Astron. Astrophys.* **25**, 363.
Eddington, A. S.: 1926, *The Interval Constitution of the Stars*, Chapter 13. (This is a more accessible reference than the 1923 paper.)
Greenberg, J. M., Pedersen, N. E., and Pedersen, J. C.: 1961, *J. Appl. Phys.* **31**, 82.
Greenstein, J. L.: 1937, *Harvard Circulars* **422**.
Henyey, L. G. and Greenstein, J. L.: 1941, *Astrophys. J.* **95**, 70.
Lindblad, B.: 1935, *Nature* **135**, 133.
Marenco, G., Schutte, A., Scoles, G., and Tommasini, F.: 1971, paper presented at the *International Vacuum Congress*, Boston.
Morgan, W. W., Whitford, A. E., and Code, A.: 1963, *Astrophys. J.* **118**, 318.
Oort, J. H.: 1932, *Bull. Astron. Inst. Neth.* **6**, 249.
Oort, J. H. and Van de Hulst H. C.: 1946, *Bull. Astron. Inst. Neth.* **10**, 187.
Spitzer, L.: 1948, *Astrophys. J.* **107**, 6.
Strömgren, B.: 1939, *Astrophys. J.* **89**, 526.
Van de Hulst, H. C.: 1949, *Recherches Astronomiques de l'Observatoire d'Utrecht* **11**, part 2.
Van de Hulst, H. C. and de Jong, T.: 1969, *Physica* **41**, 151.
Van de Hulst, H. C., Muller, C. A., and Oort, J. H.: 1954, *Bull. Astron. Inst. Neth.* **12**, 117.

DISCUSSIONS

Part I

(Lutz)

L. H. Aller: Did you examine the effects of changing the mass of the nebular shell upon the distance estimate? When Minkowski and I applied the basic principle of what is commonly called the Shklovsky method, we felt we could only give limits on the nebular distance because we thought the mass might lie anywhere between 0.05 and 0.5 solar masses. Shklovsky took the mass of the shell as 0.2 solar masses for all planetaries. Me may be right; nobody knows. The error in distance is relatively insensitive to that in mass but is *not* negligible in your treatment and may affect results as much as by changing temperature.

J. H. Lutz: If the mass chosen were too large by a factor of five the derived value of R would be too large by roughly 2.0.

Part II

(Bromage and Nandy)

Anon.: It has been shown by Williams and McIntyre that the $\lambda4430$ band could be caused by Ca°-atoms in grains, if the atoms can be inserted into holes corresponding to about the size of Ca^{++}-atoms, which is the form of Ca in interstellar space. I would like to know if there is data on the relation between the depletion of Ca and the strength of the $\lambda4430$ band?

Do you have any data on the possible anticorrelation between the strength of the $\lambda4430$ feature and the excitation of the H II regions that you have looked at?

J. M. Greenberg: An interesting anti-correlation of the sort you seem to be implying was found indirectly a number of years ago. Stoeckly and Dressler (*Astrophys. J.* **139**, 240, 1964) showed that the $\lambda4430$ feature in high velocity clouds was weakened. Lichtenstein and I (*Astrophys. J.* **68**, 74, 1963) attempted to connect this with the fact that in such cases the Ca to Na abundance ratio approached 'normal' values (Routly, P. and Spitzer, L., Jr.: 1952, *Astrophys. J.* **115**, 227).

We assumed that the $\lambda 4430$ was due to Ca imbedded in the grains and that high velocity clouds implied H II regions which evaporated the mantles of the grains. Implicit in this argument was the assumption that Ca atoms accreted more really than Na on to the grains.

W. W. Duley: I believe that McIntyre and Williams calculate that Ca atoms are not likely to give a band at 4430 Å when trapped in hydrocarbon matrices. Their calculation is in error as we see a band due to the Ca 4226 Å line at 4430 Å in solid C_6H_6.

T. Snow: Is it possible that the reddening and diffuse bands which you and Bromage have correlated in the Cygnus OB association are not necessarily circumstellar, but perhaps intracluster instead? That is, could they be occurring in dust associated with the star group as a whole, rather than dust clouds surrounding each individual star?

K. Nandy: The reddening is definitely local, but we can't say for sure whether that means close circumstellar clouds, or something less closely attached to each star.

J. M. Greenberg: In a paper Dr Wang and I gave at the Liège Symposium last year (Greenberg, J. M. and Wang, R. T.: 1972, *Mem. Soc. Roy. Sci. Liège*, 6ème, tom III, pp. 197–207). It was shown that the strength of an absorption band is amplified relative to the extinction if the dust is truly circumstellar. This, of course, assumes that the circumstellar dust is the same as the interstellar variety.

If, on the other hand, the $\lambda4430$ is contained in mantles we expect the band to be depressed if the dust in the circumstellar region consists only of the cores on which mantles form.

P. J. Treanor: You mentioned that the reddening of No. 10 of the 6 Cygnus association seems to be too small. I have also recently had some doubts about the published data on this star. A preliminary polarization measure in the red region suggested a value a good deal higher than that published for the blue. If wondered if there could be some identification error. It is quite a bright star – about $m = 9.8$.

Greenberg and Van de Hulst (eds.), Interstellar Dust and Related Topics, 575–584.

(Nandy and Seddon)

M. F. A'Hearn: I have narrow band filter measurements (in the current A. J.) of the polarization across the $\lambda 4430$ band in several stars. Despite a wavelength resolution of only 20 Å, the high polarimetric accuracy places severe limits on the possible amplitude of a dispersion-like polarization curve as described by Greenberg and by Nandy – certainly no more than a few thousandths of a magnitude.

J. M. Greenberg: In view of the importance of polarization shape of the $\lambda 4430$ band in helping to define the optical properties of grains I would like to propose that we should make more observations at better spectral resolution in order to be sure of the result.

(Hayes *et al.*)

R. K. Honeycutt: I have two comments on this interesting paper. First the knee in the reddening law at ~ 4300 Å appears to persist independently of the presence of the 'window' between 1.6 and 2.0 μ^{-1}. This conclusion is based on photoelectric scans of reddened early-type supergiants at 20 Å resolution taken at Indiana University which show the knee in the reddening curve of most stars even if one considers only the spectral region blueward of 2 μ^{-1} (that is, excluding the window).

Secondly, photoelectric scans of VI Cyg No. 12 (soon to be published by R. S. Chaldu, M. V. Penston, and myself) support the reality of the red window. The feature appears clearly in the reddening law of VI Cyg No. 12 in about the strength expected from its color excess of $E_{B-V} = 3.3$. The window looks essentially featureless at 40 Å resolution.

K. Nandy: Why use Whiteoak's data for only 28 stars, when others have published measurements on more stars at higher dispersion?

D. S. Hayes: Whiteoak's data were the first I found, and the only ones I know of yet, which have enough photometric accuracy and wide enough wavelength range. I haven't seriously tried looking for more data because Whiteoak's is adequate and this is only a preliminary analysis.

Anon. 1: In our spectrophotometry of Cyg IV No. 10, the feature you have discussed stands out quite strongly. But also, when we look at wavelengths shorter than 5000 Å, the knee at about 4300 Å is quite well defined.

D. S. Hayes: I am familiar with other attempts in the literature to locate a knee around 4300 Å while looking at a wavelength range of a couple of hundred Å on each side. In my opinion, although there may, in fact, be a knee at about this wavelength, it is not what most people mean by the 'knee', when looking at the entire interstellar extinction curve. Any 'knee' at 4300 Å must be much smaller, in the sense of having a smaller change in slope.

Anon. 2: You said 1% of the normal polarization is about the size of the polarization feature you will be looking for. Do you mean that if the polarization is 5%, that you are looking for a polarization difference of 0.05%?

D. S. Hayes: Yes. This is, as I recall, on the same order as the size of the polarization feature being searched for in the case of the 4430 Å band.

D. R. Huffman: Do you know of any material which has such a window in the absorption?

D. S. Hayes: No. Do you? Dr Greenberg and I have not yet really tried to ascribe this window to a particular material but have limited our thoughts to whether it is a possible phenomenon.

D. H. Harris: It may be that the feature you observe at $\lambda^{-1} = 1.8 \, \mu m^{-1}$ is in reality the space between two weak broad absorptions.

S. J. Shawl: What particle size and size distribution junction were used in your calculations?

D. S. Hayes: Perfectly aligned cylinders of radius a distributed according to a Greenberg distribution, $n(a) = e^{-5(a/a_0)^3}$: $a_0 = 0.225 \, \mu$.

(General)

S. van den Bergh: Could you give the arguments for and against the assumption (to M. Greenberg) that the $\lambda 4430$ feature is related to the band at $\lambda 2200$.

J. M. Greenberg: There is, to my knowledge, no *theoretical* argument which supports this hypothesis. I believe that a paper by Sneddon discusses the observational correlation. The $\lambda 2200$ feature has been variously attributed to graphite (carbon) and to silicates, but I do not recall that either of these substances has been directly used to produce the $\lambda 4430$ band.

D. H. Harris: Small quantities of impurities in graphite markedly change the strength of the 0.22 μm feature. Optical constants for less pure carbon can be found in the literature on coals.

A. N. Witt: In answer to the question whether $\lambda 4430$ is produced by scattering. I have examined spectra of various reflection nebulae in this respect. When compared with the spectra of the illuminating stars, no evidence for excess light at λ 4430 can be found in reflection nebulae. The diffuse galactic light is unfortunately much too faint to obtain sufficient spectral resolution to answer the question as to whether there is λ 4430 excess in the DGL.

Part III

(Zellner)

A. N. Witt: Do you consider that the lower polarization in the UV is due to the breakdown of the single scattering assumption, which would occur first in the UV?

B. H. Zellner: Yes. It is also true that in the UV we see to a smaller geometrical depth and hence observe a smaller range of scattering angles. It does not seem advisable, however, to invoke 2nd order mechanisms in order to explain phenomena which are already nicely reproduced by single scattering Mie calculations.

(General)

W. W. Duley: If the observations of UV surface brightness of reflection nebulae require that dielectric grains are present and that these grains are transparant to extremely short wavelengths $\lambda^1 \to 10 \, \mu^{-1}$ then the only solid material that could satisfy this requirement is solid neon which is transparent to $\lambda \simeq 900$ Å.

Part IV

(Serkowski)

Th. Schmidt: It seems that the helical interstellar magnetic field model of Mathewson and Ford can no longer be taken as the only possibility. Among other things there seems to be no real indication from the most recent faraday and pulsar mesurements for a regular magnetic field inversion between northern and southern galactic longitudes, as would be necessary for the helical field.

(Treanor)

K. Serkowski: The superiority of the ring method, as compared to taking photographs through a pair of calcite plates, is not clear to me. Are not the photographic adjacency effects affecting the results obtained with the ring method? Maybe the best solution would be to use a pair of calcite plates and a rotating tilted glass plate in front of photographic plate. This tilted plate would increase the size of Schmidt stellar images slightly. In this way the advantages of both methods may be combined.

P. J. Treanor: The question of the relative merits of the ring and calcite polarimeter is a complicated one. Undoubtedly, one has to take percautions to minimize adacency effects in the ring, particularly in relation to the calibration step I have introduced (I explain this in the full text). Also there is some loss of limiting magnitude. However, the rings present a complete analysis of the polarization and the fact that the rings extend over many grains should increase this information content in comparison with methods which rely on single or multiple star images.

(Martin *et al.*)

T. Gehrels: The circular polarization of VY Canis Majoris (T. Gehrels, *Astrophys. J.* 173, L23–L25, 1972) has the wavelength dependence as you described it, namely, with a rise from about 0.6 μm, maximum near 1 μm and down again at about 3 μm. (K. Serkowski, review in this book, p. 145). It seems therefore that you have succeeded in interpreting this phenomenon as well.

M. Harwit: If the grains in the interstellar medium produce a quarter wave or circular dichroic effect, a net angular momentum transfer takes place from the starlight to the grains. Since the starlight is preferentially directed within the galactic plane, this effect should affect grain alignment.

Anon: You have assumed that the position angle of interstellar polarization is independent of wavelength. But Coyne and Gehrels (Ref.: *Astron. J.* 71, 355, 1966) reported earlier a remarkable wavelength dependence of the position angle of interstellar polarization. How would your results be affected if there is intrinsic wavelength dependence of position angle.

(Kemp)

K. Nandy: Do you think that it is just a coincidence that the circular polarization band occurs near 4300 Å wave and a change of slope occurs in the extinction curve?

R. D. Wolstencroft: I think it is probably unrelated. The theory of Kemp requires that the wavelength dependence of circular polarization obeys a dispersion-like relation which is the first derivative of the absorption curve. This would require the maximum absorption at the point of zero circular polarization, namely at about 5500 Å rather than 4300 Å.

D. R. Huffman: I would like to report a preliminary laboratory discovery that raises a different possibility for explaining these circular polarization results. In doing so, I want to credit Dr Kemp for the use of one of his photoelastic modulators in the experiment. We have detected circular dichroism in small particles of a magnetic iron oxide oriented preferentially along the line of sight. (These are the kinds of particles suggested by the Purcell-Spitzer theory as necessary to obtain magnetic alignment in 10^{-6} G fields). This circular dichroism, which is a difference in the imaginary part (absorptive part) of the complex optical constant, would give rise to circular polarization in originally unpolarized light. I am suggesting this only as another possibility which should be considered.

(King and Harwit)

K. Serkowsky: How strong a photon alignment may be expected in the envelopes of cool stars?

J. Harwit: Perhaps you could deduce this from my article in *Bull. Astron. Inst. Czech.* **21**, 204, 1970.

(Lloyd and Harwit)

T. Gehrels: I would like to address a remark to theoreticians. About 20% of the stars observed for interstellar polarization have a remarkable wavelength dependence of the position angle. When we first found it (T. Gehrels and A. B. Silvester, *Astron. J.* **70**, 579, 1965) we made a qualitative interpretation in terms of the light traversing individual clouds that have various particle sizes as well as various orientations of the galactic magnetic field.

We now know that some of these stars may have intrinsic polarization, caused by circumstellar shells, for instance. In face, we make the rather astonishing conclusion that nearly all stars have at least some intrinsic polarization and that this is so especially when emission lines are observed.

However, the interstellar contribution to the wavelength dependence of the position angles is real (G. V. Coyne, *Astron. J.*, in preparation) and your attention is drawn to this effect.

Anon: Ireland *et al.* (Reference: *Nature* **212**, 990, 1966) tried to explain the wavelength dependence of the position angle of interstellar polarization by considering the possible rotation of the interstellar magnetic field.

(General)

M. Harwit: I would like to point out how little we actually know about the mechanism which aligns grains in interstellar space. Both the work of Spitzer and Purcell, and the calculations by King and myself show that none of the mechanisms postulated thus far align grains adequately, unless very unusual material properties are invoked. It is therefore important to look for new effects which would lead to alignment, and these effects will have to be strong, particularly in view of the existence of several strong damping mechanisms presented in the paper Leon King just read. We should also keep in mind that some alignment effects will work in certain regions of interstellar space, while others may dominate elsewhere. This will complicate the interpretation of observations to a serious extent. It also argues against interpreting the grain alignment as a reflection of the magnetic field direction; these two parameters may be quite unrelated or may be weakly related, but we have no observational evidence of a strong interrelation.

J. M. Greenberg: I am not nearly as convinced as Dr Harwit that a magnetic alignment mechanism fails, particularly since only a factor of about two in the magnetic field is at issue. In any case it is clear that no other mechanism proposed is nearly as effective (Greenberg, 1971, *Proc. of Symp. on Interstellar Molecules*, Charlottesville, Va. p. 94–124.)

T. Gehrels: I wonder if theoretical work is progressing on the effects of surface waves on the interstellar polarization and reddening phenomena. Van de Hulst devoted a chapter (Ch. 17, *Light Scattering by Small Particles*, John Wiley & Sons, Inc., 1957) to this topic. I once wrote a note (*Astron. J.* **71**, 62, 1966) in order to try a qualitative explanation of the characteristic curve of the interstellar polarization.

K. Serkowski: While the shape of the wavelength dependence of interstellar extinction around discrete absorption features such as, e.g., λ 2200 of λ 4430 seems to be more or less independent of the size of dust grains, the shape of this wavelength dependence in the red and the infrared is determined mainly by size of dust grains. Therefore it seems reasonable to represent it by plotting

$$\frac{A(\lambda) - A(\lambda_{max})}{A(\lambda_{max}) - A(k\lambda_{max})} \quad \text{vs} \quad \lambda_{max}/\lambda,$$

where λ_{max} is the wavelength of maximum polarization proportional to grain size and k is a constant such that $k-1 \ll 1$. After normalizing in this way, the wavelength dependence of extinction in Scorpius and Orion, in the red and infrared spectral regions, becomes identical with that in Perseus and the rest of the sky.

Part V

(H. M. Johnson)

P. J. Treanor: McCarthy has recently reported an analysis of the distribution of supernovae in spiral galaxies which shows a continuous increase in the frequency of supernovae per pc^3 as the nucleus is approached. Perhaps this frequency is associated with dust clouds near the nucleus as in your photograph of the M31 supernova?

(Cahn and Nosek)

L. H. Aller: In order to get improved data on interstellar extinctions for planetaries we need more reliable Hβ and radio-frequency fluxes. Many planetaries in the Perek-Kohoutek catalogue have not yet been observed. It is a matter of telescope time for the optical photometric work, particularly in the southern hemisphere where so many are located. In the radio frequency range it seems most practical to work at some wavelength near 6 cm, where you can still use a large dish (such as the Parkes instrument) and beat the confusion problem.

(Shu)

C. Heiles: (1) Observing the 1665 and 1667 MHz lines of OH allows a determination of optical depth. Mark Gordon and I have observed several points in Cloud 2, the dark Taurus cloud. We find that the OH velocity is correlated with optical depth. The sense of this correlation is what would be expected from a contracting cloud. The velocity of contraction is about 0.4 km s^{-1}.

(2) Verschuur and Turner published, several years ago, an unsuccessful attempt to detect Zeeman splitting of the OH lines in Cloud 2. They were able to place a limit of about 10^{-4} G. This is much less than would result from flux conservation during the previous contraction process. It therefore appears that magnetic fields need not be considered as inhibitors to contraction in the early stages.

(Philip)

M. P. Savedoff: On what do you (or Crawford) base the origin of your E(b-y). What is its astronomical significance?

A. D. Philip: The four-color system was calibrated by Crawford by means of photometric measures of stars in the Hyades and other open clusters for which the distance moduli are known. The color excess E(b-y) is then defined as the difference between the observed (b-y) index and the ZAMS (b-y) for a star of that effective temperature. Using ratios derived by Crawford, one can then calculate the color excesses in c_1 and m_1 and the total absorption A_v.

A. N. Witt: How do you explain the conflict between your results and those obtained from galaxy counts, which seem to indicate a much higher optical thickness of the galactic extinction layer?

A. D. Philip: The conflict between optical measures of stars at high galactic latitudes and counts of galaxies has been known for several years. I believe that the color excesses derived from photometric measures of stars are the most accurate.

(General)

C. Heiles: The Taurus dust cloud (Cloud 2) shows strong H I 21 cm self absorption. Mrs Knapp and I have observed this using the NRAO 300-ft telescope, which provides high angular resolution (HPBW = 9'). We find that the self-absorption exists not only in the cloud, but also very strongly in a narrow region *outside* of the cloud. We have not yet attempted to derive the relative amounts of cold gas inside and outside the dust cloud itself. It is not inconceivable, however, that the majority of the gas lies just outside the cloud. In this connection it is interesting to note that the CO molecule exists outside the cloud as well as inside, according to recent observations by Solomon.

F. Gardner: W44 has also been investigated with the Parkes' 64 m telescope with a 4' beam. The 43 km s^{-1} H$_2$CO absorption has been found to arise in a cloud that is centered in direction near the edge of the continuum distribution and has halfwidths of 7' $\times$ 4', when corrected for beam broadening. The location and velocity are similar to those of the OH absorption which are believed to be associated with the supernova remnant, although the OH cloud is considerably larger.

Part VI

(Huffman and Stapp)

T. Gehrels: I am rather intrigued by your remarks on the size distributions being narrow, because for the interstellar polarization we also concluded, tim eand aging (T. Gehrels, *Astron. J.* **73**, 641, 1967; K. Serkowski, T. Gehrels, W. Wisniewski, *Astron. J.* **74**, 89, 1969), that the size distribution is narrow.

(Day)

W. W. Duley: Do you have a monomolecular H$_2$ layer or does H$_2$–H$_2$ physical adsorption occur?

K. L. Day: Since it was not possible to outgas the filament, it is likely that a certain amount of H$_2$ gas was temporarily absorbed during the experiment, leading to impinging H$_2$ striking previously attached molecules of this gas. From a pure physics standpoint this is not desirable, but might actually be a better approximation to the situation in space than a scrupulously clean graphite surface.

(Watson)

H. J. Habing: I have two comments:

(1) Observations of the 21 cm line in absorption and in emission show clearly the existence of two gas components, hot rarified gas and cool clouds. Judging from the temperatures you get, your model works only for the cool clouds, although the clouds should not be too dense, because then your radiation field is too weak.

(2) You assume $n_e \approx 0.05$ cm^{-3}. If you have less electrons, then your mechanism will become less effective because the charge on the grains goes up. However, to get 0.05 in a cloud of $n_H = 10$ cm^{-3} requires $n_e/n_H = 5 \times 10^{-3}$, this means that in addition to the 912–1500 Å radiation field, another ionizing mechanism must be present. I am therefore doubtful whether your mechanism can replace the still hypothetical heating mechanisms of low-energy cosmic rays and soft X-rays. But maybe that is not what you want?

W. D. Watson: I agree with your first comment. However, I should point out that my mechanism is proposed for intermediate density clouds. There seems to be less difficulty in heating the intercloud medium. However, the conditions in denser clouds are poorly known. Hence any additional heating mechanism would be useful in pur understanding.

In reponse to point 2, I agree that electron densities which would be larger than given by ionization of carbon would imply an additional heating mechanism. However, in fact, my calculation is based on electron density due only to carbon ionization.

(Wickramasinghe)

T. Snow: I would like to make the comment that if the optical properties of grains change as you suggest when the grains reach large distances from their 'parent' star, perhaps this can at least partially explain our result that the diffuse interstellar bands do not seem to be formed in circumstellar dust shells. Grains which are still close to a star would not yet have the dislocated atoms in them which may be responsible for the diffuse bands.

N. C. Wickramasinghe: This is possible.

J. M. Greenberg: What I believe is more likely than the dislocated atom explanation for λ 4430 is that the grains which are initially ejected from the stellar atmospheres are essentially the cores on which mantles subsequently develop in interstellar space and that is the *mantles* which contain the '4430' ingredient.

M. P. Savedoff: You must be careful to include electrical charges. At such velocities (temperatures) appreciable charges may result and the electrostatic cross-section may be larger than geometric.

R. F. Willis: The statement that the optical properties of grains, which may have been bombarded by high velocity H atoms in circumstellar space, will be altered due to the high concentration of possible defects 'frozen-in' crystalline materials, is not absolutely true. For highly disordered graphite or any other electrically conducting crystal, the optical absorption spectrum will remain unaltered. For a glassy material, such as silicate, the optical spectrum is, of course, insensitive to further disorder effects. Some dielectric crystalline materials could show additional features, the strength of which will depend on both the nature of the impacting atoms (or ions) and their concentration. Nevertheless, the overall spectrum will still be closely similar to that of the undamaged crystal. The accretion of a mantle of damaged crystal plus ions plus a thin film of a dielectric would, of course, give a different optical spectrum to that of any single component grain.

Part VII

(Greenberg and Yencha)

W. W. Duley: I know of experiments which involve the trapping of radicals from microwave discharges and that in this case explosions in the condensed solid often occur. However, I know of no corresponding experiments done with UV light. Have you seen this effect experimentally?

J. M. Greenberg: In direct answer to your question: No. The UV light is used in our experiment to produce the free radicals. What triggers an explosion does not seem to me to be critical. I would expect that a warming up of the material which already has a high free radical concentration would be sufficient. I recall seeing mention (in the volume on Free Radicals by Bass and Broida) of a sudden explosion of a solid with frozen in free radicals when it is allowed to warm up. In this connection I should like to mention that this phenomenon may be the cause of the production of jets in comets as they approach the Sun.

B. N. Khare: We have performed experiments on the long wavelength UV irradiation of a frost of formaldehyde, ammonia, ethane, and water deposited on a quartz cold finger under high vacuum at 77 K. Products produced at low temperatures, as determined by gas chromatography/mass spectrometry and infrared techniques, include the organic molecules methanol, ethanol, acetone, acetylene, acetonitrile, acetaldehyde, methyl formate and possibly formic acid. Many of these product molecules have been identified in the interstellar medium. As in your case we also found a range of unidentified very high mass number products. These results, presented at the *Symposium on Interstellar Molecules,* Charlottesville, Virginia, 4 October 1971, have been published as 'Experimental Interstellar Organic Chemistry: Preliminary Findings', by B. N. Khare and Carl Sagan (in *Molecules in the Galactic Environment,* M. A. Gordon and L. E. Snyder, editors, Wiley, 1973).

J. M. Greenberg: I should point out that there are several essential distinctions between your experiment and ours. In our experiments which were first reported at the AAAS meeting in Chicago, December 1970, and at the Liège Symposium in April 1971, as well as in Charlottesville we have restricted ourselves generally to low temperatures and also to using a mixture of water, methane (or ethane) and ammonia so that we start with much smaller molecules. We believe that both of these conditions may represent better the chemical evolution of the interstellar dust. We extended our radiation down to 1400 Å in the UV. I do not believe that the mass spectrograph/gas chromatography gives the correct identification of the molecules as they exist in the *low temperature solid* but rather provides information on what molecules may be formed in the solid as it is warmed up in analogy to an interstellar grain blow-up. It may well be that somewhere during this process the grain may contain all sorts of amino acids and other very large molecules. Molecular weights of well over 350 were indicated in our gas chromatograph.

B. N. Khare: I agree that the composition of the solid may be amino acids or their precursors. Amino acids have been found in many experiments on the primitive simulation of the Earth's atmosphere, by Stanley Miller, Sidney Fox and recently by Carl Sagan and myself where we used long UV energy. However, it is now pretty well determined that whether one uses heat, shock, UV or gamma radiation, the formation of amino acids and their precursors is to be expected.

(Heiles)

M. W. Werner: I would like to suggest the possibility that molecules are not located deep inside dust clouds, where Salpeter's parameter has a large value, but are instead concentrated in a thin shell at the surface. If this were true, the edges of the cloud might show more intense molecular lines than the center does.

J. M. Greenberg: Perhaps this would be the kind of thing which would happen to grains as they migrate outward from UV shielded interior regions of a cloud and are then subjected to a higher probability of being triggered to explode and release large numbers of molecules.

S. van den Berhg: Observations of Cas A with the 200-in. show that [O II] λ 3727 has a very similar distribution to that of [O II] $\lambda\lambda$ 7320, 7331 over the surface of the remnant. This suggests that the absorption over the face of this object is probably *not* very non-uniform.

(Morris *et al.*)

T. Snow: The frequencies of the four lines of cyanoacetylene which you have observed are near the peak of the 3 K blackbody radiation field. Have you considered the blackbody radiation as a source of excitation of the lines?

M. R. Morris: This has not been considered but it should be in a complete analysis of the emission from this molecule. However, I do not think that the 3 K background would be an important excitation mechanism.

(General)

F. Gardner: The detection of the $l_{10} - l_{10}$ transition of formaldimine (C_2NH) by P. D. Godfrey, R. D. Brown, B. J. Robinson, and M. W. Sinclair is reported. Laboratory measurements showed that the two strongest peaks of the multiplet have rest frequencies of 5289.84 and 5290.78 with an intensity ratio of 1.6. In observations of Sagittarius B2 with the Parkes 64 metre telescope both these lines were detected with intensities of 0.1 K and 0.05 K with a peak-to-peak error of 0.025 K. The linewidths were 20 km s^{-1} and the radial velocity 63 $+$ 6 km s^{-1}.

R. F. Willis: Speculating on mechanisms for forming complex, large molecules there is one well-known effect involving graphite which may be of interest. Graphite consists of layers of C atoms, the spacing between which is large (~ 3.5 Å) so that molecules and atmos can easily become incorporated in the lattice. Such 'chelate complexes' are well-known. Also, the edge-faces are much more reactive due to dangling bonds and have a higher accommodation coefficient than the basal plane. It is possible that interstellar graphite grains could accrete atomic and molecular species between the atomic layers so facilitating the formation of more complex molecules. To a certain extent, such molecules would be 'shielded' from photodissociation by UV quanta. We have observed that Cs deposited on the basal plane diffuses between the layers and is trapped there. The optical properties are only slightly modified in such a system.

Part VIII

(Davies)

C. G. Wynn-Williams: One of the remarkable properties of many of the 1612 MHz OH/IR stars is that there is a symmetry between the high velocity and low velocity parts of the microwave spectrum. Does your model account for this?

(Wynn-Williams *et al.*)

M. W. Werner: It is interesting that some of your small infrared sources have no associated radio emission. This suggests that arguments based on a comparison of wide angle infrared and radio continuum measurements of H II regions may be misleading. Since much of the far infrared ($\lambda \sim 100\ \mu$) radiation may come from knots which have no radio emission, it will be necessary to compare IR and radio measurements of a given condensation before drawing any conclusions about energetic problems.

P. E. Palmer: I would like to briefly remark on the generally excellent correlation of far IR emission with millimeter-wavelength molecular line emission (HCN, CS, X-ogen, etc.). Dr Wynn-Williams kindly communicated the 20μ map of W3 to us. We observed HCN at three distinct positions in W3. Unfortunately, with the 36 foot radio telescope it is not possible to resolve all of the 20 μ sources. In NGC 7538 we also observed HCN and CS emission at the 20 μ position. NGC 7538 is also unusual in that it is one of the 6 or 7 objects in the galaxy showing maser emission in the rotationally excited states of OH.

These results are measurable if we believe that the molecular lines arise in gas mixed with the dust emitting the far IR: the large extinctions derived serves to protect the molecules and the relatively high (for H I clouds) temperatures and densities allow these molecules to be excited above the background temperature and become detectable. Alternatively, the molecules may be 'outside' the hot dust, but still in a dense region.

(General)

H. J. Habing: Dr Wynn-Williams has shown the coincidence in W3 (cont.) of the H_2O point source and an IR source that probably is a protostar. In this connection I should like to mention the work by Dr T. de Jong at Leiden, who recently constructed a model of an H_2O maser, in a protostar cloud. He considers a semi-infinite plane parallel atmosphere of uniform temperature and density. Deep in the atmosphere the H_2O molecule is in LTE, whereas close to the outer boundary LTE breaks down. The H_2O line at 1.35 cm arises in the transition of the 6_{16} to 5_{23} level, at 447 cm^{-1} above the ground state. Approaching the surface of the protostellar atmosphere from the inside one finds that a whole series of levels, among which is the 6_{16}, stays in LTE considerably longer than other, similar series, one of which contains the 5_{23} level. It follows that in the outer parts of the atmosphere a layer exists in which the 6_{16} level is inverted with respect to the 5_{23} level. A ray of 1.35 cm photons will be amplified tangentially to the stellar atmosphere.

The model has some features in common with an earlier proposal by Dr Richard Hills from Berkeley and contains an idea of Dr J. Jefferies. Today I learned from Dr W. Klemperer that Dr Phil Solomon has arrived independently at similar conclusions.

Part IX

(Woolf)

D. P. Gilra: (1) For late N-type carbon stars Gilra and Code (*Bull. Am. Astron. Soc.* 3, 379, 1971; see also my paper on 'Dust and Molecules in the Extended Atmospheres of Carbon Stars') had predicted emission due to circumstellar SiC grains in the 10–13 μ region. The infrared observations of these stars you have shown confirm our predictions very well. It seems the existence of SiC grains in circumstellar shells around these stars should be considered established.

(2) As I indicated in my paper on 'Collective Excitations and Dust Particles in Space' there are strong theoretical problems in identifying the observed features (in M supergiants etc.) in the 10 μ region with silicates. The dielectric constant of silicates is not available in this spectral region. So the absorption and scattering cross-sections cannot be calculated and we cannot investigate the shape, size and coating dependence of the strong resonances that will be present. Therefore the identification should be taken with reserve.

Part X

(Taff and Savedoff)

K. Nandy: What drives the cloud motions?

L. G. Taff: It is not our intention to account theoretically for the observed velocity distribution of the clouds. Possible mechanisms might be turbulence, galactic rotation on time scales $\gtrsim 10^8$ yr, supernovae creating expanding H II clouds.

B. D. Donn: What is the time scale of the evolution of the mass function?

L. G. Taff: Using as a basis the collision time for the smallest clouds a solution that is not an equilibrium one will decay as $e^{-t/\tau}$ to an equilibrium solution where τ is 5 to 15 times the above collision time.

L. H. Aller: Have you considered the effects of rotation?

L. G. Taff: We have not, although magnetic fields are included in some of the discussions. Rotational effects will not affect the cooling aspects of the problem because the cooling is so rapid ($\sim 10^{3.5}$ yr) but may, through shearing motions, affect the formation of very small mass fragments at the interface between clouds as they separate. For this to be very important the periods of cloud rotation would have to be much smaller than the cloud-cloud separation times.

B. J. Bok: What do you compare against?

L. G. Taff: Radio observations.

(Cameron)

J. Hackwell: We have already started to compare the infrared absorption spectra of carbonaceous chondrites with the infrared excesses of M-stars and find a surprising agreement between them. It is interesting to note that neither the M-stars nor the carbonaceous chondrites appear to show the secondary absorption bands typical of many silicates. In the case of the carbonaceous chondrites this may be due to their peculiar 'stacked' structure which is dissimilar to that found in any of the terrestrial silicates.

(McNally)

C. E. Heiles: Radio astronomers have found many molecules in regions such as the galactic center, M51, and Orion A. I believe these are just those places where matter is collapsing under its own gravity and where pressure is unimportant, similar to collapsing models published by yourself and others in past yers.

Some molecular transitions require much higher densities for their excitation than others; these always have smaller velocity widths than ones which can be excited at the lower densities. This is just what one would expect if the linewidths result primarily from large-scale collapse motions because the theoretical models show density increasing and collapse speed decreasing towards the center of the cloud. The correspondence between observation and theory can be checked to increasingly high degrees of precision in the near future by observation of strategically chosen transitions and molecules using high angular resolution obtainable with interferometers, one of which has just been constructed at the Hat Creek Observatory.